The Correspondence of John Tyndall

VOLUME 7

ACKNOWLEDGMENTS

This volume of *The Correspondence of John Tyndall* is built on the work of many colleagues whose willingness to share their time and expertise is deeply appreciated. We especially acknowledge the work of many graduate students and other colleagues in providing the original transcriptions and many initial translations of foreign-language letters. Our two research assistants, Emma Swain and Max Meulendijks, have given us huge help throughout. We greatly valued the work of James Braund and Juliette Reboul in checking the transcriptions and translations of German and French letters, respectively, and the additional assistance of Elizabeth Neswald, Efram Sera-Shriar, and Irena Brezowski. We sometimes required new scans of letters, and for this we warmly thank Jane Harrison of the Royal Institution of Great Britain. For help in creating high-quality image files of the drawings within letters in this volume, we thank Erin Grosjean. Finally, for invaluable help in answering a variety of questions about letters in this volume, and their reviews of material, we would like to acknowledge the general editors of *The Correspondence of John Tyndall*, James Elwick, Bernard Lightman, and Michael Reidy, as well as Rupert Baker (Library Manager, Royal Society), Anne Barrett (Imperial College Library), Bill Brock, Gabriel Finkelstein, Glyn Hughes (honorary archivist, Alpine Club), Caroline Lam (Geological Society), Anthony Morton (curator, Sandhurst Collection, the Royal Military Academy Sandhurst), Jennie de Protani (archivist at the Athenaeum Club), Dexter Bateman (University of Saint Andrews Library), and John Spedding.

LIST OF ABBREVIATIONS

ADB	*Allgemeine Deutsche Biographie*
The Ascent of John Tyndall	R. Jackson, *The Ascent of John Tyndall* (Oxford: Oxford University Press, 2018)
BAAS	British Association for the Advancement of Science
BL	British Library
CUL SC	Cambridge University Library, Stokes Correspondence
Faraday Correspondence	*The Correspondence of Michael Faraday*, ed. Frank A. J. L. James, vol. 5, 1855–1860 (London: Institution of Engineering and Technology, 2008)
FRS	Fellow of the Royal Society of London
Glaciers of the Alps	John Tyndall, *Glaciers of the Alps* (London: John Murray, 1860)
Heat	John Tyndall, *Heat Considered as a Mode of Motion* (London: Longman, Green, Longman, and Roberts, 1863)
Hirst Journals	William H. Brock and Roy M. MacLeod, eds., *Natural Knowledge in Social Context: The Journals of Thomas Archer Hirst FRS* (London: Mansell microfiche, 1980)
IC HP	Imperial College of Science, Technology and Medicine, College Archives, Huxley Papers
Journal	John Tyndall's Journal, 1855–72, typescript (RI MS JT/2/13c)
Kew JDH	Royal Botanical Gardens, Kew, Joseph Dalton Hooker Correspondence
LT	Louisa Tyndall, John Tyndall's widow
Life and Work	A. S. Eve and C. H. Creasey, *The Life and Work of John Tyndall* (London: Macmillan, 1945).
Mountaineering in 1861	John Tyndall, *Mountaineering in 1861: A Vacation Tour* (London: Longman, Green, Longman, and Roberts, 1862).
NLS	National Library of Scotland
ODNB	*Oxford Dictionary of National Biography*
OED	*Oxford English Dictionary*
Paris, Comptes Rendus	*Comptes rendus hebdomadaires des séances de l'Académie des Sciences*
Phil. Mag.	*Philosophical Magazine*
Phil. Trans.	*Philosophical Transactions of the Royal Society of London*

Poggend. Annal.	*Poggendorff's Annalen der Physik und Chemie*
RGO MS.RGO 6	Royal Greenwich Observatory Archives, Papers of George Biddell Airy
RI	Royal Institution of Great Britain
Roy. Inst. Proc.	*Proceedings of the Royal Institution of Great Britain*
Roy. Soc. Proc.	*Proceedings of the Royal Society of London*
RS	Royal Society of London
RS HS	Royal Society, Herschel Papers
RS RR	Royal Society of London, Referee Reports
StA JDF	Saint Andrews, Papers of James David Forbes
Tyndall Correspondence, vol. 1	Geoffrey Cantor and Gowan Dawson, eds., *The Correspondence of John Tyndall, Volume 1: The Correspondence, May 1840–August 1843* (Pittsburgh: University of Pittsburgh Press, 2016)
Tyndall Correspondence, vol. 2	Melinda Baldwin and Janet Browne, eds., *The Correspondence of John Tyndall, Volume 2: The Correspondence, September 1843–December 1849* (Pittsburgh: University of Pittsburgh Press, 2016)
Tyndall Correspondence, vol. 3	Ruth Barton, Jeremiah Rankin, and Michael S. Reidy, eds., *The Correspondence of John Tyndall, Volume 3: The Correspondence, September 1843–December 1849* (Pittsburgh: University of Pittsburgh Press, 2017)
Tyndall Correspondence, vol. 4	Ian Hesketh and Efram Sera-Shriar, eds., *The Correspondence of John Tyndall, Volume 4: The Correspondence, January 1853–December 1854* (Pittsburgh: University of Pittsburgh Press, 2018)
Tyndall Correspondence, vol. 5	William H. Brock and Geoffrey Cantor, *The Correspondence of John Tyndall, Volume 5: The Correspondence, January 1855–October 1856* (Pittsburgh, University of Pittsburgh Press, 2018)
Tyndall Correspondence, vol. 6	Michael D. Barton, Janet Browne, Ken Corbett, and Norman McMillan, *The Correspondence of John Tyndall, Volume 6: The Correspondence, November 1856–February 1859* (Pittsburgh, University of Pittsburgh Press, 2019)
Tyndall (1859)	"Note on the Transmission of Radiant Heat through Gaseous Bodies," *Roy. Soc. Proc.* 10 (1859): 37–39
Tyndall (1861)	"The Bakerian Lecture: On the Absorption and Radiation of Heat by Gases and Vapours, and on the Physical Connexion of Radiation, Absorption, and Conduction," *Phil. Trans.* 151 (1861): 1–36
Tyndall (1862)	"On the Absorption and Radiation of Heat by Gaseous Matter. Second Memoir," *Phil. Trans.* 152 (1862): 59–98

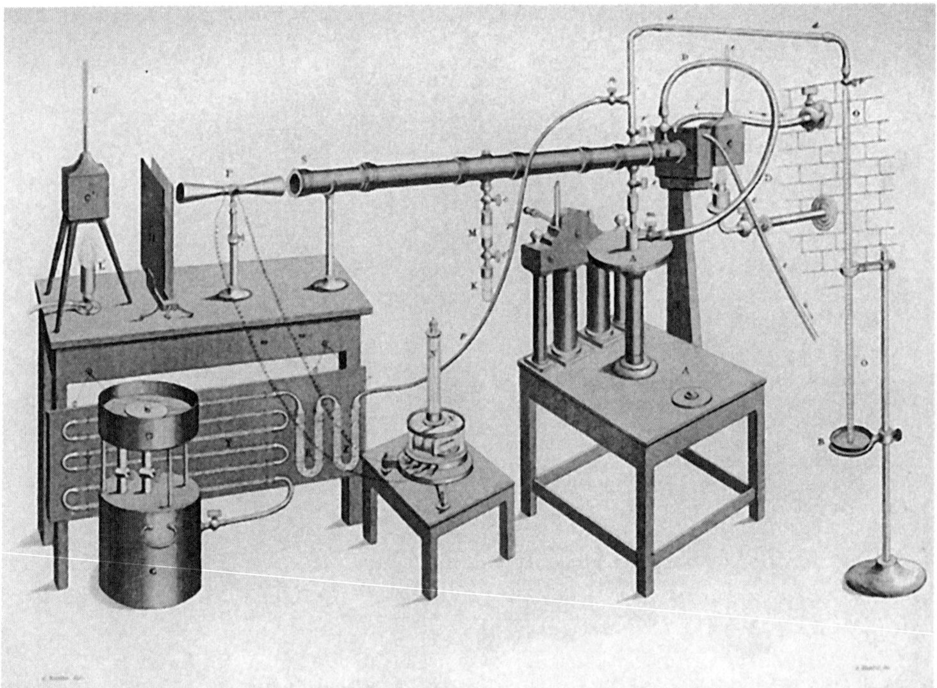

Tyndall's radiant heat apparatus. Courtesy of the Royal Institution of Great Britain.

INTRODUCTION TO VOLUME 7

The seventh volume of *The Correspondence of John Tyndall* contains 308 letters covering a period of thirty-nine months, from March 1859 to May 1862. It is a critical period in Tyndall's personal and scientific lives. The volume begins with the difficult ending of his relationship with the Drummond family, disputes about his work in glaciology, and his early seminal work on the absorption of radiant heat by gases. It ends with the start of his championship of Julius Robert Mayer's work on the mechanical equivalent of heat, which would lead inexorably to fierce arguments about priority with the "North British" physicists, a group of predominantly Presbyterian physicists including William Thomson, Peter Guthrie Tait, James Clerk Maxwell, and James Joule.[1] In between, we see Tyndall carefully establishing his own priority for his work on radiant heat, and his exchanges with the leading physicists of the day on this and other topics. In parallel, he receives offers of career opportunities. Tyndall turns down the possibility of becoming professor of natural philosophy in Edinburgh, in succession to James Forbes. He accepts the position of professor of physics at the Government School of Mines in succession to George Stokes, with its substantial additional commitment to lecturing. The lure of the Alps becomes ever stronger. In this period comes perhaps Tyndall's greatest mountaineering achievement, the first ascent of the Weisshorn, and a remarkable winter visit to Chamonix and the Mer de Glace. As his reputation grows, Tyndall continues to make his way in society. He is elected to the elite Athenaeum Club on 31 January 1860.

Tyndall's correspondents in the extant letters of this period are primarily scientific. Many encompass his network of Continental men of science, with whom about fifty of these letters are exchanged. They include Rudolf Clausius, Heinrich Gustav Magnus, Hermann Helmholtz, Robert Bunsen, and Auguste Arthur de la Rive. In England, an extensive series from William Hopkins and an exchange of about thirty letters with George Stokes give insight, respectively, into the glacier disputes with James Forbes and the refereeing process for papers for the *Philosophical Transactions*, both of Tyndall's own papers and of those he is asked to review.[2] The letters also reveal Tyndall's personal relationships with the leading older men of science, including

John Herschel, Edward Sabine, and George Biddell Airy, and the manner in which he sought to nurture them. With respect to his close friendship group, there are fewer letters with Thomas Hirst than in some previous volumes. This reflects Hirst's return to London in October 1859. He initially lived close to Tyndall in Maida Vale and they then moved in together in June 1861, renting Thomas Huxley's house at 14 Waverley Place. Heinrich Debus remained at Queenwood College during this period, and about twenty letters are exchanged with him. The largest batch of personal letters is between Tyndall and Juliet Pollock, wife of William Frederick Pollock. This close friendship, which extended to the whole Pollock family, demonstrates how Tyndall navigates his relationship with a married woman, and the way they exchange personal thoughts and beliefs.

One event of note during the period of this volume is the publication on 24 November 1859 of Charles Darwin's *On the Origin of Species by Means of Natural Selection*. Yet Tyndall's initial response to the book is hidden from us. We do know that Tyndall was present at the now famous encounter between Thomas Huxley and Bishop Samuel Wilberforce in Oxford in 1860, but only because of a report in a newspaper when Tyndall mentioned it years later when giving a lecture in Preston.[3] There are no notes in his journal and only one letter that touches upon the publication of *On the Origin of Species*—and that to Huxley, not Darwin. Huxley famously reviewed *On the Origin of Species* in the journal the *Westminster*,[4] and Tyndall wrote to Huxley after reading the published article.[5] Tyndall was particularly pleased with the last passage of Huxley's review, noting that it was "really magnificent." In this passage, Huxley poetically expressed his vision that with the adoption of Darwinian evolution the "dominion of Science" would extend into new regions of thought. It is not surprising that Tyndall would agree with this sentiment. Nevertheless, Tyndall's public statements in relation to *On the Origin of Species* date to much later than this volume.[6]

Personal and Scientific Relationships

At the start of this volume, Tyndall is extracting himself from the confusion of his relationships with Mary Drummond, her sister Emily, and their mother Maria.[7] He seeks "the advice of a lady" from Juliet Pollock[8] and relates the end of the affair to Hirst, as Michael Faraday intercedes on his behalf.[9] He appears to have no further deep attractions to potential partners until Mary Adair in the later 1860s. Tyndall's relationship with Juliet Pollock, wife of the lawyer William Frederick Pollock, who was a manager of the Royal Institution over many years, is explored in volume 6 as it started to develop.[10] Twenty-three letters between them span this volume, showing increasing fondness and intimacy. Tyndall regularly sought the advice and help of his married female friends, including Henrietta Huxley and Frances Hooker, but it is with Pollock that he seems to have had the closest relationship. He respected both her intellect and her wider advice. He asked her to review the manuscripts of both *Glaciers of the Alps* and *Mountaineering in 1861*,[11] and took her advice on the latter to tone down his potentially controversial chapter titled "Reflections," which addressed religious aspects.[12]

In the period of this volume, Tyndall started to make his first significant publication of controversial views in the newspapers, though it was not until 1865 that he published in the monthly literary periodicals.[13] His personal relationships with editors were critical to his success, and to his ability to project his ideas more widely in public.[14]

Tyndall's close relationships with many German physicists are evident in this volume, relationships that date from his time in Marburg and Berlin (see volume 3). Not only does he use those to further his own priority claims and gain support for his ideas on glaciers and radiant heat, he also highlights their contributions to British audiences. He is able to do this both through his lecturing and his editorial position at the *Philosophical Magazine*. A case in point is the Friday Evening Discourse he gave on 7 June 1861, "On the Physical Basis of Solar Chemistry,"[15] preceded by a request from Robert Bunsen for Tyndall to have a paper of Gustav Kirchhoff's published.[16] The benefits worked in both directions. Many of Tyndall's papers were published by Poggendorff in Germany;[17] a crucial help as Tyndall worked not only to claim priority but to establish his preeminence—for example, in debates about radiant heat.

In addition to his German contacts, Tyndall did not neglect to cultivate relationships with the older English men of science, whose support was important to him professionally. His relationships with George Airy and John Herschel were two he actively cultivated. When Herschel sent Tyndall a letter to complain that his work had been left out of Tyndall's Discourse "On the Physical Basis of Solar Chemistry"—"I do think it a little hard, seeing that <u>every</u> other name connected with the subject—Brewster, Talbot, Miller, Masson, Willingen, Plücker, Wheatstone, Foucault, Stokes, Thomson, is mentioned in this admirable lecture (which must have created quite a sensation and will be accepted as a résumé of the history of the subject) mine should never once have been alluded to"[18]—Tyndall quickly responded privately and in an open letter in the *Philosophical Magazine* to limit the damage.[19]

George Stokes, of Tyndall's generation, and also Irish, is a central figure. As the editor of the *Philosophical Transactions*, he had extensive interactions with Tyndall, both over Tyndall's own papers and those he asked Tyndall to review.[20] Despite their differences in religious belief, and Stokes's close connections to the North British physicists, their relationship as expressed in this correspondence is always mutually respectful. The letters also give detailed insight into Stokes's critique of Tyndall's major 1861 paper on radiant heat, and Tyndall's responses.[21]

But as this volume ends, Tyndall's links to the German physicists would soon get him into trouble in Britain. In May, he was preparing to give a Friday Evening Discourse on 6 June titled "On Force," in which he advanced the claim of Julius Robert Mayer as the originator of the theory of the mechanical equivalent of heat. In Britain, James Joule had that claim. Tyndall wrote to Helmholtz and Clausius for information and for their views,[22] and the consequences can be followed in volume 8. Tyndall found himself once more embroiled in controversial questions of priority, as he had with James Forbes and theories of glacier motion.

Glaciers of the Alps

Tyndall's first book, *Glaciers of the Alps*, was published by John Murray on 6 July 1860. He had started writing the book in October 1858, after his return from the Alps, and many letters attest to the way he sought both information and support for his ideas, challenging Forbes's viscous theory of glacier flow with his own mechanism of "fracture and regelation." As this volume opens, Tyndall gave a Friday Evening Discourse at the Royal Institution, on 4 March 1859, "The Veined Structure of Glaciers,"[23] following the presentation of his detailed views on glacier structure and flow at the Royal Society.[24] After his summer's expedition to the Alps in 1859, when he placed thermometers in several places on Mont Blanc—in an attempt, ultimately unsuccessful, to measure seasonal maximum and minimum temperatures—he based himself at Queenwood to work on his book. As his ideas crystallized, he wrote to many physicists and geologists, in Britain and abroad, to seek information, buttress his arguments, and gain support for his position. He even made a rare winter visit to Chamonix, well before the invention of a winter sports season, to measure the motion of the Mer de Glace. *Glaciers of the Alps* was in its final form around the end of April 1860, and letters show how Tyndall sought supporters to review it.

It is the Cambridge mathematician and physicist William Hopkins who plays a major role. An extensive correspondence (we do not have the corresponding letters from Tyndall) reveals Hopkins's support for, and constructive critique of, Tyndall's "fracture and regelation" theory of glacier motion over Forbes's viscous theory, and his dislike of Forbes.[25] There is only one letter from Forbes in this volume about glaciers, an open letter to Tyndall published in the *Philosophical Magazine*, in which Forbes initially defended his work.[26] Though Tyndall would not have been aware of it, behind the scenes Forbes was exchanging worried letters with William Whewell about the manner in which Tyndall was attacking his theory as well as his scientific integrity. These disputes would spill over into Tyndall's later altercations with the North British physicists over priority for discovering the mechanical equivalent of heat, which can be followed in volume 8.

Though *Glaciers in the Alps* was published by Murray, it was Longman who published most of Tyndall's subsequent books. Other correspondence reveals that the suggestion of Murray had come from Hooker.[27]

Mountaineering and the Alpine Club

As Tyndall's major work on glaciers draws to a close in the volume, his interest in mountaineering itself becomes more prominent. The summer of 1859 is perhaps a crossover point, when Tyndall climbed Mont Blanc for the second time, to place thermometers in various places, including the summit. Edward Frankland accompanied him, and they spent a night on the summit, the first time anyone had done so. He had thought then that he would not return. But his visit to Killarney in June 1860, described in a letter to Henry Bence Jones,[28] changed his mind. In a damp environment, "the mere thought of snow peaks and glaciers was an exhilaration."[29] The weather in the Alps was not good in 1860, but Tyndall, Francis Vaughan Hawkins, Johann Joseph Bennen, and Jean-Antoine Carrel reached the highest point yet attained on the Matterhorn.[30] Following this partial success, Tyndall left to inspect the site of a fatal accident on Mont Blanc, described in a letter to the *Times*.[31] He then climbed Mont Blanc for the third time, in an unsuccessful attempt to find the thermometers left the previous year. It was the next season, 1861, in which Tyndall made arguably his most noteworthy climb, the first ascent of the Weisshorn.[32] Following this season, Tyndall wrote up his exploits as *Mountaineering in 1861*. The book was published by Longman on 16 April 1862.[33] It contained a controversial chapter titled "Reflections," in which Tyndall stresses that physical processes were solely explicable on the basis of natural laws, which could be established through science alone. There was no evidence of miracles, nor need to pray to supernatural beings for protection from natural events. The mountaineer Alfred Wills approved, writing: "The old forms of faith cannot co-exist with the ascertained results of modern investigation, whether in philosophy, science, criticism or history—and those who help to prepare the human mind for what it must have to consider & for the great change in our modes of thought & of belief which will be the consequence of such consideration do a great service to their day & to their race."[34] Tyndall had intended to include a similar piece in *Glaciers of the Alps*, but was dissuaded by friends.[35] Instead, as he left for the Alps in 1860, the *Saturday Review* published his ideas as "Physics and Metaphysics."[36] It is the first substantial statement of Tyndall's "materialist" philosophy, ascribing sensations such as color and sound to the purely mechanical effects of motion impinging on eye or ear, interpreted by consciousness. Though this article appeared anonymously, as was the practice in the *Saturday Review*, and was perhaps advantageous, given its religious sensitivity, *Mountaineering in 1861* was evidently Tyndall's work. These are the first publicly visible signs of Tyndall's "scientific naturalism," which would take such provocative form in the Belfast Address of 1874.

The Alpine Club had been formed in December 1857, when Tyndall did not join and effectively declined to be the first president (see volume 6). Tyndall joined a year later, but never found it congenial. He might have become the second president, after John Ball, had the altercations with Forbes not caused some sensitivity or offense.[37] He was elected vice-president in 1861, but resigned shortly afterward.[38]

Though Tyndall was increasingly drawn to the Alps—which he visited every summer from 1856 until his death in 1893, apart from 1891, when he was too ill to travel—he did not neglect mountains and cliffs closer to home. Letters in this volume refer to visits to the Lake District in England,[39] a winter climb of Snowdon,[40] and walking and cliff climbing in Devon and Cornwall.[41] Tyndall frequently commented that getting out of London and into the countryside was important for maintaining and restoring his physical and mental health.

Radiant Heat

The motivations for Tyndall's exploration of the interaction of radiant heat with gases were multiple.[42] He started his experiments in the spring of 1859, having designed a differential spectrometer that allowed the absorption of very small amounts of heat, previously undetectable, to be measured by difference. Initial success came quickly, so that he could write on 18 May, after just a couple of weeks, that "the subject is completely in my hands."[43] He announced his early results to the Royal Society on 26 May,[44] and in a Discourse at the Royal Institution on 10 June, chaired by Prince Albert.[45] Tyndall had made the seminal discovery that many gases, including coal gas, water vapor, and carbon dioxide, could absorb heat.[46] It was important that Tyndall establish his priority, but he was equally careful not to reveal too much. Even a year and a half later, when Tyndall was carrying out his more detailed work, he could write to Debus: "Do not mention these things to anybody. Faraday has warned me against speaking of the subject at all."[47] Tyndall sent brief notes to *Cosmos* and *Il Nuovo Cimento*,[48] but a much longer letter to de la Rive, which was published in the *Bibliothèque Universelle*.[49] It is only in this letter that he specifically mentions his discovery of the absorption of heat by water vapor and carbon dioxide. On 1 June 1859 he wrote to Clausius, also documenting his priority, and describing his new instrument.[50] Although Tyndall made some further measurements in June and July, there is a gap between this 1859 work and the next phase, starting in the autumn of 1860, caused by a combination of ill health, the demands of finishing *Glaciers of the Alps*, and his lecture commitments at the Royal Institution and the Government School of Mines in the early part of 1860.

It was the experiments in the autumn of 1860 that led to his first significant paper on radiant heat, given as the Royal Society's Bakerian Lecture in early 1861. In this paper Tyndall made the connection between the absorption of heat by gases in the atmosphere, and climate change on geological timescales, writing: "[I]f, as the above experiments indicate, the chief influence be exercised by aqueous vapour, every variation of this constituent must produce a change of climate. Similar remarks would apply to the carbonic acid [carbon dioxide] diffused through the air."[51] Tyndall had this paper published in Germany too, as a letter of 14 February 1861 from Poggendorff attests.[52] Poggendorff's letter also mentions the fact that Heinrich Gustav Magnus, whom Tyndall greatly admired and in whose laboratory he had worked, was himself experimenting on radiant heat. Magnus wrote to Tyndall a month later,[53] challenging both his priority and his claim that water vapor was a strong absorber of heat. This was the start of a dispute, friendly and mutually respectful (unlike some of Tyndall's other controversies), that ran until Magnus's death in 1870. Tyndall replied immediately, documenting his priority in considerable detail.[54] He took care also to reinforce his priority in another letter to Clausius.[55] The challenge from Magnus over the absorption of heat by water vapor would exercise Tyndall for several years. He carried out experiments throughout the autumn of 1861, focused on the absorption of heat by water vapor, which he wrote up for publication in 1862.[56] As this volume ends, Poggendorff writes in May 1862 with news of the translation of his Tyndall's paper for the *Annalen*, and to assure him, concerning Magnus, that "you do not need to worry; he is far from taking the dispute with you personally."[57] Tyndall also sought views and information from Sabine, his important early patron, to support his work on radiant heat and related physics, though we possess only the responses from Sabine.[58]

Career Opportunities

At this point in his career, Tyndall had built a formidable reputation as a popular lecturer, both at the Royal Institution and the London Institution. As the volume opens, Tyndall is giving his final lectures at the Royal Institution in a series of twelve on the Force of Gravity. His 1860 lecture series was on light, and his 1861 series on electricity, with his first set of Christmas Lectures in 1861 on light. The lecture series in 1862, on heat, would become his successful popular book *Heat as a Mode of Motion*, published in 1863.

In late 1853, Tyndall had been approached by the School of Mines to lecture in physics, but he had declined the offer, as he felt that the commitment to serve on the council would cut across his obligations to the Royal Institution.[59] George Stokes had been appointed instead. Stokes resigned in 1859 and Tyndall told Hirst that Roderick Murchison in the "handsomest manner" had offered it to him.[60] Urged on by Huxley,[61] who was instrumental in setting up the opportunity, he accepted after checking with Faraday, and wrote to John Barlow arguing that the lecturing was routine and would not affect his activities at the Royal Institution.[62] That may have been disingenuous. Lecturing was always something that Tyndall took seriously, and this new responsibility did affect the time he could dedicate to research. It is possible that his insecurity about his financial situation was a factor in his decision, though there is no direct evidence of this. Certainly this post was welcome financially, as Tyndall could write immediately to his mother increasing her annual allowance.[63]

After James Forbes had been appointed principal of Saint Andrew's College in succession to Brewster, Lyon Playfair wrote in November 1859 to see if Tyndall would be a candidate for the chair of physics at Edinburgh that Forbes was vacating.[64] Tyndall could see the advantage of founding a true school of experimental physics, which he thought the country lacked, but had to judge it against the advantage of London and his proximity to Faraday. He exchanged a series of letters with Playfair, and sounded out friends including Faraday,[65] Frederick Pollock, and Juliet Pollock.[66] Though Frederick Pollock advised him to accept, Juliet advised the opposite, and Tyndall recorded in his journal, "This advice I shall follow."[67] He turned down the opportunity.[68] Murchison wrote that he would never forget Tyndall's magnanimous gesture of staying in London "to support the advancement of physical science in this great Metropolis."[69] Faraday had advised that he would not have swapped Edinburgh for London,[70] and Forbes himself complained in a private letter about how he was cut off from events in the metropolis.[71] This was a decision that would shape the rest of Tyndall's career as a central scientific figure in London.

Conclusion

The period of just over three years covered by this volume is crucial to Tyndall's scientific career and life trajectory. The discovery of the absorption of heat by gases in 1859, for which he was to win the Royal Society's Rumford Medal in 1864 (see volume 8), is a major advance in physics, underpinning our current understanding of the greenhouse effect, climate change, weather, and meteorology. It set a direction for his own research program over many years. Even in the 1880s, with his work on the action of intermittent beams of heat on gases, he would return to the topic to address continuing opposition to his findings and conclusions. Tyndall's decision to stay in London, rather than move to Edinburgh, located him in the metropolis for the rest of his career.

The ascent of the Weisshorn in 1861 is one of the pioneering first ascents of the Alpine "Golden Age," and it alone would establish Tyndall's reputation as a leading figure. But this is also the period in which Tyndall finally appreciated the deep relationship he had with the Alps. After his visit to Killarney in 1860, when he realized how much he missed them, he was to return every year until his death, apart from 1891, when he was too ill to travel.

This period also sees the first signs of the emergence of Tyndall as a voice in wider cultural debates, with his article "Physics and Metaphysics" in the *Saturday Review*. He is now an establishment figure, elected to the Athenaeum Club in 1860, and increasingly confident about making potentially controversial public statements, even though they might conflict with the beliefs of Faraday, his revered mentor.

Yet beneath all this, Tyndall's emotions bubble. Many of these letters reveal Tyndall's inner feelings and the passions that drive him. His commitment to work, a Carlylean duty but perhaps also a financial necessity, put him under great pressure. His multiple lecturing roles encroached on his time for research—hence the delay between his initial spring 1859 discoveries and his detailed work in the autumn of 1860—and his attraction, as an Irishman of relatively humble origins, to heady society gatherings added to the strain. To mental and social exhilaration were added mental and physical exhaustion leading to short visits to the country if he could find the time. He revealed his feelings to his friends, and they to him. In these letters we see Tyndall's deepest thoughts conveyed to Juliet Pollock, the poignancy of Debus's feelings for Tyndall, Tyndall's deep respect for Faraday and Herschel, Hopkins's disdain for Forbes, and Magnus's careful approach to avoiding any personal controversy with Tyndall. Tyndall often wore his heart on his sleeve, at least to his friends. These letters show us his real character.

Notes

1. See C. Smith, *The Science of Energy: A Cultural History of Energy Physics in Victorian England* (Chicago: University of Chicago Press, 1998).
2. For an extensive discussion of the wider Tyndall/Stokes correspondence, see M. Baldwin, "Tyndall and Stokes: Correspondence, Referee Reports and the Physical Sciences in Victorian Britain," in *The Age of Scientific Naturalism*, ed. B. Lightman and M. S. Reidy (London: Pickering & Chatto, 2014), 171–86. For recent work by Aileen Fyfe and colleagues on reviewing cultures and the Royal Society, see N. Moxham and A. Fyfe, "The Royal Society and the Prehistory of Peer Review, 1665–1965," *Historical Journal*, FirstView, https://doi.org/10.1017/S0018246X17000334.
3. "Professor Tyndall on Scientific Research," *Leeds Mercury*, 5 December 1884. These

remarks are also reported in W. T. Jeans, *Lives of the Electricians: Professors Tyndall, Wheatstone, and Morse* (London: Whittaker, 1887), 97–98.

4. T. H. Huxley, "Art.VIII.—Darwin on the Origin of Species," *Westminster Review* 17 (April 1860): 541–70.
5. Letter 1697.
6. Notable examples include his lecture in Liverpool in 1870, "Scientific Use of the Imagination," and the Belfast Address in 1874.
7. The story of Tyndall's relationship with the Drummond family, and his approaches to two of the three daughters, is related in *The Ascent of John Tyndall*, chapters 7 and 8.
8. Letter 1597.
9. Letters 1604, 1630, and 1631.
10. See the introduction to *Tyndall Correspondence*, vol. 6, for a summary.
11. Letters 1642, 1682, and 1810.
12. Letter 1866.
13. "The Constitution of the Universe," *Fortnightly Review* 3 (December 1865): 129–44.
14. Letters 1707, 1735, and 1743.
15. J. Tyndall, "On the Physical Basis of Solar Chemistry," *Roy. Inst. Proc.* 3 (1861): 387–96.
16. Letter 1700.
17. See, for example, letter 1900.
18. Letter 1796.
19. Letters 1799 and 1815.
20. On Tyndall as referee for Stokes, see Baldwin, "Tyndall and Stokes," 171–86.
21. Letters 1780, 1781, 1782, 1783, 1784, and 1787.
22. Letters 1877, 1891, 1892, 1893, 1896, and 1898.
23. J. Tyndall, "The Veined Structure of Glaciers," *Roy. Inst. Proc.* 3 (1859): 72–78.
24. J. Tyndall, "On the Veined Structure of Glaciers; with Observations upon White Ice-Seams, Air-Bubbles and Dirt-Bands, and Remarks upon Glacier Theories," *Phil. Trans.* 149 (1859): 279–307.
25. Letters 1703, 1708, 1709, 1731, 1741, 1746, 1751, 1765, 1774, 1839, 1840, and 1845.
26. Letter 1594.
27. Joseph Dalton Hooker to Tyndall, undated, RI MS JT/1/TYP/8/2552.
28. Letter 1717.
29. *Life and Work*, 362.
30. F. Vaughan Hawkins, "Partial Ascent of the Matterhorn," in *Vacation Tourists and Notes of Travel in 1860*, ed. F. Galton (Cambridge: Macmillan, 1861), 282–304.
31. Letter 1727.
32. On Tyndall, risk, and masculinity, see R. D. Eaton, "In the 'World of Death and Beauty': Risk, Control, and John Tyndall as Alpinist," *Victorian Literature and Culture* 41, no. 1 (2013): 55–73; F. O'Gorman, "'The Mightiest Evangel of the Alpine Club': Masculinity and Agnosticism in the Alpine Writing of John Tyndall," in *Masculinity and Spirituality in Victorian Culture*, ed. A. Bradstock et al. (New York: Palgrave, 2000), 134–48; B. Hevly, "The Heroic Science of Glacier Motion," *Osiris* 11 (1996): 66–86; and M. S. Reidy, "Mountaineering, Masculinity, and the Male Body in Victorian Britain," *Osiris* 30 (2015): 158–81.
33. Advertisements, *Saturday Review*, 5 April 1862, 401.
34. Letter 1888.
35. *Life and Work*, 83.
36. J. Tyndall, "Physics and Metaphysics," *Saturday Review*, 4 August 1860, 140–41.
37. Letters 1749 and 1750.
38. Letter 1849 gives Tyndall's ostensible official reasons for resignation. See *The Ascent of John Tyndall*, 224–45, for a more extensive discussion of this episode.

39. Letter 1606.
40. Letters 1682 and 1743.
41. Letters 1885 and 1886. See also Journal, RI MS JT/2/13c/1205–11.
42. For an outline of the stimuli to Tyndall's work on radiant heat, see *The Ascent of John Tyndall*, 155.
43. Journal, 18 May 1859, RI MS JT/2/13c/1158.
44. J. Tyndall, "Note on the Transmission of Radiant Heat through Gaseous Bodies," *Roy. Soc. Proc.* 10 (1859d): 37–39.
45. J. Tyndall, "The Transmission of Heat of Different Qualities through Gases of Different Kinds," *Roy. Inst. Proc.* 3 (1859): 155–58.
46. There is no evidence that he, or any other European involved in this work, was aware that the American Eunice Foote had discovered the absorption of heat (from sunlight) by water vapor and carbon dioxide in 1856. See E. Foote, "Circumstances Affecting the Heat of the Sun's Rays," *American Journal of Science and Arts* 22 (1856): 382–83, and R. Jackson, "Eunice Foote, John Tyndall, and a Question of Priority," *Notes and Records* (2019, doi .org/10.1098/rsnr.2018.0066).
47. Letter 1733.
48. "Sur la transmission de la chaleur de qualité différente à travers les diverses espèces de gaz, par M. John Tyndall," *Cosmos* 15 (1859): 321–25; "Della trasmissione del calore di diverse qualita' attraverso al diversi gas, di T. Tyndall," *Il Nuovo Cimento* 10 (1859): 196.
49. Letter 1617, "Sur la diathemansie des gaz, Lettre de M. John Tyndall a M. le prof. Auguste de la Rive," *Bibliothèque Universelle de Genève* 5 (1859): 232.
50. Letter 1619.
51. J. Tyndall, "The Bakerian Lecture: On the Absorption and Radiation of Heat by Gases and Vapours, and on the Physical Connexion of Radiation, Absorption, and Conduction," *Phil. Trans.* 151 (1861): 1–36.
52. Letter 1767.
53. Letter 1771.
54. Letter 1773.
55. Letter 1793.
56. J. Tyndall, "On the Absorption and Radiation of Heat by Gaseous Matter. Second Memoir," *Phil. Trans.* 152 (1862a): 59–98.
57. Letter 1900.
58. Letters 1661, 1681, 1689, 1759, 1760, 1764, 1777, and 1790.
59. See *Tyndall Correspondence*, vol. 4, letter 0863.
60. Letter 1637.
61. Letter 1638.
62. Letter 1644.
63. Letter 1649.
64. Letter 1645.
65. Letters 1652 and 1655.
66. Letter 1656.
67. Journal, 19 November 1859, RI MS JT/2/13c/1163.
68. Letter 1658.
69. Letter 1659.
70. Letter 1655.
71. In a letter to Alfred Wills, Forbes wrote that he was poorly situated for the scientific world of London. James David Forbes to Alfred Wills, 12 April 1859, the Alpine Club Library.

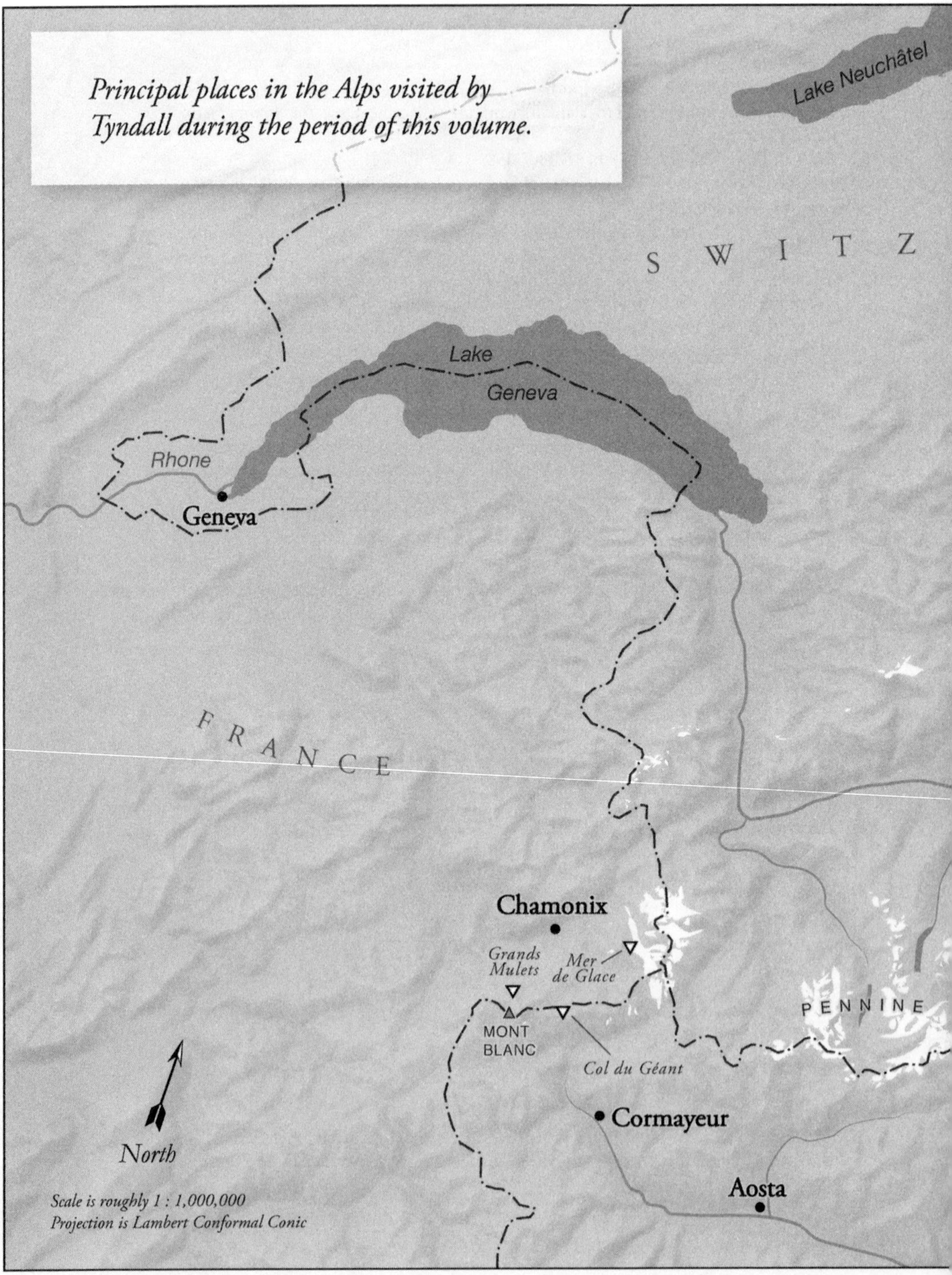
Principal places in the Alps visited by Tyndall during the period of this volume.
Lake Neuchâtel
S W I T Z
Lake
Geneva
Rhone
Geneva
F R A N C E
Chamonix
Grands
Mulets
Mer
de Glace
MONT
BLANC
Col du Géant
PENNINE
Cormayeur
Aosta
North
Scale is roughly 1 : 1,000,000
Projection is Lambert Conformal Conic

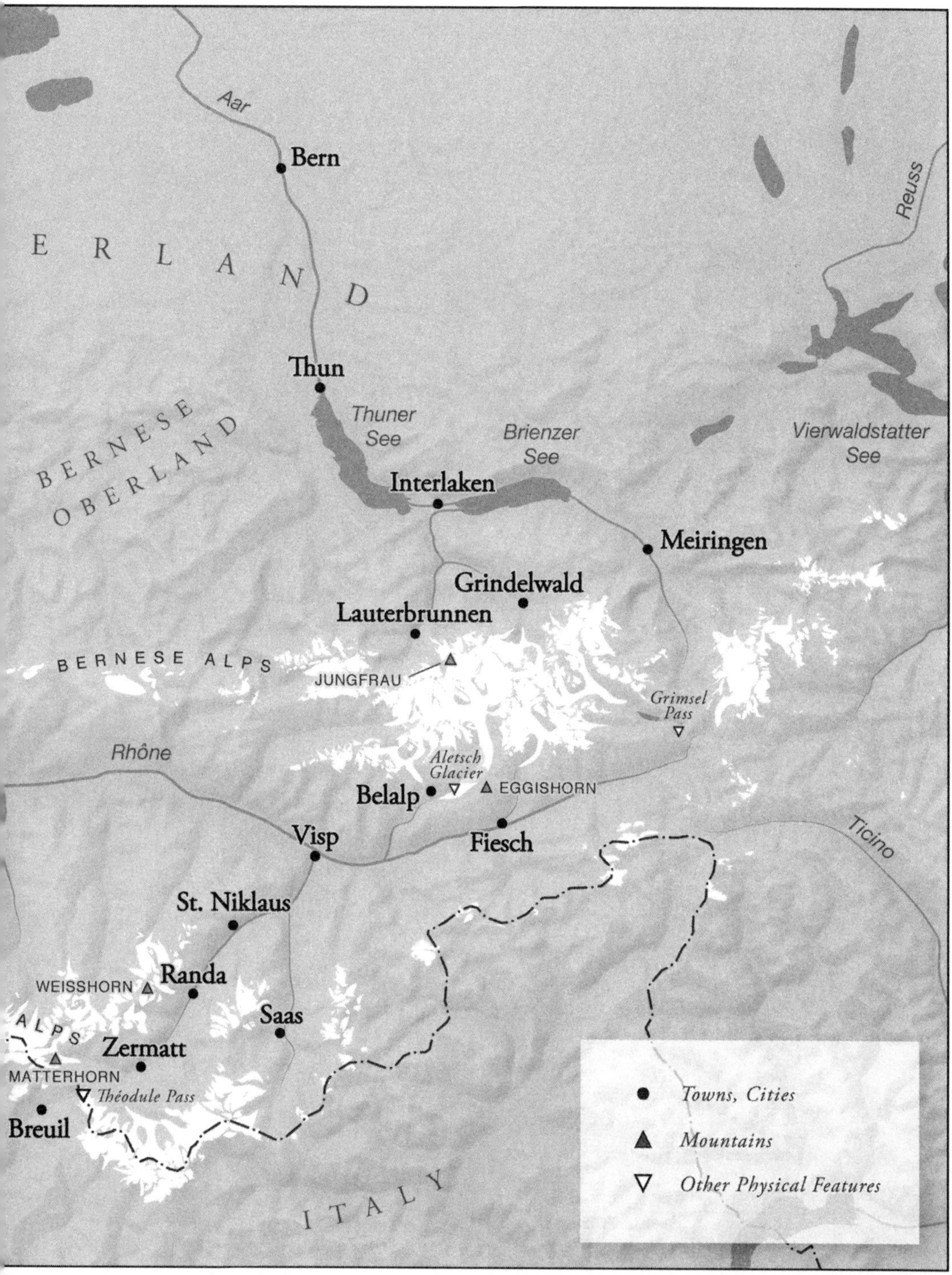

Credit: Benjamin Meader

EDITORIAL PRINCIPLES

Our aim is to include all letters to and from John Tyndall that are currently extant. Some of the letters are from published sources, such as letters written by Tyndall to newspapers or journals, but the majority are either the originals or typescripts produced by Louisa Tyndall from originals. One central editorial principle has been to transcribe, wherever possible, from original letters rather than Louisa's typescripts. This was because of Louisa's conscious and unconscious editorial interventions, a point discussed in more detail below. We have also aimed to reproduce, as accurately as possible, the text of the letters as they were written. Spelling mistakes that have appeared in the handwritten letters have therefore been preserved; mistakes appearing where only the typescript letter has survived are silently corrected, as in our judgment they are almost always typographical errors. But, in general, nonstandard spellings (e.g., the use of dialect) have been retained even when in a typescript-only letter (though there is the possibility that the spelling is Louisa's, not her husband's). Where the transcription is based on a typescript transcription by Louisa (or one of her assistants), this will be indicated by "LT Transcript Only." For those cases in which our transcription comes from typescript letters not done by Louisa, or it is unclear if she was involved, this will be indicated by "Typed Transcript Only." For the rare case when only a handwritten transcript of a letter exists, this will be indicated by "Transcript Only."

We have included illustrations that are part of the letter. In cases where it was possible to reprint the original handwritten image, we have done so. In other cases, volume editors have reconstructed drawings because it was not possible to reprint an image from the handwritten letter (due, for example, to deterioration of the original document).

While quotations within some handwritten and typescript letters use double quotation marks, for consistency we have used single marks in the transcriptions. However, if our source for a letter was a publication such as a journal, newspaper, or book, we have retained the quote marks as they appear in that publication.

The letters are presented in chronological order of writing. Where more than one letter has the same date, letters from Tyndall (in alphabetical order of addressee) come first, followed by letters to Tyndall (in alphabetical order of writer), with the exception that if the order of writing can be determined, that order takes precedence. Where letters cannot be dated precisely they are placed at the latest likely date of writing. (In some cases, if it is possible, but unlikely, that they were written later, notes indicate such possibilities.) To indicate the beginning of a new year, the date of that year has been inserted in a large font before the first letter from that year. Readers should note that although volumes are arranged in chronological order, letters that are discovered from years already covered in published volumes will appear in the final volume of the correspondence.

For non-English language letters, both the transcription of the original and the English translation are presented. The transcription of the non-English letter appears first, followed by the English translation in a smaller font. Editorial notes are inserted in the English translation. Some foreign letters are in an original hand, while others exist only as transcriptions by Louisa or others who often did not have a full command of the language. In the latter case these letters contain frequent spelling, grammatical, and interpretation errors that do not seem to have been committed by the original author. In these typescript-only cases, and only in these cases, spelling and grammatical errors have been corrected, as have poor interpretations where a better word choice was obvious from the context.

There is a standard format for each letter. The first line lists the author, if it is to Tyndall, or the recipient, if it is from Tyndall; the date; and the letter number. The second line lists any information supplied by the writer in their salutation, such as the place, date, or time of day when the letter was written. (In some cases, though, where the writer put the date at the foot of the letter, we have left it there in order to reproduce the text accurately.) The opening salutation, the text of the letter, and closing then follow. Then come the source and editorial notes, the latter indicated in the text by Arabic numerals.

Editorial notes are intended to provide the contextualization necessary to understand the contents of each letter. There are basically six types of editorial notes. Many of them are informative notes on persons mentioned in the letters. If a person referred to in a letter is not described in a note, they have been mentioned more than twice in the volume and they will have an entry in the biographical register at the back of the volume. For the rest—those mentioned once or twice—biographical information will appear in the editorial note appended to the letter in which he or she is first mentioned. When there is a second reference to that person, a note will refer back to the letter in which he or she was first mentioned. Biographical notes on well-known figures such as Faraday or Huxley stress what that person was doing at the time when the letter was written, rather than providing a full overview of the person's life.

A second type of editorial note identifies allusions and quotations. In the case of quotations, where possible, the first edition of the work is cited unless a different edition is mentioned in the letter. If a quote is not in English a translation is given. Glosses on obscure places, abstruse words, and words used in an unfamiliar sense constitute the third type of editorial note. The fourth type provides information about the context of the letter, drawing on such sources as contemporary publications, scholarly sources, Tyndall's personal journals and field notes, and other letters in the *Tyndall Correspondence*. The fifth type of editorial note is cross-references to other letters in the correspondence, referring to the letter number. The last type is information on significant textual changes that Tyndall made to the original that is thought to be relevant to the meaning of the letter.

There are a series of conventions for indicating when editors have given additional information about the text of a letter:

When Louisa Tyndall has made an annotation or marginal insertion that the editors have deemed important enough to retain, we have included her annotation in an editorial footnote.

When editors other than she have made insertions, they are indicated by unitalicized square brackets; for example, [word].

When certain words are ambiguous and we have had to conjecture their meaning, they are indicated by italicized square brackets; for example, *[*words*]*; when we have considered such words illegible the number of illegible words is indicated in full italicization and square brackets; for example, *[3 words illeg]*.

If a word or series of words are not present or have been destroyed—by inkblots, holes, or mold, for example—they are indicated as missing, italicized and in angular brackets; for example, *<3 words missing>*.

In extremely rare cases where it is clear that Louisa herself has intentionally destroyed a few words, they are indicated as excised italicized and angular brackets; for example, *<3 words excised>*.

Finally, a brief word about Louisa. She was responsible for collecting together many of Tyndall's letters in order to write a biography of her late husband. This biography was not completed. It is important to recognize that Louisa was not always reliable as a transcriber. She often threw out the manuscript letter when she had made her transcription, but some original manuscripts remain, which indicate her editing processes. Like any transcriber, she was sometimes inaccurate, but corrections show that she proofread her transcriptions. However, she also edited Tyndall's writing to make his letters seem more consistent or grammatically correct. For example, she made punctuation more formal, corrected grammar, and, occasionally, corrected quotations. She consistently corrected Tyndall's usual (deliberate) lower case for "god" to "God." Her typewriter was unable to insert the superscripts that Tyndall used in abbreviations, or the "&c" symbols that stood for an address at the foot of the letter. (Modern typefaces are unable to deal with all the variant scrawls that the "etc." symbol took, and they are all formalized here to "&c.") Louisa also left out details that she considered too private or too salacious for public consumption. She equated postmark and date of writing and overlooked the time taken to write some letters. When she inferred dates from internal evidence, her reasoning was sometimes faulty. Similarly, her identifications of people are sometimes in error. Independent evidence has always been sought for her dates and identifications. Lastly, the patterns of punctuation differ between LT sources and manuscript sources. Tyndall himself used dashes more often than commas, semicolons, and stops, and he used dashes of many different lengths, which are all standardized here to em-dashes. It is often difficult to distinguish between very short dashes and commas or very short dashes and stops. Similarly, capitalization is often ambiguous; distinguishing S/s, for "science," for example, is a particular problem. Therefore, scholars for whom punctuation and capitalization are significant will need to consult the manuscripts. Users should also be aware that throughout this project the transcribers and editors have generally worked from scans and photographs rather than the original manuscripts.

NOTE ON MONEY

In the nineteenth century the currency used in England was the pound sterling, usually abbreviated to £ (or, occasionally, to L). A pound consisted of 20 shillings (abbreviated *s*) and each shilling contained 12 pennies (*d*). A guinea was 21 shillings; that is, £1 1s. A crown was 5 shillings (5s), and half a crown equaled two shillings and sixpence (2s 6d). The smallest coin was a farthing, one-quarter of a penny (¼d).

Several forms of representation were in contemporary use. Sometimes the numbers for pounds, shillings, and pence were separated by dots, forward slashes, or spaces; thus, for example, 3 pounds, 4 shillings, and 5 pennies could be written as £3.4.5, £3/4/5, or £3 4 5. A zero could be represented by a dash; for example, 7 shillings was often written as 7/- or just 7/. A half crown was often written as 2/6.

This volume contains just one mention of currency from other countries (letter 1841). The Reichsthaler was a silver coin issued by Prussia from 1750 that became a common currency throughout much of Germany in the mid-nineteenth century before being replaced by the Vereinsthaler in 1857. In the mid-nineteenth century the exchange rate between the German and British currencies was approximately 7 Thaler per pound.[1]

1. H. Vizetelly, *Berlin under the New Empire: Its Institutions, Industry, Monuments, Social Life, Manners*, 2 vols. (London: Tinsley Brothers, 1879), 1:166.

TIMELINE OF JOHN TYNDALL'S LIFE

Year	Event
c. 1822	John Tyndall born at Leighlin Bridge, County Carlow, Ireland
c. 1836	Attends John Conwill's National School in Ballinabranna
1839	Begins employment as civil assistant with Ordnance Survey of Ireland in Carlow
1840	Becomes civil assistant in the Ordnance Survey Office at Youghal, Cork County
1841	First appearance in print (poem in *Carlow Sentinel* under pseudonym "W[alter] S[nooks]")
1842	Joins English Ordnance Survey in Preston
1843	Leads written criticism of Ordnance Survey in *Liverpool Mercury*; dismissed November
1844	Home in Ireland, mostly unemployed, until obtaining a position with the firm of Nevins and Lawton, Surveyors, of Manchester
1845–47	Works for Richard Carter, Surveyor, of Halifax
1845	Meets Thomas Archer Hirst in Halifax
1847	Begins teaching mathematics and surveying at Queenwood College, Hampshire
1848–50	At University of Marburg, working for PhD with Robert Bunsen and Friedrich Stegmann
1850	Returns to England; meets Michael Faraday and William Francis and attends his first British Association meeting, before returning to Marburg
1851	Moves from Marburg to Berlin (April–June); then returns to Queenwood
1852	Elected fellow of the Royal Society
1852	Friendship developed with Thomas Henry Huxley
1853	First lecture at Royal Institution (RI)
1853	Meets Herbert Spencer
1853	Appointed professor of natural philosophy at RI
1853	Awarded Royal Medal for magnetic work; declines when controversy arises over award
1856	Meets Thomas Carlyle
1857	Climbs Mont Blanc for first time with Hirst
1859	Demonstrates existence of greenhouse gases
1859	Joins Government School of Mines as professor of natural philosophy
1860	*The Glaciers of the Alps* published
1860	Meets John Lubbock and his wife, Ellen Lubbock
1861	First ascent of the Weisshorn

Year	Event
1861	Delivers his first Christmas Lectures at RI
1862	*Mountaineering in 1861* published
1862	Begins to champion Mayer's priority in discovery of conservation of energy
1863	*Heat Considered as a Mode of Motion* published
1864	Awarded Royal Society's Rumford Medal
1864	First meeting of X Club
1865	*On Radiation* published
1865	Engages in public controversy over the efficacy of prayer
1866	Succeeds Michael Faraday as scientific adviser to Trinity House
1867	Faraday dies; Tyndall becomes superintendent of the RI
1867	*Sound: A Course of Eight Lectures* published
1868	Successfully climbs Matterhorn
1868	Discovers the cause of light scattering, to be known as "Tyndall Effect"
1868	*Faraday as a Discoverer* published
1868	Lecture on "Scientific Materialism" at the British Association
1868	Resigns from Royal School of Mines
1869	Joins Metaphysical Society
1870	"On the Scientific Use of the Imagination" lecture at British Association
1870	*Three Scientific Addresses* published
1870	*Researches on Diamagnetism and Magne-Crystallic Action* published
1871	Meets Louis Pasteur for first time while in Paris
1871	*Hours of Exercise in the Alps* published
1871	*Light and Electricity* published
1871	*Fragments of Science* published
1872	Debates over how to measure the efficacy of prayer, aka the "Prayer-Gauge Debate"
1872	*Contributions to Molecular Physics in the Domain of Radiant Heat* published
1872	*The Forms of Water in Clouds and Rivers, Ice and Glaciers* published
1872–73	USA lecture tour
1873	*Six Lectures on Light* published
1874	As president of the BAAS, delivers the "Belfast Address"
1876	Marries Louisa Charlotte Hamilton
1877	Develops "Tyndallization" (lengthy sterilization process to destroy heat-resistant pores)
1877	Tyndall and Louisa build summer cottage, Alp Lusgen, at Belalp, northern side of Valais, above Brig
1877	*Fermentation and its Bearings on Phenomena of Disease* published
1879	*Fragments of Science*, which had gradually expanded, first published in two volumes
1881	*Essays on the Floating Matter of the Air* published
1885	Tyndall and Louisa build Hindhead retreat, Surrey Downs
1887	Resigns from RI
1890	Fight with Gladstone in the *Times* over Irish Home Rule
1892	Hirst dies
1892	*New Fragments* published
1893	Dies at Haslemere, Surrey, accidentally poisoned

TIMELINE OF EVENTS IN JOHN TYNDALL'S LIFE SPECIFIC TO VOLUME 7

Month and year	Biographical details	Notable social, political, and cultural events
March 1859	Gives a Friday Evening Discourse at the RI, "On the Veined Structure of Glaciers" (4th) The managers of the RI decide to increase Tyndall's salary to £300 per annum (7th) At a social event also attended by Mary and Maria Drummond in late March or early April (Journal, RI MS/JT/2/13c/1156)	The Society for Promoting the Employment of Women is founded on an unidentified date in 1859 Hugh Miller's *Sketchbook of Popular Geology* is published posthumously (19th)
April 1859	Finishes the series of twelve lectures on the "Force of Gravity" (12th) Travels to the Lake District in northwest England (15th–c. 21st) with Frankland. Leaves London for Windermere (15th), rows to Ambleside (16th), climbs Helvellyn and surrounding areas, before traveling to Keswick (16th–19th) Receives notice of election as honorary member of Société de Physique et d'Histoire Naturelle de Genève (c. 17th) Visits Richard Dawes in Hereford for "a few days" (c. 26th) (RGO Archives/MS.RG0.6/378)	The Second Italian War of Independence, also known as the Austro-Sardinian War, breaks out (29th). The war involved the French Empire, under Napoleon III, and the Kingdom of Piedmont-Sardinia, under Camillo Benso, Count of Cavour, and Victor Emmanuel II of Italy, against the Austrian Empire, under Franz Joseph I Construction of the Suez Canal begins (25th) Charles Dickens starts his new magazine *All the Year Round*, wherein he publishes *A Tale of Two Cities* in serial form
May 1859	Moves to 31 Aberdeen Place, Maida Vale (6th) Large break in journal until early November (4th) Communicates his research on radiant heat to the Royal Society (26th)	Several conflicts during the Second Italian War of Independence take place, including the Battle of Montebello (20th), the Battle of Varese (26th), the Battle of San Fermo (27th), and the Battle of Palestro (30th–31st) with a French and Sardinian victory over the Austrian Empire

Month and year	Biographical details	Notable social, political, and cultural events
May 1859		The 1859 United Kingdom general election (18th) sees the Whigs under Lord Palmerston maintain their majority in the House of Commons. In Ireland the Irish Conservative Party wins a majority in the under the Earl of Derby The Prussian geographer and explorer Alexander von Humboldt dies aged 86 (6th)
June 1859	Gives a Friday Evening Discourse at the RI, "On the Transmission of Heat of Different Qualities through Gases of Different Kinds" (10th)	The Second Italian War of Independence continues, with the Battle of Magenta (4th), and the Battle of Solferino (21st–24th). Although the Battle of Solferino results in a French and Sardinian victory over the Austrian Empire, both sides suffer large losses.
July 1859	Receives notice of election as honorary member of the Marburg Philosophical Society (22nd) Faraday intervenes on Tyndall's behalf on the situation with the Drummonds (c. 29th)	Napoleon signs an armistice with Austria in the Conference of Villafranca (11th). While this ends the Franco-Piedmontese war against Austria, unrest continues in Italy. The terms of the treaty of Villafranca are not agreed with Piedmont-Sardinia, causing Cavour to resign as president Big Ben, the Great Bell in the Palace of Westminster, chimes for the first time (11th) George Eliot publishes *The Lifted Veil*
August 1859	Travels to Chamonix (8th–c. 29th) via Dijon (9th) and Geneva (10th). Climbs Mont Blanc (c. 19th–20th). Writes a will prior to departing, and tells Hirst that he has bequeathed him his journals and to "never let any thing unkind be heard after me: There is not much of it but whatever there is suppress it" (RI MS JT/1/T/652, RI MS JT/1/HTYP/518)	

Month and year	Biographical details	Notable social, political, and cultural events
September 1859	Accepts the chair of physics at the Royal School of Mines (c. 10th) Spends time at Queenwood (c. 7th–21st) noting, "I wrote a good deal while in the country" (RI MS JT/1/T/653)	A solar storm takes place between 28 August and 2 September, including a large geomagnetic storm (1st–2nd) which results in bright auroras and the failure of telegraphic systems Big Ben cracks and is out of order for four years The annual meeting of the BAAS is held in Aberdeen (14th–21st)
October 1859	Assists Hirst with finding lodgings at 27 Aberdeen Place (6th)	A very severe storm, the Royal Charter Storm, hits the Irish Sea (25th–6th), causing an estimated eight hundred deaths. The storm takes it name from a steam clipper traveling from Melbourne, Australia, that wrecks close to Liverpool
November 1859	Decides against becoming a candidate for the chair of physics at the University of Edinburgh. Explains to Playfair that he "live[s] in hope that men in power will eventually become sufficiently alive to the importance of the subject, to grant to London the school of physics which the country so much needs." Researches Big Ben, the Great Bell in the Palace of Westminster, which had recently cracked, occasionally recording his progress in his journal Due to his improved financial circumstances increases his annual allowance to his mother, Sarah (13th) Attends the christening of Huxley's daughter, Marian (18th)	Charles Darwin publishes *On the Origin of Species* (24th) The terms of the Treaty of Villafranca are formally confirmed (10th), though unrest continues
December 1859	Continues research on Big Ben, contributing to a report ordered by the House of Commons, John Tyndall, "REPORT on the Great Bell of Westminster," *Bell (New Palace, Westminster): Copy of reports on the condition of the great bell of the new Palace of Westminster* (1860), p. 553	Queensland officially separates from New South Wales (10th) with Brisbane as the capital

Month and year	Biographical details	Notable social, political, and cultural events
December 1859	Dines at the Alpine Club (8th) Travels to the French Alps (22nd–1 January) via Paris (22nd) and Geneva (24th–25th) where he dines with de la Rive (24th) Experiments with glaciers and snow by Chamonix and Montanvert (25th–30th). Returns via Geneva and Paris (30th–1 January)	
January 1860	Starts his lecture series on Light at the RI (12th) to a large audience Gives the Friday Evening Discourse at the RI, "On the Influence of Magnetic Force on the Electric Discharge" (20th) Large break in journal (28th) until late October	Having resigned as prime minister of Piedmont-Sardinia in protest over the Treaty of Villafranca in July 1859, Cavour returns to office (21st)
February 1860	Investigates the relationship between organic and inorganic matter, and queries if Berthelot was the first to produce an organic compound from an inorganic (23rd) Determines to avoid social events for 6 weeks (ending c. 20 March) to finish *Glaciers*.	Hirst is offered an appointment to University College teaching mathematics (28th) Gustav Kirchhoff and Robert Bunsen announce their discoveries on the spectral patterns of chemical elements using a spectroscope as "Chemische Analyse durch Spectralbeobachtungen," *Poggend. Annal.* 186, 6 (1860): 161–189
March 1860	Finishes lecture series on light at the RI (29th) Signs petition against removing the natural history collections from the British Museum (29th) Works on completing the final stages of *Glaciers* Period of self-imposed avoidance of social events ends (c. 20th) Orders an unidentified scientific instrument from Duboscq, who specialized in optical and photographic equipment	The First Taranaki War in New Zealand begins (17th), a year-long violent conflict between Māori and the New Zealand government

Month and year	Biographical details	Notable social, political, and cultural events
April 1860	Sends *Glaciers* manuscript to the printers, William Clowes and Sons (c. 30th)	George Eliot publishes *The Mill on the Floss* (4th)
May 1860	Spends time in Queenwood (c. 13th–21st) and is unwell for a longer period	The conflicts in Italy continue, with the Expedition of the Thousand that last throughout the year with the Battle of Calatafimi (15th) and the siege of Palermo by General Giuseppe Garibaldi's forces Following the annexation of Savoy by France, de la Rive appointed to be a Swiss special ambassador to England The architect Charles Barry dies (12th)
June 1860	Gives Friday Evening Discourse at the RI, "Remarks on Some Alpine Phenomena" (1st) Takes a short holiday in Ireland (14th–26th) Leaves for BAAS meeting in Oxford (28th)	The annual meeting for the BAAS is held in Oxford (26th–3 July)
July 1860	Returns from BAAS meeting in Oxford (4th) *Glaciers of the Alps* published (6th)	The Nightingale Training School and Home for Nurses opens in London (9th) A total solar eclipse occurs (18th), particularly visible in Spain The Great Meteor Procession occurs (20th), a procession of Earth-grazing meteors observed in North America General Giuseppe Garibaldi's forces defeat the Neapolitan in Milazzo during the Expedition of the Thousand (20th)
August 1860	Leaves London for Switzerland (2nd) "Physics and Metaphysics" published in the *Saturday Review* (4th) Forces a route over the Lawinen Thor (10th) Attempts the Matterhorn with Bennen and Vaughan Hawkins (c. 20th)	

Month and year	Biographical details	Notable social, political, and cultural events
August 1860	Inspects the site of the accident on the Col du Géant (c. 23rd) Climbs Mont Blanc (c. 30th)	
September 1860	At the funeral of Huxley's son Noel (18th)	The Meteorological Office in London begins collecting weather reports (1st), spearheaded by Robert Fitzroy Large geomagnetic storm (1st) The Karlsruhe Congress, an international conference of chemistry, is held in Karlsruhe, Germany (3rd–5th). A key topic of discussion during the conference is the question of standardization of chemical formulae and atomic weights Garibaldi's forces take control of Naples during the Expedition of the Thousand (7th). The Kingdom of Piedmont-Sardinia takes control of central Italy in the Battle of Castelfidardo (18th). The Piedmont-Sardinia forces join with Garibaldi's.
October 1860	Finds that "there is no sensible thickness of olefiant gas that does not absorb an appreciable quantity of heat" (26th) Notes that the ability of a gas to radiate heat is proportional to its ability to absorb: "that quality, whatever it is, which enables them to accept the motion of the ether, enables them also to put in motion" (27th)	The conflicts in Italy continue, with the Battle of Volturnus (1st) between Garibaldi's expedition and Neapolitan forces, with an initial defeat for Garibaldi
November 1860	Challenges Thackeray about spirit-rapping (8th) Subscribes to the Tyndale monument (10th) Member of the Council of the Royal Society for a two-year term (30th)	Abraham Lincoln wins the US presidential election (6th) for the Republican Party with Hannibal Hamlin as vice-president The Siege of Gaeta begins (5th), marking the final stages of this period of unrest in Italy Robert Bunsen is awarded the Copley Medal and James Clark Maxwell is awarded the Rumford Medal, announced at the anniversary meeting of the Royal Society (30th)

Month and year	Biographical details	Notable social, political, and cultural events
November 1860		John Barlow resigns as secretary of the RI, and is succeeded by Henry Bence Jones
December 1860	Examination commitments (1st–8th) Dines at the Alpine Club and proposes Kennedy's health as incoming president (13th) Walking expedition to Snowdonia with Huxley and Busk (26th–31st).	Bence Jones is elected secretary of the RI (3rd) A brief ceasefire is ordered during the Siege of Gaeta (8th–13th)
January 1861	Gives the first of ten Working Men's evening lectures on "Magnetic and Electrical Phenomena" at the School of Mines (8th) Gives Friday Evening Discourse at the RI, "On the Action of Gases and Vapours on Radiant Heat" (18th) Gives the first of twelve weekly lectures on electricity at the RI (24th), ending on 25 April	A brief ceasefire is ordered during the Siege of Gaeta (9th–22nd) First round of general elections are held in Italy (27th)
February 1861	Gives the Bakerian Lecture at the RS, "On the Action of Gases and Vapours on Radiant Heat" (7th)	The Siege of Gaeta ends (13th), concluding the war between the Kingdom of Sardinia and the Kingdom of the Two Sicilies, with a Sardinian victory Second round of general elections are held in Italy (3rd) with a victory for Cavour and the Historical Right Political unrest in the United States grows, and the Peace Conference of February 1861 (4th) is held in Washington, DC, in an unsuccessful effort to reach agreements on issues that are dividing the union, in particular the slavery trade. President-elect Lincoln supports a ban on slavery in all states The Meteorological Office in London issues their first storm warning (5th)

Month and year	Biographical details	Notable social, political, and cultural events
March 1861	Awarded £200 from the Government Grant Committee at the RS for his work on radiant heat (20th) Leaves with Vaughan Hawkins and Francis Galton for a walking trip in Devon and Cornwall (23rd)	Lincoln inaugurated as president of the United States (4th), followed by a ten-day tour of the country Princess Victoria, Queen Victoria's mother, dies aged seventy-four (16th) Victor Emmanuel is declared King of Italy (17th) with Cavour as the first prime minister of Italy.
April 1861	Returns from walking trip in Devon and Cornwall (2nd) Gives the first of thirty-six lectures on physics at the School of Mines (15th)	The United Kingdom Census is recorded (7th), giving Tyndall's age as thirty-nine and occupation as "Professor of Philosophy" The American Civil War begins (12th), starting a four-year bloody conflict with the Baltimore Riot, also known as the Pratt Street Riot (19th). Lincoln drafts seventy-five thousand troops from Northern states (15th)
May 1861	Visits Sir Benjamin and Lady Brodie at Broome Hall with Grove (4th–6th)	Conflicts break out between Union and Confederate forces in Virginia as part of the American Civil War Queen Victoria issues a proclamation of neutrality (13th) on the American Civil War
June 1861	Gives Friday Evening Discourse at the RI, "On the Physical Basis of Solar Chemistry" (7th) Visits the Isle of Wight (16th–19th) Moves to 14 Waverley Place (Huxley's house) with Hirst (19th)	Cavour, the prime minister of Italy, dies (6th) aged fifty
July 1861	Leaves London for Switzerland (31st–early September)	The first major battle during the American Civil War, the First Battle of Bull Run (21st), concludes with a Confederate victory
August 1861	Large gap in journal (31st) until mid-November	

Month and year	Biographical details	Notable social, political, and cultural events
August 1861	Travels to Switzerland through Paris (1st–2nd). Reaches Basel (2nd–3rd), Meiringen (2nd–3rd), Hof (3rd–4th), and area around Grimsel (4th–8th). Continues to Fiesch (8th) and climbs the Eggishorn (10th). Trip to Belalp (13th–16th). Continues to Randa (16th), and ascends to summit of Weisshorn (19th). He remains in the area for the rest of the month (20th–31st)	
September 1861		Ellen Wood publishes the novel *East Lynne* (19th) in a three-volume version following its earlier publication in serial form in the New Monthly Magazine Annual meeting of the BAAS in Manchester (4th–11th)
October 1861	Visits Bence Jones in Folkstone (c. 29th) and Lubbock in Kent (c. 6th)	Charles Dickens publishes *Great Expectations* in three volumes, following its publication in serial form in the periodical *All the Year Round*
November 1861		The United Kingdom becomes more directly involved in the American Civil War, in a diplomatic crisis known as the Trent Affair (8th–14 January) Louis Agassiz is awarded the Copley Medal, announced at the anniversary
December 1861	Is unwell for several days (c. 15th). Gives first Christmas lectures, on light, in six installments (26th, 28th, 31st, and three dates in January)	Death of Prince Albert (14th)
January 1862	Accepts and then resigns the vice-presidency of the Alpine Club Gives the second half of his first Christmas lectures, on light (2nd, 4th, 7th) RI lectures on heat begins, in a series of 12 lectures (23rd–10 April)	The Union has its first significant victory over Confederate forces during the American Civil War in the Battle of Mill Springs (19th–1 May) in Kentucky Resolution of Trent Affair (14th)

Month and year	Biographical details	Notable social, political, and cultural events
January 1862	Gives the Friday RI Discourse, "On the Absorption and Radiation of Heat by Gaseous Matter" (17th)	
February 1862		The Hartley Colliery disaster, a large coal mining accident, kills 204 and prompts discussion on the safety of mining (16th)
March 1862		
April 1862	Completes his series of RI lectures on heat (10th), which began in January *Mountaineering in 1861* published (16th) Spends time in Dorset and Devon (c. 29)	New Orleans is captured by the Union (25th), an important event during the American Civil War The explorer and naval officer James Clark Ross dies (3rd)
May 1862	Travels to Isle of Wight (18th)	The Great London Exposition, or the International Exhibition of 1862, held in South Kensington, begins (1st) Westminster Bridge, designed by Thomas Page, opens (24th)

The Correspondence of John Tyndall

1859

From James David Forbes [late February 1859][1] 1594

XXXI. *Remarks on a paper "On Ice and Glaciers" in the last Number of the Philosophical Magazine. In a letter to* Prof. Tyndall. *By* Prof. J. D. FORBES.

MY DEAR SIR,

IN the part of your short paper "On Ice and Glaciers," in the Philosophical Magazine for February,[2] in which my writings are referred to, I think that my meaning has been misapprehended. Perhaps I can very briefly state what, in each of two cases, I conceive to be a needless source of difficulty unintentionally entertained by you with reference to my theoretical views or explanations.

The explanations which I now offer will tend, I hope, towards a reconcilement of our conclusions, and not to controversy.

I. With reference to the theory of "Regelation,"[3] you say, on p. 98 of the Phil. Mag. For February,—"Why is the 'rapid pounding' necessary in the experiment of Prof. Forbes? Doubtless in order that the ice may be brought into contact with the thermometer *before* its temperature has risen to 32°.[4] But give the ice time to rise to 32°; let its last residue of cold be abolished—the mass thus warmed, and in which the finest thermometer will not show the smallest fraction of a degree below 32°, may, with the utmost facility, be converted by pressure into *solid ice.*"

And again, "Let the thawing surface of a mass of ice be scraped away, so as to obtain a fine ice-powder possessing the temperature of that surface. Let not the alleged magazine of cold *within the ice* be at all called upon; such a powder, or more properly fine slush, the temperature of which no thermometer can show to be below 32°, may, as in the former case, be converted by pressure into solid slabs of ice."

According to my view, in the former of these cases the thermometer can only show the temperature of the *ice* when the latter is freshly pounded. If kept at a thawing temperature, each icy fragment becomes invested with a film of water which has the temperature of *water,* not of *ice.* You say, "Give the ice time to rise to 32°." I answer, that it is not in the nature of ICE to do so. The heat communicated becomes latent in the water produced, not one iota of it reaching the ice, which, whilst it remains Ice, is incapable of having its temperature raised. Precisely the same consideration applies to the second paragraph. Granulate the ice as we will, whilst any Ice remains, it has the temperature of Ice, not of Water. The powder or fine slush which you describe has the temperature of 32°, because the thermometer is in contact with the water alone, which forms so large a part of is composition. A thermometer is, in short, incapable of taking the temperature of ice unless that ice be *dry*. What, then, is the effect of the pressure which you describe as converting the "slush" into "solid slabs"? You admit, with me, that pressure is not necessary in the general case for regelation (p. 94). The effect of the pressure, then, is merely to banish nearly all the perfect water (that whose temperature the thermometer had shown) from the mass, leaving merely films of water (more or less viscid) between the still icy particles which constituted the slush. These particles, small as they are, contain each its magazine of cold, and being numerous as they are small, this cold is adequate to the consolidation of the interposed films of water, precisely as when the experiment is made on larger masses.

In this explanation I admit your view, that water is absolutely frozen in the process; but I believe also in the cohesive aggregation or *welding* under pressure of surfaces of ice softened by imminent thaw, though not yet reduced to water.

I think that, on consideration, it will appear clear to you that the preceding explanations render the facts you mention inevitable deductions from the doctrine that ice absorbs latent heat *gradually**.

II. As regards the theory of the Veined Structure of Glaciers[5] (p. 94 of your paper), there would perhaps be some inconvenience in my dictating the interpretation which certain passages of my writings more than twelve years old must bear, and declaring that no other interpretation is admissible. Such an *affidavit* of a pleader in his own cause might be received with distrust. It is better that others should study what I have written, and that they should conclude my meaning from an impartial examination of *the whole* documents. To facilitate this examination, I am about to publish, in a collected form, all my minor papers on the Theory of Glaciers[6] which can throw any light on this question; and I hope that *then,* when the difficulty of fairly and fully examining what I have said on the subject has been removed, my meaning will be found to be both definite and intelligible.

In the present instance, if, as stated in your paper, more than one writer has attached a certain significance to the terms of my Thirteenth Letter on Glaciers[7]—and if, of these writers, Professor William Thomson be one,—it is not, I think, too much to affirm that it is probable that their concurring, though independent, inference as to my meaning is correct. I have not myself any doubt of its correctness; but I repeat that I do not cite myself as a witness in a case of interpretation which affects my own credit. Professor Thomson, or some one who maintains the same opinion which he does, will, I have no doubt, be ready and willing to give grounds for his judgement.

Why I very reluctantly appear in the matter is for this reason—that I believe you have not altogether understood, perhaps have overrated, what I claim to have established, particularly in connexion with the change admitted by all parties to have occurred in some of my ideas, in consequence of my journey in 1846,[8] described in the Thirteenth Letter before mentioned. Several passages at pages 95 and 96 of your paper seem definitely to indicate that you ascribe to me an abandonment of a previously elaborate theory of the cause of the Veined Structure, and the substitution of a theory subversive of, and superseding in all points the old one. I will cite only one passage, including a quotation from myself, as given at p. 95 of your paper. It is as follows:—

"'Mr. Hopkins,' writes Professor Forbes in 1845, in reference to an experimental proof, 'denies that the ribboned (veined) structure is produced by differential motion ..... *No person who has seen the model made, or even been told how it was made, and inspects the ribboned structure upon its surface, can, I think, unless influenced by previous theoretical views, entertain any other opinion.'* Is it to be supposed that convictions thus strongly uttered, based upon years of observations, and established, according to the above quotation, by the testimony of the senses themselves, are meant to be reversed by a single observation, which, after all, is essentially defective, involving, in reality, not a *fact,* but an *opinion?*"

The italics are yours. Here I am represented as stating a theory of the veined structure in 1845, which, it is argued, I could not (as I am represented to have done) have relinquished in merely one or two sentences of a letter of 1846. I shall probably give you satisfaction when I assure you that the doctrine of the origin of the veined structure, so far as described in the preceding citation, and in several other passages of those pages of your article already referred to (pp. 95 and 96), was neither given up by me in 1846, nor in any subsequent year—nay, that I hold it still. I also entertain the very same views with reference to Mr. Darwin's theory of banded lavas which I did in 1845,[9] which at page 96 you understand me to have abandoned, or at least that I am represented to have abandoned them, in 1846*.

The difficulty, which until 1846 did not become thoroughly plain to me, was this:—How is the structure of glassy ice induced in the glacier without previous fusion and fresh congelation? In 1842 I believed that the plates of hard blue ice, which by their interposition between spaces of a less compact character, compose the Veined Structure, could only be due to the freezing of water in the glacier. I considered them as veins of infiltrated water frozen during winter. To the same cause I then attributed the conversion of the granular mass of the *névé*[10] into glassy ice. It was always a reluctant admission on my part: for the congelation of infiltrated water to any great extent, even in winter, was a relic of the Dilatation theory of glacier motion;[11] and against that theory I had entered an earnest protest. Between the printing of my Fourth Letter on Glaciers in 1842,[12] and the publication of my 'Travels' in 1843,[13] I had already abandoned the use of this reluctantly admitted fact—the winter congelation of the *depths* of the glacier—so far as I had employed it to explain the recovery of the level of the glacier between autumn and the following spring; and I felt my position to be stronger when I obtained another explanation*. A similar but more effectual deliverance from a felt difficulty arose in 1846, when I arrived at a clear persuasion of what (as will appear form a comparative study of my writings) had already dawned upon me some time previously, namely, that the glassy structure of ice is attainable by the cohesion under pressure (especially if aided by motion with friction, or *kneading*) of the semi-opake[14] and porous material of the glacier.

Although I had previously expressed several times my growing belief that the renewed cohesion of the bruised surfaces of the ice in the glacier proper, under the mutual pressure of their parts, might account for the facts both of motion and of structure there, I had not until the summer of 1846 disembarrassed myself of the complication arising from the conversion of granular snow in the *névé* into pellucid ice. I had, indeed, come very near to it; for already, in 1844, I had approximated the two phænomena in the following passage, in which, after speaking of the differential motion which tends to produce the veined structure in the glacier proper, I added—"I believe that it is during the progress of the glacier thus subjected to a new and peculiar set of forces depending upon gravity, and which remodel its internal constitution by substituting hard blue ice in the form of veins for its previous snowy texture, that the horizontal stratification observed in the higher part of the glacier or *névé* is gradually obliterated*." You see that even at this date the formation of the hard blue ice of veined structure, and the conversion of the *névé* into pellucid ice, appeared to me to be very intimately connected,—so much so, that until the last could be explained without interior congelation, little was gained in point of simplification by rejecting it in the former instance. It will be found, as I have said, by a consecutive perusal of what I wrote between 1844 and 1846, how gradually I approached the conviction (only obtained on the spot in the latter year) that the phænomena might be *all* explained by pressure and cohesion, the latter arising from the softening of snow or ice when thaw is imminent or in progress. In 1846, then, I abandoned no part of the theory of the veined structure on which, as you say, so much labour had been expended, except the admission, always yielded with reluctance, and got rid of with satisfaction, that the congelation of water in the crevices of the glacier may extend in winter to a great depth†.

I remain, dear Sir, | Yours faithfully, | JAMES D. FORBES.

Prof. Tyndall, | Royal Institution.

P.S. —It is foreign to the purpose of this letter to express an opinion as to whether any theory which you have formed as to the "veined structure" is independent, or otherwise, of the fundamental idea of differential motion arising under conditions of unequal pressures combined with friction. I reserve it until your matured views are published. But it did appear to me, some time ago, that in Mr. Sorby's most ingenious researches on slaty cleavage,[15] and in your extension of them to homogeneous bodies such as wax and dough, the efficiency of differential motion was tolerably present. —J. D. F.

[* See Mr. Faraday's remarks on Regelation in our present Number, p. 162,—ED.][16]

[* For the sake of clearness, I hope I may be permitted to state that the assumption to which I referred is expressed by Mr. Darwin in the following words:—"In the ice the porous laminæ are rendered distinct by the subsequent congelation of infiltrated water;"[17] and my statement is, that Prof. Forbes does not use a word which would lead us to suppose that he wished to modify that assumption.[18]—Phil. Mag. 1845, vol. xxvi. p. 354[19] . —J. T.]

* 'Travels,' 1st ed. (1843), p. 384, or 2nd ed. p. 386; and note the limitations under which congelation in the interior of the glacier is admitted, at pages 232, 360, 372 of both editions. In so far as the *glacification* of the *névé* is concerned, it was no assumption of mine. I accepted the universal opinion of the time as stated by De Saussure, as well as by MM. de Charpentier and Agassiz.

* Sixth Letter on Glaciers, 1844.[20]

† That this is a question of some nicety will be readily perceived; and my caution in overtly rejecting its influence altogether may be understood. For winter congelation must act to some depth; and certain phænomena cannot be explained without it. Such are the lenticular frozen cavities which you have described in your earliest paper on glaciers, which I take to be the same as those longitudinal infiltrated cracks previously described by me as visible in the glacier of Bossons,[21] and other torrential glaciers. Such icy infiltrations are apparently too wide to be produced except by direct congelation. But then they are always near the side or surface of a glacier.

J. D. Forbes, 'Remarks on a paper "On Ice and Glaciers" in the last Number of the Philosophical Magazine. In a letter to Prof. Tyndall. By Prof. J.D. Forbes', *Phil. Mag.,* 17 (1859), pp. 197–201.

---

1. *[late February 1859]:* the month and year are given based on relation to letter 1588 (12 February 1859) in *Tyndall Correspondence,* vol. 6, in which Forbes informed Tyndall that he would be writing this for the *Phil. Mag.* It appeared in the March issue, so was probably written at the end of February.

2. *short paper . . . for February:* J. Tyndall, 'Remarks on Ice and Glaciers', *Phil. Mag.*, 17 (1859), pp. 91–96. See also 'On the Veined Structure of Glaciers; with Observations upon White Ice-Seams, Air-Bubbles and Dirt-Bands, and Remarks upon Glacier Theories', *Phil. Trans.*, 49 (1859), pp. 279–307.
3. *the theory of "Regelation":* Tyndall explained glacier motion by the *fracture* of ice (similar to his theory of slaty cleavage, which he lectured on at the RI on 6 June 1856, 'Comparative view of the Cleavage of Crystals and Slate Rocks') and *regelation* (the refreezing of water between two pieces of ice). In *Glaciers of the Alps,* Tyndall described regelation as when 'Two pieces of ice at 32° Fahr., with moist surfaces, when placed in contact freeze together to a rigid mass'. He continued, 'When the attachments of pressed ice are broken, the continuity of the mass is restored by the regelation of the new contiguous surfaces. Regelation also enables two tributary glaciers to weld themselves to form a continuous trunk; thus also the crevasses are mended, and the dislocations of the glacier consequent on descending cascades are repaired. This healing of ruptures extends to the smallest particles of the mass, and it enables us to account for the continued compactness of the ice during the descent of the glacier' (p. 423).
4. *before its temperature has risen to 32°:* much of the debate surrounding glacial movement centered around the nature of ice at 32° F. See, for example, J. D. Forbes, 'Prof. Forbes on some properties of ice near its melting point', *Athenaeum,* 1608 (21 August 1858), p. 238.
5. *As regards the theory of the Veined Structure of Glaciers:* Tyndall applied his theory of regelation to explain the veined structure and movement of glaciers. See 'On the Veined Structure of Glaciers; with observations upon White Ice-seams, Air-bubbles and Dirt-bands, and remarks upon Glacier Theories', *Phil. Trans.,* 149 (1859), pp. 279–307.
6. *I am about to publish, in a collected form, all my minor papers on the Theory of Glaciers:* J. D. Forbes, *Occasional Papers on the Theory of Glaciers* (Edinburgh: Adam and Charles Black, 1859).
7. *my Thirteenth Letter on Glaciers:* J. D. Forbes, 'Thirteenth Letter on Glaciers. Addressed to Professor Jameson', *Edinburgh New Philosophical Journal,* 42 (1847), pp. 136–54. First read at the Royal Society of Edinburgh on 21 December 1846. It also appeared in his collection *Occasional Papers* (1859), pp. 186–205.
8. *my journey in 1846:* Forbes visited several glaciers in 1846, including the Unteraargletscher (formed by the confluence of the Lauteraargletscher and Finsteraargletscher) and the Mer de Glace.
9. *I also entertain the very same view with reference to Mr. Darwin's theory of banded lavas which I did in 1845:* Forbes drew comparisons between the structure of glaciers and some volcanic rocks, and communicated with Darwin on this matter. See 'Extracts from Letters to the General Secretary, on the Analogy of the Structure of some Volcanic Rocks with that of Glaciers. By C. Darwin, Esq., F.R.S. Specimens were exhibited. With Observations on the same subject, made by Professor Forbes', *Edinburgh New Philosophical Journal,* 38 (1845), pp. 369–73.
10. *névé:* compacted or hardened snow, as found on the upper part of a glacier (*OED*).
11. *Dilatation theory of glacier motion:* Forbes described the dilatation theory of glacier

motion in his *Occasional Papers* (1859): 'The glacier being traversed by innumerable capillary fissures, and being in summer saturated with water in all its parts, it was natural to invoke the freezing action of the night to convert this water into ice, and by the amount of its expansion to urge the glacier onwards in the direction of its greatest slope' (p. 249). First proposed by the Swiss scholar Johann Jakob Scheuchzer (1672–1733) after he visited the Rhône and other glaciers of the Alps in 1705, in *Itinera per Helvetia Alpinas Regiones* (1723), the theory was also promoted by de Charpentier in his *Essai sur les glaciers et sur le terrain erratique du bassin du Rhône* (Lausanne, 1841), and Agassiz in his Études sur les Glaciers (Neuchâtel: Gent et Gassmann, 1840).

12. *Fourth Letter on Glaciers in 1842:* J. D. Forbes, 'Fourth Letter on the Glacier Theory, to Professor Jameson', *Edinburgh New Philosophical Journal,* 24 (1843), pp. 1–10. It also appeared in his collection *Occasional Papers* (1859), pp. 26–35.
13. *publication of my 'Travels' in 1843:* J. D. Forbes, *Travels through the Alps of Savoy and Other Parts of the Pennine Chain with Observations on the Phenomena of Glaciers* (Edinburgh: Adam and Charles Black; London: Longman, Brown, Green, and Longmans, 1843).
14. *semi-opake:* semi-opaque, only partially transparent (*OED*).
15. *Sorby's most ingenious researches on slaty cleavage:* Sorby undertook research on slaty cleavage through the 1850s. See H. C. Sorby, 'On Slaty Cleavage, as exhibited in the Devonian Limestones of Devonshire', *Phil. Mag.,* 11 (1856), pp. 20–37; and H. C. Sorby, 'On the Theory of the Origin of Slaty Cleavage', *Phil. Trans.,* 12 (1856), pp. 127–29. In the latter, Sorby offered a correction to Tyndall's representation of his views on slaty cleavage. Tyndall's response, in the same issue, is J. Tyndall, 'Observations on the preceding Paper', *Phil. Trans.,* 12 (1856), pp. 129–35.
16. *[* See Mr. Faraday's remarks on Regelation in our present Number, p. 162*: refers to Faraday's article 'XXIV. On regelation, and on the conservation of force', *Phil. Mag.,* 17 (1859), pp. 162–69.
17. *"In the ice the porous lamina are rendered distinct by the subsequent congelation of infiltrated water;":* this quotation is from Darwin's book *Geological Observations on Volcanic Islands* (1844) p. 70.
18. *Prof. Forbes does not use a word which would lead us to suppose that he wished to modify that assumption*: Tyndall, wrote 'In his reciprocal reference to Mr. Darwin's theory, Prof. Forbes endorses the comparison of that eminent naturalist, and *does not use a word which could lead us to suppose that he wished to modify Mr. Darwin's assumption*', in 'Remarks on Ice and Glaciers', *Phil. Mag.*, 17 (1859), p. 96.
19. *Phil. Mag. 1845, vol. xxvi. p. 354:* The article 'Notice respecting New Books:—Darwin on the Volcanic Islands visited during the Voyage of the Beagle.', *Phil. Mag.,* 26 (1845), pp. 342–58, includes a discussion of the links between Darwin's observations of volcanic islands and Forbes' views of the structure of glaciers and their motion on pp. 354–55.
20. *Sixth Letter on Glaciers, 1844:* J. D. Forbes, 'Sixth Letter on Glaciers. Addressed to the Right. Hon. Earl Cathcart', *Edinburgh New Philosophical Journal,* 37 (1844), pp. 231–49. It also appeared in his collection *Occasional Papers* (1859), pp. 43–56.
21. *the glacier of Bossons:* a large glacier on the north side of Mont Blanc.

From Julius Plücker 7 March 1859 1595

My dear Sir!

I thank you for your very kind letter of the 14th of February.[1] Prof. Faraday informed my self that my last letter[2] did not reach you in town. To answer old letters is not at all amusing, therefore I thank you the more.

I am very glad to see that, with infatigable zeal, you continue your researches about glaciers[3] and that finally you succeeded to reduce all principal phenomena to their physical causes.[4]

I am working always on electric discharges through rarefied gases. Lately I finished that part of these researches which regards the action of the Magnet on such currents. Two months ago, after having sent my paper to Poggendorff,[5] I gave in a letter to Prof. Faraday[6] a short account of some of the most striking new phenomena[.]

In the general case the positive light, if acted upon by the magnet, moves in spirals towards the negative wire where heat or chemical action are produced. In all cases the law of these spirals and their modifications easily derives from the well known magnetic action on an electric particle moving towards the negative wire. Properly speaking, the negative light, exhibiting the magnetic surfaces, has nothing to do with the current itself; I think it a secondary phenomenon.

I have before me some papers of mine, but I should not like to send them over to my English friends, but joined to the best and longer paper provided now.[7]

Recently all my attention is absorbed by the splendid phenomena of electric gas spectra.[8] Generally in the case of such spectra all the light is concentrated into a certain number of bright luminous lines, of an absolutely constant refrangibility.[9] These lines will in most cases replace with great advantage Fraunhofers dark lines;[10] in my experimental researches I continually made use of the three beautifull lines of hydrogen gas. —I procured an excellent instrument in order to measure the positions of the questioned lines with greatest accuracy. I examined already a certain number of gas spectra—where principally, I meet with chemical questions.

Nearly every week I get new wonderfull appearances, not unworthy to be schown in Roy. Institution. So I sit lately in examining the vapour of the chlorides of tin and titane. The discharge through the vapour of the feros chloride in an larger tube is of a deep blue colour, which, when the discharge enters a very narrow tube, becomes a bright one of gold. The Magnet, when acting upon the larger tube provokes beautifull golden lightnings within the blue gas.

I was happy to know that Prof. Faraday continues very well. Pray present to him my compliments.

Most truly | Yours | Plücker | Bonn 7/3 59

RI MS JT/1/P/128

---

1. *your very kind letter of the 14th of February:* Tyndall sent Plücker a letter dated 13 February 1859 (see *Tyndall Correspondence,* vol. 6, letter 1589).
2. *my last letter:* letter missing (see also *Tyndall Correspondence,* vol. 6, letter 1589, n. 1).
3. *you continue your researches about glaciers:* Tyndall had been researching the nature and behaviour of glaciers since the mid-1850s. In 1859 Tyndall was working on his book *The Glaciers of the Alps* (see n. 4).
4. *finally you succeeded to reduce all principal phenomena to their physical causes:* likely refers to Tyndall's proposed theory of glacial motion. Tyndall believed that glaciers moved by perpetually fracturing and relegating, and he elaborated on this theory in his book *The Glaciers of the Alps: Being a Narrative of Excursions and Ascents, an Account of the Origin and Phenomena of Glaciers and an Exposition of the Physical Principles to which They are Related* (London: John Murray, 1860). It was published on July 6, 1860.
5. *I am working always . . . having sent my paper to Poggendorff:* likely refers to J. Plücker, 'Ueber die Constitution der elektrischen Spectra der verschiedenen Gase and Dämpfe', *Poggend. Annal.,* 107 (1859), pp. 497–539.
6. *I gave in a letter to Prof. Faraday:* Plücker wrote to Faraday on 3 July 1859, where he elaborated on his experiments using Ruhmkorff's Apparatus (letter 3612, *Faraday Correspondence,* Vol. 5, 1855–60, pp. 530–31).
7. *I have before me some papers of mine . . . joined to the best and longer paper provided now:* the enclosed paper is missing, but may refer to J. Plücker, 'Abstract of a Series of Papers and Notes Concerning the Electric Discharge through Rarefied Gases and Vapours', *Roy. Soc. Proc.,* 10 (1860), pp. 256–69, which was read to the society on 19 January 1860.
8. *the splendid phenomena of electric gas spectra:* the light produced by a gas-charge lamp.
9. *refrangibility:* refers to the ability of light to be refracted (*OED*).
10. *Fraunhofers dark lines:* the dark lines in the solar spectrum, caused by substances in the solar atmosphere absorbing radiation in specific parts of the continuous solar spectrum. They were noticed by Wollaston, investigated by Joseph Fraunhofer (1787–1826) and named after him (*OED*).

From Juliet Pollock [9 March] 1859[1] 1596

My dear M^r^. Tyndall

I write to tell you with what true, heartfelt satisfaction I heard from Sir Charles Hamilton[2] of the last resolution passed at the R.I. concerning you.[3]

Sir Charles came at once (with his usual benevolence) to tell it to me knowing the pleasure that I should derive from the intelligence. —

I trust that you are well and happy, and that I shall see you soon to shake hands with you, and tell you better than I can write it, how glad I must always be to hear of every augmentation to your welfare.

—If you want an interesting, clever, admirably written novel; get Adam Bede.[4]—

Yours most truly | Juliet Pollock

59 Montague Squ[are][c] W. | Ash Wednesday—59/

RI MS JT/1/P/183
RI MS JT/1/TYP/6/1953

1. *[9 March] 1859:* the letter is signed Ash Wednesday '59, which fell on 9 March.
2. *Charles Hamilton:* Sir Charles Hamilton was a Manager of the RI.
3. *the last resolution passed at the R.I. concerning you*: the Managers of the RI decided to increase Tyndall's salary to £300 per annum. This was proposed by Jones and seconded by Webster (RI Manager's Minutes, p. 266, 7 March 1859). See also letters 1597, 1602, and 1649.
4. *Adam Bede:* a novel written by Mary Anne Evans (1819–80) under the pen-name George Eliot.

To Juliet Pollock 9 March 1859 1597

9th March | 1859

My dear Mrs. Pollock,

Your letter[1] is just like you—you have always been my kind & warm friend and you are so now!

Nothing could be nicer than the spirit in which this kind deed[2] has been done, nothing more grateful to the feelings of a man like myself than the manner in which it has been communicated to me.

I have been thinking of you very much of late, and nearly a year ago you were in my thoughts in connexion with the same subject.[3] As we and Walter[4] sat together watching the performance of Mr Spedding[5] in the archery ground you little knew the soil upon which some of the words which then fell from you at that time were falling.

It would have been perhaps well for me if I had then and there asked your counsel on the subject which occupied my thoughts, I thought of doing so often afterwards, but I finally thought it better and manlier to keep it to myself.

I do not know whether I shall speak of it even now, but if I should require your advice on a matter which affects me personally very deeply, though it is of absolute unimportance to the rest of the world, I want to know whether you feel sufficient interest in me to listen to a round unvarnished tale of half an hour's duration, and to give me the benefit of your advice afterwards.

It is a question which requires the consideration of a lady, and I think I could tell it better to you alone; you can repeat it to Mr. Pollock afterwards—but it is a matter which must rest sacred between ourselves. I want simply to check my own conduct, to see whether what my feelings tell me to be right is right—I want to act truly and to avoid acting unkindly if I can.

Ever yours | John Tyndall.

RI MS JT/1/T/1132
RI MS JT/1/TYP/6/1952

1. *Your letter:* letter 1596.
2. *this kind deed:* Tyndall's salary at the RI had been increased, see also letters 1596, n. 3 and 1602.
3. *I have been thinking of you very much of late, and nearly a year ago you were in my thoughts in connexion with the same subject:* this likely refers to an ongoing conversation regarding Tyndall's past relationship with Mary Drummond. Tyndall recorded an accidental encounter with a woman presumed to be Mary in his journal for March 4 1859 (Journal, RI MS JT/2/13c/1155): 'Probably that clever woman thinks that she will again succeed in twisting me round her finger—doubtless she is capable of trying to revive it all and give me another draught which should make me hollow her like another *[circe]*, but if she knew how matters stand she would spare her clever head the trouble.' See also letter 1604.
4. *Walter:* Walter Herries Pollock.
5. *Mr Spedding:* likely the editor and biographer James Spedding (1808–81) particularly known for his work on Francis Bacon (*ODNB*).

From George Gabriel Stokes 12 March 1859 1598

Lensfield Cottage | Cambridge | March 12th 1859

My dear Dr Tyndall,

My wife charged me most particularly to thank you for the ticket, or rather order, you were so kind as to send her for your lecture.[1] She would have been very glad to have gone if it had been in her power. I intended to have thanked you viva voce but both last Thursday and the Thursday before other things put it out of my head when I saw you.

I have received a paper from Dr Robinson[2] 'On the stratification of the electric light' for the Phil. Mag.[3] It will occupy 7 pages in print. I mention it now that you may expect it. I keep it for the present to read more carefully as it is a subject I have thought a good deal about. I intend to post it so as to arrive in London on Monday Morn[ing]g. I will address it to you at the Royal Institution.

Yours very truly | G. G. Stokes.

RI MS JT/1/S/220
RI MS JT/1/TYP/4/1363

1. *your lecture:* Tyndall gave a series of 12 lectures on the Force of Gravity at the RI. Stokes is likely referring to the lecture Tyndall gave on 3 March rather than 10 March 1859, given that Stokes noted that he previously intended to thank Tyndall for the ticket.
2. *Dr Robinson:* Thomas Romney Robinson (1793–1882), astronomer and physicist. The Robinson anemometer is named after him (*ODNB*).
3. *a paper from Dr Robinson 'On the stratification of the electric light' for the Phil. Mag:* this paper was published as T. R. Robinson, 'XLII. On the Stratification of Electric Light', *Phil. Mag.*, 17 (1859), pp. 269–74.

To The Editor of the *Morning Post* 14 March 1859 1599

SUBMARINE TELEGRAPHS.
TO THE EDITOR OF THE MORNING POST[1]

SIR,—With reference to the notice on the submarine telegraph which appears in your impression of to-day,[2] I hope you will permit me to say that my knowledge of the subject[3] at present extends simply to the fact that I have submitted india-rubber and gutta percha[4] covered wires to a test of extreme severity, and have found that insulating powers of both substances excellent. I do not now know which of them is best, and I have never made an experiment with either of them in which hydraulic pressure was applied.[5]—I am, sir, your obedient servant,

14th March, 1859. | JOHN TYNDALL.

Morning Post, Wednesday, March 16, 1859

1. *TO THE EDITOR OF THE MORNING POST:* Algernon Borthwick (1830–1908). Borthwick was appointed editor of the *Morning Post* following the death of the previous editor, his father Peter Borthwick (1804–52), in 1852. The editorial line of the paper was conservative (*ODNB*).
2. *the notice on the submarine telegraph which appears in your impression of to-day:* 'Submarine Telegraph', *Morning Post,* 14 (March 1859), p. 3.

3. *my knowledge of the subject:* Tyndall was cited in the article as having given assurance to the company producing a new process of insulation for the telegraph wire (Messrs. Silver, of Cornhill and Bishopsgate-street) that india-rubber would be able to withstand the hydraulic pressure under water.
4. *gutta percha:* a substance that resembles rubber, derived from a type of Malaysian tree of the Palaquium genus (*OED*).
5. *I have never made an experiment with either of them in which hydraulic pressure was applied:* according to the article, Tyndall had stated that he had performed tests on more sensitive wiring with no problems with the coating. See also letter 1755, n. 2.

To Heinrich Debus [16 March 1859][1] 1600

My dear Heinrich,

Your paper[2] has reached me & I have sent it to the Royal Society—

It is very nicely prepared and your abstract is quite in order.

Before Easter I do not expect to be able to leave town—but if you could then join Frankland and myself for a week's trip somewhere we should both be delighted.

I am right glad to hear that your numbers are augmenting.[3]

Ever yours | John Tyndall | Wednesday

RI MS JT/1/T/339

---

1. *[16 March 1859]:* The letter is dated based on the reference to Debus's paper, which is assumed to be 'On the Oxidation of Glycol and on some Salts of Glyoxylic Acid', *Roy. Soc. Proc.,* 9 (1859), pp. 711–16. This article included a note, 'Communicated by Dr. Tyndall. Received March 16, 1859'. 16 March 1859 was a Wednesday, which corresponds with the note 'Wednesday' in this letter.
2. *Your paper:* see n. 1.
3. *I am right glad to hear that your numbers are augmenting:* refers to Queenwood; see also letter 1685, n. 5.

To Sir James (Unknown) 19 March 1859 1601

My dear Sir James

It would give me very great pleasure to accept the invitation with which you have honoured me[1] were it not that I am already pledged to Lady Brodie[2] for Wednesday the 23rd.

Most truly yours | John Tyndall

19th March 1859

Bonham's, Knightsbridge, Auction, 15 Nov 2017, lot 93

1. *the invitation with which you have honoured me:* invitation missing.
2. *I am already pledged to Lady Brodie:* unknown event with Lady Brodie, Anne Brodie, neé Sellon (1797–1861). Lady Brodie was married to the physiologist and surgeon Sir Benjamin Collins Brodie, first baronet (1783–1862) (*ODNB*).

To Managers of the RI[1] 19 March 1859 1602

To the Managers of the Royal Institution

Gentlemen,

You have recently, without solicitation on my part, but animated solely by your own kindness, augmented my income by one hundred pounds a year.

This advance is no doubt of importance to me, but I should be untrue to my own feelings if I omitted to state that I derive a higher pleasure from the manner in which the act has been done and communicated to me, than from the addition which it makes to my material possessions.

Believe me, gentlemen, the kindness, the delicacy, the perfect gentlemanly spirit, which you have manifested towards me, are duly appreciated by me.

I take it for granted that it is not on account of my poor lectures[2] that this deed has been accomplished; that it is your desire to encourage and enable me to follow, according to my ability, that course of original investigation to which the World owes so much of its scientific knowledge and the Royal Institution so much of its renown. In this regard I would only ask your patience, and I cherish the humble trust that the future will not quite defeat the strong desire which I now feel to render myself worthy of the kindness so freely & spontaneously bestowed.

I remain Gentlemen | Your obedient Servant | John Tyndall.

19th March | 1859.

RI MS AD_03_04

1. *To Managers of the RI:* The Managers of the RI in March 1859 were, as listed in the RI Managers' Minutes XI, p. 277 (5 April): Lord Ashburton, Warren De la Rue, George Dodd, Sir Charles Fellows, William Grove, Sir Charles Hamilton Bt CB, Sir Henry Holland Bt, H. Bence Jones, Sir R Murchison, James Rennie FRS, Robert Roupell MA QC, Rev William Taylor, John Webster, Charles Wheatstone, Col Philip Yorke. See also letter 1596, n. 3.
2. *my poor lectures:* in 1859 Tyndall gave a series of twelve lectures on the Force of Gravity at the RI.

To George Gabriel Stokes 23 March 1859 1603

March 23rd, 1859

My dear Sir

I return you Mr. Gassiot's[1] paper on Electrical Discharges.[2]

The paper gives evidence of great care, great skill, and, above all great labour, in the devoting & carrying out of the experiments.

It contains a great number of valuable facts which in my opinion must have an important bearing upon the final solution of the question: many of them are of a highly suggestive character.

On this account I would recommend the paper for publication in the Philosophical Transactions.

I have communicated to Mr. Gassiot my belief[3] that the paper might be condensed with advantage, and he expresses his entire willingness to conform to this suggestion should the decision of the committee regarding it be favourable to its publication

I am dear Sir | Yours very faithfully | John Tyndall

Prof. Stokes | Sec. R.S.

RS RR/4/104

1. *Mr. Gassiot's:* John Peter Gassiot.
2. *I return you Mr. Gassiot's paper on Electrical Discharges:* this article was published as 'On the Stratifications in Electrical Discharges, as observed in Torricellian and other Vacua—Second Communication', *Phil. Trans.*, 149 (1859), pp. 137–60, and built upon his previous article, 'On the Stratifications and Dark Band in Electrical Discharges, as observed in Torricellian Vacua', *Phil. Trans.*, 148 (1858), pp. 1–16.
3. *I have communicated to Mr. Gassiot my belief:* letter missing.

To Thomas Archer Hirst 3 April 1859 1604

Royal Institution | Albemarle Street London | 3rd April 1859

My dear Tom

I tried to commence two or three things this morning. I tried various portions of my work, but I do not find that my brain possesses its maximum power and clearness, and as I am unwilling to write a book which does not express my whole strength my project this morning was unsatisfactory. It is no compliment to you to say that I turn to you under these circumstances—still I know you will not be unwilling to read a line from John though only a fraction of John's spirit may pervade it.

I would have written to you long ago had it not been for circumstances which neither you nor I could have anticipated; and which disturbed me for a time as a tempest shakes the trees of autumn. I waited until I should right my boughs and wash my draggled leaves before appearing in your presence. Sufficient calm at lengths has come, and I will sketch for you a swift picture of what has taken place, waiting until we meet for the full development & filling in of the outline

You know what my feeling were when I wrote to you last[1]—It is my lot to inform you that they are very different now. You have some idea of my relationship to the people[2] whom I mentioned to you at Saas,[3] but you have a very imperfect idea of the affection and the tenderness with which I saw, or thought I saw myself treated. Every thing went as merry as a marriage bell—I was incessantly with them. This bearing towards me was the subject of common remark—on the benches of the Royal Institution the audience talked to each other regarding their demeanors towards me there—strange to say the lady[4] so behaved as to leave the general impression that they made advances to me and not I to them—which indeed was substantially the correct impression. At length things went so far that some of the lady's friends[5] warned her regarding her conduct. Her replies to them were circulated and were highly unsatisfactory to me—I wrote to her on the subject,[6] stating however my belief that it was all false—she replied denying the proper part of the rumours that were circulated, and at the same time repeated to me with consummate coolness the selfsame tale, regarding the preoccupation of her daughter's affections which I mentioned to you at Saas.[7] For a month previous I had noticed many things which displeased & pained me. I noticed a love of amusement which I had not observed before, and some little differences occurred between us. But still the evidences of affection were so strong that I overlooked all, and should as soon have doubted the light of the sun as their truth and goodness—But here came the reiteration of the old tale; I was expected to remain with them for ever as their dear and valued friend—and it was assumed that I should continue to write to them and continue to visit them just as usual.

Well this letter[8] let light in upon my mind, but it was the light of lightning which reveals us our bad work. At all events—I saw in this woman whose open affection for me had stupefied me, a being unworthy of my slightest regard. My course was to maintain a dignified silence and to have no more to do with them. She wrote to me—I sent her letter back unopened. Her daughter[9] wrote to me[10] entreating me to send her mother one line of kindness, but her letters[11] revealed to me at the same time what I do not like to mention. Still I wished to give the woman an opportunity of clearing herself if she could, and I complied with her daughters request. She did not clear herself—but revealed still more to me the rottenness of the seed upon which I had based my trust. Again and again she wrote to me—she came herself to my lodgings, but I was not at home. At length I told her sufficient of my mind[12] to let her understand clearly how I regarded her conduct. I have not heard from her since.

The storm as I have said is passed—and my friends rejoice in my liberation, for I now find that they long dreaded the consummation. I am getting on with my work, and enjoy a solid peace of mind, and a stronger hope of being useful in the world than I have entertained for the last two years. I will now pause—When I see you I will tell you all, and you will then see that my mistake was natural

Yours ever affectionately | J. Tyndall

RI MS JT/1/T/649

---

1. *I wrote to you last:* letter missing.
2. *The people:* this likely refers to Maria Drummond and three daughters, in particular Mary Drummond, as well as Mrs Barlow. Tyndall courted Mary for a period in 1858 (see for example letter to Hirst: Volume 6, letters 1564 and 1576). In a journal entry for Sunday 3 April 1859, Tyndall recalled that he had attended a gathering hosted by Mrs Barlow the previous Friday, which was also attended by the Drummonds. It is unclear if it is Friday 1 April or Friday 25 March. In his journal entry for 3 April 1859, Tyndall recorded that an incident had taken place at 'Mrs Barlow's on Friday evening last' (Journal, RI MS/JT/2/13c/1156), between himself and the Drummonds. See also letter 1597.
3. *Saas:* Tyndall travelled to Saas-Fee, the main village of the Saas Valley in the Swiss canton of Valais in August 1858 (*Tyndall Correspondence*, vol. 6, letter 1489, n. 33).
4. *the lady:* likely refers to Maria rather than Mary Drummond.
5. *the lady's friends:* not identified.
6. *I wrote to her on the subject:* letter missing.
7. *regarding the preoccupation of her daughter's affections which I mentioned to you at Saas:* Maria Drummond wrote to Tyndall about his relationship with her daughter Mary Drummond, and Tyndall recorded their exchange in letters to Hirst (*Tyndall Correspondence*, vol. 6, letters 1564 and 1576) where Tyndall seemingly expressed an uneasiness about the tone taken by Maria towards him.
8. *This letter:* letter missing.
9. *Her daughter:* Mary.
10. *wrote to me:* letter missing.
11. *her letters:* letter missing.
12. *At length I told her sufficient of my mind:* this letter is missing and the date of the meeting is not identified.

From Heinrich Wilhelm Dove 15 April 1859 1605

My dear Sir

Ich schreibe deutsch, was Sie wohl nicht so vergessen werden haben, als ich mein Englisch und sende Ihnen beifolgend 5 Abdrucke zweier kleiner Aufsätze in Pogg. Ann nebst dazugehörigen Tafeln, eine Mittheilung enthaltend, welche ich im März der hiesigen Akademie gemacht. Es betrifft die Anwendung des Stereoskops um falsche Banknoten von echten, überhaupt eine Copie von ihrem Original zu unterscheiden. Ich lasse eben darüber eine etwas ausführlichere Brochure drucken von deren ich *[ihr]*, wenn sie erschienen ist ein Exemplar zusenden werde. Von den mit gesandten Abdrucken und Tafeln sind Sie vielleicht so freundlich 1 an die Herrn Faraday, Grove Miller und General Sabine zu geben und 1 an *[Hr.]* Stokes. Ich habe noch einige andre Anwendungen des Stereoskop[s] gemacht, von denen das folgende eine. Da die Elasticität der Metalle verschieden, so erhalten bekanntlich Dräthe, welche man durch dasselbe Loch im Drathzug gezogen nicht genau dieselben Durchmesser, da sie sich nach dem Heraustreten aus der Pressung ungleich ausdehnen. Es schien mir wahrscheinlich, dass dies auch bei dem Prägen der Medaillen stattfinden müsse. Kleine Unterschiede müssen sich im Stereoskop nach Art eines Nonius wie bei Maasstäben dadurch kundgeben dass das in einer Ebene liegende gekrümmt *[erscheint]*. Legt man z.B. die Pariser Ausstellgs Medaille, die Seite wo die Wappen sind nach oben ins Stereoskop, die eine von Silber die andre von Bronze, so erhält man eine schön aus beiden Farben combinirte Medaille als hohles Schild. Unterschiede der spannenden Gewichte bei elastische[n] Substanzen auf welchen gleich weit abstehende Striche gezogen, Temperatureinflüsse und die beim *[Erstarren]* der Körper eintretenden *[Erschngen]* können auf diese Weise bemerkbar gemacht werden. Da in *[unsern]* Zeitungen sehr unklare Berichte über die von mir gefundene Anwndg des Stereoskops *[erschienen]* sind, so wird es Ihnen vielleicht recht *[1 word illeg]* sein, etwas *[1 word illeg]* darüber zu erfahren.

Ihren schönen die Gletscher und das Eis überhaupt betreffenden Untersuchungen bin ich mit dem lebhaftesten Interesse gefolgt. Leider habe ich zum Experimentiren selbst gar keine Zeit und kann mich daher nur an den Arbeiten meiner jüngern Freunde erfreuen. Ich bin jetzt Rector der Berliner Universität habe dabei 3 Professuren und dirigiere das Meteorologische Institut von jetzt 70 Stationen. Einen grossen Folioband über die Ergebnisse der ersten 10 Jahre habe ich im vorige[n] Jahre veröffentlicht und drucke jetzt den *[6ten]* Theile der recht periodischen Veränderungen auch wohl 40 Bögen. Ausserdem war ich lange Zeit unwohl. Bei der Inspectionsreise vor 2 *[Jahren]* kam ich bei Besteigung der Beskiden in ein furchtbares Hagelwetter und bekam als Folge ein Rheuma im rechten Knie, so dass ich vorigen Sommer ein warmes Bad besuchen musste, Warmbrunn in Schlesien zum Glück jetzt mit Erfolg[.]

Ich bin so vielen meiner Freunde Briefe schuldig, dass ich am /liebsten/ mich mündlich entschuldigen möchte. Wie gern wäre ich zu der Meeting in Aberdeen gekommen, aber als Rector ist es mir nicht möglich dies Jahre zu reisen. Wenn Sie General Sabine sehen, so sagen Sie ihm doch, dass er /nächstens/ eine[n] langen Brief erhalten wird, wo ich wegen meines langen Schweigens Verzeihung bitte, und ihn zugleich ersuche, mein Sachwalter bei Admiral Fitzroy zu sein, dem ich für die besorgte Uebersetzung meiner Law of Storms zu grösstem Danke verpflichtet bin.

Nun habe ich noch eine Bitte. Von den Proceedings der Royal Society deren Mittheilung ich der Güte der /Societät/ verdanke ist mir vol VIII No. 26 nicht zugegangen. Kann diese Lücke vielleicht ergänzt werden. Sie würden mich dadurch sehr /verbinden/.

Von Ihren Berliner Bekannten ist aller wohl. Professor Dirichlet in Göttingen aber schwer wohl hoffnungslos erkrankt. Knoblauch habe ich vor einigen Tagen hier gesehen. Es ist wohl aber sein Vater krank.

Voriges Jahr hatten wir in Carlsruhe die Hoffnung Sie zu treffen da Sie in der Schweiz waren aber Sie kamen nicht. Von England war nur Miller da nebst Frau. Die Meeting war sehr heiter und nachher in München das Jubilaeum der 700 jährigen Gründung der Stadt prächtig, nur Bier /zu/ viel.

Ich gebe die Hoffnung nicht auf Sie /wieder/ einmal hier zu sehen. Berlin ist freilich nicht London, aber Berlin ist doch Berlin und Sie wissen, wie aller sich freut, wenn alte Bekannte /wiederkehren/[.]

Meine freundlichsten Grüsse an Hr. Faraday.

Mit alter Freundschaft | Ihr | H.W. Dove

$\frac{15}{4}$ 59 Berlin

My dear Sir

I am writing in German, which you will probably not have forgotten as I have my English, and am sending you enclosed 5 offprints of two short essays in Pogg. Ann[1] together with the plates that belong with them, containing a communication which I made to the Academy here[2] in March. It concerns the use of the stereoscope to distinguish fake banknotes from real ones, or in general a copy from its original. I am just now having a somewhat more detailed brochure about it printed, of which I shall send a copy /to it/, when it has appeared. Of the enclosed offprints and plates, will you possibly be so kind as to give 1 to Herr Faraday, Herr

Grove, Herr Miller[3] and General Sabine and 1 to *[Herr]* Stokes. I have made some other uses of the stereoscope, of which the following is one. As the elasticity of metals is different, then wires which one has pulled through the same hole in a draw plate are known not to get exactly the same diameters, as they expand unevenly after emerging from pressing. It seemed probable to me that this would also have to take place with the striking of medals. Small differences will have to be revealed in the stereoscope in the manner of a vernier[4] as with scales through the fact that anything lying in a plane *[appears]* curved. If one places, for example, the Paris Exposition Medal[5] into the stereoscope with the side where the coats of arms are facing up, one side being made of silver, the other of bronze, then one gets a medal beautifully combined of both colours as a hollow shield. Differences in tensile weights with elastic substances on which lines have been drawn at equal distances, temperature influences, and *[phenomena]* that occur when the bodies *[solidify]* can be made noticeable in this way. As only very unclear reports about the use of the stereoscope found by me have *[appeared]* in *[our]* newspapers, then perhaps it will be quite *[1 word illeg]* for you to learn something *[1 word illeg]* about the matter.

I have followed with the liveliest interest your nice studies concerning glaciers and ice in general.[6] Unfortunately, I have no time whatsoever to do any experimenting myself and can therefore only take pleasure in the works of my younger friends. I am now Rector at the University of Berlin; hold three professorships into the bargain[7] and direct the Meteorological Institute, comprising now 70 stations[8]. Last year I published a large folio volume about the results of the first 10 years and am now printing the *[6th]* part of the quite regular changes also, probably 40 sheets. In addition to this, I was unwell for a long time. On the inspection tour 2 *[years]* ago, I was caught in a terrible hailstorm while climbing the Beskids Mountains[9] and as a consequence got rheumatism in my right knee, with the result that I had to visit a hot spa, Warmbrunn in Silesia[10], last summer, fortunately now with success.

I owe letters to so many of my friends that I would *[prefer]* to apologise in person. How gladly I would have come to the meeting in Aberdeen[11], but as Rector it will not be possible for me to do any travelling this year. If you see General Sabine, then do tell him that he will *[shortly]* receive a long letter, in which I shall ask him to pardon my long silence and at the same time request him to be my advocate with Admiral Fitzroy, to whom I owe the greatest debt of gratitude for taking care of the translation of my Law of Storms.[12]

Now, I have another request. Of the Proceedings of the Royal Society, *[whose]* communication I owe to the kindness of the *[Society]*, I have not received vol VIII No. 26[13]. Can this gap possibly be filled. In doing so, you would greatly *[oblige]* me.

All of your acquaintances in Berlin are well. Professor Dirichlet[14] in

Göttingen, though, is seriously, probably hopelessly ill. I saw Knoblauch[15] here some days ago. He is well, but his father ill.[16]

Last year we hoped to meet you in Karlsruhe[17] as you were in Switzerland, but you did not come. From England, only Miller was there, together with his wife. The meeting was very cheerful and afterwards in Munich, the celebration of the founding of the city 700 years ago was splendid, only there was *[too]* much beer.

I am not giving up hope of seeing you here once *[again]*. Of course, Berlin is not London, but Berlin is Berlin, though, and you know how everyone is pleased when old acquaintances *[return]*.

My kindest regards to Herr Faraday.

With old friendship | Your | H.W. Dove

$\frac{15}{4}$ 59 Berlin

RI MS JT/1/D/138

---

1. *enclosed 5 copies of two short essays in Poggend. Annal.:* the inserted copies are missing, but it was likely copies of the two articles Dove published in 1859, 'Stereoskopische Darstellung eines durch einen Doppelspath binocular betrachteten Typendrucks', *Poggend. Annal.,* (1859) pp. 655–57, and 'Beweis, dass die Tartinischen Töne nicht subjektiv, sondern objektiv sind', *Poggend. Annal.,* (1859) pp. 652–54.
2. *the Academy here:* The Royal Prussian Academy of Sciences located in Berlin.
3. *Miller:* possibly the crystallographer Willam Hallowes Miller (1801–80).
4. *vernier:* a vernier scale.
5. *Paris Exposition Medal:* The International Exhibition, Exposition Universelle, was held in Paris in 1855 where Dove acted as a judge.
6. *your nice studies concerning glaciers and ice in general:* during the 1850s and early 1860s Tyndall was engaged in controversy with James David Forbes over the nature and motion of glaciers. Tyndall proposed a theory of fracture and regulation to account for glacial motion, see letter 1594, n. 2–3.
7. *I am now Rector at the University of Berlin; hold three professorships into the bargain:* Dove was elected Rector at Berlin's Friedrich–Wilhelms–Universität for the year 1858–59. Dove became an ordinary professor at Berlin's Friedrich–Wilhelms–Universität in 1845, during which time he also continued his work at the Institute of Technology and at the Military school in Berlin.
8. *direct the Meteorological Institute, comprising now 70 stations:* Dove became director of the Meteorological Institute of Prussia in 1849.
9. *Beskids Mountains:* likely refers to the mountain range Silesian Beskids, or the mountain range Moravian-Silesian Beskids.
10. *Warmbrunn in Silesia:* the town Warmbrunn in Silesia was home to a wellknown spa (Walter White, *A July Holiday in Saxony, Bohemia, and Silesia,* Chapman and Hall, 2013).

11. *the meeting in Aberdeen:* the annual meeting of the BAAS was held in Aberdeen in 1859.
12. *the translation of my Law of Storms:* Refers to Dove's book, *The Law of Storms Considered in Connection with the Ordinary Movements of the Atmosphere* published in 1861, was translated by Robert Fitzroy from the original German Über das Gesetz der Stürme, first published in 1857.
13. *vol VIII No. 26:* likely refers to *Roy. Soc. Proc,* 8 (1856).
14. *Professor Dirichlet, in Göttingen:* Peter Gustav Lejeune Dirichlet (1805–59), a German mathematician at the University of Göttingen.
15. *Knoblauch:* Karl Hermann Knoblauch (1820–95), a German physicist at the University of Halle, with whom Tyndall had worked on diamagnetism. Knoblauch's research focused particularly on radiant heat.
16. *He is well, but his father ill:* Knoblauch's father, the politician and entrepreneur Carl Friedrich Wilhelm Knoblauch (1793–1859), passed away later that year, see letter 1662, n. 2.
17. *Karlsruhe:* city in what was then the Grand Duchy of Baden, in the present-day federal state of Baden-Württemberg in Germany. The association *Versammlung Deutscher Naturforscher und Aerzte* (f. 1822) met in Karlsruhe in September 1858.

To Juliet Pollock 17 April 1859 1606

Keswick Sunday | 17th. Ap. 1859

My dear Mrs. Pollock

I am going to write to you as much as a bad head and half time will let me. I started from London on Friday morning—in the carriage with me were a gentleman and his grown up daughters; a Bolton manufacturer, who was a ranter and his little daughter, and a woman who kissed an old man several times with great affection before entering the carriage. The Bolton man sang hymns, and whistled hymn tunes when his voice was tired—it was very unpleasant, and I had half a mind to tell him so. Had the man sung common songs I would have listened to him with pleasure but I suppose it was the impudent selfishness which rendered the man regardless of the comfort of others, coupled with the implied assertion of a more than average sanctity that exasperated me. Had he been under a showerbath I should have pulled the string and thus incurred the charge of religious persecution.

I reached Windermere[1] at 6 oC and before dinner walked to Orrest Head[2] and saw Wordsworth's 'beautiful romance of Nature'[3] Next morning at 9 oC. we rowed from Windermere to Ambleside[4]—the lake was smooth and sunny when we departed, but it became black and rough, and to shelter ourselves from one shower we pushed ashore and landed on a wooded knoll. Near the place stood a granite cross marking the spot where two young men had been drowned the year before within 18 feet of the shore. Remember Walter must be taught to swim. The storm howled over us spitting heavy snow flakes against us. The little lake looked quite grand, black as ink under the black scowl of heaven with the crests of its little sputtering waves rendered doubly white by contrast. The slate rocks all over the knoll were carved and sculptured by some local madman, who had a passion for chiseling—there were Wordsworth, Davy, and a hundred other names, some deeply cut, others in relief, the rock around the letters being chiseled away. We landed at Ambleside, walked thence to Rydal Mount; and looked round in Wordsworth's nest.[5] Thence to Grassmere, and stood for some minutes beside his grave.[6] A clear stream rushes near it, a few trees are at hand; one black cypress gives character to the scene, and there the poet sleeps, while the plainest slab of black slate which contains only his name, marks his resting place. If I can manage it I will be buried in a country churchyard. I hate town burials.

From Grassmere we turned into what Frankland called a gully of the mountains,[7] having first obtained a little information regarding Helvellyn.[8] All day the snow storms assailed us at intervals, and here we were told that we should have wild work upon the mountain. We however steadily climbed the gorge, Griesdale Tarn[9] being the landmark at which we first aimed: once we saw a cascade precipitating itself over a barrier which stretched across the gorge, and we thought the tarn must lie behind that barrier: we surmounted it, there was no tarn, but the sunniest little mountain basin imaginable: clasped all round by the hills, sheltered from the storm, and so cosy and warm that the <u>comforts</u> of the spot heightened our estimate of its <u>beauty</u>. It seemed to have been once a lake where waters had withered away, leaving its bed all smooth and grassy. And the barrier was in front of us which we surmounted in due time, and here the keen north wind met us again. Below us was the Tarn,[10] the most lovely, and earnest looking piece of water imaginable: beyond it was spur from the summit of Helvellyn, the mountain top however being shut out from view. We descended to the Tarn—the hillside down which we went being covered with miniature glaciers. The snow had melted, and again coagulated so as to form little patches exactly resembling glaciers of the second order. All the higher summits were white, a definite snow line drew itself across the spur of Helvellyn, in fact a little imagination only was needed to cause us to fancy ourselves in the Alps. And indeed I know not that higher enjoyment or better feelings are excited by the loftiest of mountains than those called up by our own little hills. They are very picturesque. The slate rock of which they are in a great measure composed give a roughness to the eminences which renders them bolder than higher mountains with smoother outlines. They are knobbed and warty and very black under the black clouds. We were soon at the foot of Helvellyn, and very soon afterwards upon his sides. We reached the summit of the spur, each pursuing his own way; and our ways happened to be highly divergent. But we closed again upon the top of the spur, and went along the ridge towards the summit. The north was before us filled with black clouds; it sent its wind against us, and we concluded that it would empty its reservoir of snow upon us in due time. The flat ridge along which we marched was covered with snow, and over the precipices to our right a beautiful cornice was folded, which reminded me of cornices which I had seen elsewhere, the blackness in front augmented–dark as the gloom of the darkest night, when the light of stars is cut off by clouds. But this storm did not yet smite us, it moved right and left as if diverted by the mountain, and showered down its snow in the flanking valleys. A grander phenomenon of the kind I have never seen. The snow as, it fell from the black mass overhead, was drawn into vast vertical striae,[11] these grouped themselves into distinct columns which stalked like giants over the mountain heads. The blackness at length closed in upon us, and the blast augmented to a storm. It was dreadfully cold, and the pain of the face so great that I pinched mine at intervals to assure myself that it was not frostbitten. The falling snow soon dwindled to a thin hail, and we emerged in sunlight and calm at the other side of the storm. The summit was before us with the sunshine resting on it, scarcely a breath of air. In fact the character of the storms throughout the entire day was that of a series of undulations, first a rush, then a lull, the middle point between two atmospheric billows, being absolutely without motion. We enjoyed the perfect calm for some minutes upon the mountain head, and leaving Frankland and his sandwiches there, I descended to inspect Stridin Edge & the Red Tarn.[12] The place is awkward, & in foggy weather might be dangerous. Some parts are very steep, and the whole scene is very solemn. The mountain on that side is scooped into a vast crater-like hollow. Stridin Edge bends round the Red Tarn like a vast ridge of scoria,[13] cast up from the hollow now occupied by the Tarn; and indeed the bubbled lava under one's feet, and which composes the upper part of the mountains gives strength to the fancy. On the other side the tarn is bounded by Swirrel Edge,[14] another wedgelike ridge which widens at its extremity to a bold conical mass, and forms Catchedecam.[15] Beyond and around were the more distant lakes and mountains—the sea, the Isle of Man[16] & Morecombe Bay.[17] Black clouds again rose in the north, but this time they bore on directly towards us. There was no shelter save a mound of stones bearing the stump of a flagstaff which had once served as a signal for the ordinance surveyors: each of us danced for 10 minutes on the shale of the summit to get his feet warm and when the fronts of the storm reached us, we crunched down in shelter of the mound. a snow drift was our cushion; and there we sat and enjoyed the atmospheric war. The sun sometimes broke through the clouds, sending vast fanlike rays on all sides. Again columns of light and shade would diverge in one directions only–the columns forming a series of conical props for the sun itself which seemed planted at their place of apparent intersection. But the blackness was above all description grand, and the contrasts wonderful.

The boundary of the storm was marked with perfect definition: outside of it rocks and fells, and lakes lay bathed in sunlight; in front of the gloom the hills were of a grimness which suggested a kind of supernatural horror. Frankland said that if a little nitrate of *[strontium]* were ignited on Scawfell Pike,[18] so as to redden the clouds, we should have a fair representation of a certain locality. Here and there through a hole in the inky mass the sunbeams escaped, gilding the hills in small patches, and once they fell upon a little tarn which reflected the light upwards from amid the deepest gloom, with the effulgence of burnished silver. On hitting the knots and prominences of the mountain the wind was broken into eddies which caught the snow and twirled it round and round—mingled with this was the snow of the clouds themselves, the flakes of which as they crossed the more illuminated portions of the heavens seemed like a multitudinous flight of insects. We waited a good while, but still the storm continued—it was the most persistent we had yet encountered. It subsided a little and we commenced a descent in a direction opposite to that of our ascent. The tail of the storm contained its chief energy, it smote us with wild fury, tore up the crust that had formed upon the snow, and seizing upon the lighter and drier mass underneath beat it against us with pitiless animosity—there was a kind of unreasoning madness in this portion of the storm which suggested the fury of a woman with strong emotions and little intellect—a wild wasteful expenditure of strength, which speedily finished in complete exhaustion.

We went down to the base of the mountain, and marched along the high-way towards Keswick.[19] During half the journey snow fell heavily. The hedgerows were laden, the carol and brine & furse laden, the laurel leaves especially laden, and the whole landscape was white. Towards sunset all cleared away except a storm which strode through the valleys to our right. You know the sun's beams when he is setting are red—well these stalking snow columns became finally illuminated with this red light, and presented a most wonderful appearance. Turning towards Helvellyn the effect was lively beyond description—the mountain was quite clear of cloud, and a patch of heaven for some distance above the summit was equally free; but there was something in the nature of the light which gave the mountain a light and unsubstantial appearance, as if indeed it consisted of ether in a slightly more condensed form than that above and around it. The clouds about the west were magnificent. I cannot describe them, and the calm delight that they imparted suggested a relationship between them and the human soul.

<Hereafter LT Transcript Only>

There is something in this universe different from the spinning of cotton, different also from the investigation of mere physical law. But I wont moralize

Good bye | Always yours | J. Tyndall

If you have the time and mind to write me a line I shall call at the post office Windermere on Friday afternoon. | I was very sorry indeed to find myself unable to accept Lady Martin's[20] invitation for yesterday.[21] It was so kindly

<Handwritten Letter Resumes>

given that I felt more than usual regret at being unable to accept it.

BL, Add MS 63092, ff. 16–227v
RI MS JT/1/TYP/6/1954–58

---

1. *Windermere:* the name refers both to a town and lake in Cumbria.
2. *Orrest Head:* a hill by the shores of Windermere.
3. *Wordsworth's 'beautiful romance of Nature':* this refers to William Wordsworth's poem 'On the Projected Kendal and Windermere Railway' from 1844, which described Orrest Head. It was published as, William Wordsworth, 'On the Projected Kendal and Windermere Railway', *Morning Post,* 16 October, 1844.
4. *Ambleside:* a town in Cumbria.
5. *walked thence to Rydal Mount; and looked round in Wordsworth's nest:* a Rydal Mount is a house in the village Rydal, where the poet William Wordsworth lived until his death in 1850.
6. *Thence to Grassmere, and stood for some minutes beside his grave:* Wordsworth is buried in St Oswald's Church in the village of Grasmere.
7. *into what Frankland called a gully of the mountains:* from the description, Tyndall and Frankland may have walked up Tongue Gill, arriving at Grisedale Hause, which overlooks the tarn.
8. *Helvellyn:* a mountain in the Lake District.
9. *Griesdale* Tarn: Grisedale Tarn, near Helvellyn in the Lake District.
10. *the Tarn:* Grisedale Tarn.
11. *vertical striae:* striae here refers to bands in the snow.
12. *Stridin Edge & the Red Tarn:* Striding Edge is a rocky ridge leading onto Helvellyn, with Red Tarn below it.
13. *scoria:* mass formed by cooling of molten lava (*OED*).
14. *Swirrel Edge:* Swirral Edge is a ridge leading onto Helvellyn, on the opposite side of Red Tarn to Striding Edge.
15. *Catchedecam:* Catstyecam, a mountain linked to Helvellyn by Swirral Edge.
16. *Isle of Man:* island in the Irish Sea.
17. *Morecombe Bay:* Morecambe Bay, the estuary in the south west coast of Cumbria.
18. *Scawfell Pike:* Scafell Pike, the highest mountain in England, located in the Lake District in Cumbria.
19. *Keswick:* town in Cumbria.
20. *Lady Martin's:* may refer to Helena Faucit Saville, Lady Martin (1814–98), an actress (*ODNB*).
21. *Lady Martin's invitation for yesterday:* invitation missing. The nature of the event is not known.

From Auguste Arthur de la Rive 17 April 1859 1607

Genève, le 17 Avril, 1859.

Mon cher Monsieur,

J'ai chargé un ami de vous porter un diplôme de Membre honoraire de notre Société de Physique. Dans le même pli il y a en avait un pour M. Airy que vous aurez bien la bonté de lui faire parvenir.

Notre Société ne se fait aucune illusion sur la valeur de ses diplômes; elle les considère bien moins comme un honneur qu'elle *[confère]* que comme un témoignage d'estime et d'affection qu'elle donne aux savants qui ont fait quelques visites à Genève, et qui se sont occupés de questions scientifiques dont la nature touche à notre pays et l'intéresse. C'est à ce double titre que nous avons tenu à vous compter dans le nombre de nos Membres honoraires, & nous espérons que vous voudrez bien considérer cette nomination comme l'expression de notre attachement pour vous.

J'espère un peu, si du moins les affaires du continent s'arrangent, pouvoir vous faire une visite au mois de Mai. J'ai quelques affaires en Angleterre & je désire vivement revoir quelques uns de mes amis de Londres. Vous êtes en tête ainsi que votre excellent et digne maître Faraday. Rappelez-moi à son bon souvenir en attendant que *[j'aie]* le très grand plaisir de lui serrer moi-même la main.

Je ne vous parle pas science; c'est un sujet que nous traiterons quand je vous verrai; je vous apporterai j'espère, quelques petites choses nouvelles, & surtout j'en apprendrai beaucoup auprès de vous.

En attendant croyez-moi; très cher Monsieur et ami,

Votre tout dévoué et bien affectionné | <u>Aug. de la Rive</u>.

Geneva, April 17th, 1859

My dear Sir,

I asked a friend[1] to bring you a diploma of Honorary Member of our Society of Physics.[2] In the same envelope there was one for Mr Airy, to whom you would be kind to send it.

Our Society is under no illusions about the value of their diplomas; it considers it much less as an honour that it confers than as a mark of esteem and affection that it gives to scholars who have made some visits to Geneva, and who took an interest in scientific questions whose nature affects our country. It is for these two reasons that we wanted to count you among our Honorary Members, & we hope that you will consider this nomination as an expression of our attachment to you.

I hope, if at least the affairs on the continent get settled,[3] to be able to make a visit in the month of May. I have some business in England & I am *[eager]* to visit again some of my friends in London. You are at the top of my list, as well as your excellent and worthy teacher Faraday. Remind him about me until I have the great pleasure of shaking his hand myself.

I am not talking to you about science; it is a topic that we will discuss when I see you. I hope that I will bring you some new things, & especially that I will learn a lot from you.

Meanwhile believe me, dear Sir and friend,

Your devoted and affectionate | Aug. de la Rive.

RI MS JT/1/TYP/1/358
LT Transcript Only

---

1. *a friend:* not identified.
2. *our Society of Physics:* Société de Physique et d'Histoire Naturelle de Genève.
3. *the affairs on the continent get settled:* likely refers to the ongoing unrest in Europe in this period. In April 1859, the month this letter was written, Austria declared war on Sardinia.

To Francis Fox Tuckett 19 April 1859 1608

Keswick 19th Apr. 1859 | Royal Institution of Great Britain

My dear Sir

I believe I am the sole member who is carrying out the principles of the Alpine Club[1] of this season. My presence among the Cumberland mountains[2] will deprive me of the pleasure of seeing you on Tuesday. This I regret as I should like to see the apparatus.[3]

With regards to Mr Barnard[4] he would probably gladly avail himself of your offer—I was so busy when your last letter[5] reached me that I omitted to send it to him as I intended: his address is Charlotte Street Fitzroy Square, but I forget the number:—the porter at the Royal Institution[6] knows it and would tell you if you cared to enquire.

I hope you will have a pleasant meeting on Tuesday—pray commend me to our *[*President*]*[7].

Ever truly | Yours J. Tyndall

F. F. Tuckett *[*Esq*]*

Science Museum, Wroughton (MS 2108/79a)

---

1. *the principles of the Alpine Club:* refers to him being in the Cumbrian mountains at a time when members of the Alpine Club, which was founded 1857, were not likely to be there.
2. *Cumberland mountains:* the Lake District in North West England.
3. *the apparatus:* not identified.

4. *Mr Barnard:* likely refers to George Barnard (1807–90), a landscape artist who was a member of the Alpine Club.
5. *your last letter:* letter missing.
6. *the porter at the Royal Institution:* unidentified. There were three porters at the time: Thomas Smith (door porter, who slept at the RI), James Eed (hall porter), and Henry Lacey.
7. *[President]:* the first President of the Alpine Club was the politician and naturalist John Ball (1818–89), who held the post until 1860.

From William Frederick Pollock 18–20 April 1859[1] 1609

59 Montagu Square | 18th April 1859

My dear Tyndal

I have shared the pleasure of reading your very interesting letter from Keswick[2]—& in imagination I have trodden with you the mountain paths you describe between Ambleside & Keswick via Helvellyn,[3] with which I am pretty familiar—That crater like hollow at the bottom of which lies Red Tarn,[4] with its twin guardian ridges, is to me, unacquainted as I am with mountain scenery on a larger scale, a grand recollection—and I should like to have visited it again in your company—I hope you will have seen Scaw Fell[5] before you get this—the whole of Borrowdale[6]—Sty Head[7]—& the summit itself of Scaw Fell are extremely fine—I rather envy you the sight of that country in time of snow—My own experience of the English Lakes has always been in Autumn—probably on the whole—the best season for seeing them—But once, when I was on Ullswater[8] as late as the end of October I saw the mountain-tops in snow—& it added so much of a special beauty to them, that I have ever since longed to be among them some time in winter or early spring—Wordsworth always maintained that the scenery among which he lived was the finest in winter—& I have heard him say so to myself, standing in his garden at Rydal Mount[9]—But notwithstanding this, I cannot help thinking that the opinion was in great measure due to the wish of a resident on the spot to convince an occasional autumn visitor that the best of the land was not so easily to be seen—& to secure for himself the enduring advantages—or the reputation of them, of being always in attendance at Nature's receptions, where he was certainly by position & by other qualifications, so well entitled to officiate as perpetual gentleman usher in daily waiting—I once spent part of two very pleasant days in Wordsworth's[10] company at Rydal[11]—The morning of a Saturday spent in walking round Rydal Lake—& a Sunday afternoon in taking tea with the Poet at the Mount[12]—I had previously made his acquaintance at Lowther Castle,[13] where it was my good fortune to be almost the only fellow guest with him during 2 or 3 days—&

I saw the most beautiful part of that fine domain under his guidance—The last time I past through Ambleside[14] the man who was driving us, had never heard of Wordsworth—I suppose he was a stranger imported to meet the increased wants of the road, under the system of railways & excursion tickets, but it grated harshly on the ear, did that avowal of utter ignorance of the name of the time honoured genius WW[15]—& I thought the man should have been better taught—or at least better crammed, to discharge his duties—for in such a country, it is clearly the function of the driver to answer tourists' questions as well as to drive—& I gave him an unloving guerdon when he parted from us—

I met *[Poeris]*[16] at breakfast a day or two since—He is not a man of striking appearance—but able & agreeable, & I should like to see more of him—

We had a good lecture from Sir Cha*[rle]*s Lyell on Friday last at the R.I.[17] In manner of delivering it was better than any previous one I have had the pleasure of hearing from him—and as criticism on this is all to which he is open, you may conceive the evening was a very successful one—His facts, as to the actual angles of descent at which lava streams on Etna have cooled & remained as now seen, were very interesting—& his deductions from them in favour of his theory of craters of eruption as against Von Buch's[18] theory of elevation,[19] were overwhelming—Indeed I almost grudged the exercise of such able championship in favour of a cause of such intrinsic strength—& I wished Sir Charles a foeman more worthy of his steel—It will however be amusing to see what answer is attempted by the continental geologists who may feel themselves pledged to support their own case—when once personal reputation has become involved, it takes a long time before error is acknowledged even in the face of the strongest array of adverse facts—

I took Fred[20] on Saturday to see Henry V[21] at the Princess' Theatre[22]—& I think I never heard so much noise within four walls as prevailed from seven o'clock to eleven—Bells, guns, shouts, trumpets maintained a continual din on the stage—In front of the curtain where a certain amount of occasional sound in the shape of applause is generally thought desirable during a theatrical performance, there was not so much breach of silence—& most people seemed to come away satiated rather than satisfied with the many appeals made to their senses of sight and hearing—Of practical feeling in the representation—or of passion in the acting, there is none—I went in a very unprejudiced frame of mind, & not expecting that Charles Kean's[23] acting would be worse than on former occasions—but I must admit that I was disappointed—& this distinguished performer surpassed himself in badness—Scene painters, upholsterers, wardrobe-men & women, property makers, pyrotechnists, stage drillers had done their utmost and their best—But the individual human interest of the piece was overwhelmed by all their material circumstances—I doubt whether even much better acting could have struggled successfully against them—The sensual faculties alone were addressed—the mind is never reached—Shakespeare appears between the acts, as depicted on the drop-scene—but during the acts he was altogether out of sight—

I am going down to Brighton[24] at the end of the week, when I shall have a holiday, to fetch up Walter[25]—who will I hope be with us for a fortnight—during which time I trust you may have many opportunities of seeing him—

The best ham I ever encountered was at the hotel[26] at Windermere Station[27] last autumn—I fancy you may be reading this over breakfast over a similar piece of excellence—& if the letter has tired you, I hope you may find consolation in the porcine dainty—

I am yours ever truly | W. F. Pollock

20th April 1859.

RI MS JT/1/P/231
RI MS JT/1/TYP/6/1959–61

---

1. *18–20 April 1859:* The letter appears to have been started on 18 April and finished on 20 April.
2. *your very interesting letter from Keswick:* Tyndall wrote to Julia Pollock from Keswick, a town in Cumbria, see letter 1606.
3. *Ambleside & Keswick via Helvellyn:* see letter 1606, n. 4, 8, and 19.
4. *Red Tarn:* see letter 1606, n. 12.
5. *Scaw Fell:* Sca Fell, or Scafell, is a mountain in the Lake District.
6. *Borrowdale:* valley in the Lake District.
7. *Sty Head:* the Sty Head Pass links Borrowdale with Wasdale, and can be used as a route on to Scafell.
8. *Ullswater:* lake in the Lake District.
9. *Rydal Mount:* see letter 1606, n. 5.
10. Wordsworth's: the poet William Wordsworth (1770–1850), see also letter 1606, n. 3.
11. *Wordsworth's company at Rydal:* see letter 1606, n. 3 and 6.
12. *the Poet at the Mount:* refers to William Wordsworth, who lived on the Rydal Mount, see letter 1606, n. 5.
13. *Lowther Castle:* a castle in the Lake District, previously the seat of the Earls of Lonsdale, now a visitor attraction with a 130-acre garden.
14. *Ambleside:* see letter 1606, n. 4.
15. *WW:* William Wordsworth.
16. *[Poeris]:* not identified.
17. *good lecture from Sir Cha[rle]s Lyell on Friday last at the R.I:* Charles Lyell gave the Weekly Evening Meeting at the RI on Friday 15 April 1859, which was published as C. Lyell, 'On the Consolidation of Lava on Steep Slopes, and on the Origin of the Conical Form of Volcanoes', *Roy. Inst. Proc.,* 3 (1859), pp. 125–31.

18. *Von Buch's:* the German geologist Leopold von Buch (1774–1853).
19. *his deductions from them in favour of his theory of craters of eruption as against Von Buch's theory of elevation:* Lyell presented his research on Mount Etna to the Royal Society on 10 June 1858, later published as *On the Structure of Lavas which have consolidated on Steep Slopes: with Remarks on the Mode of Origin of Mount Etna and on the Theory of Craters of Elevation* (1858).
20. *Fred:* Frederick Pollock.
21. *Henry V:* Play by William Shakespeare.
22. *Princess' Theatre:* Theatre in Oxford Street, London.
23. *Charles Kean:* Charles Kean (1811–68) was an actor and theatre manager (ODNB).
24. *Brighton:* city on the south coast of England.
25. *Walter:* Walter Pollock.
26. *the hotel:* not identified.
27. *Windermere:* town in Cumbria.

To Auguste Arthur de la Rive 22 April 1859 1610

Royal Institution, 22nd, April, 1859.[1]

My dear Sir

I have just received your letter[2] on the banks of the Lake of Windermere in Westmoreland.[3] It has been sent to me in company with other pleasant letters: but I can assure you that none of them gave me greater satisfaction than yours.

Accept my warmest thanks for this new proof of your kindness towards me, and I beg of you also to convey to those concerned in my election[4] my grateful acknowledgments for the honour which they have done me, tell them that I shall endeavour to render myself more worthy than I now am of the distinction which they have conferred upon me.

Nevertheless apart from this subject I confess that your note has a peculiar charm for me—it is the spirit of friendship which it breathes which renders it chiefly precious to me: and surely few have more reason to be grateful than myself for having such friends given to me as you and Faraday. And I know that in your case as in his the longer we know each other the more I shall respect and esteem you.

I know it will give him delight to see you once more in London,[5] and to me also it will be a source of the liveliest gratification—I sincerely hope that nothing may occur to prevent you from carrying out your intention.

The diploma shall be duly transmitted to Mr Airy.[6]

Believe me my dear Sir | Ever most faithfully Yours | John Tyndall

RI MS JT/1/TYP/1/359
LT Typescript Only

---

1. Typescript letter included the handwritten note, possibly by Louisa Tyndall, 'written from Windermere'.
2. *I have just received your letter:* letter 1607, dated 17 April 1859.
3. *the Lake of Windermere in Westmoreland:* Tyndall was spending time in the Lake District, see also letters 1606 and 1609.
4. *my election:* Tyndall had been elected as an Honorary Member of Société de Physique et d'Histoire Naturelle de Genève, see letter 1607.
5. *see you once more in London:* De La Rive was planning a visit to London, see letter 1607.
6. *The diploma shall be duly transmitted to Mr Airy:* Airy was also elected Honorary Member of Société de Physique et d'Histoire Naturelle de Genève, see letter 1607.

To Juliet Pollock [22 April 1859][1] 1611

Good Friday

My dear M^rs^. Pollock

It is very kind both of you and Mr Pollock to write to me, & read both your letters[2] with very great pleasure. I have just received a bundle of letters[3] and they are almost all pleasant ones. De la Rive writes to me[4] from Geneva in a most affectionate style—enclosing me a diploma of membre honoraire of the Genevese Socête de physique.[5] Dove writes to me[6] very pleasantly from Berlin, other letters came from other places more or less distant, and on the whole they diffuse around me quite an intellectual fragrance, which it is very pleasant to enjoy.

I have just completed my lake excursion,[7] and a most pleasant & successful one it has been. I had many grand days, though perhaps no day so grand as that one upon Helvellyn.[8] On Sunday last I was strongly tempted to trespass upon the grounds of your friends at Greta Bank[9]—it would have shortened our way up Skiddaw.[10] However we reached the summit without doing any thing illegal. It was a fine day, the proportion of sunshine to cloud greater than on the day preceding, but there was no such concentrations of atmospheric energy. There was a sleepy sunny effect diffused over the landscape, and sometimes dark cloud shadows moved over it, but on the whole it was a dilution of the grandeur of Saturday. On the summit there is a post and from the post long streamers of snow were drawn out like the filaments of a gray beard to windward. Many of these were of the most beautiful structure: exactly like a complicated moss bleached to whiteness and frozen to ice.

From Keswick[11] we went through Borrodale[12]—exploring the action of its ancient Glaciers, and afterward[13]

RI MS JT/1/T/1210
RI MS JT/1/TYP/6/1962

---

1. *[22 April 1859]:* dated to April 1859 based on when Tyndall and Frankland went on a Lake District excursion, and a letter from W. F. Pollock, 18 April 59 (letter 1609). Good Friday fell on 22 April in 1859.
2. *your letters:* Juliet Pollock's letter is missing; see also letter from W. F. Pollock (letter 1560).
3. *I have just received a bundle of letters:* likely the letters had been bundled because Tyndall had been travelling through the Lake District with Frankland.
4. *De la Rive writes to me:* letter 1607.
5. *enclosing me a diploma of membre honoraire of the Genevese Socête de physique*: According to the *Memoires de la Société de physique et d'histoire naturelle de Genève,* 15: VIII (1859), Tyndall and George Airy became honorary members on 1 August 1859.
6. *Dove writes to me:* letter 1605.
7. *I have just completed my lake excursion:* Tyndall and Frankland made a Lake District excursion in April 1859.
8. *Helvellyn:* see letter 1606, n. 8.
9. *your friends at Greta Bank:* at the time, Greta Bank under Skiddaw was probably occupied by John James Spedding (John Spedding, personal communication).
10. *Skiddaw:* mountain in the Lake District.
11. *Keswick:* town in the Lake District underneath Skiddaw.
12. *Borrodale:* misspelling of Borrowdale, a valley near Keswick.
13. Fragment, letter ends here.

From George Biddell Airy 23 April 1859 1612

Royal Observatory Greenwich | 1859 April 23

My dear Sir

I am much obliged for your sending me your lecture upon Glaciers,[1] which has been here three or four days.

In the interesting Lower Glacier of Grindelwald[2] to which you allude with so much pleasure, there is one appearance of which I should be glad to have the explanation. In descending from the Eis Meer, you probably remember how that first the ice is broken by enormous cross-crevasses: that then the crevasses become a little curved, then convexly upwards, that the great vertical laminæ[3] then begin to crack across a little, that then the *[crosses]* break the ice into blocks, that these blocks become pyramidal pinnacles, to all appearance perfectly homogenous, and that this state is succeeded by a plain solid glacier state completely riddled with horizontal veins.

Now the thing which I want to have is, how these horizontal veins are formed, because, in the state which immediately precedes them, there is not (as far as my eyes could see) a single vein.

I wish I could have two photographers at command, who, acting under my instructions, and separated to such distances as I should indicate, would take pairs of stereoscopic plates[4] at different parts of the path and at different places on the Glacier.

I am, my dear Sir, | yours truly | GB Airy

RGO Archives/MS.RG0.6/378

---

1. *your lecture upon Glaciers:* likely refers to Tyndall's paper, 'On the Veined Structure of Glaciers; with Observations upon White Ice-Seams, Air-Bubbles and Dirt-Bands, and Remarks upon Glacier Theories', *Phil. Trans.,* 149 (1859), pp. 279–307.
2. *Lower Glacier of Grindelwald:* a glacier in the Swiss Alps, below the Eiger.
3. *laminæ:* or laminae, a thin layer (*OED*), in this context referring to a layer in the ice.
4. *stereoscopic plates:* referring to stereoscopic photography.

From George Adolf Erman[1] 23 April 1859 1613

LXVII. *Some occasional Observations on the Structure, the Melting, and the Crystallization of Ice,* made *in Siberia. By* A. ERMAN, *Prof. Univ. Berol.*[2] | *To John Tyndall, Esq., F.R.S.*

DEAR SIR,

YOUR important experiments and observations on the physical properties of ice, of which I have read the report in the Philosophical Magazine, No. 108,[3] remind me of some remarks on the same subject which occurred to me during my travels in Siberia.[4] It seems that at present these remarks may become useful; for some of them agree perfectly with the results which you have already obtained, and therefore confirm them, and some others, like all phaenomena that are still waiting for an explanation, may be an inducement to new experiments. I think, therefore, that either yourself in the second part of your researches, or other observers who may participate in them, would like to take notice of my observations. If you are of this opinion, I should wish the description of what I have given to become known to English philosophers, by your publishing it in the Philosophical Magazine. I have endeavoured to have these observations quite verbally translated from the German volumes, entitled *Reise um die Erde, u.s.w.,*[5] which I published in the years

1835 to 1849; and I hope you will convince yourself of this agreement by comparing these with the passages quoted from the German work. I may add that my translation of many of the passages in question agrees fully with that of Mr. W. D. Cooley,[6] in his somewhat abridged English edition of part of my book entitled "Travels in Siberia", &c., by A. Erman: London, 1848.[7] The remainder of the annexed remarks, however, will be found in full accordance with the German only, whereas some particulars have been omitted in the English abridgement, perhaps as not being of interest to the public in general. I have nevertheless quoted at each of these passages the pages of the English, as well as of the German edition, when the phaenomena in question are either merely alluded to, or fully described.

Believe me, dear Sir, to be with the sincerest regard, | Your obedient Servant, | A. ERMAN.

Berlin, 122 Friedrichstrasse, | April 23, 1859.

A.Erman, 'LXVII. Some occasional Observations on the Structure, the Melting, and the Crystallization of Ice, made in Siberia', *Phil. Mag.,* 17 (1859), p. 405.

---

1. *From George Adolf Erman:* George Adolf Erman (1806–77) was a German physicist and explorer.
2. *LXVII. Some occasional Observations on the Structure, the Melting, and the Crystallization of Ice,* made *in Siberia. By A. ERMAN, Prof. Univ. Berol:.* Erman's letter was published as part of the article 'LXVII. Some occasional Observations on the Structure, the Melting, and the Crystallization of Ice, made in Siberia', *Phil. Mag.,* 17 (1859), pp. 405–13, p. 405.
3. *the report in the Philosophical Magazine, No. 108:* Likely refers to Tyndall's paper 'Remarks on Ice and Glaciers', *Phil. Mag.,* 17 (1859), pp. 91–96.
4. *during my travels in Siberia:* Erman travelled through Siberia and North Asia in 1828, 1829, and 1830.
5. *Reise um die Erde, u.s.w., which I published in the years 1835 to 1849:* Erman's book, *Reise um die Erde durch Nord-Asien und die beiden Oceane in den Jahren 1828, 1829 und 1830 ausgeführt,* first appeared in 1833.
6. *Mr. W. D. Cooley:* the geographer William Desborough Cooley (1795?–1883). Cooley was part of the early formation of the Hakluyt Society and edited several of the society's publications (*ODNB*).
7. *his somewhat abridged English edition of part of my book entitled 'Travels in Siberia', &c., by A. Erman: London, 1848:* It was published in English in 1848, as *Travels in Siberia: including excursions northwards, down the Obi, to the Polar Circle, and southwards to the Chinese frontier.*

To George Biddell Airy 26 April 1859 1614

Deanery Hereford. | April 26th 1859

My dear Sir

Your note by post[1] reaches me at Hereford[2] where I have been staying for a few days with my excellent friend the Dean.[3]

My absence from London has prevented me from carrying out a duty towards you imposed upon me by Prof. De la Rive of Geneva from whom a note reached me[4] a few days ago at Windermere[5] charging me to transmit to you a diploma of the Societe de Physique of Geneva.[6]

With regards to the glaciers the question[7] is so large that one can only attack it piecemeal and after each attack many questions will remain. I know the place to which you allude[8] very well; and as far as I have been able to look into the matter the following is the explanation that appears to be correct.

The ice near the bed of the glacier is forcibly compressed by the mass above it, and I should undoubtedly expect to find horizontal veining there if one could get at the place. The portions of the glacier above this (supposed) horizontally veined ice is undergoing continual waste, and near the terminus of the glacier has so far disappeared as to permit of the horizontal veining to show itself. I do not think the homogenous ice of the foreground has been converted into the veined ice lower down; but the veined ice lower down comes to view by the wasting of the pyramids.

I omitted in the lecture[9] to state that I had a few observed cases of slipping like those to which you referred as having seen upon the places of the Rhone[10]—The Allalin Glacier[11] shews numerous such cases with great distinctness

I am now throwing my thoughts upon the subject of glaciers into a more compacted form, and hope to present them to you thus before the expiration of another year.

Believe me dear Sir | very faithfully yours | John Tyndall

RGO Archives/MS.RG0.6/378

---

1. *Your note by post:* letter 1612, dated 23 April 1859.
2. *Hereford:* city in Herefordshire.
3. *the Dean:* Richard Dawes.
4. *from whom a note reached me:* letter 1607.
5. *at Windermere:* Tyndall spent time in the Lake District during April 1859, see also letters 1606 and 1609.
6. *a diploma of the Societe de Physique of Geneva:* Tyndall and Airy had been elected Honorary

Members of Société de Physique et d'Histoire Naturelle de Genève.

7. *With regards to the glaciers the question:* Airy asked Tyndall about the formation of the horizontal veins in glaciers.
8. *the place to which you allude:* an unspecified place on the Lower Glacier of Grindelwald, see also letter 1612, n. 2.
9. *I omitted in the lecture:* see letter 1612, n. 1.
10. *the Rhone:* The Rhone is a major river in Europe, originating in the Rhone Glacier.
11. *The Allalin Glacier:* the Allalin Glacier in the Pennine Alps, under the Allalinhorn, which Tyndall explored.

To Unidentified[1] 27 April 1859 1615

My dear Sir

Mr. Gassiot's note will form a valuable appendix to his paper, and I have no hesitation in recommending it for publication in the Philosophical Transactions[2]

I am dear Sir | very much yours | John Tyndall

27th April 1859

RS MM/769/106

1. *To Unidentified:* the recipient was either Walter White or G. G. Stokes, based on the review for *Phil. Trans.*, but has not been further identified.
2. *Mr. Gassiot's note will form a valuable appendix to his paper, and I have no hesitation in recommending it for publication in the Philosophical Transactions:* Gassiot's article 'On the stratifications in electrical discharges, as observed in torricellian and other vacua. Second communication', *Phil. Trans.*, 149 (1859), pp. 136–60, included two additional notes which were added after the communication of the original article. The first note was dated as received on 11 January 1859, and the second note was dated as received on 17 February 1859.

To [George Gabriel Stokes][1] 25 May 1859 1616

25th May 1859

My dear Sir

I think it very desirable that Dr Radcliffe[2] should have an opportunity of fully stating his news (which are in many respects very ingenious) of muscular action—This opportunity would be afforded by a full abstract of his paper appearing in the Proceedings of the Royal Society;[3] but in the present state of our Knowledge regarding the process[4] described in the paper. I think it would be hazardous to publish it in the Philosophical Transactions.

Believe me dear Sir | Very since[rel]y yours | John Tyndall

RS RR/4/215

---

1. *To [George Gabriel Stokes]*: Stokes is the presumed recipient of this letter, as the letter concerns the peer review process of an article for the *Roy. Soc. Proc,* and included the later addition '1859 | Radcliffe | by | Tyndall' in what appears to be Stokes's handwriting.
2. *Dr Radcliffe:* Charles Bland Radcliffe (1822–89) was made full physician to the Westminster Hospital in 1857. Radcliffe research focused in particular on the nervous system, in particular on the electrical physiology of muscle and nerve (*ODNB*).
3. *I think it very desirable . . . Proceedings of the Royal Society:* Radcliffe's research on muscular action was published as Charles Bland Radcliffe, 'An inquiry into the muscular movements resulting from the action of a galvanic current upon nerve', *Roy. Soc. Proc.,* 10 (1860), pp. 347–59.
4. *the process*: that is, the effect of electrical stimuli on muscles. See also Charles Radcliffe's earlier publication, 'On Muscular Action from an electrical point of view', *Roy. Soc. Proc.,* 9 (1859), pp. 690–97.

To Auguste Arthur de la Rive [May 1859[1]] 1617

Lettre de M. John Tyndall à M. le prof. Auguste de la Rive.

Mon cher Monsieur,

Je viens d'envoyer à la Société royale une note sur des recherches relatives à l'influence d'enveloppes diathermanes pareilles à notre atmosphère sur le passage de la chaleur rayonnante; elles seront, je crois, de nature à vous intéresser. Vous n'ignorez pas que M. Pouillet, dans son travail remarquable sur le rayonnement solaire, attribue la température proportionnellement plus élevée des couches inférieures de l'atmosphère à l'absorption par ces couches d'une portion de la chaleur obscure émise par la terre. M. Hopkins a fait aussi un bon travail sur le même sujet, mais il manquait aux recherches théoriques de l'un et l'autre de ces savants une base expérimentale : savoir des données exactes sur la manière dont la chaleur rayonnante se comporte dans son passage à travers un milieu gazeux. C'est l'espoir de combler cette lacune qui m'a conduit à étudier sous un point de vue général la transmission de la chaleur rayonnante à travers les différents gaz.

Melloni a démontré que la loi de l'inverse des carrés était applicable au passage de la chaleur rayonnante à travers l'air, et il conclut de ses expériences à ce sujet "qu'entre les limites de ses observations, c'est-à-dire pour un intervalle de cinq ou six mètres, l'air n'exerce aucune absorption sensible sur le rayonnement des corps chauds." Vous aurez probablement répété plus d'une fois l'expérience du célèbre physicien italien, et vous l'aurez trouvée, comme moi, parfaitement exacte.

Le pouvoir absorbant extraordinaire des composés d'hydrogène dont j'ai déjà eu l'occasion de vous entretenir, m'a suggéré l'idée atmosphère de ce gaz agirait peut-être plus énergiquement que l'air pour arrêter le calorique rayonnant. Pour m'en assurer, je me suis servi d'un tube long de quatre pieds et large de trois pouces, fermé à chacune de ses extrémités par un disque de sel gemme. Ce tube est muni à l'une de ses extrémités d'un robinet au moyen duquel il peut être visse à la machine pneumatique, de manière à y faire ou un autre gaz quelconque. La propriété bien connue du sel gemme de transmettre également tous les rayons calorifiques, quelle que soit leur qualité, m'a permis de faire passer à travers mon tube des rayons calorifiques de toutes les espèces, sans crainte de les altérer.

Le tube étant dispose horizontalement et plein d'air, une source de chaleur fut placée à l'une de ses extrémités, et à l'autre extrémité une pile thermoélectrique. Le calorique rayonnant, qui passait alors à travers l'air renfermé dans le tube, produisait sur la pile un effet qu'accusait aussitôt un galvanomètre était revenue au repos, j'ai fait le vide dans le tube pour constater si l'absence de l'air exerçait une influence sensible sur la quantité de chaleur transmise. Je n'en aperçus aucune. Il en a été de même lorsque, à la place de l'air, j'ai introduit dans le tube du gaz hydrogène. Ces résultats sont de nature à confirmer pleinement les conclusions de Melloni déjà citées.

L'expérience ci-dessus me suggéra pourtant quelques réflexions sur la méthode employée. Je m'étais aperçu que lorsque la source de chaleur était passablement intense, tel qu'un boulet de cuivre chauffé presque au rouge, la déviation de l'aiguille devenait très-considérable. Or, chacun sait que les degrés élevés du galvanomètre représentent une force beaucoup plus considérable que les degrés inférieurs. Il en résulte que, pour produire une diminution sensible sur une forte déviation de l'aiguille, il devient nécessaire d'opérer une réduction comparativement beaucoup plus considérable dans la force du courant. De là l'inconvénient de se servir d'une source calorifique intense; d'autre part, si cette source est trop faible, la quantité de chaleur rayonnante qui passe à travers le tube, devient assez minime pour que la fraction absorbée par la colonne gazeuse doit naturellement être insensible.

J'ai donc cherche à appliquer une source de chaleur assez forte pour que la portion de cette chaleur absorbée restât une quantité appréciable, et en même temps à faire en sorte que la chaleur rayonnante, émise par la source, put agir sur l'aiguille du galvanomètre dans la position la plus sensible de celle-ci. Voici le procédé que j'ai employé. Les deux fils d'un galvanomètre différentiel furent mis en communication avec deux piles thermoélectriques, de telle sorte que les courants provenant de l'une et de l'autre pile dussent faire le tour de l'aiguille dans des directions opposées. Une source de chaleur étant placée en face de l'une des extrémités du tube, l'aiguille se trouva promptement déviée par l'action du courant provenant de la pile située à l'extrémité opposée. Je rapprochai alors la seconde pile de la source calorifique, jusqu'à ce que son courant neutralisât complétement celui de la première, et aussitôt on vit l'aiguille redescendre au zéro. Voilà donc que, par cette disposition, nous avons obtenu deux forces puissantes en état d'équilibre parfait; il ne s'agit plus que de s'assurer si un changement dans la nature du milieu traversé par la chaleur rayonnante aura pour effet de déranger cet équilibre. Or, voici le résultat général que j'ai obtenu dans le cas de l'air atmosphérique. Je suppose que l'aiguille soit au zéro lorsque le tube est plein d'air; dès qu'on a fait le vide, on voit l'aiguille se mettre en mouvement, indiquant ainsi que le courant provenant de la pile située à l'extrémité du tube l'emporte sur l'autre; ou en d'autres termes, que, par suite du déplacement de l'air, une plus grande quantité de chaleur rayonnante a pu passer à travers le tube. En laissant rentrer l'air dans le tube, on voit l'aiguille revenir aussitôt au zéro. J'ai obtenu dans quelques cas des résultats tout à fait inattendus; l'épuisement de l'air de tube ayant plus d'une fois fait descendre l'aiguille du galvanomètre de 40 à zéro, pour remonter à 40 du côté opposé, tandis que la réadmission de l'air parfaitement pur et sec; je me bornerai à vous assurer qu'elles étaient telles que vous pouvez regarder le fait ci-dessus comme parfaitement établi.

J'ai opéré de la même manière sur plusieurs autres gaz, tels que l'oxygène, l'hydrogène, l'azote, l'acide carbonique, le gaz oléfiant, le gaz d'éclairage, l'ammoniaque, etc., et aussi sur les vapeurs d'éther, d'alcool, d'eau, de sulfure de carbone, etc., et j'ai obtenu pout résultat, que le pouvoir diathermane des différentes substances gazeuses varie probablement tout autant que celui des corps solides et liquides. Prenons pour exemple le gaz d'éclairage; le pouvoir diathermane de ce gaz, compare à celui de l'oxygène ou de l'azote, est à peu près dans le même rapport que le pouvoir diathermane de l'alun est à celui d'un des corps solides qui se laissant le plus facilement traverser par la chaleur rayonnante. L'introduction, dans le tube vide, d'une quantité du gaz d'éclairage équivalente à un pouce de mercure seulement, produit une forte déviation dans l'aiguille du galvanomètre, et je ne doute pas de pouvoir arriver à apprécier l'effet absorbant de ce gaz à la tension d'un millimètre. J'ai aussi remarqué que le pouvoir absorbant de la vapeur de sulfure de carbone.

En poursuivant ces recherches, j'ai aussi réussi à démontrer que le même gaz est plus ou moins perméable à des rayons de chaleur de qualité différente, et même que tel rayon de chaleur peut passer à travers le tube sans être sensiblement affecte par la présence du gaz qui s'y trouve. Je dois ajouter que ces recherches doivent être regardées comme seulement à leur début, et qu'il me faudra beaucoup de temps pour les compléter; mais j'ai cru que, même au point où elles en sont, la méthode employée et les résultats déjà obtenus seraient de nature à vous intéresser.

Recevez, etc. | John Tyndall

PS. Je me sers maintenant d'un galvanomètre à un seul fil, de deux sources de chaleur et des deux faces de la même pile.

Letter from Mr. John Tyndall to Prof. Auguste de la Rive.

My dear sir,

I have just sent to the Royal Society a note[2] on researches into the influence of diathermal envelopes[3] similar to our atmosphere on the passage of radiant heat; they will, I think, be of interest to you. You do know that M. Pouillet, in his remarkable work on solar radiation,[4] attributes the proportionately higher temperature of the lower layers of the atmosphere to the absorption by these layers of some dark heat emitted by the earth. Mr. Hopkins has also done good work on the same subject,[5] but the theoretical researches of both scientists lacked an experimental basis: to know the exact manner in which radiant heat behaves in its passage through a gaseous medium. It is the hope of filling this gap which led me to study from a general point of view the transmission of radiant heat through the different gases.

Melloni has demonstrated that the law of inverse squares[6] was applicable to the passage of radiant heat through the air, and he concludes from his experiments on this subject "that between the limits of its observations, that is to say, for an interval of five or six meters, the air does not exert any sensible absorptive force on the radiation of hot bodies." You will probably have repeated more than once the experiment of the celebrated Italian physicist, and you will find it, like me,[7] perfectly exact.

The extraordinary absorptive power of the hydrogen compounds, of which I have already spoken, has suggested to me that an atmosphere of this gas would perhaps act more energetically than the air to stop the caloric radiation. In order to ascertain this, I have employed a tube four feet long and three inches wide, closed at each end by a disk of rock salt. This tube is fitted at one of its ends with a tap by means of which it can be screwed to the pneumatic machine, so as to empty it; a second tap makes it possible to introduce air or any other gas into it. The well-known property of rock-salt to transmit all calorific rays, whatever their quality, has permitted my calorific rays of all types to pass through my tube, without fear of altering them.

An air filled tube was horizontally arranged, a source of heat placed at one end, and at the other end a thermoelectric cell. The calorific ray, which was then passed through the air enclosed in the tube, caused the battery cell to produce an immediate effect on a very sensitive galvanometer[8] placed in the circuit. As soon as the galvanometer's needle had returned to rest, I emptied the tube to ascertain whether the absence of the air exerted a sensible influence on the quantity of heat transmitted. I did not notice any. It was the same when, in instead of air, I introduced into the tube hydrogen gas. These results are such as to fully confirm the conclusions Melloni already cited.

The above experiment made me reflect on the methods employed. I realized that when the source of heat was quite intense, such as a ball of copper heated almost to red, the deflection of the needle became very considerable. Everyone knows that the higher degrees of the galvanometer represent a force much greater than the lower degrees. As a result, in order to decrease the deflection of the needle, it is necessary to reduce the force of the current. Hence the disadvantage of using an intense heat source; on the other hand, if this source is too weak, the quantity of radiating heat passing through the tube becomes small enough that the fraction absorbed by the gaseous column must naturally be negligent.

I have, therefore, endeavored to apply a sufficiently strong source of heat for the portion of this absorbed heat to remain in substantial quantity, and at the same time to cause the radiant heat emitted by the source to act upon the needle of the galvanometer in its most sensitive position. This is the procedure I used. The two wires of a differential galvanometer were connected to two thermoelectric cells,[9] so that the currents coming from one and the other would have to go around the needle in opposite directions. A heat source was placed opposite one of the ends of the tube, the needle was promptly diverted by the action of the current from the battery at the opposite end. I then brought the second battery cell closer to the calorific source until its current completely neutralized that of the first, and at once the needle was seen to fall back to zero. Here, therefore, by this arrangement we have obtained two powerful forces in a state of perfect equilibrium; it is only a question of ascertaining whether a change in the nature of the medium passed through by radiant heat will have the effect of disturbing this equilibrium. Now, here is the general result I obtained in the case of atmospheric air. I suppose the needle is at zero when the tube is full of air; as soon as the vacuum has been emptied, the needle is seen to move, indicating that the current from the battery at the end of the tube carries it over the other; or in other words, that as a result of the displacement of the air a greater quantity of radiant heat may have passed through the tube. By allowing the air to enter the tube, the needle is immediately returned to zero. In some cases, I have obtained quite unexpected results; the exhaustion of the air of the tube having more than once made the needle of the galvanometer descend from 40 ° to zero, to rise to 100 ° on the opposite side, while the readmission of the air made it immediately return to its first position. I shall not here enter into the details of the precautions taken to obtain perfectly pure and dry air; I will confine myself to assuring you that they were such that you can look at the fact above as perfectly established.

I have experimented in the same manner on several other gases, such as oxygen, hydrogen, nitrogen, carbonic acid, olefin gas, lighting gas, ammonia, etc., and also on the vapors of ether, alcohol, water, carbon disulphide, etc., and I have concluded,[10] as a result, that the diathermic power of the different gaseous substances probably varies as well as that of the solid bodies, and liquids. Take for example the lighting gas; the diathermic power of this gas, when compared with that of oxygen or nitrogen, is nearly in the same ratio as the diathermanal power of alum is to one of the solid bodies which are more easily passed through by radiant heat. The introduction, into the empty tube, of a quantity of the lighting gas equivalent to one inch of mercury alone produces a great deflection in the needle of the galvanometer, and I have no doubt, absorbing the effect of this gas at the voltage of one millimeter. I have also observed that the absorbing power of ether vapor is infinitely greater than that of carbon disulfide vapor.

In continuing these researches, I have also succeeded in demonstrating that the same gas is more or less permeable to heat rays of different quality, and even that such a heat ray may pass through the tube without being sensibly affected by the presence of the gas therein. I must add that these investigations must be regarded as being at their beginning, and that it will take me a long time to complete them; but I have thought that, even at the point where they are, the method employed and the results already obtained would be of interest to you.

Yours sincerely, etc. | John TYNDALL.

PS. I now use a single-wire galvanometer, two heat sources and two sides of the same cell.

Bibliothèque Universelle revue suisse et étrangère, 5 (1859), pp. 232–36.

---

1. *[May 1859]:* this is a conjectured date, based on the publication of the letter in the July volume of *Bibliothèque Universelle revue suisse et étrangère.* Tyndall likely wrote this letter in May (or possibly early June), as he was circulating information about his experiments on radiant heat through gases in this period (J. Tyndall, Contributions to Molecular Physics in the Domain of Radiant Heat (London: Longmans, Green and Co., 1872), p. 62.
2. *I have just sent to the Royal Society a note:* likely refers to John Tyndall, 'Note on the Transmission of Radiant Heat through Gaseous Bodies', *Roy. Soc. Proc.,* 10 (1859): pp. 37–39.
3. *diathermal envelopes:* a concept from thermodynamics, a surface which allows passage to rays of heat (*OED*).
4. *M. Pouillet, in his remarkable work on solar radiation:* refers to Pouillet's 'Memoir on the Solar Heat, on the Radiating and Absorbing Powers of the Atmospheric Air, and on the Temperature of Space', *Scientific Memoirs,* 4 (1846), pp. 44–90.
5. *Mr. Hopkins has also done a good job on the same subject:* it is unclear what part of Hopkins's research Tyndall was referring to. Hopkins published extensively on the temperature of the earth such as: W. Hopkins, 'Presidential Address', BAAS Report, 23 (1853), pp.

xli–lvii; W. Hopkins, 'On the Causes which may have produced Changes in the Earth's Superficial Temperature', *Quarterly Journal of the Geological Society,* 8 (1852), pp. 56–92. It is likely that Tyndall was referring to Hopkins's article 'XXIX On the External Temperature of the Earth, and the other Planets of the Solar System', *Phil. Trans.*, 9 (1856), pp. 628–72. In this article, Hopkins examined various modes of the transmission of heat, and discussed the Dynamical Theory of Heat (see also letter 1623, n. 4).

6. *Melloni has demonstrated that the law of inverse squares:* see M. Melloni, 'Proposal of a New Nomenclature for the Science of Calorific Radiations', *Scientific Memoirs,* 3 (1843), pp. 527–36 and M. Melloni, 'Memoir on the Free Transmission of Radiant Heat through Different Solid and Liquid Bodies', *Scientific Memoirs,* 1 (1837), pp. 1–5. See also letter 1619, n .8.
7. *like me:* refers to Tyndall's article 'Note on the Transmission of Radiant Heat through Gaseous Bodies', *Phil. Mag.,* 19 (1860), pp. 60–61. See also letter 1619.
8. *galvanometer:* a device for determining existence and direction of a galvanic current (*OED*), when the thermopile is used in conjunction with a galvanometer it is called a thermo-multiplier (*OED*).
9. *thermoelectric cells:* see also letter 1619 n. 9.
10. *I have concluded:* see n. 7, and letter 1619.

To Heinrich Debus [May or June 1859][1] 1618

Monday.

My dear Heinrich.

Es thut mir Leid nicht gegenwartig zu sein als Du Deinen Vortrag hieltest.[2] But as you say I needed Erfreschung[3]—and found it among the mountains. We had fine weather in Cumberland[4]—fine I say though we were sometimes in heavy snow. Sometimes in thick fog. but sometimes also in unclouded sunshine—the atmospheric changes were wonderful, and we climbed all the highest peaks.

I am now working hard at my book,[5] and hope to make it a respectable one: I do not want it to pass away in a year, and hence while I try to write in a pleasant style and in a way which shall interest the general reader my aim also is to concentrate in it a solid body of Scientific truth.

Your estimate of Carlyle I think is a very correct one, and I am half tempted to send your letter[6] to Mrs Carlyle, without letting her know any thing about its origin. I have looked over the first volume[7] only and not the whole of that—but I shall read all someday.

I was sorry to hear of your cold and glad to hear of your industry. I have left south Audley Street & now live in a more open place near St. John's Wood.[8] I hope I may be able to pay you a visit during the spring or summer. Believe me it will give me pleasure to do so; but my book is now my first care.

This is a glorious day and it makes me long for the country—what a weight our civilization and social experiments lay upon our shoulders—and how I sometimes long to rid myself of the load, and to return to the free joyous life which I lived when a boy. Still it is only when the mind is weary that one feels them—Assuming the heart is strong enough to cope with, and were to enjoy the passages of civilization—

Goodbye Heinrich | Ever yours. J Tyndall

RI MS JT/1/T/327

1. *[May or June]:* This letter has been dated to May or June 1859 based on the reference to Tyndall's recent visit to Cumberland and his move to lodgings in 31 Aberdeen Place, Maida Vale.
2. *Es thut mir Leid nicht gegenwartig zu sein als Du Deinen Vortrag hieltest:* I am sorry for not having been present when you gave your lecture.
3. *Erfreschung:* refreshment.
4. *Cumberland:* Tyndall visited the Cumberland mountains in April, see also letter 1608 for more details on this trip.
5. *my book:* likely refers to *Glaciers of the Alps* published in 1860.
6. *your letter:* letter missing.
7. *the first volume:* likely refers to Carlyle's *History of Friedrich II of Prussia: Called Fredrick the Great* which was published in 6 volumes between 1858 and 1865.
8. *I have left south Audley Street & now live in a more open place near St. John's Wood:* Tyndall moved to Maida Vale in May 1859.

To Rudolf Clausius 1 June 1859 1619

1st June 1859

My dear Clausius,

Many thanks to you for the paper on the Potential function,[1] which though it is beyond me is still precious as a proof of your friendly feeling. The second copy has been duly presented to the Royal Society.[2]

I have been at work for some time upon a new subject which promises to give good results. Melloni, you know, proved the law of inverse squares to be true of radiant heat passing through air, and he inferred from this that "autre les limites d'observation, C'est à dire, pour une intervalle de cinq ou six metres, l'air n'exerce aucune absorption sensible sur le rayonnement des corps chauds." This is the only experimental conclusion on this subject which we possess.[3]

But you know that M. Pouillet,[4] and Hopkins[5] have considered the influences of a diathermic envelope[6] like our atmosphere to be such that while the rays of the sun can reach us freely, a great portion of them are stopped when they have been changed by the absorption of the earth into obscure heat. Pouillet you know also explains the greater warmth of the lower atmospheric strata by supposing that the radiant-heat emanating from the earth is there principally absorbed.[7]

Reflecting on all this I felt that our knowledge as to the action of gaseous bodies upon radiant heat was absolutely nil and I thought it very desirable to investigate the transmission of radiant heat through gases in general.

I had often verified the experiments of Melloni regarding the law of inverse squares;[8] but I thought that our atmosphere's hydrogen might act with greater energy than air, because all the components of this element are too opaque to radiant heat. I therefore had a tube constructed 4 feet long 3 inches wide, and by suitable means contrived to stop its end securely by plates of rocksalt. A stopcock was attached to the tube which could be screwed on to an airpump intended to exhaust the tube. There was also a passage devised for the admission of any gas, and also a second arrangement by which the vapour of any liquid might be admitted into the exhausted tube.

I first placed a source of heat at one side of this tube and a thermoelectric pile at the other. When the needle of the associated galvanometer had come to rest I exhausted the tube, and watched whether the removal of the air had any sensible influence on the radiation—it had not—the needle remained perfectly steady, and was equally so when the air was readmitted. The experiment was repeated with hydrogen and with the same result.

It was evident therefore that some other mode of experiments must be devised. I saw that the experiment just recorded has this disadvantage. When great heat was made use of the needle of the galvanometer /pushed/ to the high degrees, and you know it requires a far greater amount of force to move the needle in this position than when it is near to zero. Then, on the contrary, a feeble source of heat was used, and the needle came to rest in a sensitive position, the fractional part of this, a small quantity absorbed by the gas in the tube might well escape observation.

My object then was to operate with quantities of heat so large that the percentage climbed would be a sensitive quantity, and to apply this quantity to the needle in its sensitive position. You will understand my way of proceeding in a moment. I took two thermo-electric piles;[9] had a galvanometer[10] constructed with two wires so that the currents from the two piles could be sent round the instrument in opposite directions. The heat falling in one pile had to pass through the tube, and hence the medium traversed by its /path/ could be varied; while the medium traversed by the heat falling upon the other pile remained unchanged.

Imagine these two powerful currents looping round the galvanometer in opposite directions. They neutralize each other, and the needle stands at zero.

I exhaust the tube, and find immediately that the current from the pile at this end of the tube now predominates, or in other words by the removal of the air a greater quantity of heat has been permitted to pass. When the tube is quite exhausted & the needle quite at rest in its new position let air be admitted. The needle then shows a diminution of the current which, when the tube was empty, was excited in the pile.

Effects which would surprise you have been obtained in this way; for example with air I can easily obtain a deflection from 40° on one side of zero to 40° on the other; the results moreover are perfectly constant, and the amount of absorption varies with the density of the air in the tube.

I have operated in the same manner with several other gases and vapours, and have obtained very remarkable results. It is quite probable that gases vary as much as solids & liquids in this department to radiant heat. Some of these bear the same relation to others that alum does to the better diathermic solids. As similar diversity reigns among the vapours. I believe that I shall be able to measure the absorption of plates of some gases not more than a millimetre in thickness.

The experiments have also proved that the selfsame gas operates differently at different qualities of heat. Yesterday I sent through the tube a powerful wave of heat which was not sensibly affected by the most energetic gas that I have as yet discovered. The same gas operating upon heat of a different quality causes the needle to move 80° on one side of zero to 80° on the other.

I am now in the middle of the investigation, and hope that when it is complete it will be found a good piece of work.

And now I have several favours to ask of you:—the first is that you would present my kind remembrances to M. Mousson, and ask him to be good enough to send me a copy of the little paper on the present state of the glacier question[11] which he gave to me in Zürich & which I have mislaid. It refers among other things to the application of Faraday's experiment on regeneration to the receding of the crevasses;[12] leaving the question of structure untouched, because it was thought that it still remained to be solved.

I also want you to ascertain for me how I may obtain Rendu's Essay upon glaciers.[13] I would give a good deal for the perusal of this Essay if you should have a copy of it, and would lend it to me I should be careful to return it to you safely. It may be possible to obtain a copy of it among the booksellers of Zürich.

Thirdly, I want you to give me your opinion on that theory of the coldness of the upper strata of the atmosphere which refers it to the fact that the attenuated air has a much higher specific heat than that which is more dense. I should take it very kind of you if you would write to me soon on the three points each mentioned.

Believe me ever | most truly yours | John Tyndall

RI MS JT/1/T/176

1. *Many thanks to you for the paper on the Potential function:* the letter from Clausius containing the paper is missing. Clausius's work on the Potential function was published as *Die Potentialfunction Und Das Potential: Ein Beitrag Zur Mathematischen Physik* in 1859.
2. *presented to the Royal Society:* a notice about Clausius's book *Die Potentialfunction Und Das Potential* was published as 'Presents received by the Royal Society with the Names of the Donors', *Phil. Trans.,* 149 (1859), p. 9.
3. *Melloni, you know . . . which we possess:* Tyndall was referring to Melloni's book *La thermochrôse ou la coloration calorifique* published in 1850. The quote on page 136 appeared slightly different to the version in this letter, 'qu'entre les limites de l'observation, c'est-à-dire, pour un interval de cinq à six mètres, l'air n'exerce aucune absorption sensible sur le rayonnement des corps chauds'.
4. *M. Pouillet:* the French physicist Claude Servais Mathias Pouillet (1790–1868). Pouillet's research on heat and atmospheric phenomena included *Eléments de physique expérimentale et de météorologie,* vol 1–4 (Paris/Brussels, 1827–32) and 'Memoir on the Solar Heat, on the Radiating and Absorbing Powers of the Atmospheric Air, and on the Temperature of Space', *Scientific Memoirs,* 4 (1846), pp. 44–90.
5. *Hopkins:* Hopkins undertook extensive research on radiant heat, which, together with the research of the physicist Jean Baptiste Joseph Fourier (1768–1830), Pouillet (see n. 4), and de Saussure (see letter 1741, n. 18) was significant for Tyndall's experimental work on radiant heat (Bernard Lightman and Michael Reidy (eds), *The Age of Scientific Naturalism: Tyndall and His Contemporaries,* (London: Pickering & Chatto, 2014, pp. 39–46).
6. *diathermic envelope:* see letter 1617, n. 3.
7. *Pouillet you know . . . there principally absorbed:* see n. 4.
8. *I had often verified the experiments of Melloni regarding the law of inverse squares:* Tyndall published these considerations of radiant heat and the law of inverse squares as 'Note on the Transmission of Radiant Heat through Gaseous Bodies', *Phil. Mag.,* 19 (1860), pp. 60–61.
9. *thermo-electric piles:* a thermopile is a device for conversion of heat into electricity (*OED*).
10. *galvanometer:* see letter 1617, n. 8.
11. *the little paper on the present state of the glacier question:* this may refer to Mousson's book *Die Gletscher des Jetztzeit* published in 1854.
12. *It refers among other things to the application of Faraday's experiment on regeneration to the receding of the crevasses:* likely refers to Faraday's experiments on ice, presented at the RI in June 1850 and published as an abstract, 'Royal Institution', *Athenaeum,* (June 1850), p. 641. Faraday's research on ice was also published as 'On Regelation, and on the conservation of force', *Phil. Mag.,* 17 (1859), pp. 162–69.
13. *how I may obtain Rendu's Essay upon glaciers:* Louis Rendu's *Théorie des Glaciers de la Savoie* proposed that a glacier flows like a river, and became a centrepiece in the dispute about priorities for theories of glacier motion between Tyndall, Agassiz, Forbes and others: L. Rendu, *Théorie des Glaciers de la Savoie* (Chambéry: Chez Puthod, Imprimateur-Libraire, 1840). Rendu's paper was notoriously difficult to obtain in Britain at the time.

From David Brewster 1 June 1859 1620

June 1 1859

My Dear Mr Tyndall,

Having occasion to write an Article on Glaciers, I have cursorily read your papers,[1] and those of Mr Huxley on the subject,[2] and also Prof[esso]^r^ Forbes's Occasional Papers just published.[3]

I do not fully comprehend what the Professor (p. XVI) calls your theory,[4] and therefore you would oblige me much if you could, in a few sentences, describe it, in order that I may study the evidence of it as given in your paper.

I am anxious to know the optical structure of the ice which forms the blue veins, in order to be able to judge whether it is formed from the nevé[5] by pressure, or from the slow action of corporeal forces, as takes place in the decomposition of bodies such as glass & circular crystals which, in the course of time, return into their elements;—or by traction, as explained in one of my late papers in the Edin. Transactions.[6]

I would like also to know what is the optical structure of the solid Ice produced by pressure in your experiments. Is the axis of Double refraction parallel to the direction of the force? or does the ice exhibit the same effect that is produced by the compression of glass, and crystals that have a feeble double refraction.

I observe that Prof^t^. Forbes regards your researches as confirming his theory. I am,

my Dear Mr Tyndall, | Ever Most Truly yrs. | D Brewster

Edin. 42 Charlotte Square | June 1st 1859.

P.S. My address after today is St. Andrews.

RI MS JT/1/B/128
RI MS JT/1/TYP/1/159

---

1. *read your papers:* see for example John Tyndall and Thomas Huxley, 'On the Structure and Motion of Glaciers', *Phil. Trans.*, 147 (1857), pp. 327–46; John Tyndall, 'On some Physical Properties of Ice', *Phil. Trans.*, 16 (1858), pp. 333–56.
2. *those of Mr Huxley on the subject:* likely refers to John Tyndall and Thomas Huxley's co-written article, see n. 1.
3. *Prof[esso]^r^ Forbes's Occasional Papers just published:* Forbes's book *Occasional Papers on the Theory of Glaciers* (Edinburgh: Adam and Charles Black, 1859).
4. *what the Professor (p. XVI) calls your theory:* refers to Forbes's *Occasional Papers,* where he outlined his perception of Tyndall's theory of glacial motion as one of fracture and regelation in comparison with his own, the viscous theory of glacial motion. See also letter 1594.
5. *nevé:* the hardened snow found on glaciers (*OED*).

6. *as explained in one of my late papers in the Edin. Transactions:* 'On the Production of Crystalline Structure in Crystallised Powders, by Compression and Traction', *Transactions of the Royal Society of Edinburgh,* 20 (1853), pp. 555–59.

From David Brewster 7 June 1859 1621

My dear M[r] Tyndall,

As you are occupied with the interesting subject of Glaciers, I may mention to you some observations of mine connected with the subject

1. Altho' Prof. Forbes (Travels p. 29)[1] mentions that I and Col. Sabine and Elie de Beaumont had seen the veined structure of the Glacier ice, yet he appears as the discoverer, and first publisher of the fact. I discovered it on the 10[th] Sept[em]b[e]r 1814 and described it in my Journal, and in the Article Glaciers in the Edin[b[u]r[gh]]. Encyclopaedia[2*3] published soon afterwards. I copy the description in my Journal thus—

"In many places there is something like a vertical stratification in the Ice masses stretching in the direction of the valley, and the surface of the Ice has an appearance as if intersected by veins."[4]

In my Journal I mention the similarity of the veins to those of rocks.

On the subject of the optical structure of Ice to which you have more than once referred, you will find my observations in the Article Ice in the Edin[b[u]r[gh]]. Encyclopaedia vol. XI p. 637;[5] and in reference to the regelation of the water in Ice cavities you will find a remarkable expt., in the Edin[b[u]r[gh]].Transactions 1826, vol. X. p. 426–27.[6]

I am persuaded that the existing controversy respecting the Glaciers—the cause of your hexagonal flowers and liquid hexagon, and the molecular condition of Ice after pressure, cannot be settled till the optical state of the Ice as shown by polarised light has been thoroughly examined.

I am confident that you will never produce either hexagonal Flowers or hexagonal spaces in Ice perfectly crystallised, that is in Ice in which the optical structure is throughout uniaxial, or rather in which the optical axes parallel in every point of the Mass.

I consider the flowers are as the result of the decomposition of a mass in which the molecules have been prevented from taking their proper position, and I have discovered similar phenomena in the decomposition of solid bodies.

I have just received your note of Saturday,[7] and I am,

My dear M[r] Tyndall, | Ever most truly yours | St. Leonard's College | D. Brewster

S[t]. Andrews | June 7[th] 1859.

RI MS JT/1/B/129
RI MS JT/1/TYP/1/160

---

1. *Altho' Prof. Forbes (Travels p. 29):* refers to Forbes's book *Travels through the Alps of Savoy* from 1843.
2. *the Article Glaciers in the Edinb[u]r[gh]. Transactions*: D. Brewster, 'Glaciers', *Edinburgh Encyclopaedia,* 10 (1830), pp. 284–87. In the review [Anon.], 'Forbes's *Travels through the Alps of Savoy*', *Edinburgh Review,* 80 (1844), pp. 135–63, it was noted that 'The earliest observation of this structure was made on the Glacier de Bois, by Sir David Brewster in 1814, as stated by Professor Forbes; and was published in 1816 or 1817 in the article Glacier in the 'Edinburgh Encyclopedia" (p. 141).
3. **:* Brewster added a marginal note, *Vol. X p. 285.
4. *"In many places . . . an appearance as if intersected by veins.":* see Brewster, 'Glaciers', *Edinburgh Encyclopaedia,* 10 (1830), p. 285.
5. *the Article Ice in the Edinb[u]r[gh]. . Encyclopaedia vol. XI p. 637:* D. Brewster, 'Ice', *Edinburgh Encyclopaedia,* 11 (1830), pp. 636–42.
6. *the Edinb[u]r[gh]. Transactions 1826, vol. X. p. 426–27:* refers to D. Brewster, 'On the Refractive Power of the Two New Fluids in Minerals, with Additional Observations on the Nature and Properties of these Substances', *Transactions of the Royal Society of Edinburgh,* 10 (1826), pp. 407–27.
7. *your note of Saturday:* note missing.

From Rudolf Clausius 9 June 1859 1622

Zürich 9. Juni 59

Lieber Tyndall,

Ich beeile mich Ihren freundlichen Brief zu beantworten, und die gewünschte Sendung beizufügen.

Von Hern Mousson schicke ich Ihnen den Aufsatz über die Gletscherfrage und einen anderen in Pogg. Ann. erschienenen Aufsatz, zugleich mit freundlichen grüssen von Mousson.

Der Aufsatz von Rendu ist in den Memoiren der Savozischen Academie gedruckt. Das einzige Exemplar in Zürich von dem ioch erfahren konnte, gehört Prof. Escher von der Linth. Da dieser gegenwärtig verreist ist, so hat Mousson den Band ohne sein Wissen aus seinem Zimmer genommen, so dass ich ihn Ihnen ebenfalls schicken kann. Abr ich möchte Sie bitten, ihn später, wenn Sie ihn nicht mehr gebrauchen, zurückzuschicken.

Was die Frage betrifft, ob sich die Kälte der höheren Luftschichten aus der Thatsache erklären lasse, dass die verdünte atm. Luft eine viel grössere specifische Wärme habe als die dichtere, so glaube ich dass die Erkläurng in dieser

Form, in welcher sie sich allerdings in vielen Lehrbüchern befinet, unzulässig ist. Die frühere Annahme, dass die specifische Wärme der Luft von der Dichtigkeit abhänge, ist nicht richtig. Ich habe schon in meinem ersten Aufsatze über die mech. Wärmetheorie nachgewiesen dass die specifische Wärme der permanenten Gase sehr wahrscheinlich von der Dichtigkeit unabhängig sei, und dieses ist durch die späteren Versuche von Regnault vollständige bestätigt. Jene Ansicht über die specif. Wärme war aus der Thatsache entstanden, dass die Luft sich bei der Ausdehnung abkühlte aber diese Abkühlung erklärt sich nach der mechanischen Wärmeth. aus der bei der Ausdehnung durch Überwindung des Druckes gethanen Arbeit.

Wenn man nun aber die Frage in der Form stellt, ob die Kälte der oberen Luftschichten sich aus der Thatsache erklären lässt, dass die Luft sich bei der Ausdehnung abkühlt, so glaube ich, dass diese Thatsache, wenn vielleicht auch nicht die einzige, so doch eine der Ursachen jener Kälte ist. Da es nämlich in der Atmosphäre vielfach vorkommt, dass untere Luftmassen aufwärts und obere Luftmassen abwärts strömen, und jedesmal im ersteren Falle Ausdehnung und daher Abkühlung, im letzteren Falle Verdichtung und daher Erwärmung stattfindet, so scheint mir dadurch eine Temperaturdifferenz zwischen den unteren und oberen Regionen nothwendig bedingt zu sein.

Ihre Versuche über die strahlende Wärme haben mich und auch Mousson, ausserordentlich interessirt, und ich habe darin wieder Gelegenheit gehabt, Ihre Geschicklichkeit und Ihren Scharfsinn in der Ausführung und Anordnung der Experimente zu bewundern, wodurch Sie eine Feinheit der Messung erreicht haben, die man von vorn herein für unmöglich gehalten haben würde. Ich bin sehr gespannt darauf die näheren Details dieser Untersuchungen zu erfahren, sowie auch auf Ihre Abhandlung über die Gletscher, welche ich schon lange erwartet habe. Wie ich aus den 'Proceedings' sehe, haben Sie schon Mittehilungen über Ihre neueren Gletscher-Untersuchungen an die Royal Society gemacht, und hoffentlich werden dieselben bald gedruckt werden.

Mit freundlichen Grüssen, | Ihr | Clausius

Zurich, 9. June 59

Dear Tyndall,

I am hurrying to reply to your friendly letter[1] and to enclose the mailings you requested.[2]

I am sending you the paper on the glacier question from Mr. Mousson[3] and another one that was published in Pogg. Ann.,[4] with kind regards from Mousson.

The paper from Rendu[5] is in the Memoirs of the Savoyen Academy. The only copy that I could find in Zurich belongs to Prof. Escher von der Linth. Since he is currently out of town, Mousson took the volume out of his room without his knowledge, so that I could send it on to you. But I would ask you to please send it back later, when you no longer need it.

In regard to the question whether the cold of higher layers of the atmosphere can be explained from the fact that the rarified atmospheric air has a much greater specific heat than the more condensed, I think that the explanation in the form in which it is found in many textbooks, is inadmissible. The earlier assumption, that the specific heat of the air is dependent upon its density, is incorrect. Already in my first paper on the Mechanical Theory of Heat[6] I proved that the specific heat of the permanent gases very likely is independent of density, and this was fully confirmed by the later experiments of Regnault. That view of the specif[ic] heat came from the fact that air expands when it cools, but this cooling can be explained according to the mechanical th[eory] of heat from the work performed during expansion by overcoming the pressure.

When one formulates the question this way, whether the cold of the upper layers of the air can be explained from the fact <u>that the air cools during expansion</u>, I think that this fact, while probably not the only one, would clearly be one of the causes of the cold. Since namely it often happens in the atmosphere that lower air masses flow upward and upper air masses downward, and every time in the first case expansion and thus cooling, in the latter case condensation and thus warming take place, it seems to me that in this way a temperature difference between the lower and upper regions would be conditioned out of necessity.

Mousson and I found your experiments on radiant heat[7] to be extremely interesting, and I once again had the opportunity to marvel at your dexterity and your insightfulness in conducting and arranging experiments, through which you have achieved a delicacy of measurement that one would have considered beforehand to have been impossible. I am very much looking forward to hearing more details of these experiments, as well as to your paper on glaciers, which I have been expecting for some time. As I see in the 'Proceedings', you have already communicated your recent glacier studies to the Royal Society,[8] and hopefully they will soon be published.

With kind regards, | Your | Clausius

RI MS JT/1/TYP/7/2234–35
LT Typescript Only

---

1. *your friendly letter:* letter 1619, dated 1 June 1859.
2. *the mailings you requested:* Tyndall asked Clausius to have Mousson send him a copy of his book *Die Gletscher des Jetztzeit* (see letter 1619, n. 10), which Clausius then included with this letter. Tyndall also asked if Clausius knew where to procure a copy of Rendu's *Théorie de glaciers de la Savoie* (see letter 1619, n. 13).
3. *the paper on the glacier question from Mr. Mousson:* see n. 2.

4. *another one that was published in Pogg. Ann.*: likely refers to 'Einige Thatsachen betreffend das Schmelzen und Gefrieren des Wassers', *Poggend. Anal.*, 181 (1858), pp 161–74.
5. *The paper from Rendu:* see letter 1619, n. 12.
6. *my first paper on the Mechanical Theory of Heat:* likely refers to 'Ueber die bewegende Kraft der Wärme und die Gesetze, welche sich daraus für die Wärmelehre selbst ableiten lassen', *Poggend. Anal.*, 155 (1850), pp. 500–524.
7. *your experiments on radiant heat:* see for example letter 1619, n. 8.
8. *As I see in the 'Proceedings', you have already communicated your recent glacier studies to the Royal Society:* likely refers to Tyndall's 'Observations on the Mer de Glace—Part 1', *Roy. Soc. Proc.*, 9 (1858), pp. 245–47, or Tyndall's 'On Some Physical Properties of Ice', *Roy. Soc. Proc.*, 9 (1857), pp. 76–80.

To Rudolf Clausius 17 June 1859 1623

17th June | 1859

My dear Clausius,

The books[1] have reached me quite safely and I return both you and Professor Mousson my very best thanks for your kindness—As soon as I have made an abstract of Rendu's paper[2] I will return the volume.

Might not the precise lowering of temperature through the cause you mention[3] be calculated from the dynamic theory?[4]

Ever yours | John Tyndall

There are two of my papers on the glaciers in the hands of the Royal Society[5]—this year I hope to finish a little book upon the subject[6]

RI MS JT/1/T/177

1. *The books:* see letter 1622, n. 3–5, and letter 1619, n. 10 and 11.
2. *Rendu's paper:* see letter 1619, n. 13.
3. *the cause you mention:* Clausius suggested that the cooling can be explained according to the mechanical theory of heat, see also letter 1622.
4. *the dynamic theory:* relating to the science of energy, in particular the relationship between heat and mechanical energy. Both Clausius and Tyndall published on the subject, see also letter 1622.
5. *two of my papers on the glaciers in the hands of the Royal Society:* likely refers to Tyndall, 'On the Physical Phenomena of Glaciers—Part I. Observations on the Mer de Glace', *Phil. Trans.*, 149 (1859), pp. 261–78, and Tyndall, 'On the Veined Structure of Glaciers; with Observations upon White Ice-Seams, Air-Bubbles and Dirt-Bands, and Remarks upon Glacier Theories', *Phil. Trans.*, 149 (1859), pp. 279–307.
6. *a little book upon the subject:* refers to Tyndall's book *The Glaciers of the Alps,* see also letter 1595, n. 4.

To James David Forbes[1] 17 June 1859 1624

17th June 1859

My dear Sir

I hope you will excuse my tardiness in replying to your last note.[2] Through a change of home which compelled me to pack up all my books the notes of my lectures in which the experiments you refer to[3] were made have gone out of my reach, and I have been hunting through various periodicals since, but as yet without success; to find the paper containing the original experiment. I hope however to find it for you before the week ends and will at once send it to you.

most truly yours | John Tyndall[4]

University of St Andrews Special Collections, msdep7: Incoming letters 1859, n0.76.

1. Recipient determined through letters 1625 and 1627.
2. *your last note:* note missing.
3. *the experiments you refer to:* likely refers to Tyndall's experiments on ice and glacial motion, see also letter 1594, n. 2 and 3.
4. A note was added at the end of the letter, likely by Forbes, 'Dr Tyndall | June/59 | About Page's | Exp[erimen]t on Vibration'.

To James David Forbes[1] 20 June 1859 1625

Royal Institution of Great Britain

My dear Sir

The paper of Page to which I referred is contained in Silliman's American Journal for 1850. vol. 9. page 105.[2]

most truly yours | John Tyndall

20th June 1859

University of St Andrews Special Collections, msdep7: Incoming letters 1859, n0.78.

1. *To James David Forbes:* Recipient determined from letter 1627, see also letter 1624.
2. *Silliman's American Journal for 1850. vol. 9. page 105:* refers to Chas G. Page, 'Vibrations of Trevelayan's bars by the Galvanic Current', *The American Journal of Science and Arts*, 9 (May 1850), pp. 105–7.

To Unidentified 05 July 1859 1626

Royal Institution of Great Britain | 5th July 1859

Dear <*1–2 words excised*>[1]

I feel truly obliged to you, and I know the best way of proving to you how I appreciate your kindness is to show you that it has not been misapplied—I shall endeavour therefore to make these plates of rocksalt[2] perform a piece of work which I trust when completed will not be unworthy of your attention. —With many thanks believe me

Most truly yours | John Tyndall

The Waller Manuscript Collection, Uppsala University, alb-13:143

1. *Dear <1–2 words excised>:* Recipient name excised or never written. It is possible that this is a draft letter, given the lack of recipient. In the article, 'The Bakerian Lecture. —On the Absorption and Radiation of Heat by Gases and Vapours, and on the Physical Connexion of Radiation, Absorption, and Conduction', *Phil. Trans.*, 151 (1861), pp. 1–36, Tyndall wrote that he had received one plate of rocksalt from the Trustees of the British Museum and one from Mr Harlin, and received a piece of salt which was later turned into two plates from Mr. Lettsom and Mr. Darker. It is likely that this letter was intended for one of them.
2. *plates of rocksalt:* this was likely the plates Tyndall used for his research for his Bakerian Lecture. In this article (see reference in n. 1), he noted that, 'Rock-salt immediately suggests itself as the proper substance; but to obtain plates of suitable size and transparency was exceedingly difficult.' (p. 3).

From James David Forbes 15 July 1859 1627

Pitlochry | N.B. | 15 July 1859

My dear Sir

I thank you for giving me the reference to Mr Page's paper in Silliman's Journal[1] & especially for your interesting Paper on the Absorption of Radiant Heat by the Gases.[2] I shall look with much interest to the fuller account of your results which I hope you will soon publish. I wish you had mentioned the length of tube you used, & the estimated percentage stopped. No doubt it would occur to you to try the effect of condensed gases & to see whether it follows the degree of condensation, or a less rapid law. I confess I am surprised at the magnitude of the effects you have obtained.

I remain dear Sir | Yours faithfully | James D. Forbes

Prof. Tyndall | F. R. S.

RI MS JT/1/F/33
RI MS JT/1/TYP/12/3843

1. *Mr Page's paper in Silliman's Journal:* see letter 1625, n. 2.
2. *your interesting Paper on the Absorption of Radiant Heat by the Gases:* likely refers to Tyndall's article 'Note on the Transmission of Radiant Heat through Gaseous Bodies', *Roy. Soc. Proc.,* 10 (1859), pp. 37–39.

To Rudolf Clausius 19 July 1859 1628

My Dear Clausius,

I was very much rejoiced to receive the intimation[1] of your Verlobung[2] and I sincerely trust that it is the commencement of a life of true & solid happiness <on> the part both of yourself and your intended Bride.[3]

It is now my turn to think of uniting my fate to somebody willing to share it; for if I do not quickly do so I shall be left behind by all my friends.

Will you present my respects & congratulations to Miss Rimpham,[4] and I trust at one future day that you will allow me the pleasure of making her personal acquaintance.

In the /hearts greetings/ | Believe me dear Clausius | ever truly yours | John Tyndall

July 19/1859

RI MS JT/1/T/178

1. *receive the intimation:* letter missing.
2. *Verlobung:* engagement.
3. *your intended Bride:* Clausius married Adelheid Rimpham on 13 November 1859. See letter 1674, n. 12, and 1802, n. 5.
4. *Miss Rimpham:* Adelheid Rimpham.

To Thomas Archer Hirst 22 July 1859 1629

Tyndall | 22nd July 1859

My dear Tom

I have received[1] a diploma for you & another for myself making us both Ehren Mitgliede[2] of the Marburg Philosophical Society.[3] The enclosed is from Schell,[4] but until I came to the 'Attraction' I did not know whether it was intended for you or for me.

affectionately yours | John.

RI MS JT/1/T/650
RI MS JT/1/HTYP/517

1. *I have received:* the letter with the diplomas is missing.
2. *Ehren Mitgliede:* Honorary Member.
3. *Marburg Philosophical Society:* both Tyndall and Hirst had close connections to Marburg, as they obtained their PhD at the University of Marburg.
4. *The enclosed is from Schell:* a missing note from Schell. Hirst recorded receipt of Schell's note through Tyndall's letter in his journal for 24 July, noting that it informed him and Tyndall about their election as honorary members of the Marburg Philosophical Society (*Hirst Journal*, RI MS JT/2/32c/1495). See also letter 1630, n. 4.

From Thomas Archer Hirst 24 July 1859 1630

Broughton[1] | July 24th 1859

My dear John

Wright's[2] bright little boy[3] entered my room this morning and awoke me by presenting your letter. I read Schell's little note[4] with great pleasure and no little surprise. It was a graceful act on the part of our old friends and tutors at Marburg to elect us members of their Society.[5] We may accept the act as expressive of their approval of the use we have made of the knowledge which they imparted to us.[6] Schell mentions having sent me a mathematical memoir[7] if you could send me it together with the diploma[8] I should like to look at both. If you can do so by return of post I shall still be here but on Wednesday I leave for Bristol (19 Small Street)[9] I am enjoying my visit at Broughton greatly after being tossed about for three years it is almost like coming home.[10] It is true that every object around me still awakens memories which though sad enough to dwell upon and still such as I do not wish entirely to escape from.[11] I can now look back upon the bright position of my life which was spent here with a sort of subdued & chastened pleasure & without danger of paralyzing my future action with vain regrets. In short that happy day dream[12] is beginning to lose its undue prominence and to reveal to me its meaning and true value. I would gladly remain here longer for I enjoy perfect liberty and can use any day as I think best, but I must hasten on and get all my necessary visits paid so that I may settle down once more to solid work. I find Mrs Wright[13] looking better and younger than she did three years ago. She is the same gentle graceful wife she always was contented with her work and her home and hallowing both as true women only can. I know not why but sometimes the tears rise in my eyes as I watched her nursing her baby and playing with her boy. The three are so perfectly harmonious so beautiful.

I have thought frequently about you since I came and of our conversation on the last three evenings. I am still utterly at a loss how to account for what has happened to you or what to say about it.[14] I should like much to know Faraday's reply for the more I think of it the more I am convinced you have done well to place all in his hands. You will write to me again before you leave for Switzerland will you not?

The post is just leaving, good bye dear John; run away soon to your favourite mountains and come back to me strong & healthy.

Yours affectionately | T A Hirst

RI MS JT/1/H/224

---

1. *Broughton:* Hirst travelled to Broughton, a village close to Queenwood College, to visit Richard Wright (see n. 2) and his family.
2. *Wright's:* Richard Sears Wright (1829–92), mathematician and teacher at Queenwood College between 1851 and 1863. The name of his wife is not identified. In this period, Richard and his wife had two surviving children, Frances Mary Wright and Henry Wright.
3. *Wright's bright little boy:* refers to Henry Wright.
4. *I read Schell's little note:* note missing. Hirst recorded receipt of Schell's note through Tyndall's letter in his journal, noting that it informed him and Tyndall about their election as honorary members of the Marburg. See also letter 1629, n. 4.
5. *to elect us members of their Society:* see letter 1629.
6. *the knowledge which they imparted to us:* both Hirst and Tyndall studied in Marburg, see also letter 1629, n. 3.
7. *Schell mentions having sent me a mathematical memoir:* letter missing.
8. *the diploma:* see n. 3 and 4.
9. *I leave for Bristol (19 Small Street):* Hirst was visiting his elder brother, the architect John Henry Hirst (1826–82) at his family home in Bristol.
10. *after being tossed about for three years it is almost like coming home:* Hirst was referring to the time he spent teaching at Queenwood College, which was near the village of Broughton.
11. *I do not wish entirely to escape from:* Hirst resigned from Queenwood in 1856 when his late wife Anna became ill. Anna passed away in 1857.
12. *that happy day dream:* in his journal entry for 22 July 1859, Hirst wrote 'As I look on them sometimes seated in their garden with Henry playing around them I remember old times when the same sight awakened in me the desire to be like them and also I remember too how harshly cruelly I have been taught that their lot was not to be mine'. (*Hirst Journal*, RI MS JT/2/32c/1495).
13. *Mrs Wright*: see n. 2.
14. *about it:* presumably this refers to Faraday intervening in the fallout from Tyndall's relationship with Mary Drummond, to manage the rumors spread by her mother Maria Drummond. See also letters 1597, 1604, and 1631.

To Thomas Archer Hirst 29 July 1859 1631

29th July 1859

I am aiming to get off by the 6th. of Aug. Frankland accompanies me![1]

My dear Tom

I also was pleased with the good feeling of our Marburg friends:[2]—Though small there is something specific in the honour[3]—it is like that of a prophet in his own country. I cannot however explain the omission of Frankland which Mrs F. appears to feel rather keenly[4]—probably they intend to send him a diploma direct.

Faraday looked into my /Punch & Judy den/ on Thursday, and we arranged that I should go out to them on the next day. I have just returned from Hampton Court.[5] I joined him at 2. Gathered some apricots from his garden walls had a wholesome plain dinner at 3, and wholesome talk afterwards. After tea we went upon the river. Mrs Faraday[,] Faraday, their niece & myself—Sweet and earnest spread the glassy water in the evening light. We spent then a most delightful hour and a half, and then returned to his beautiful little house upon the green.

That matter is in his safe keeping. he called upon the Lady promptly after receiving my note,[6] she was not at home and he meets her by appointment on Thursday—I doubt not he will advise both her and me for our benefit.[7]

I was much pleased with the spirit of your note Tom,[8] & particularly gratified by your allusion to Mrs. Wright & her children.[9] She is a most excellent little woman: and long may you my boy preserve a fresh heart to find beauty in a mother's love, and delight in its contemplation

ever affectionately | John

RI MS JT/1/T/651

---

1. *I am aiming to get off by the 6th. of Aug. Frankland accompanies me!:* this was included as a postscript.
2. *our Marburg friends:* see letter 1629 and letter 1630.
3. *the honour:* Tyndall and Hirst were made Honorary Members of the Marburg Philosophical Society, see letter 1629.
4. *which Mrs F. appears to feel rather keenly:* according to Tyndall, Sophie Frankland expressed her disappointment in the absence of a diploma for Edward Frankland, who was not elected to the society.
5. *Hampton Court:* Michael and Sarah Faraday's residence.
6. *my note:* note missing, but presumed to be relating to Miss Drummond, see letters 1597, n. 3, 1604, n. 2, and 1630, n. 14.

7. *That matter is in his safe keeping. he called upon the Lady promptly after receiving my note . . . I doubt not he will advise both her and me for our benefit:* see letter 1630 n. 14.
8. *your note Tom:* see letter 1630.
9. *your allusion to Mrs. Wright & her children:* see letter 1630, n. 1 and n. 2.

To Thomas Archer Hirst 07 August 1859 1632

7th. Aug. | 1859.

My dear Tom.

I am off tomorrow[1]—the 8th—I made my will yesterday, but forgot to mention in it my journals. This is to say that they are yours in case I do not return. Never let any thing unkind be heard after me: There is not much of it but whatever there is suppress it

Ever affectionately | John Tyndall (over.)[2]

I have made you one of my executors. The policy has been charged for extra on account of risks. 50 per cent the clerk told me.

RI MS JT/1/T/652
RI MS JT/1/HTYP/518

1. *I am off tomorrow:* Tyndall left for Dijon on 8 August 1859 to travel through the French Alps for the month of August.
2. *over:* indicating to turn the page.

From Auguste Arthur de la Rive 16 August 1859 1633

Genève, le 16 Août, 1859.

Mon cher Monsieur,

Je viens d'écrire à Balmat à Chamounix pour le charger de vous dire la même chose que je vous écris maintenant à Londres, afin que, soit que vous soyez à Chamounix, soit que vous soyez à Londres, vous sachiez que nous désirons vivement votre présence à Genève le 23, le 24 et 25 Août prochain. Nous venons d'organiser une réunion scientifique inofficielle & toute spontanée en l'honneur d'Agassiz qui avait le désir de rencontrer ses amis de Suisse. Vous comprenez combien votre présence nous serait précieuse & je suis convaincu que vous trouveriez vous-même quelque intérêt à cette réunion.

Tâchez donc, très cher Monsieur, de nous y rencontrer. On se réunira le mardi soir 23 Août chez moi & les séances auront lieu le mercredi, 24 & le Jeudi 25 à 10h du matin. On dînera ensuite ensemble &c, &c.

J'espère donc que nous vous verrons à notre réunion.

Votre tout dévoué et bien affectionné | Aug. de la Rive.

Pardonnez-moi si je vous écris si tard, mais ce n'est que ce matin que la convocation a été définitivement décidée.

Geneva, August 16th 1859

My dear Sir,

I just wrote to Balmat[1] in Chamounix[2] to ask him to tell you the same thing that I am writing to you now to London, so that, whether you are in Chamounix, or if you are in London,[3] you will know that we are eager to see you in Geneva on the 23rd, 24th and 25th of next August. We have just organized an unofficial & spontaneous scientific reunion, in honour of Agassiz, who wished to meet with his friends in Switzerland. You understand how precious your presence will be to us, & I am convinced that you will find some interest in the reunion as well.[4]

Try, dear Sir, to meet us there. We will meet on Tuesday evening, August 23rd, at my home & the sessions will take place on Wednesday the 24th & Thursday the 25th at 10 o'clock in the morning. Then we will have dinner together &c, &c.

So I hope we'll see you at our reunion.

You devoted and affectionate | Aug. de la Rive.

Forgive me for writing to you so late, but it is only this morning that the meeting has been finally decided on.

RI MS JT/1/TYP/1/360
LT Typescript Only

1. *I just wrote to Balmat:* letter missing.
2. *Chamounix:* a village under Mont Blanc, a common starting point for climbing the mountain.
3. *whether you are in Chamounix, or if you are in London:* Tyndall was travelling through the French Alps in August 1859, and had been in Chamonix since 11 August 1859. He returned to London around 29 August 1859.
4. *I am convinced that you will find some interest in the reunion as well:* see also letter 1634.

To Auguste Arthur de la Rive [18 August 1859][1] 1634

My dear Sir

Balmat has just shewn me a letter[2] in which you have been kind enough to invite me to meet M. Agassiz at Geneva on the 25th.[3]—I have long desired to make the acquaintance of M.Agassiz, and it would give me great delight to do so in your company.

Tomorrow we start on an expedition up Mont Blanc.[4] I intend to divide the mountain into 6 stations at each of which I intend to place registering thermometers both in the air and in the ice. The Summit, the Corridor, the Grand Plateau, the Grand Mulets and two other stations on the glacier des Bossons[5] will I hope teach us something regarding the atmospheric temperature, and the depth to which the cold of winter penetrates.

We purpose lodging two nights upon the summit. If we can get our work sufficiently advanced it will give both Balmat and myself the greatest pleasure to join you, but we have a heavy piece of work before us—however we shall do all in our power to reach you by the 25th.

If I cannot come so soon it will give me great pleasure to call upon you on my return through Geneva—it was always my intention to do so.

If the weather be favourable I intend to make several other observations of which I shall render you an account in due time.

Ever Yours Most truly | John Tyndall

Chamouni[6] | Thursday night.

RI MS JT/1/TYP/1/361
LT Typescript Only

---

1. *[18 August 1859]:* this is a conjectured date. Tyndall left London on 8 August, a Monday and returned around 29 August. Assuming that Tyndall would not be able to climb Mont Blanc on Thursday 11 August, the date of this letter would be 18 August. That also allows time for Tyndall to have seen the letter concerning the gathering in Geneva.
2. *Balmat has just shewn me a letter:* De la Rive sent a letter to Balmat, as well as to Tyndall (addressed both to Chamonix and London), inviting both to an event in honour of Agassiz. See letter 1633.
3. *on the 25th.*: De la Rive invited Tyndall to an event that extended from 23–25 August 1859, see letter 1633.
4. *Mont Blanc:* the tallest mountain in the French Alps. Tyndall first ascended the mountain in 1857.
5. *glacier des Bossons:* Bossons Glacier, a large glacier on Mont Blanc.
6. *Chamouni:* Chamonix. See letter 1633, n. 2–3.

To Thomas Archer Hirst 29 August 1859 1635

29th Aug. 1859

My dear Tom.

Here is a terrible paper[1] for you. It was sent under cover to me, and I am just arrived in time to receive it. I reached London on Monday morning having completed more swiftly than I anticipated my work in the Alps.[2] I am now hard at work trying to get at portion of my book off my hands.

I went to the top of Mont Blanc,[3] and we planted scopes & thermometers at several stations between top & bottom—We remained a night at the summit—I suffered awfully from headache, for I left Chamouni[4] with a splitting head. Frankland acquitted himself very creditably—the snow was in first rate condition and the climb altogether an easy one.[5]

I am undecided about Aberdeen,[6] but the probability is that I shall remain here and get a lot of my book off hands.[7]

Ever affectionately yours | John

Monday

Stokes has resigned his Professorship at the school of Mines[8]—and Sir Roderick Murchison in the handsomest manner offers it *[to me]*. [9] I do not know the motive of Stokes—perhaps he was too mathematical.[10]

RI MS JT/1/HTYP/518

1. *terrible paper:* not identified.
2. *my work in the Alps:* Tyndall travelled through the French Alps during August 1859, where he for example set up experimental stations on Mont Blanc.
3. *Mont Blanc:* see letter 1634, n. 4.
4. *Chamouni:* as part of his visit to the French Alps, Tyndall spent time in Chamonix, see also letter 1634, n. 4.
5. *Frankland [acquitted] himself . . . the climb altogether an easy one:* This was omitted in Lousia Tyndall's typed version.
6. *Aberdeen:* see letter 1605, n. 11. See also letters 1638, n. 11 and 1639, n. 5.
7. *I am undecided about Aberdeen . . . get a lot of my book off hands:* this was omitted in Louisa Tyndall's typed version. LT's typed letter included instead 'When are you coming to London. I am thinking of taking a run into Hampshire for a few days'. LT omitted and changed several parts in her transcription. See also n. 10.
8. *Stokes has resigned his Professorship at the school of Mines:* Stokes taught at the Royal School of Mines during the 1850s to supplement his income as Lucasian Professor of Mathematics, until the Lucasian Chair was assigned a higher pay (George Stokes and Joseph Larmor (ed), *Memoirs and Scientific Correspondence of the late Sir George Gabriel Stokes,* vol. 1 (*Cambridge University Press:* 1907, 2010), p. 142). See also letters 1638, 1639, 1640, 1641, and 1642.
9. *to me:* the letter from Murchison is missing. Tyndall accepted a post as Professor of Natural Philosophy at the Royal School of Mines in London in 1859, which he held for nine years.
10. *Stokes has resigned . . . perhaps he was too mathematical:* this part was omitted in Louisa Tyndall's typed version. The end paragraphs of LT transcript of letter 1637 belonged to the handwritten version of this letter, but were excluded in the LT transcript. See also letter 1637, n. 5.

To William Frederick Pollock [4 September 1859][1] 1636

Queenwood | near Stockbridge | Hants. | Sunday

My dear Pollock,

Your note reached me[2] as I was stepping into a cab to quit London for a week. Since I saw you I have had very little rest, nor can I now afford myself such, but I thought a little pure air combined with my writing might do me good. I have therefore located myself on the Hampshire downs for a week or so, perhaps longer, if I am not tempted to go to Aberann.[3]

I finished my work in the Alps, much sooner than I expected[4] and tho' work at home called me too urgently to permit of my allowing myself much pleasure. I took a tent up to the summit of Mont Blanc and remained a night upon the top. A strong north wind swept over the summit, but we dropped down a few yards to leeward, and thus found the mountain itself our protector. Our tent was 10 feet wide and 11 people were crammed into it—this kept us warm.

The sunrise I cannot describe—a hundred elements conspired to give it indescribable magnificence. I have heard it said that Mont Blanc was too high for a good view, too high certainly for a view of the pine trees and chalets of the valley of Chamouni[5] but not too high for the grand congregation of mountains which surrounded him. All were clear, and of a majesty not to be described. When the sun rose the transmitted light which is red, fell upon one side of the mountain, and caused it to blush but the other was illuminated by the pure firmamental blue: and the contrast of both was astonishing. As day advanced the wind veered and attacked us covering us with showers of the light dry snow.

We planted a cross twelve feet high upon the summit, fixing 6 feet in snow and leaving 6 feet in air. Between the top and bottom we fixed 6 stations at each of which a cross is planted, and maxima and minima thermometers attached to each.[6] This done, and as I had a pressing invitation from De la Rive to join a scientific reunion at Geneva[7] I turned my steps homeward.

Frankland accompanied me and acquitted himself most creditably.

I spent an extremely pleasant day at Geneva, made the acquaintance of Agassiz, *<1 word excised>*, */*Plantamon*/*,[8] Escher and many others, we all dined together, and afterwards went to the country house of */*M.*/* Favre[9] where we were entertained with the most liberal hospitality—the grounds commanded a view of the lake of Geneva, and we saw the evening red upon the snows of Mont Blanc. As darkness fell all the trees were overhung with lamps of various colours which had a most charming effect. Tho' much pressed to remain for some days I left them with these pleasant memories laid up in store. I have written a little note to Walter.[10] With kindest regards to Mrs Pollock and to Fred

Believe me | most truly yours | J. Tyndall

RI MS JT/1/TYP/6/1915–16
LT Typescript Only

1. *[4 September 1859]:* this letter is dated based upon Tyndall's return to London on or around 29 August 1859. He was at Queenwood by 7 September (letter 1639). The Sunday before was 4 September 1859.
2. *Your note reached me:* letter missing.
3. *Aberann:* likely a mistaken transcription of Aberdeen, the location of the BAAS meeting, or of Abbots Anne, referenced in letter 1775, n. 1.
4. *I finished my work in the Alps, much sooner than I expected:* Tyndall visited the French Alps during August 1859, and returned to London around 29 August 1859.
5. *Chamouni:* see letter 1633, n. 3.
6. *We planted a cross twelve feet high upon the summit . . . and maxima and minima thermometers attached to each:* see letters 1634 and 1635.
7. *I had a pressing invitation from De la Rive to join a scientific reunion at Geneva:* see letter 1633.
8. *Plantamon:* not identified.
9. *Favre:* not identified.
10. *I have written a little note to Walter:* Pollock's son, note missing.

To Thomas Archer Hirst [September 1859][1] 1637

Sep. 1859.

My dear Tom

Right glad I am to hear that England continues to agree with you: with regard to lodging[2] the most practical plan appears to be that we should live close at hand. There are many fit places near my place, and we should find no difficulty in getting you quarters.[3] Perhaps if it were thought advisable we might induce my hostess to give an old lodger[4] notice to quit, but I am dubious about this as the said lodger has been with her for 7 years and has paid her punctually. Still he does not appear to be over amiable, but, I do not like to incur the responsibility of dislodging him. When you come to town we shall arrange all. You can sleep at my place until you get a nest for yourself—perhaps there is some advantage in being separate it gives each of us the run of two houses instead of one.[5]

RI MS JT/1/T/653
LT Typescript Only

1. *[September 1859]:* uncertain dating indicated on the typescript with a handwritten '?5 September'. Based on the discussion of Hirst's lodgings, it is likely that the letter was written in late September, see n. 3. See also letter 1635.
2. *with regard to lodging:* Tyndall was assisting Hirst with finding a place upon his return to England.
3. *we should find no difficulty in getting you quarters:* Hirst noted in his journal on 6 October 1859, that Tyndall found him lodgings at 27 Aberdeen Place (*Hirst Journal*, RI MS JT/2/32c/1504).
4. *my hostess to give an old lodger:* not identified.
5. *instead of one:* the end paragraphs of the LT transcript of this letter belonged to letter 1635, see also letter 1635, n. 10.

From Thomas Henry Huxley 5 September 1859 1638

Aberdour | Fife, N.B. | 5th. September 1859.

My dear Tyndall,

I met Faraday on Loch Lomond[1] yesterday[2] and learned from him that you had returned,[3] whereby you are a great sinner for not having written to me. Faraday told me you were all sound wind and limb, and had carried out your object,[4] which was good to hear.

Have you had any letter from Sir Roderick?[5] if not pray call in Jermyn St.,[6] and see Reeks[7] as soon as possible.

The thing I have been hoping for for years past has come about—Stokes having resigned the Physical Chair in our place, in consequence of his appointment to the Cambridge University Commission.[8] This unfortunately occurred only after our last meeting for the session, and after I had left town, but Reeks wrote to me about it at once. I replied as soon as I received his letter and told him that I would take upon myself the responsibility of saying that you would accept the chair if it were offered you. I thought I was justified in this by various conversations we have had; and, at any rate, I felt sure that it was better that I should get into a mess than that you should lose the chance.

I know that Sir Roderick has written to you[9] but I imagine the letter has gone to Chamonix,[10] so pray put yourself into communication with Reeks at once.

You know very well that the having you with us at Jermyn St., is a project that has long been dear to my heart, partly on your own account, but largely for the interest of the school. I earnestly hope that there is no impediment in the way of your coming to us. How I am minded towards you, you ought to know by this time, but I can assure you that all the rest of us will receive you with open arms. Of that I am quite sure.

Let me have a line to know your determination. I am on tenterhooks till the thing is settled.

Can't you come up this way as you go to Aberdeen?[11]

Every yours faithfully | T. H. Huxley.

P.S. I thought I might mention the Jermyn St matter to Faraday privately, and did so. He seemed pleased that the offer had been made.

IC HP 8.39
RI MS JT/1/TYP/9/2893–94

1. *Loch Lomond*: a lake in Scotland.
2. *yesterday*: 4 September 1859.
3. *you had returned*: Tyndall travelled through the French Alps during August 1859, see letters 1634 and 1635.
4. *carried out your object*: Tyndall undertook experimental research while mountaineering, see letters 1634 and 1635, n. 3.
5. *Have you had any letter from Sir Roderick?*: the letter is missing, but Tyndall mentioned it in a letter to Hirst (see letter 1635, n. 10), and Tyndall copied parts of it into letter 1644. See also letters 1639, 1640, 1641 and 1642.
6. *Jermyn St*: the Museum of Practical Geology and the Royal School of Mines was located on Jermyn Street.
7. *Reeks*: Trenham Reeks (1823–79), geologist and curator of the Museum of Practical Geology and registrar of the Royal School of Mines.
8. *The thing I have been hoping for for years past has come about . . . his appointment to the Cambridge University Commission*: see letter 1635, n. 9–11.
9. *has written to you*: letter missing.
10. *Chamonix*: see letter 1633, n. 2–3.
11. *Aberdeen*: see letter 1605, n. 11. See also letters 1635, n. 8, and 1639, n. 5

To Thomas Henry Huxley 7 September 1859 1639

Queenwood, | near Stockbridge, | Hants | 7th. Sep. 1859.

My dear Huxley,

On my descent from <u>the mountain</u>[1] I had a letter from DelaRive[2] urging me to come to Geneva to see Agassiz and to attend a réunion of the Societé Helvétique.[3] My main work being accomplished, and my desire to return to the work calling for me at home being very strong, I accepted the invitation, and afterwards continued my journey homewards.

I have been minded to write to you since, and simply waited until I could throw a few sentences together for you to communicate to the Physical Section.[4] I do not think I can afford the time to go to Aberdeen;[5] it would cost me 10 days, and only think of my book and my investigation![6] I shall never get that terrible book off my hands if I do not cling to it.

Sir Roderick wrote to me,[7] and in the handsomest manner. I have replied[8] that if my inclinations are to decide the matter I shall accept the post[9] that I wished however to speak to Faraday previous to deciding but that I hoped he would see nothing in my connexion with the school of mines subversive of my allegiance to the Royal Institution.

You are a good fellow and of course it will be a delight to me to be the fellow worker of you and the rest of them. The only thing which prevents me form saying yes to the offer at once is the feeling that the people at the R.I. have been extremely kind and liberal towards me, and I should like to pay them the respect of letting them know how matters stand before the thing is finally decided.

I am glad that you spoke to Faraday[10]—I will now write to him it will expedite matters—From Sir Roderick's letter I inferred that a quick decision was not necessary and I therefore purposed to wait until I could talk to Faraday on the subject: But I will now write to him and let you know the result

Good luck to you and all around you | Ever Yours | J. Tyndall

IC HP 8.40
RI MS JT/1/TYP/9/2895

---

1. *the mountain*: Mont Blanc, see letter 1634, n. 4.
2. *I had a letter from DelaRive*: see letter 1633.
3. *Societé Helvétique*: Société Helvétique des Sciences Naturelles.
4. *the Physical Section*: Section for Mathematical and Physical Sciences at the British Association for the Advancement of Science.
5. *Aberdeen*: see letter 1605, n. 11. See also letters 1635, n. 8, and 1638, n. 11.
6. *my book and my investigation*: Tyndall was in the final stages of writing his book, *Glaciers of the Alps*. For further details on his investigations, see letter 1595, n. 3–4.
7. *Sir Roderick wrote to me*: Tyndall copied parts of these letters which concerned the sudden availability of a position at the Royal School of Mines into letter 1644. See also letters 1635, 1638, 1640 and 1641.
8. *I have replied*: letter missing.
9. *the post*: see letter 1635, n. 9 and 10.
10. *I am glad that you spoke to Faraday*: Huxley wrote in letter 1638 that he had met with Faraday at Loch Lomond, and spoken privately regarding the position at the Royal School of Mines.

From Michael Faraday [10 September 1859][1] 1640

Dundee Saturday

My dear Tyndall,

I heard of the offer to you on Loch Lomond.[2] I hope you will accept it[3] have no time to say more or I shall lose the post.

Ever yours | M. Faraday

RI MS JT/1/TYP/12/4157
Faraday Correspondence 5: 3639

---

1. *[10 September 1859]:* dated based on the reference to Loch Lomond, letters 1638 and 1639, and *Faraday Correspondence* 5: 3639.
2. *I heard of the offer to you on Loch Lomond:* Tyndall had been offered a post at the Royal School of Mines, and Faraday first discovered this through Huxley, see letter 1638. See also letters 1635, 1638, 1639, 1641, and 1642.
3. *I hope you will accept it:* Tyndall expressed to Huxley, that he wanted to ensure that the RI, and in particular Faraday, approved of him taking the position at the Royal School of Mines, before he accepted it. See letter 1639.

To Thomas Archer Hirst 21 September 1859 1641

21st. Sept 1859

My dear Tom.

11d to pay!—Mr. Tardy[1] deserves to be fined 2/6. But perhaps he is sending you a diploma which may be worth the money.[2]

I have just come in from Queenwood, and tomorrow commence my experiments. I wrote a good deal while in the country. My book is getting on but it is horrible work and I hate it.[3]

I think I shall never undertake to write a book again. If one were a scamp the work would be easy enough; but for an honest man it is dreadful.

I have accepted the Chair in the School of Mines.[4] All the good fellows wrote to me about it.[5] It has long been a project with Huxley, and now it is accomplished—nothing could be nicer than the tone of Murchison's letters[6]—they do not sound in the least as if he were conferring a favour upon me.

When are you coming to London. I am now fixed here & the sooner you come the better. Don't you think we might keep a horse & have a gallop out in the country sometimes?

ever affectionately | John.

RI MS JT/1/T/653
RI MS JT/1/HTYP/518

1. *M^r^. Tardy:* not identified.
2. *M^r^. Tardy deserves to be fined 2/6. But perhaps he is sending you a diploma which may be worth the money:* not identified.
3. *My book is getting on but it is horrible work and I hate it:* Tyndall was working on his book, *Glaciers of the Alps* (see letter 1595, n. 4).
4. *I have accepted the Chair in the School of Mines*: following Stokes's resignation, a position opened up at the Royal School of Mines which was offered to Tyndall; see letters 1635, 1638, 1639, 1640, and 1642.
5. *All the good fellows wrote to me about it:* see letters 1638, 1639, n. 7, and 1640.
6. *Murchison's letters:* while Murchison's letters to Tyndall are missing, Tyndall mentioned the correspondence in letter 1587, n. 7.

To Juliet Pollock [24 September 1859][1] 1642

Saturday night

My dear M^rs^. Pollock

I was in the country when I last heard from you but now I am in Babylon[2] once more. Your letter told a sad tale,[3] it was very simple, but very affecting. I had one or two days previous suddenly lost an old friend:[4] he was one of four of which I also formed one, who edited a volume of scientific memoirs together.

I was glad to hear such good accounts of your driving. Depend upon it you are enacting a wise philosophy in practising these exercises; but the best of it I suppose is that you do not do it through philosophy, but simply because you like it.

I spent a few gallant days in the country, galloping over downs, across pastures and through wood. My companion also would be ready to give me a good character he was loud in the praise of the firmness which which I held my seat—in fact I liked the thing so much, and found it so good for me that I think I shall buy a horse and have a gallop every day.

One or two of the days that I thus spent in the country are still sunny to my thoughts. The woods of Norman Court; the seat of Mr Baring[5] are most lovely; fine beeches and fine oaks, and the branches and foliage throw themselves together in a wild free fashion quite peculiar to themselves. With the sunlight dripping like gold from the leaves of the beech, and the trunks half illuminated at the loveliness of these woods was inexpressable.

I am writing hard, very hard, that is very constant, but terribly slow. How some men write so many books I know not—To write any thing passable is to me a work of immense labour, and the most labour is unfortunately expended on the worst writing whose final doom is the wastepaper basket. I have got the proofs of about 160 pages of my book,[6] I am writing it in rather a harum scarum way, purposing to arrange it when it is all fairly off my hands. Did you not say you would look at some of the proofs for me? If you would do so, just here & there in order to give me the benefit of your opinions upon the style I should feel truly indebted to you—I think the introduction, one or two of the ascents, and a little bit of the science would enable you to form a judgment of the whole thing, and if you condemned it I should feel strongly disposed to commit it to the fire.

When I called upon you last I wanted to talk to Mr Pollock about an offer which had been just made to me by Sir Roderick Murchison.[7] He wrote to me in the most handsome terms offering me the chair of Physics in the School of Mines, and assuring me that nothing should be required of me which would loosen my allegiance to the Royal Institution. Many of the Professors there, my personal friends, wrote to me also,[8] in short the thing was so kindly done that I felt strongly disposed to accept the offer. I wrote to Mr Faraday[9] telling him this—he advised me to accept it[10] & I have done so—My lectures there will be mere lessons compared with those of the R.I. I hope to able to get over four of them next week.

M^r. & M^rs. Barlow[11] have returned, both well, I saw them a few days ago.[12]

Poor Miss Moore I regret to hear is ill; but I suppose you know more about her than I do. Your favourite and you no doubt correspond.

I was at Hampton Court yesterday and spent several pleasant hours with our friends. M^rs. Faraday beat me twice over at Russian tactics.[13] In fact I had forgotten the lessons that you gave me. If I remember aright I used to beat you!.

I am very tired, writing is terribly tiresome, so I purpose taking a dash into the country tomorrow and restoring my vigour. I was at Aldershott last Sunday with a friend[14] who parted from me when we were both boys—We lived together in the south of Ireland, and used to break all our furniture once a fortnight. It seems but yesterday since we parted, but he has been a good portion of the time since in India, where [h]is health suffered and his hair partially fell off. So the ardent little fellow with whom I used to struggle 18 years ago has changed to a bald headed man without my seeing the gradation—

Good bye—I dont know where you are, but I will send this to Montagu Square. Kind regards to Mr Pollock & love to the boys.

Most sincerely yours | John Tyndall

RI MS JT/1/T/1217
RI MS JT/1/TYP/6/1923–24

1. *[24 September 1859]:* Tyndall returned from Queenwood on or just before 21 September (see letter 1641). The Saturday just following this was Saturday 24 September 1859.
2. *Babylon:* likely meaning that he was back in London.
3. *Your letter told a sad tale:* letter missing.
4. *lost an old friend:* refers to Arthur Henfrey, (1819–59) who died 7 September. Henfrey edited with Huxley *Scientific memoirs, selected from the transactions of foreign academies of science, and from foreign journals: Natural History* (London: Taylor and Francis, 1853), and Tyndall coedited the other volume on *Natural Philosophy* with William Francis (also 1853).
5. *woods of Norman Court; the seat of Mr Baring:* Norman Court in Salisbury was one of many properties owned by the Baring family.
6. *my book:* likely refers to *Glaciers of the Alps* (see letter 1595, n. 4).
7. *offer which had been just made to me by Sir Roderick Murchison:* referring to the Chair of Physics in the School of Mines, see letters 1635, 1638, 1639, 1640, 1641, and 1644, n. 3–4
8. *Many of the Professors there, my personal friends, wrote to me also:* only some of this correspondence is known, see letters 1638, 1639, n. 7, 1640, 1641, n. 6, and 1644, n. 3–4.
9. *I wrote to Mr Faraday:* letter missing.
10. *he advised me to accept it:* see letter 1640.
11. *M^r. & M^rs. Barlow:* John Barlow (1798–1869) and Cecilia Anne Barlow (née Law, d. 1868).
12. *I saw them a few days ago:* dates not identified.
13. *Russian tactics:* presumably a board game, not otherwise identified.
14. *with a friend:* likely Thomas Charles Higginson (*c.* 1821–98), a surveying colleague of Tyndall in Ireland and England. See also letter 1769.

From Auguste Balmat 13 October 1859 1643

Chamonix le 13 8bre, 59

Monsieur,

D'après le rapport que vous me fîtes par votre lettre, sur les poteaux que nous avions placés sur le glacier des Bossons, je m'empressais accompagné de Tairraz, d'aller les visiter. Nous avons trouvé celui que nous avions placé sur le plateau supérieur un peu poussé par la glace contre la *[surface]* et légèrement tourné, nous l'extrayâmes et le replaçâmes comme il devrait l'être; les thermomètres exposés à l'air étaient de 12 et 13 degrés de chaleur au dessus <de> zéro. Celui dans la glace nous l'y avons laissé et sans le visiter. Il est replacé solidement. Nous descendîmes ensuite au plateau inferieur. Les thermomètres ont été volés et le poteau a été jeté dans un ruisseau, que nous trouvâmes à quelques pas de l'endroit où nous l'avions placé.

Comme nous avons toujours un temps d'été, nous ne les replacerons qu'au moment qu'il ne passera plus personne dans cet endroit. Le Monsieur Anglais qui vous a fait le rapport est probablement dans l'erreur car à la dernière ascension que je fis au Mont Blanc avec Mr. M*[essu]* je trouvai celui du Mont Blanc à 21 degrés et demi le 9 7bre au dessous de zéro, et les autres poteaux n'étaient pas dérangés.

Conséquemment je ne puis comprendre comment ce Monsieur les a compris.

Nous avons encore aujourd'hui à Chamonix 29 degrés de chaleur au moment du Soleil.

Nous croyons trouver le voleur sous peu. Nous en fîmes part secrètement à l'autorité supérieure qui en a été fortement vexée. Elle travaille ainsi que nous en secret à le *découvrir.*

Quand nous en connaitrons quelques nouvelles nous vous en ferons part.

Balmat Auguste | Tairraz

Chamonix, October[1] 13th '59

Sir,

From the account you made in your letter,[2] on the posts[3] we had placed on the Bossons glacier,[4] I hastened to take a look at them in the company of Tairraz.[5] We found the one placed on the upper plateau somewhat pushed by the ice against the *[surface]*, and slightly turned, we dug it up and put it back as it should be; the thermometers exposed to the open air displayed a temperature of 12 and 13 degrees above zero. We left the one in the ice as it was without a taking a look. It is solidly replaced. Afterwards, we climbed down to the lower plateau. The thermometers have been stolen, and the post thrown in a brook that we discovered a few steps away from the place where we placed it.

Since the weather is still summery, we will not replace them until after no one can walk by this location. The English gentleman[6] who gave you this account is probably mistaken because in the last ascent of the Mont Blanc I made with Mr. M*[essu]*,[7] I found the[8] Mont Blanc[9] at 21 degrees and a half on September the 9th and the other posts had not been disturbed.

As a result, I cannot understand the manner in which this English gentlemen understood them.

The temperature today in Chamonix[10] is up to 29°C when the sun is high.

We believe we will shortly find the thief. We shared this with the higher authority[11] which was greatly upset by it. Just like us, it works in secrecy to find him.

When we will have more details about this, we will share them with you.

Balmat Auguste | Tairraz[12]

RI MS JT/1/B/5

1. The French 8bre is an abbreviation of October.
2. *in your letter:* letter missing.
3. *the posts:* as part of his expedition to the French Alps, Tyndall set up experimental stations. See letters 1634, 1635, n. 3, and 1645.
4. *Bossons glacier:* see letter 1634, n. 5.
5. *Tairraz:* Joseph Tairraz, a mountain guide.
6. *The English gentleman:* not identified.
7. *with Mr. M[essu]*: likely François Messat, also a mountain guide working near Chamonix.
8. *I found the:* that is, the temperature.
9. *Mont Blanc:* see letter 1634, n. 4.
10. *Chamonix:* see letter 1633, n. 2–3.
11. *We shared this with the higher authority:* a local representative of the Sardinian state.
12. *Tairraz:* may refer to Hotel de la Couronne, kept by Joseph Tairraz (*The European Indicator, Or Road-book for Travellers on the Continent* (London, Paris, Brussels: Félix de Monnier, 1841), p. 71).

To John Barlow 4 November 1859 1644

4th November 1859.[1]

My dear Mr. Barlow.

Might I request you to lay the following statement before our managers at their next meeting?

It is now more than five years since a chair of physics in the Government School of Mines was offered to me by the late Sir Henry De la Beche.[2] I then feared that the acceptance of the position would interfere with my duties at the Royal Institution, and accordingly declined it.

Toward the end of last August I received a communication[3] from Sir Roderick Murchison which contained a similar offer.[4] it was accompanied by the following words: 'You may perhaps feel some hesitation in accepting any offer which might weaken your allegiance to the Royal Institution; but I feel assured that you could carry out all our behests without in any way distracting from your other essential duties.' Letters from many of my personal friends[5] connected with the School of Mines, all of them highly gratifying to me, reached me at the same time, or soon [afterwards].

Had this offer been made to me at a period of the year when our managers held their meetings, my first act would have been to lay the subject before them and ascertain their wishes regarding it. Under the circumstances I wrote to Mr. Faraday[6] who was then in Scotland:[7]—I also endeavoured to see two of my friends[8] who are managers of the Institution, but they were out of town. M^r^. Faraday advised me to accept the appointment,[9] and my two friends[10] have since expressed opinions quite in harmony with that advice.

After the receipt of Mr. Faraday's note[11] I wrote to Sir Roderick Murchison[12] accepting the Chair offered to me. This subsequent note to me contains the following words, which show the sentiments under which the offer was made and accepted:—'Your acceptance of the office of Professor in our establishment gave me and all concerned the utmost satisfaction. Rely upon it that I have too great a regard for the Royal Institution and every one of its officers to hamper you in any way as to duties which should or could interfere with the working of that admirable body.'

With regard to my duties at The School of Mines I may remark that they are quite different from those at the Royal Institution. In the former case my sole aim is to instruct a class of young men who are unacquainted with the rudiments of physics. This I do in the simplest manner possible, making a selection from the knowledge I possess, and putting that with life and vigour before them, but aiming at nothing new, either in the matter delivered, or in the form of the experiments. Unlike our audience of the R.I. the class renews itself almost yearly, so that the same course of lectures, is to be given every year.

Considering the honourable and friendly manner in which I have been treated by our managers, I should not willingly neglect any act which my respect or duty towards them might suggest. I think it is right that they should be made acquainted with the above facts, and if they come further information I shall be happy to supply it to the best of my ability.

Believe me | ever faithfully yours | John Tyndall

Rev[eren]^d^. John Barlow | &c &c &c | Hon. Sec. R.I.

Rev[eren]^d^. John Barlow | Hon. Sec. R.I. | Royal Institution | Albemarle Street.[13]

RI MS AD/03/D/04

---

1. The letter included a note 'Tyndall', which appears to have been added later.
2. *Henry De la Beche:* the geologist Henry De la Beche (1796–1855) was a central figure in establishing the Royal School of Mines, and was the first director of the Museum of Practical Geology.
3. *I received a communication*: letter missing.
4. *a similar offer:* Tyndall was offered a post at the Royal School of Mines, see letters 1635, 1639, 1640, 1641, and 1642.
5. *Letters from many of my personal friends:* only some of this correspondence is known, see letters 1638, 1639, n. 7, 1640, and 1641, n. 6.

6. *I wrote to Mr. Faraday:* letter missing.
7. *who was then in Scotland:* Faraday first learned about the offer through Huxley while in Scotland, see letters 1638, and 1640, n. 2.
8. *two of my friends:* not identified.
9. *M[r]. Faraday advised me to accept the appointment:* see letter 1640.
10. *my two friends:* not identified.
11. *Mr. Faraday's note:* see letter 1640.
12. *I wrote to Sir Roderick Murchison:* letter missing.
13. *Rev[eren][d]. John Barlow | Hon. Sec. R.I. | Royal Institution | Albemarle Street:* on envelope.

From Lyon Playfair 9 November 1859 1645

9th Nov/59. | Confidential[1]

My dear Tyndall

Prof. James Forbes is appointed to St Andrews as Principal[2] to succeed Sir D. Brewster. The Chair of Physics here will therefore soon be declared vacant.[3] It is only worth about 450£ I believe at present, but it is certain to be endowed by the University Commission in March next & will then be worth probably 200£ more if they act rightly 100£ more if they act shabbily.

Is there any chance of your becoming a Candidate? If so I think that you could be brought forward under the most favourable chances for success. The Patrons are the University Court

Yours truly | Lyon Playfair

RI MS JT/1/P/117
RI MS JT/TYP/3/985

1. *Confidential:* Tyndall recorded receipt of this letter in his journal for Thursday 10 November 1859, 'Playfair has written me a confidential note about Edinburgh—he asks me whether I could be prevailed upon to become a candidate—as my chances of success are most favourable'. (Journal, RI MS JT/2/13c/1161).
2. *Forbes is appointed to St Andrews as Principal:* Forbes took up the position as Principal of the United College, St. Andrews in 1859, which he held until his death in 1868.
3. *The Chair of Physics here will therefore soon be declared vacant:* Forbes held the post at the University of Edinburgh between 1833 and 1860. Tyndall deliberated whether to become a candidate after Forbes (see letters 1646, 1648, 1651, 1652, 1654, 1655, and 1657) but ultimately decided against it (see letters 1656, 1658, 1659, and 1660).

To Lyon Playfair [10–11 November 1859][1] 1646

Royal Institution

My dear Playfair, I am very much obliged to you for your kind and flattering letter. The subject[2] was mentioned to me some days ago by a mutual friend of ours:[3] I will tell you my feelings on the matter. In this first place it would be strongly against my inclination to become a candidate for the post[4]: the influences must be very cogent which would overcome this repugnance on my part. —I dislike election contests of all kinds. In the next place the income, even in its amended form, would be less than what I am *[now]* making in London. They give me £300 a year for 12 lectures at the Royal Institution:[5] my income from the School of Mines will, even at the commencement, be £250 a year; my military examinerships are another 100, and an additional hundred may be set down from other sources. Set this beside the sum you mention, and above all take into account my connexions in London, the cordial relationship in which I stand with Faraday, and I think when you take a balance you could hardly advise me to become a candidate for the Edinburgh Chair. My strongest inducement would be the hope of being able to found a school of experimental physics there, but, this I hope to do eventually in London

With very many thanks | believe me dear Playfair | most truly yours | John Tyndall

RI MS JT/1/T/1126
RI MS JT/1/TYP3/986

---

1. *[10–11 November 1859]:* there is a presumed draft version of this letter which included further details on Tyndall's considerations for not taking up the post (RI MS JT/1/T/1127, RI MS JT/1/TYP/3/987), 'My dear Playfair, The subject of the note which you have so kindly written to me was communicated to me a few days ago by a mutual friend of ours. —There is a great deal in the position which is attractive to me: it is a whole and single thing & would do away with the distraction of incidental work. There is also the possibility of founding a school of true experimental physics in the country which it as yet lacks. There is on the other hand my connexions here in London many of which are near and dear to me. Above them all stands the cordial relationship which subsists between Faraday and me & which has grown only firmer as years have advanced. Sir David Brewster did me the honour to offer me the Chair of St. Andrews some months ago, but the above considerations induced me to decline it. Since that time Sir Roderick Murchison has written to me the handsomest terms, offering me a Chair of Physics in the School of Mines—this I have accepted. With regard then to income the matter stands thus: They allow me at the

Royal Institution £300 for twelve lectures annually:—thus cutting off 7 of my lectures and augmenting my salary by £100—The latter act of kindness was performed without my knowing that it was at all in contemplation. My income from the school of mines will, even at the commencement be between two and three hundred a year. My examinerships are worth at least £100 more. Add 100 to this, which I cannot help making from other sources, even if *[I tried]* to do so, and you have a total of between seven & eight hundred a year. In the most favourable case the chair at Edinburgh appears to be worth no more than £650. The whole case is now before you, and I think you will agree with me that the balance of influence is on the side of my remaining where I am.'

2. *the subject:* see letter 1645.
3. *has written to me:* likely refer to a missing letter from Murchison, which Tyndall discussed in letter 1635 to Hirst.
4. *the post*: the Chair of Physics at the University of Edinburgh (see letter 1645, n. 3).
5. *[the Royal Institution]:* Tyndall's financial situation had recently improved, see letter 1596, n. 3.

From Richard Owen 12 November 1859 1647

Brit. Mus. | Nov. 12. /59.

My dear Professor,

I have no objection to the publication of my letter on the subject of the relaxed restrictions at Chamouni;[1] and indeed, feel happy that any communication of mine should be deemed worthy of insertion in your forthcoming work[2]

Sincerely yours. | R[ichar]^{d}. Owen

BL, Add MS 63092, ff. 16–227v

1. *the publication of my letter on the subject of the relaxed restrictions at Chamouni:* refers to a letter dated 18 February 1859 (Volume 6, letter 1590), which was later reprinted in *Glaciers of the Alps,* pp. 193–94. For Chamouni, or Chamonix, see letter 1633, n. 2–3.
2. *your forthcoming work:* refers to Tyndall's book *Glaciers of the Alps* which was published in 1860 (see letter 1595, n. 4).

From Lyon Playfair 12 November 1859 1648

Confidential | 12 Nov/59[1]

My dear Tyndall.

Your reasons are very strong;[2] but pause a little before you decide finally. To day Simpson[3] & Wilson pledged themselves to support you & we three could certainly influence the majority of the Court in your favour, as this majority consists of Town Councillors who have always acted with us & ask our opinions. Besides I am sure your name would secure you nearly an unanimous Election. I therefore wish you to think of it as an office you could get.[4]

The emoluments which I give you are those made by Forbes (deducting payment of the assistants which lecture for him). But his health has made the Chair run down, and it is not what it was or could be made. As an instance of this, when I came here, Gregory had run down this Chair to £450 yearly. My fees are already £1250 but as I have an ambition to found a Laboratory & to introduce the Tutorial System into the Scotch Colleges as a supplement to the Professorial System,[5] I make very little profit out of the fees. But the rise shows you what may be done in a Chair fairly worked.

You would have only six months Lectures, one hour daily & six months entire Holiday, for investigations & travel.

Therefore still think of it. I may mention to you privately that no official resignation can be given in till the 2nd Dec[emb]r & that probably Forbes will keep the Chair all winter.[6] But Candidates become so thick that you must decide promptly, say within 8 days, whether your friends should put you forward or not.

Brewster & Forbes have made me their umpire as to the period of their mutual resignations, so that there will be no sudden Election, but at the same time Candidates are already actively canvassing.

Yours truly | Lyon Playfair

RI MS JT/1/P/118
RI MS JT/TYP/3/988

---

1. Tyndall recorded the receipt of this letter in his journal for Monday 14 November, 'Another 'confidential' letter from Playfair. Simpson and Wilson have pledged for me—my name he says would secure for me a nearly unanimous elections, and therefore he wishes me to think of it as an office which I could get'. (Journal, RI MS JT/2/13c/ 1162)
2. *Your reasons are very strong:* letter 1646. See also 1645, n. 3.
3. *Simpson:* likely the physician James Simpson, based on a reference in letter 1651.
4. *we three could certainly influence . . . I therefore wish you to think of it as an office you could get:* Playfair was referring to the hiring process for the Chair of Physics at the University of Edinburgh, detailed in letter 1657.
5. *introduce the Tutorial System into the Scotch Colleges as a supplement to the Professorial. System:* mainly practiced at Cambridge and Oxford, tutorials were regular meetings of small groups of students with a tutor, who trained their students with exercises related to the subjects discussed in the lectures. Playfair thought that lectures had too many students for teaching to be effective, and therefore adopted the tutorial system. Other Scottish colleges

would follow Playfair in adopting tutorial systems. (W. P. Doyle, 'Lyon Playfair (1818–1898)'. University of Edinburgh, School of Chemistry. Accessed May 23, 2018. http://www.chem.ed.ac.uk/about-us/history/professors/lyon-playfair).

6. *keep the Chair all winter:* see letter 1645, n. 2.

To Sarah Tyndall 13 November 1859 1649

13th, November, 1859.

My dear mother

I am about to propose an alteration in the annual remittance which I have hitherto made to you. My position as regards money has become better[1] than it was when I last wrote to you,[2] and I think it might increase your comfort and allow you more freedom of action if I augmented your allowance by £20 a year.

If then you would let me know when the next quarterly remittance will be needed I will commence by sending you the increased sum

If anything should occur to render an addition to the sum I mention necessary, you have only to write to me and I will send you whatever you require

Your affectionate son | John

RI MS JT/5/16a

---

1. *My position as regards money has become better:* Tyndall's financial situation had improved in two main ways. His income from the RI had increased by 100 pounds per year; see letter 1596. In addition, Tyndall had accepted the post as Professor of Natural Philosophy at the Royal School of Mines in London in 1859; see letter 1635.
2. *when I last wrote to you:* letter missing and date uncertain, but Tyndall thanked the RI for increasing his pay in a letter dated 19 March 1859; letter 1602.

From Richard Dawes 14 November 1859 1650

Deanery, Hereford, | Nov. 14th 1859.

My dear Tyndall,

Don't think because I have not written to thank you for a copy of your paper on Glaciers[1] before this, that Deans[2] are an overworked class of men and have not time to write to their friends, but I happened to receive it during our audit which is rather a busy week, and I wished to read it before writing to you. It has given me a very great deal of pleasure the doing so and although I don't pretend to be versed in such matters, yet the way in which you set about to ascertain the movement of these masses, the obstructions of different kinds which you have to contend with, and the attempt to ascertain the points of maximum movement have really interested me—not merely on account of the glaciers in Switzerland but the course of the river Wye[3] and a walk on its banks after a flood will interest me more in future than they have done.

I shall watch the bending of the river and ascertain the velocity of the stream in different parts of the same section, and I shall look at the dirt beds left after a flood with greater interest than I have hitherto done; I wish the dirt beds of the Wye were the only ones we had here, they do good, but we have many dirt beds of another kind which I would like to examine (with the eye of a Tyndall) but I dare not.

Last week Mr and Mrs Philipps[4] were staying with us. Mary Best that[5] was, and she wished to be most kindly remembered to you.

Mrs Dawes begs me to give you her kindest regards, also to Frankland and Mrs Frankland when you see them,

and Believe me, my dear Tyndall, | Always sincerely yours | R. Dawes.

RI MS JT/1/TYP/1/319A
LT Typescript Only

1. *Don't think because I have not written to thank you for a copy of your paper on Glaciers:* it is uncertain which paper Tyndall sent Dawes. It likely either refers to a draft of *Glaciers of the Alps,* or to one of the following three papers: J. Tyndall, 'On the Physical Phenomena of Glaciers—Part I. Observations on the Mer de Glace', *Phil. Trans.,* 149 (1859), pp. 261–78; J. Tyndall, 'Remarks on Ice and Glaciers', *Phil. Mag.,* 17 (1859), pp. 91–96; J. Tyndall, 'On the Veined Structure of Glaciers; with Observations upon White Ice-Seams, Air-Bubbles and Dirt-Bands, and Remarks upon Glacier Theories', *Phil. Trans.,* 149 (1859), pp. 279–307.
2. *Deans:* Richard Dawes was the Dean of Hereford.
3. *river Wye:* fifth longest river in the UK, running along the Anglo-Welsh border. Tyndall drew comparison between the flow of glaciers and rivers in his research.
4. *Mr and Mrs Philipps:* not identified.
5. *Mary Best:* possibly a daughter of Samuel Best, see letter 1775.

From [William Fullarton] Cumming[1] 15 November 1859 1651

9 Athol Street, Edinburgh | Nov. 15th 1859.[2]

My dear Dr Tyndall

After a journey of 12 hours, I arrived here ½ frozen last night from Lemington.

Doctor Simpson[3] (of chloroform renown)[4] joined me at the Carstairs station,[5] and as there were none else in the carriage, we had a talk 'de omnibus rebus'[6] all the way to Edinburgh.

'Whom are you to have in room of James Forbes' said I 'Oh Tyndall, I hope' said he. We then had a long discussion which ended by our deciding that in the Chair of Nat. Philosophy in our[7] 'Alma Mater', you would be 'the right man in the right place'. S.[8] declared with confidence his belief that if you would only accept, he and Geo. Wilson[9] and Lyon Playfair would be able to secure your election without any trouble on your part. I asked him if I might tell you so, to which he unhesitatingly assented, adding he did not think you would be required to produce testimonials even.

Arrangements are made for the present session, so you would have nearly a whole year of preparation, and I would really urge you, at all events, to let it be understood that you would accept the Chair of Office[.] No doubt the candidates will be 'Legion',[10] but I have a strong notion you would distance them all.

Simpson has a great deal to say in these matters, and I am sure, from what passed between us yesterday, he did put his shoulder in earnest to the wheel that should roll you into the Nat. Phil. Chair. I found my sister[11] lamenting her widowhood—the Commissioner[12] having been absent for nearly a fortnight and not expected home until tomorrow.

Dinner comes on the table, and I conclude in haste hoping to have an early line to say you are ready to enter the field.

Ever most sincerely Yours | [W. F.] Cumming

Professor Tyndall | &c. &c.

I have just written to Arygle[13] the substance of my conversation with Prof. Simpson.

RI MS JT/1/TYP/1/278
LT Typescript Only

---

1. Louisa Tyndall added the note 'W. G. Cumming | Nov. 4. 60', but the sender is presumed to be W. F. Cumming.
2. Tyndall recorded the receipt of this letter in his journal for Wednesday 16 November 1859, 'A letter from Dr Cumming about the Edinburgh Chair—he seems confident of my election' (Journal, RI MS JT/2/13c/1162).
3. *Simpson:* likely refers to James Simpson (see letter 1648, n. 3).
4. *(of chloroform renown):* Simpson advocated obstetric anaesthesia, in particular by use of chloroform (*ODNB*).
5. *Carstairs station:* station in Sought Lanarkshire, Scotland.
6. *de omnibus rebus:* a Latin saying, often used in the longer version 'de omnibus rebus et

quibusdam aliis' which can loosely be translated to mean 'on everything and anything', or simply 'to ramble on'.

7. *Chair of Nat. Philosophy:* Tyndall considered taking up the newly available Chair of Physics at the University of Edinburgh. See letter 1645, n. 3.
8. *S.:* likely refers to Simpson, see n. 3.
9. *Geo. Wilson:* George Wilson.
10. *No doubt the candidates will be 'Legion':* means that Cumming expected there to be multiple candidates for the posts.
11. *my sister:* Mary Coxe (née Cumming, 1811–c. 1875). Their mother was the subject of Robert Burns's 'Bonnie Lesley' as noted in the novelist Charles Gibbon's dedication of *In Honour Bound,* vol. 1 (London: Richard Bentley and Son, 1874) to Lady Coxe. In 1841 she married the Scottish physician James Coxe. She was a correspondent of Tyndall's and he claimed to have toasted her on the summit of Mont Blanc in 1857. See also letter 1739.
12. *the Commissioner:* James Coxe (1811–78), Scottish physician who studied at Göttingen, Heidelberg, Paris, and received his MD in 1835 from the University of Edinburgh, where he then practiced medicine. In 1841 he married Mary Anne Cumming. In 1855 he was appointed to the Royal Commission that examined the condition of the lunatic asylums in Scotland and, following the 1857 Act that established the General Board of Lunacy for Scotland, he served as one of its two paid commissioners. He was knighted in 1863 for his services. ('Sir James Coxe'. *Illustrated London News,* 18 May 1878, p. 471).
13. *Argyle:* George Campbell, 8th Duke of Argyll (1823–1900), politician, and writer on science and religion. A regular correspondent with Tyndall.

To Michael Faraday 16 November 1859 1652

Royal Institution | 16 Nov 1859

My dear Mr. Faraday,

Would you think for a moment at your leisure over the subject of this letter.[1] I have already had one from Playfair,[2] to which I have replied[3] in the terms indicated in the first lines of his present letter.[4] I have also been called upon by people of influence from Edinburgh, unconnected with the University.[5]

I want as much light as possible on the question before deciding[6] and therefore I am anxious to secure the thoughts of yourself and one or two more of my friends—my own feelings are distinct enough. Still a case of the kind is too important to be decided without reflection

Ever yours | J. Tyndall

RI MS JT/1/TYP/12/4092
Faraday Correspondence 5: 3674

1. *the subject of this letter:* it is uncertain if 'this letter' refers to this specific letter, or a missing enclosed letter. Based on Tyndall's wording regarding Playfair's letters (see n. 2), Tyndall is likely referring to letter 1648 where Playfair outlined in more detail the circumstances of the Chair of Physics at the University of Edinburgh which would soon be vacant. See also letter 1645.
2. *had one from Playfair:* likely letter 1645; see also letter 1648.
3. *to which I have replied*: likely letter 1646.
4. *his present letter*: likely letter 1648, see n. 1.
5. *I have also been called upon by people of influence from Edinburgh, unconnected with the University*: see for example letter 1651.
6. *before deciding*: Tyndall was deliberating whether he would become a candidate for the post in Edinburgh, see letter 1645, n. 3.

From Auguste Balmat 16 November 1859 1653

Chamonix le 16 Novembre 1859

Monsieur le Professeur,

Je m'empresse de répondre à votre lettre du 10 par laquelle vous me demandez des renseignements sur vos projets. Pour ce qui concerne la possibilité de placer des jalons sur la mer de glace et rester au Montanvert, je le crois tout à fait probable moyennant des précautions sur la mer de glace. Quant au temps, c'est plus difficile à résoudre, pour le moment il est beau, mais d'ici à la fin de décembre il peut changer plusieurs fois. Conséquemment, s'il est beau quand vous serez ici, je crois pouvoir vous dire que vous réussirez, mais je crois que, si vous voulez, il serait mieux que je vous écrivisse vers le 8 ou 12 décembre.

Concernant le temps il pouvrai quelques fois tombé une grande quantité de neige qui empecherais même d'arriver jusqu'au Montanvert.

Il y a aussi une communication depuis Genève à Sallanches, tous les jours, et de Sallanches à Chamonix. Le courrier descend journellement cette année avec une petite voiture ou un traineau suivant le temps, de sorte que c'est très facile.

Si vous désiriez par example que j'allât vous chercher à Sallanches, vous sauriez qu'a m'en informer sur votre réponse (que j'attends avant de vous *[*réagir*]*) en m'indiquent le jour.

En attendant je suis | votre serviteur | Auguste Balmat

Chamonix the 16th of November 1859

Sir the Professor,

I hasten to answer your letter of the 10th[1] in which you asked me for information relating to your projects. Concerning the possibility of putting stakes on the Mer de Glace,[2] and staying at Montanvert,[3] I believe it is quite possible with precautions on the Mer de Glace. As for the weather, it is difficult to determine, for the moment it is pleasant, but from now to the end of December it can change several times. Consequently, if it is pleasant when you are here, I believe I can tell you that you will succeed, but I think, if you want, that it would be better if I wrote to you around the 8th or the 12th of December.

As far as the weather is concerned, a great quantity of snow could fall several times, which could even prevent reaching Montanvert.

There is also a communication between Geneva and Sallanches, every day, and from Sallanches to Chamonix. The mail courier travels daily this year with a small carriage or a sleigh depending on the weather, of the sort which is very simple.

If you desired for example that I came to meet you at Sallanches,[4] you would only have to let me know in your response (which I await before *[responding]*) letting me know the day.

I remain | your servant | Auguste Balmat

RI MS JT/1/C/35

---

1. *your letter of the 10th:* letter missing.
2. *putting landmarks on the Mer de Glace:* as part of his research on glaciers, Tyndall set up experimental stations. See letters 1634, 1635, n. 3, 1643, n. 3, and 1645.
3. *that remains at Montanvert:* there was an issue that the posts set up for making measurements in the glaciers were moving. Balmat outlined these problems in letter 1643.
4. *I came to meet you at Sallanches:* it appears Tyndall did not meet Balmat at Sallanches, but rather changed vehicle there (Journal, RI MS JT/2/13c/1168–70).

From Lyon Playfair 16 November 1859 1654

16 Nov 59

My dear Tyndall

My black ink bottle is broken so Excuse red ink.

I write in reference to what I said in my last[1] in regard to my own fees, which last year increased from £450 to £1250 & this year will likely be more. But you are to recollect that my Class[2] is imperative with Medical Students—or rather the Subject of Chemistry—for the Class may be taken anywhere. But Physics[3] is only imperative for Students in Arts & not for those in Medicine. The total number of students in the two faculties is about the same; but hitherto there have been much fewer degrees taken out in the Arts faculty, so that the same number of Students are not compelled to take the Subject.

But a great impulse will be given to degrees in Arts by our new University

Constitution.[4] We have also made a rule that Students are to be Examined in Physics before they are admitted into the Medical Faculty i.e. Physics is one of the Subjects included in the preliminary Examinations w[hi][c[h]] we are to require in future. This will force Medical students to study the Science.

With all this however, supposing two Professors of Equal efficiency, there is more chance of fees for the Prof. of Chemistry than for the Prof. of Physics, because the former has Summer practical & Winter Analytical classes which the latter has not.

Only therefore take my increase of fees as showing the Elasticity of the University & not as a measure of your amount.

Yours truly | Lyon Playfair.

RI MS JT/1/P/119
RI MS JT/TYP/3/989

1. *what I said in my last:* letter 1648.
2. *my Class:* Playfair held the Edinburgh Chair of Chemistry.
3. *But Physics:* the person appointed to the Chair of Physics at the University of Edinburgh would be teaching physics. Tyndall was deliberating whether he would become a candidate for the post, but ultimately decided against this; see letter 1645, n. 3.
4. *But a great impulse will be given to degrees in Arts by our new University Constitution:* this appears to be a reference to the 'Universities (Scotland) Act 1858', in which Scottish universities were expected to set up 'University Courts', which would become the governing body of the institutions. This meant that the University of Edinburgh had the power of appointing its staff, and deciding on its curriculum. These responsibilities had previously been in the hands of the Town Council.

From Michael Faraday [17] November 1859[1] 1655

Private | 15 Nov 1859

My dear Tyndall,

I really cannot advise you:[2] I can only tell you what I should do, and what I did do, under like circumstances. When the chair of Chemistry was offered me under the strongest private assurances of the authorities, I declined it. It was all a matter of feeling with me, but the feeling was that, if I had a sufficient moderate income in London, nothing would make me change London for Edinburgh. Others might reverse the terms, and say nothing should bring them to London, so that really I have no advice to give; for I suppose I may now assume that you have a competency in London, and that all beyond will come under the points of honour, prosperity, or pleasure.

Ever, my dear Tyndall | Truly yours | M. Faraday

RI MS JT/1/TYP/12/4150
Faraday Correspondence 5: 3675

1. *[17] November 1859:* the letter was dated 15 November, but is believed to have been finished two days later (*Faraday Correspondence,* 5: 3675, p. 592). Depending on its relation to letter 1652, this letter could also be dated to 16 November.
2. *I really cannot advise you*: Tyndall was considering if he wanted to be a candidate for the newly vacant Chair of Physics at the University of Edinburgh; see letter 1645, n. 3.

To Juliet Pollock [17 November 1859][1] 1656

Thursday.

My dear Mrs Pollock,

I send you a German exercise,[2] written by a man of fine serene, philosophic mind. Von Ettingshausen[3] of Vienna. I read some time ago a memoir from a young man whose acquaintance I made in Vienna in 1856. One of the most <*1–2 words excised*> intellects of the day—and united with it a most manly and transparent character. I wrote to him to express the pleasure his memoir gave me, the more so as I had heard that he was ill. This is the reply I received.[4]

The attraction towards Edinburgh is very strong[5]—I have 5 days remaining to consider the subject and then must give my final answer.[6] Looked at from the point of pure reason, there is no doubt of the preponderating advantages of Edinburgh—'One hour a day for 6 months and six months entire holiday' thus the question is put to me—much could be done—the same lectures to be repeated year after year being merely kept at the level of our knowledge, and the hope of founding a school of true experimental physics which would have a lasting influence upon the country. The reasons pro are very strong. Still reason I fear will not decide it—this entre nous[7]

very sincerely yours | J. Tyndall.

You can give me Ettingshausens's exercise when I come over Saturday.

If I cannot get home to dress I hope you will excuse me. I take it for granted you will be alone

RI MS JT/1/TYP/6/1963
LT Typescript Only

1. *[17 November 1859]:* dated based on the discussion surrounding the Chair of Physics at the University of Edinburgh. In letter 1648, Playfair stated that Tyndall had eight days to decide. As Tyndall notes he now has five days to give a final answer, this letter is likely a few days prior to 24 November 1859 when Playfair replied to Tyndall's letter letting him know that he had finally decided against becoming a candidate. Tyndall recorded in his diary for

19 November that he dined with the Pollocks. This letter has therefore been dated to 17 November 1859, which was a Thursday.

2. *I send you a German exercise:* letter missing.
3. *Von Ettingshausen:* Andreas von Ettingshausen (1796–1878), German mathematician and physicist at the University of Vienna. He was appointed director of the University's physics institute in 1852.
4. *I wrote to him to express the pleasure his memoir gave me, the more so as I had heard that he was ill. This is the reply I received:* this exchange is missing.
5. *The attraction towards Edinburgh is very strong:* referring to the availability of the Chair of Physics at the University of Edinburgh, see letter 1645, n. 3.
6. *I have 5 days remaining to consider the subject and then must give my final answer:* Tyndall decided against applying for the position. Tyndall outlined these reasons in a letter to Playfair on 11 November 1859 (letter 1646), and finally let Playfair know in a letter recorded in Tyndall's journal for Saturday 19 November 1859 (letter 1658).
7. entre nous: in confidence.

From Lyon Playfair 19 November 1859 1657

Confidential | 19 Nov/59.

My dear Tyndal

I have been over to St Andrews today & saw Sir David Brewster.

The Court of Patrons of the University Chairs consists of Seven Curators—Four appointed by the Town Council, three by the University Court[1]—

The three latter will probably be Sir D. Brewster[2] & the Assessor of the Lord Chancellor[3] & Rector.[4] I feel tolerably sure of getting the Town Council Men[5] in your favour & went to see Sir David to find how far he was to be depended upon. He cordially promised to support you if you become a Candidate. The Assessor of the Lord Chancellor will be Sir John Melville[6] & as he will owe his nomination to the representation of Sir David Brewster & myself, I think he will be inclined to support our Candidate. The Assessor of the Lord Rector is not yet known.

Your chances are therefore as good as they can be if you choose to become a Candidate[7]

Yours truly | Lyon Playfair

I do not think Forbes would much promote your Candidateship but I do not think he would actually oppose it. Sir David will resign about the 28th & Forbes take his Commission on the 30th.[8] But no new Professor w[ould]d require to commence his labours before next November.

I should mention that there is a first rate Museum of Physical Apparatus attached to the Chair.

RI MS JT/1/P/120
RI MS JT/TYP/3/990

1. *University Court:* for the first meeting of the new University Court at Edinburgh University in 1859, the University Court consisted of: William Ewart Gladstone (1809–98), David Brewster (1781–1868), Francis Brown Douglas (1814–85), John Brown (1810–82), John Melville (1802–60), R.S. Grieve (N.D.), Edward Francis Maitland (1803–70), and Robert Christison (1797–1882). (http://ourhistory.is.ed.ac.uk/index.php/First_Meeting_of_University_Court,_1859 accessed June 22 2018).
2. *Brewster:* Brewster was acting in the role as Principal.
3. *the Assessor of the Lord Chancellor:* John Melville (see n. 6)
4. *Rector:* William Ewart Gladstone.
5. *the Town Council Men:* as part of the Universities (Scotland) Act, the role of the Town Council changed and they exercised less control over the university, as the university became self-governing.
6. *Sir John Melville:* John Melville (1802–60) was a prominent Scottish lawyer, and Fellow of the Royal Society of Edinburgh.
7. *if you choose to become a Candidate:* Tyndall was considering if he wanted to be a candidate for the newly vacant Chair of Physics at the University of Edinburgh (see letter 1645, n. 3.), and Playfair was considering whether Tyndall would be appointed if he applied for the post.
8. *Sir David will resign about the 28th & Forbes take his Commission on the 30th:* see letter 1645, n. 2.

To Lyon Playfair 19–21 November 1859[1] 1658

My dear Playfair,

I have carefully thought over the subject[2] of your recent letters[3] to me, and my conclusion is that I ought not to leave London.

In this deciding I see clearly the responsibility of an act by which I virtually decline the best position which this country has to offer to a man like myself; for I feel that no other Chair of Natural Philosophy in the kingdom possess advantages equal to those of the Edinburgh one, whether for the investigation of the science, or its diffusion among the community at large

But I am here close to the heart of England, and in the midst of my personal friends. I cannot break my connexions, and I live in hope that men in power will eventually become sufficiently alive to the importance of the subject, to grant to London the school of physics which the country so much needs.

With regard to the relation in which you have placed yourself to me in this matter,[4] believe me I feel truly grateful to you; and I shall avail myself of the very earliest opportunity to visit Edinburgh for the purpose of personally thanking you and the other gentlemen who have honoured me by the spontaneous offer of their support

Believe me dear Playfair | Most faithfully yours | John Tyndall.

RI MS JT/2/13c/1163

1. *[19–21 November 1859]:* the letter was written on 19 November and posted on 21 November (Journal, RI MS JT/2/13c/1163).
2. *the subject:* Tyndall was deliberating whether he would become a candidate for the Chair of Physics at the University of Edinburgh. See letter 1645, n. 3.
3. *your recent letters:* see letters 1645, 1648, 1654, and 1657.
4. *With regard to the relation in which you have placed yourself to me in this matter:* Playfair had been lobbying on behalf of Tyndall in anticipation that he would become a candidate for the position; see for example letter 1657.

From Robert Murchison 21 November 1859 1659

Nov. 21. 1859.

My dear Dr. Tyndall

I cannot say how much gratified I am by being made acquainted by yourself of your resolution to stand by and support the advancement of physical science in this great Metropolis.[1]

Be persuaded that such conduct (which I consider to be truly magnanimous) will never be forgotten by me, and will ever prompt me to urge upon the Government the necessity of rendering the material in your lecture room[2] worthy of the directing spirit.

Yours most truly | Roderick Murchison

RI MS JT/2/13c/1164

1. *your resolution to stand by and support the advancement of physical science in this great Metropolis:* this refers to Tyndall decision to not make himself a candidate for the Chair of Physics at the University of Edinburgh, and instead stay in London (see letter 1645, n. 3).
2. *material in your lecture room:* refers to Tyndall's appointment at the Royal School of Mines; see also letter 1635, n. 10.

From Lyon Playfair 24 November 1859 1660

24 Nov/59.[1]

My dear Tyndall

You have probably judged wisely[2] but I am sorry for our University. Do not think it necessary to come here to thank any body. We only did our duty in trying to secure the Man we thought best for the interests of the University. But if you ever wish to come here for pleasure, I have a large house & ample accommodation & welcome for my friends—It will give me the greatest pleasure to see you. Mrs Playfair desires me on her part to press on you this invitation.

Poor George Wilson[3] died yesterday. He had a soul fit for a better world & his poor diseased body was quite unfitted for this one. With diseased lungs, liver & spleen he has only been kept alive for many years by his genial happy & graceful mind.

His Professorship, as things go, is a good one, the Endowment being £400, independent of fees. It is the only Chair of Technology in the Kingdom & was founded at my recommendation when in office, so that I am anxious to see it worthily filled[4]

Yours Sincerely | Lyon Playfair

RI MS JT/1/P/121
RI MS JT/TYP/3/991

---

1. Tyndall noted in his journal for Friday 25 November 1859, 'Heard of George Wilson's death—Playfair has written me a kindly pleasant note'. (Journal, RI MS JT/2/13c/1164).
2. *You have probably judged wisely:* refers to Tyndall's decision not to become a candidate for the newly vacant Chair of Physics at the University of Edinburgh (see letters 1593, 1595, 1598, 1599, 1601, 1602, 1604, and 1607).
3. *George Wilson:* the chemist George Wilson passed away on 22 November 1859 after a long period of illness. Wilson was one of several Edinburgh figures who supported the idea (referenced in letters 1648 and 1651) of Tyndall becoming a candidate for the vacant Chair at Edinburgh. See also letters 1645, 1646, 1652, 1654, 1655, 1656, 1657, 1658, and 1659.
4. *His Professorship . . . I am anxious to see it worthily filled:* the Regius Chair of Engineering was abolished after Wilson's death. The botanist Thomas Croxen Archer (1817–85) succeeded Wilson as director of the Industrial Museum.

From Edward Sabine 24 November 1859 1661

13. Ashley Place | Nov. 24 | 59

Dear Tyndall,

If you will kindly take the trouble to read the 1st page of my paper in the Phil Trans. for 1852. Art. VIII (page i the vol. 103)[1] you will see that it was the circumstance of my having ascertained the existence of a decennial period in the frequency of occurrence and aggregate amount of deflection proved by the magnetic storm, (or disturbance) in different years, which led me to connect the discovery which was wholly due to my own researches, with the decennial period observed by Schwabe[2] in the Solar Spots—and if you will further kindly take the trouble (and I am sure you will not grudge the trouble where the object is to do justice to one who pursues the same path of inductive inquiry as yourself) to refer to p. 127 in the Vth Art. of the Phil. Trans. for 1851[3] near the bottom, you will see that my discovery of the decennial period in the magnetic storm was no accidental circumstance;—but was the result of a premeditated and previously announced mode of enquiry, on which I had been engaged for some years before, with the express purpose of ascertaining whether any period of the kind existed or not.

You will see in page 118 of my paper in 1852 that I have distinctly noticed the priority of publication of Lamont's[4] table of the periodical variation[5] of the range of the Solar Diurnal variation of the Declination.

My paper was delivered to the Secretary of the Royal Society on March 18. 1852, & contains 4 similar tables to the one which Lamont printed in Pogg.[6] on the preceding Dec 23rd (1851) for Toronto & Hobarton, and also tables of the Decennial variations of the Solar Diurnal variations of the Inclination & Total Force at Toronto & Hobarton. No doubt Lamont's publication of the one element on the one station preceded mine of the 3 elements at the two stations by nearly 3 months: But neither Lamont, nor any other magnetician here or elsewhere,[7] has ever to this day ascertained the existence or non-existence of the decennial period of the magnetic storms. The object & method by which the object was attained were my own; the connexion of the decennial period so discovered with the of the Sun Spots was my own: & when I published them I added the decennial period of the Solar-diurnal variation far more extensively described than it had been, or has been since by any one.

Sincerely yours | Edward Sabine

RI MS JT/1/S/17
RI MS JT/1/TYP/4/1312

1. *Phil Trans. for 1852. Art. VIII (page i the vol. 103):* 'On Periodical Laws discoverable in the mean effects of the larger Magnetic Disturbances—No 2', *Phil. Trans.*, 124 (1852), pp. 103–24.
2. *Schwabe:* Samuel Heinrich Schwabe (1789–1875), a German astronomer known for his solar observations. He proposed that sunspots followed a ten-year periodicity.
3. *p. 127 in the Vth Art. of the Phil. Trans. for 1851:* 'On periodical laws discoverable in the mean effects of the larger magnetic disturbances', *Phil. Trans.*, 141 (1851), pp. 123–39.
4. *Lamont's:* see letter 1777, n. 2.
5. *table of the periodical variation:* see Lamont, 'Ueber den allmäligen Kraftverlust der Magnete, mit besonderer Rücksicht auf die Bestimmung der Variationen der erdmagnetischen Intensität', *Poggend. Annal.*, 84 (1851), pp. 440–44.
6. *Lamont printed in Pogg:* see n. 5.
7. *But neither Lamont, nor any other magnetician here or elsewhere:* Sabine further discussed the differences between his and Lamont's observations in letter 1777, dated 17 April 1861.

From Karl Hermann Knoblauch[1] 28 November 1859 1662

Halle, 28th. Nov. 1859.[2]

My dear Friend,

I have many things to report to you to day. First of all I must with much grief mention the decease of my dear father.[3] I already wrote to you[4] of his grave illness in last winter. Though several attacks of the sickness had gone over, his health came no more back to its usual state. Sleep had left him entirely and so his forces became weaker and weaker. It was a great joy for him as well as for me that it was possible to him in summer to come over to Halle[5] and see my little Johannes.[6] He lived in Wittekind,[7] a little bathing-place in our neighbourhood.[8] First the change of air and the joy seemed to have a good influence and he felt his state of health much better than in Berlin. But the excessive heat of last summer and new attacks of the ill (it was an organic defect in the heart) made it worse again, the *[forces]* were more and more consumed and so he died in my arms the 4th of August. You know how intimately our lives were connected with each other: I have lost the most conscientious father and the most faithful friend. I had communicated to you this wistful news earlier, had I not been afraid that a letter would not meet you at home, as you certainly have been abroad.[9]

After this sorrowful time I spent some weeks in Thüringen,[10] especially in *[Liebza]*,[11] where I had spent already my vacation time last summer. By chance I met both times two sisters from Warrschau,[12] one a widow Mrs. *[Fliegner]* with two children,[13] the other Miss Winter.[14] The frequent commerce and the various opportunities which the little bathing place offered to make a nearer acquaintance have made this meeting a more intimate one. I saw the ladies once more at Jena,[15] where M^rs^. */Fliegner/* is living now, then I met Miss Winter at Dresden (where she has relations, among other: the well known Schnorr of Carolsfeld, actual Director of the Dresden Gallery of pictures).[16] There, on the returnal to Warsaw she became my bride. I hope from her conscientious character not only to get a faithful wife for myself but also a true mother for my dear child[17]—both of which I had possessed in so high a degree in my Elizabeth,[18] who will never be forgotten in my memory.

It was a peculiar misfortune that even this happy event was followed by a new affliction. I received in Dresden at the same time the news of the death of my aunt,[19] the same, who after the early decease of my amiable mother[20] had taken her place in the full sense of the word for my sister and me, so that I considered her always as my second mother and bore to her the full veneration and love of a son.

You see, my dear friend, that in the last time Heaven has sent me much grief and that He begins to send me new joy!

Referring to my scientific occupation I send to you several papers: one on the influence of different media in */lenses/* another on the Interference of Heat. You may imagine that I lay a certain weight on this discovery, as I best know the extreme difficulties which occur in proving experimentally the Cold not in the usual sense as a want but as a */real/* fact of Heat itself. Finding the cold bands which separate the warm ones in the diffracted heat, and showing their existence with a sureness, which makes it possible to observe them even for the most unexercised one, I have believed to fill up an essential void in the Doctrine of Radiant Heat. —I have proved by the thermo–multiplier the existence of the Interference of Heat in all possible principles of producing them: difference of ways; difference of substances; different reflexions; different velocity of rays. So I have repeated in the Chapter of Radial Heat[21] all experiments of Thomas Young with Light.[22]

Reading the Paper, you will find another fact, as a secondary observation: the two groups of rays which Heat offers after the reflexion from Crystals etc. I hope you will to a certain degree be satisfied by the mode of experiments. The Interference–Colours (qualitative differences) of Heat have also been shown by the inquiry. I take the liberty to send several copies of the papers for you, Prof. Faraday and our common friend Hirst, whose address is unknown to me.

I add a paper for the London Royal Society[23] as a peculiar mark (sign) of my high esteem for this, one of the first Academies in Europe. I have joined some other papers which are of a peculiar interest.

You receive:

1) 6 cards for you, Dr Hirst, Prof. Frankland,[24]
2) for you:
   a) Memoir on the Interference of Heat,[25] } quite new
   b) " " Optical Experiments[26]
   c) " " Diffraction and double Refraction of Heat[27] } not yet in your possession
   d) " " Influence of Electricity on Crystals[28]
   e) a copy of my water rays.[29] } new
3) No a and b for Dr Hirst
4) " " " " " Prof. Faraday (with my respectful compliments)
5) for the Royal Society
   a) Memoir on the Interference of Heat[30] referring to:
   b) " " on the Diffraction of Heat.[31]
   c) Memoir on the influence of Metals on Radiant Heat,[32] referring to:
   d) De calore radiante disquisitions[33] etc.
   e) Two Papers on the influence of Crystals on Rad. Heat. Pleochroismus.[34]
   f) On the Influence of Electricity on Crystals[35]
   g) The Physical Qualities on Woods.[36]

—

A great part of my Experiments of last Summer is not yet published.

My best thanks for your very interesting memoirs on Glaciers.[37]

Good bye my dear friend! My Johannes sends his kindest remembrances to you, as his father's friend.

Do not forget us!

Yours | most affectionately | Herm. Knoblauch

I am sure you will be kind enough to present my Memoirs to the Royal Society in one of the Sections with my fullest respects, and add some words about the contents.[38] My best thanks for this favour before! | H. K.

I have the honour to add in the name of the naturforschende Gesellschaft von Halle[39] a diploma,[40] which creates you a member of this respectable society, which in the course of its existence since almost a hundred years contains the most eminent and celebrated names among their members.

Wishing to make Prof. Faraday also a fellow of it, I heard that he were so already since a longer time.

Receive this little mark of my high esteem for your scientific labours at the same time as an homage of the unaltered friendship in which I remain

Ever yours | most faithfully | Herm. Knoblauch. | Halle, | 7. December | 1859.

RI MS JT/1/K/22
RI MS JT/1/TYP/7/2502–5

1. *Knoblauch:* see letter 1605, n. 15.
2. Tyndall recorded a brief summary of the content of this letter in his journal for Thursday 15 December 1859, 'Knoblauch's father is dead—He is also verlobt. I was glad to hear it, for his life in Halle seemed a lonely one.' (Journal, RI MS JT/2/13c/1167). Knoblauch's father had been ill for a longer period, see also letter 1605.
3. *my dear father:* the politician and entrepreneur Carl Friedrich Wilhelm Knoblauch (1793–1859).
4. *I already wrote to you:* letter missing.
5. *Halle:* city in Germany, home of the University of Halle.
6. *my little Johannes:* Knoblauch's son with his late wife, Elizabeth (see n. 17), Johannes Knoblauch (1855–1915).
7. *Wittekind:* not identified.
8. *in our neighbourhood:* Knoblauch worked at the University of Halle from 1853 onwards.
9. *as you certainly have been abroad:* Tyndall travelled through the French Alps during August 1859. See letters 1634, 1635, and 1643.
10. *Thüringen:* a region in central Germany.
11. *[Liebza]:* not identified.
12. *Warrschau:* Warsaw, capital of Poland.
13. *widow M^rs^. [*Fliegner*] with two children:* not identified.
14. *Miss Winter:* not identified, presumed to be Knoblauch's second wife.
15. *Jena:* city in Germany, home of the University of Jenna. It is unclear if Knoblauch was undertaking research while in Jenna.
16. *Schnorr of Carolsfeld, actual Director of the Dresden Gallery of pictures:* refers to Julius Schnorr von Carolsfeld (1794–1872).
17. *my dear child:* see n. 6.
18. *Elizabeth:* Knoblauch's late wife, Elizabeth (1827–55).
19. *my Aunt:* not identified.
20. *the early decease of my amiable mother:* Knoblauch's mother Emilie Henriette died very young, in 1821.
21. Chapter of Radial Heat: it is uncertain which of Knoblauch's articles on radiant heat is referred to; it is possibly H. Knoblauch, 'Über die Polarisation der strahlenden Wärme durch einfache Brechung', *Poggend. Annal.*, 161:2 (1852) pp. 169–88.
22. *all experiments of Thomas Young with Light:* The physician Thomas Young (1773–1829) carried out research on light, including the famous double–slit experiment which showed that light behaved as a wave.
23. *I add a paper for the London Royal Society:* Knoblauch outlined several gifts to the Royal Society, see n. 31–37.
24. *6 cards for you, Dr Hirst, Prof. Frankland:* cards missing.
25. *Memoir on the Interference of Heat:* it is uncertain which of Knoblauch's articles on Heat is referred to. Given the timing of this letter, it may refer to Knoblauch's article 'Ueber die Interferenz der Wärmestrahlen', *Poggend. Annal.*, 184:12 (1859), pp. 610–15.

26. *Optical Experiments:* may refer to Knoblauch's article 'Ein Paar optischer Vorlesungs-Versuche', *Poggend. Annal.,* 183:6 (1859), pp. 323–26.
27. *Diffraction and double Refraction of Heat:* likely refers to Knoblauch's article 'Über die Doppelbrechung der strahlenden Wärme', *Poggend. Annal.,* 150:5 (1848), pp. 1–9.
28. *Influence of Electricity on Crystals:* H. Knoblauch, 'Über das Verhalten krystallisirter Körper zwischen elektrischen Polen', *Poggend. Annal.,* 159:6 (1851), pp. 289–99.
29. *my water rays:* not identified.
30. *Memoir on the Interference of Heat:* see n. 26
31. *on the Diffraction of Heat:* see n. 27.
32. *the influence of Metals on Radiant Heat:* H. Knoblauch, 'Über den Einfluss welchen Metalle auf die strahlende Wärme ausüben', *Poggend. Annal.,* 177:6 (1857), pp. 161–213.
33. *De calore radiante disquisitions:* H. Knoblauch, *De Calore Radiante: Disquisitiones Experimentis Quibusdam Novis Illustratae,* (Berolini, 1846).
34. *Two Papers on the influence of Crystals on Rad. Heat. Pleochroismus:* H. Knoblauch, 'Über die Abhängigkeit des Durchgangs der strahlenden Wärme durch Krystalle von ihrer Richtung in denselben', *Poggend. Annal.,* 161:2 (1852), pp. 169–88, and likely refers to H. Knoblauch, '1. Gangunterschied durch ungleiche Wegeslängen in einem und demselben Medium', *Monatsberichte der Königlichen Preuss. Akademie der Wissenschaften zu Berlin,* (1859) pp. 565–66.
35. *Influence of Electricity on Crystals*: see n. 28.
36. The Physical Qualities on Woods: H. Knoblauch, *Ob bei verschiedenen Holzarten ein gewisser Zusammenhang zwischen den un ihnen beobachteten Eigenschaften* (Halle: 1858).
37. *memoirs on Glaciers:* Refers to Tyndall's book *The Glaciers of the Alps,* which was published in 1860.
38. *I am sure you will be kind enough to present my Memoirs to the Royal Society in one of the Sections with my fullest respects, and add some words about the contents:* it is uncertain if and when Tyndall did this, and it does not appear that Knoblauch's papers were published in the *Roy. Inst. Proc.* Several of his articles did appear in the *Phil. Mag.,* including: H. Knoblauch, 'LVI. On the Connexion between the Structure and the Physical Properties of Wood', *Phil. Mag.,* 115:17 (1859), pp. 348–52.

H. Knoblauch, 'XVI. On the Interference of Heat', *Phil. Mag.,* 125:19 (1860), pp. 126–28; and H. Knoblauch, 'XXI. Intelligence and Miscellaneous Articles. Optical Lecture-Experiments', *Phil. Mag.,* 125:19 (1860), 162–64. Tyndall frequently reference Knoblauch, for example in his Friday evening lecture at the RI 'On the Action of Gases and Vapours on Radiant Heat' (January 18, 1861).

39. *naturforschende Gesellschaft von Halle:* scientific society founded in 1779.
40. *a diploma:* Tyndall recorded receipt of this diploma in his journal for Thursday 15 December 1859, 'He sent me a diploma from Halle. His letter is full of warm friendship.' (Journal, RI MS JT/2/13c/1167).

From Heinrich Daniel Ruhmkorff 4 December 1859 1663

Paris den 4ten Decemb | 1859

Hoch geehrtester Herr Thyndal

Die beyden instrumente welche si mir die Ehre machen zu bestellen werde im Monath februar fertig werden ich werde alles mögliche thun damit sie Ihnen gefallen. Ich verbleibe mit der grosste Hochachtung

Ihr | ganzergebenster | Ruhmkorff

Paris 4th Decemb | 1859

Most dear Mr Tyndall

The two instruments[1] which you do me the honour of ordering will be ready in the month of February. I will do everything possible to ensure they meet your approval. Respectfully

Yours | sincerely | Ruhmkorff

RI MS JT/1/R/42

---

1. *The two instruments:* it is uncertain what instrument Tyndall had ordered from Ruhmkorff, but Jules Célestin Jamin mentioned a device being sent to Tyndall in letter 1721.

To Rudolf Clausius 7 December 1859 1664

7th Dec. 1859

My dear Clausius.

Will you have the kindness to assist me in the following matter.

I am anxious to put Agassiz's labours on the glaciers[1] in a fair light before the English public as I think he has been neither justly nor generously treated.

I am very anxious to learn when he first spoke to Mr Wild[2] about engaging him to make a survey of the glacier of the Aar.[3]

I know that Herr Wild made measurements upon the glacier in 1842: but I want to know whether Agassiz did not bespeak his Services in 1841, or at all events early in 1842.

Would you ask Mr Escher von d. Linth[4] whether he and Agassiz did not make an arrangement in 1840 to determine the velocity at various points across the Aletsch[5] (and across the Aar glacier,) I believe they actually set up posts, which however on account of not being sunk sufficiently deep were of no use.

In 1841 Agassiz set up a series of posts on the Aar glacier and in 1842 their motion was determined by Mr Wild.[6] About 3 weeks previously Forbes had determined the motion of the Mer de Glace: but I believe notwithstanding the speed with which he communicated his results to the Editor of the Edinburgh Philosophical Journal,[7] that Agassiz was beforehand with him in publication.

I tell you frankly my impression that judged of by the laws tacitly adopted by scientific men, the proceeding of Forbes in this matter[8] was not generous; and Agassiz has suffered from it in England up to the present time. I am anxious to do him justice in my little book & therefore ask you to be kind enough to obtain from M. Wild & M. Escher the above information for me.

Would you also kindly inform me where the papers of Mayer on the dynamical theory of heat are to be found.[9]

Finally would you ask M. Mousson whether any other writer besides Agassiz ever asserted that the sides of the glacier move more quickly than the centre. People say here that this was the general opinion, before the researches of Forbes.[10]

Ever yours most truly | John Tyndall

RI MS JT/1/T/179

---

1. *Agassiz's labours on the glaciers:* Aggasiz undertook research on the movement and nature of glaciers from the 1830s onwards. His publications on glaciers include J. L. R. Agassiz *Études sur les glaciers* (1840); J. L. R. Agassiz, 'Observations on the glacier of the Aar', *Edinburgh New Philos. J.,* 38 (1842), pp. 339–402; J. L. R. Agassiz, 'Observations sur le glacier de l'Aar', *C.R. Acad R. Sci. Paris.* 15 (1842), pp. 284–88; J. L. R. Agassiz, 'The glacial theory and its recent progress', *Edinburgh New Philos. J.,* 33 (1842), pp. 217–83.
2. *Mr Wild:* Johannes Wild (1814–94), a Swiss engineer, cartographer, and professor in Zurich.
3. *glacier of the Aar:* glacier system in the Bernese Oberland in Switzerland.
4. *whether he and Agassiz did not make an arrangement in 1840 to determine the velocity at various points across the Aletsch (and across the Aar glacier):* Arnold Escher von der Linth sent Tyndall a letter outlining his and Agassiz' work, though requesting that he did not use the information in 'war of priority', see letter 1670.
5. *Aletsch:* the largest glacier in the Alps.
6. *their motion was determined by Mr Wild:* see for example Wild's measurements in Louis Agassiz, *Système glaciaire ou recherches sur les glaciers* (Paris: Victor Masson/Leipzig: L. Voss, 1847).
7. *Forbes had determined the motion of the Mer de Glace . . . communicated his results to the Editor of the Edinburgh Philosophical Journal:* see 'Professor Forbes' Account of his recent Observations on Glaciers. Communicated in the following Letters to the Editor, Professor Jameson', *Edinburgh New Philosophical Journal,* 33 (1842), pp. 338–52.

8. *the proceeding of Forbes in this matter:* Tyndall was referring to the accusations levelled against Forbes that he had been dishonest in his academic dealings with Agassiz. See F. F. Cunningham, *James David Forbes: Pioneer Scottish Glaciologist* (Edinburgh: Scottish Academic Press, 1990).
9. *Would you also kindly inform me where the papers of Mayer on the dynamical theory of heat are to be found:* this is significant, as Tyndall would later champion Mayer's priority in discovery of conservation of energy. See also letter 1674.
10. *before the researches of Forbes:* see for example, Forbes's *Travels through the Alps of Savoy* (see letter 1594, n. 13).

From Charles Dufour[1] 10 December 1859 1665

Morges le 10 Décembre 1859.

A Monsieur le Professeur Tyndall à Londres

Monsieur!

Vous vous souvenez peut-être de la conversation que j'eus le plaisir d'avoir avec vous à Genève l'été dernier, lors de votre retour du Mont Blanc, conversation dans laquelle il fut question de l'observation de la scintillation dont je m'occupe depuis plusieurs années. Vous me dites alors que cette question vous intéressait beaucoup aussi, & que vous désiriez vivement que je vous communique ce que je publierai sur ce sujet. Je viens ici Monsieur remplir ma promesse & vous envoyer une copie des instructions que je publie actuellement pour l'observation de la scintillation dans le Bulletin de la Société vaudoise des Sciences naturelles, & que j'ai déjà envoyées l'année dernière à M. *[*Haemitz*]* à Dorpat, & à M. Poey à la Havanne, qui m'avaient prié de les leur communiquer. Seulement comme j'ai vu Monsieur que la langue française ne vous était pas familière, j'ai prié une excellente relation que j'ai en Angleterre (Madame *[*Cugny Trice*]* Morrill Place Tunbridge Wells) de bien vouloir me traduire ce manuscrit en anglais (*[*du*]* français). Madame *[*Cugny*]* vous expédiera en même temps le premier mémoire que j'ai publié sur la scintillation en 1856 & dont Mr Piazzi Smyth a déjà publié il y 2 ou 3 ans un résumé dans les Monthly Notices de la Société astronomique de Londres.

Maintenant Monsieur, si comme on peut le supposer, & comme je le désire dans l'intérêt de la science vous vous proposez d'entreprendre encore une fois l'ascension du Mont Blanc, & de passer la nuit dans les hautes régions de l'air, oserais-je vous prier de bien vouloir aussi prêter quelque attention au phénomène de la scintillation. Les résultats que vous obtiendrez seraient précieux Monsieur, car il n'est pas ordinaire de trouver des savants disposés à passer la nuit sur le sommet du géant des Alpes, & j'*[1 word illeg]* Monsieur, qu'avant d'apprendre le résultat de votre entreprise, je croyais que l'on aurait plutôt des observations faits au *[*pôle*]* même que sur la cîme du

Mont Blanc.

Agréez Monsieur l'assurance de l'estime & de la haute considération | de votre dévoué Ch. Dufour | Professeur à Morges (Suisse)

Morges,[2] December 10th 1859.

To Professor Mr. Tyndall in London,

Sir!

Perhaps you remember the conversation that I had the pleasure of having with you in Geneva last summer—at the time of your return from Mont Blanc, a conversation regarding the question of observing scintillation, as I have dealt with this subject for several years. You told me then that this question greatly interested you as well, and that you would much desire to see what I published on this subject. I now, Sir, fulfill my promise, and send you a copy of the instructions that I am currently publishing for the observation of scintillation in the Bulletin de la Société vaudoise des Sciences naturelles,[3] and which I already sent last year to Mr *[Haemitz]*[4] in Dorpat[5] and to Mr Poey[6] in Havanne,[7] who asked me about it. But as I saw, Sir, that the French language was not familiar to you, I asked an excellent acquaintance that I have in England (Madame *[*Cugny Trice*]*[8] Morrill Place Tunbridge Wells[9]) to be kind enough to translate this manuscript for me into English (from French).[10] Madame *[*Cugny*]* will send you[11] at the same time the first memoir that I published about scintillation in 1856[12] and which Mr Piazzi Smyth[13] had already published 2 or 3 years ago his summary in the Monthly Notices of the Society of Astronomy in London.[14]

Now, Sir, if, as may be supposed, and as I wish in the interest of science, you propose to undertake once more the ascent of Mount Blanc, and to spend the night in the upper regions of the air; can I also ask that you please pay attention to the phenomenon of scintillation? The results that you will obtain will be precious, Sir, since it is not usual to find savants inclined to spend the night on the summit of the giant of the Alps.[15] And I *[1 word illeg]*, Sir, that before receiving the result of your venture, I believe that one could make many observations at *[*perch*]* itself as well as at the top of Mont Blanc.

Please accept my highest esteem and best wishes | your devoted | Ch. Dufour | Professor at Morges (Switzerland)

---

1. *From Charles Dufour:* Charles Dufour (1827–1902), Swiss mathematician and astronomer. Dufour was professor in mathematics in Lausanne between 1855 and 1869, after which he was professor in astronomy until his death in 1902.
2. *Morges:* a city in Switzerland.
3. *the instructions that I am currently publishing for the observation of scintillation in the Bulletin de la Société vaudoise des Sciences naturelles:* published as Charles Dufour, 'Instructions

pour l'observation de la scintillation des étoiles', *Bulletins des Séances de la Société Vaudoise des Sciences Naturelles,* 6 (1858–61) pp. 363–71.

4. *Mr [Haemitz]*: not identified.
5. *Dorpat:* refers to the University of Dorpat (now University of Tartu) in Estonia.
6. *Mr Poey*: the Cuban meteorologist and physicist Andrés Poey Aquirre (1825–1919), who founded the Physical–Meteorological Observatory in 1856 (see Angelo Baracca et. al. (eds) *The History of Physics in Cuba,* (Springer: Heidelberg, New York and London, 2014) p. 76.
7. *Havanne*: Havana in Cuba.
8. *Madame [Cugny Trice]*: not identified.
9. *Tunbridge Wells:* town in Kent, England.
10. *translate this manuscript for me into English (from French):* not identified.
11. *will send you*: letter missing.
12. *the first memoir that I published about scintillation in 1856:* refers to Charles Dufour, 'Sur la scintillation des étoiles', *Bulletins des Séances de la Société Vaudoise des Sciences Naturelles,* 5 (1856–58), pp. 17–26.
13. *Mr Piazzi Smyth*: the Italian-born astronomer Charles Piazzi Smyth (1819–1900), Astronomer Royal for Scotland and professor of astronomy at the University of Edinburgh.
14. *published 2 or 3 years ago his summary in the Monthly Notices of the Society of Astronomy in London:* published as Charles Dufour, 'Notes on the Scintillation of the Stars (Extracts of Letters to Professor Piazzi Smyth)', *Monthly Notices of the Society of Astronomy in London,* 18 (1857–58), pp. 51–54.
15. *it is not usual to find savants inclined to spend the night on the summit of the giant of the Alps:* refers to Tyndall's ascent of Mont Blanc.

To Alfred Wills[1] 11 December 1859 1666

11th Dec. 1859

My dear Wills,

I was just preparing for bed when accident drew my eyes to the following passage in your article in the National Review:[2]—

"The theory commonly called De Saussure's[3]—now more than a century old—was based upon the supposed immobility of glaciers during winter. De Charpentier[4] and Agassiz in our own day have supported the Dilatation theory[5] by an appeal to the equally erroneous notion that glaciers move faster at the sides than at the centre."[6]

I would take it as a great favour if you would refer me to the passage in De Saussure's works in which he bases his theory upon the supposed immobility of glaciers during the winter, and also to the passage in Charpentier's works in which he appeals to the quicker motion of the sides in support of the theory of Dilatation.

Believe me | very sincerely yours | John Tyndall.

Wellcome Library, MS. 7777/02

---

1. Alfred Wills (1828–1912), judge and mountaineer. Wills was very involved in the Alpine Club, and was its third President between 1863 and 1865 (ODNB).
2. *your article in the National Review:* Alfred Wills, 'Glaciers and Glacier Theories', *the National Review,* 17 (July 1859), pp. 1–30.
3. *De Saussure's:* Horace-Bénédict de Saussure proposed what was called the sliding theory of glacier movement, which assumed that glaciers were rigid bodies that moved by sliding in its entirety. He published his ideas in his four volume *Voyage dans les Alpes* (Lyon: Fauche-Borel, 1779–96).
4. *De Charpentier:* Jean de Charpentier.
5. *the Dilatation theory:* the 'dilatation theory of glacier motion' is associated with De Charpentier; see his book *Essai sur les glaciers et sur le terrain erratique du bassin du Rhône,* (Lausanne: 1841). The theory was further developed by Agassiz. Dilatation means expanding, or widening out (OED); in relation to glaciers it means the flow of water permeating the ice thereby expanding and moving the glacier.
6. *'The theory commonly called De Saussure's . . . than at the centre':* passage appears on p. 3 of Wills's article, see n. 2.

To Unidentified[1] 13 December 1859 1667

Hochgeehrte Freundin.

Alles was Sie geschrieben haben ist so gut und Ihre Einladung auf Deutsch ausgedruckt, klingt so freundlich dass ich es kaum wiederstehen kann 'Ja' zu sagen.

Ich werde kommen wenn es nur möglich ist. 'Ich hoffe, ja ich glaube auch' daß dieß der Fall sein wird.

Ich schicke Ihnen eine Verlobungscarte welche ich gestern erhielte; es wird Ihnen zeigen wie man solche sachen in Deutschland offentlich macht.

Ich schicke Ihnen auch einen Englischen Brief von einem Deutschen geschriben; Er ist sehr nett und artig.

Mit aller Hochachtung | Ihr ergebenster | John Tyndall

13th Dec. 1859

Auch schicke ich Ihnen zu gleicher Zeit eine *[*Todken Karte*]*—nein ich will es nicht. Ja doch ich will.

Highly esteemed friend.[2]

Everything that you wrote[3] is so good and your invitation[4] expressed in German sounds so friendly that I can hardly resist to say 'yes'.

I will come as soon as it is possible. 'I hope and I also believe'[5] that this will be the case.

I am sending you an engagement card,[6] which I received yesterday; it will show you how these things are made public in Germany.

I am also sending you an English letter written by a German;[7] it is very pleasant and proper.

Respectfully | Your most devoted | John Tyndall

13th Dec. 1859

I also send you a *[*condolence*]*[8] card at the same time[9]—no, I do not want to. But yes, I do want to.[10]

RI MS JT/1/T/1133
RI MS JT/1/T/6/1964

---

1. *[To Unidentified]:* a later archival note assigned this letter to Juliet Pollock, however this has not been verified.
2. *friend:* the German Freundin implies a woman.
3. *Everything that you wrote:* letter missing.
4. *your invitation:* invitation missing.
5. *'I hope and I also believe':* it is uncertain what Tyndall is referring to by use of quotation marks, presumably it references the wording in the invitation.
6. *I am sending you an engagement card:* card missing.
7. *an English letter written by a German:* not identified, letter missing.
8. *[condolence]:* Tyndall wrote 'Todten Karte', or 'Todtenkarte', which could either mean a condolence card or a death certificate.
9. *I also send you a [condolence] card at the same time:* card missing.
10. *But yes, I do want to:* presumably this, and the rest of the letter, was intended as a joke.

From Auguste Balmat — 18 December 1859 — 1668

Chamonix le 18 Décembre 1859

Monsieur le Professeur,

Je vous prie de me pardonner du retard que j'ai mis à répondre à votre lettre. J'en suis grandement fâché, je ne l'ai reçu qu'aujourd'hui, on la gardée il parait par oubli je crois. Je voie *[*qu'elle est*]* arrivée à Chamonix le 11 du courant au bureau de la poste. C'est peut-être un *[1 word illeg]* encore car le temps c'est grandement changé depuis quelques jours, hier nous avions un froid de 24 degrés de centigrades et avec ça de la neige et du vent. Conséquemment, il serait impossible pour le moment d'effectuer votre travail aujourd'hui. Surtout nous avons de mauvais pronostics du temps, dans ce moment il neige très fortement avec un grand vent, aussitôt que le temps sera remis je vous écrirai. Je crois réellement qu'il serait mieux d'attendre encore quelques temps si vous le pouvez.

Quant aux *[jalons]*, ils seront prêts si en tous cas vous arriviez.

En attendent de vos nouvelles, ayez Monsieur Tyndall les sentiments de ma haute considération

Votre Serviteur | Auguste Balmat

Chamonix 18 December 1859

Professor,

Please forgive me for my tardy response to your letter.[1] I am truly furious that I have not received it until today. It looks like it has been kept out of forgetfulness. I see it arrived at Chamonix on the 11th of this month at the post office. Perhaps it is a *[1 word illeg]* again because the weather greatly changed in the past several days, yesterday we had –24 degrees Celsius and with it snow and wind. As a result, it would be impossible to do your work today. We have an especially poor weather forecast; right now it is snowing a lot with strong wind. As soon as the weather improves I will write to you. I honestly believe that it will be better to wait for some time, if you can.

As for the stakes,[2] they will be ready when you arrive in any case.

While waiting for your news, please accept, Mr. Tyndall, my highest regards Your servant | Auguste Balmat

RI MS JT/1/B/6

1. *your letter*: letter missing.
2. *As for the stakes:* as part of his research on glaciers, Tyndall set up experimental stations and used stakes to measure the motion of the glaciers. See also letters 1634, 1635, n. 3, 1643, n. 3, and 1645, 1653.

To First Commissioner of Works[1] 20 December 1859 1669

REPORT[2] on the Great Bell of Westminster,[3] by *John Tyndall*, F.R.S., | dated 20th December 1859.

SECTION 1. —PRELIMINARY OBSERVATIONS.[4]

1. IF a circular plate of metal be fixed at its centre, and a violin bow drawn across its edge, it will, as is known, divide itself into an even number of vibrating segments, each separated from its neighbor by a line of rest, commonly called a nodal line. If the plate be horizontal, and if fine sand be strewn upon it, the sand will be ejected from the vibrating parts, and will arrange itself along those radii with which the nodal lines coincide.

2. When a bell is struck it divides itself in a similar manner. Thus, as regards the great bell of Westminster, we should conclude that the point on which the

hammer falls lies in the middle of a vibrating segment, while two other points on the soundbow,[5] the one 45 degrees to the right and the other 45 degrees to the left of the point struck by the hammer, would be nodes. At a further distance of 90 degrees from each of these a node would occur, thus making four in all. The circumference of the soundbow being, in round numbers, 28 feet, it follows that the distance between two adjacent nodes is seven feet.

3. Knowing, however, that, in practice, the conditions of theory are rarely strictly fulfilled, I thought it advisable to examine, by direct experiment, the condition of the bell while in a state of vibration. I therefore divided the circumference of the soundbow into 28 equal parts, and ascertained the comparative amplitudes of the vibrations at the several points of division.

4. This examination shewed that while the middle of each vibrating segment gave the maximum amplitude of oscillation, the nodal lines were, by no means, lines of absolute rest, but shared, to a considerable extent, the general agitation.

These remarks are made partly to give those unacquainted with the principles of acoustics an intelligent conception of the mode of the bell's vibration, and partly to assist those conversant with the subject to draw their own conclusions, should they desire to do so, regarding one of the principal defects of the bell.

SECTION 2.—ON THE CAVITIES IN THE BELL.

1. On the soundbow, and diametrically opposite to the hammer, a number of cavities occur, forming a group which is included in a space of 3½ in. by 1½ in. The centre of this group (which is the principal one) is about 12½ in. distant from the edge of the bell. The cavities (six in number, excepting one or two very small ones), vary from three–tenths of an inch to less than one–tenth of an inch in diameter, and the depth of the deepest of them is two–fifths of an inch. For the purpose of future reference, I will denote this group by the symbol (G_1).

2. To the left of this group (*i.e.,* to the left of an observer facing the group), and at a distance of about eleven inches from it, two cavities occur, the largest of which measures two–tenths of an inch across, at the surface, and is two–tenths of an inch deep. The diameter of this cavity underneath the surface augments to three–tenths of an inch.

3. At a distance of a foot further to the left, and 16 inches from the edge of the bell, is a group of pits, the largest diameter of the largest of them measuring, at the surface, one–fourth of an inch, the largest inner diameter being four–tenths of an inch: the depth of this cavity is one–eighth of an inch. All the others are small; and with the exception of one little outlier, the whole group would be covered by an area of a square inch.

4. Between this group and the point struck by the hammer; at a distance

of four feet nine inches from the latter, and 14 inches from the edge of the bell, a cavity occurs which measures five-eighths of an inch across: it is filled with a foreign metal, and I am therefore unable to state its depth.

5. At a distance of 10 inches to the *right* of the group G_1, a cavity occurs one-fourth of an inch in diameter, and two-tenths of an inch deep.

6. At a distance of six inches above the latter, and in a line which inclines from it to the east, a cavity occurs, measuring, in its longest direction, one-eighth of an inch at the surface, but running under the surface until it attains about three-eighths of an inch in diameter.

7. Finally, at a distance of three feet four inches to the right of G_1 and at 14 inches from the edge of the bell, a cavity occurs, measuring, in its longest direction, one-fourth of an inch across, and having a depth of two-tenths of an inch.

8. All these cavities appear to have been produced by the rising of air bubbles through the metal while in a state of fusion; and their occurrence along the soundbow is, I think, due to the fact that the slope of the mould, at this place, formed a kind of roof over the bubbles and prevented their escape.

SECTION 3. —ON THE CRACKS IN THE BELL.

1. The principal cracks are two in number; the first, which I will call C_1, is diametrically opposite to the hammer, and, hence, in the middle of a vibrating segment. It is nine inches long, and runs transverse to the soundbow. It does not reach the edge of the bell, but terminates two inches above the edge.

2. The second crack, C_2, occurs at a distance of two feet four inches to the left of C_1. It also runs transverse to the soundbow; measures 13 inches in length, and terminates at a distance of five inches from the edge of the bell. This crack occurs within 15 inches of a node, and the amplitude of the vibrations immediately adjacent to the crack was found by experiment to be less than that of the vibrations at the node itself.

3. Besides these two principal cracks there are three minor ones, varying from two and-a-quarter to two inches in length. Were the bell closely examined with a good light, additional small cracks might possibly be detected.

4. To the naked eye the sides of the cracks appear to be in mechanical contact; but when a wet cloth is dabbed against the larger ones, the water is forced into them, or soaks in by capillary attraction; and when the bell is struck, subsequently, bubbles issue from the cracks throughout their entire length.

SECTION 4. —ON THE CONNEXION OF THE CRACKS AND CAVITIES.

1. The pits above described, and other smaller ones which have been disclosed by the removal of a portion of one of the ornamental rings of the bell, must undoubtedly be regarded as defects; but how far such defects are

unavoidable in the casting of large bells I am not in a position to state. My duty at present is to estimate the possible influence of the cavities in the production of the cracks.

2. The first principal crack, $C_{1,}$ certainly originates (or terminates) among the lower members of the group of cavities denoted by G_1. Two of the lowest holes, both of them small ones, are about three–fourths of an inch apart, and occur nearly in the same vertical line. The crack commences midway between both, intersects the lower cavity, and afterwards passes through an impitted portion of the bell.

3. Hence, though the superior end of this crack is close to the cavities of the group $G_{1,}$ the fact that it reaches less than half an inch higher than the lowest member of the group, and that the main body of the group is not intersected by the crack, tends, in my opinion, to invalidate the assumption that this crack has been caused by the cavities.

4. But setting this apparently doubtful case aside, a reference to the other cracks appears to me quite fatal to the notion that the cracks are due to the cavities. The second principal crack, denoted by the symbol C_2, is the most formidable of all, and it runs from beginning to end through an impitted portion of the metal. As for the minor cracks, no cavity occurs near to any one of them.

5. With regard to the manner in which the bell is suspended, I do not think that it had any influence in producing the cracks.

6. Thus, I think the elements of the question reduce themselves to two, viz., the quality of the metal of the bell, and the magnitude of the blows to which it has been subjected. Each of these elements must be considered in relation to the other, and when this is done, the conclusion appears to me unavoidable that the hammer and the bell are not suited to each other. If the bell be of the proper composition and right tenacity, the blows of the hammer must have been too heavy; if, on the contrary, the blows have been no heavier than they ought to have been, it necessarily follows that the metal of the great bell of Westminster is too brittle.

CONCLUDING REMARKS.

1. In the foregoing report I have confined myself to the physical portion of the question. With the chemical and metallurgical portion I have nothing to do; nevertheless I may be permitted to suggest, as possible, that, owing to the imperfect mixture of the constituents, or through a defect of cooling, the external portions of the soundbow may vary in composition, or tenacity from the average composition, or tenacity, of the bell; thus forming a brittle casing round a less brittle mass. The truth or fallacy of this conjecture experiment alone can decide; but if it should prove true, the cracks might, with proper care, be confined to the brittle envelope, and thus the bell might be saved.

2. The depth of the cracks is unknown; but they appear to penetrate the soundbow to only a small fraction of its thickness. If they were sufficiently deep the grating of their surfaces would injure the tone of the bell, but, as yet, the tone appears to be unaffected.

3. It seems to me worthy of consideration whether the cracks at present existing ought not to be cut out of the soundbow, the striking permitted to proceed as formerly, and the bell subjected to a daily examination. The result *might* be, that the metal, being partially relieved from strain at the places now intersected by the cracks, would yield no further.

4. It might also be desirable to shift the position of the hammer so as to produce a different system of vibrating segments from that now existing. If, under these circumstances, new cracks should appear, or the old ones show a tendency to deepen, the recasting of the bell would seem to be the only resource.

(signed) *John Tyndall.*

To the First Commissioner | of Works, &c.

John Tyndall, 'REPORT on the Great Bell of Westminster', *Bell (New Palace, Westminster): Copy of reports on the condition of the great bell of the new Palace of Westminster,* (1860) p. 553.

1. Until his death on December 17, 1859, the First Commissioner of Works was Henry Fitz-Roy (1807–59). He was succeeded by William Cowper (1811–88) on February 9, 1860 (*ODNB*).
2. REPORT: this was part of a larger report ordered by the House of Commons, *Bell (New Palace, Westminster): Copy of reports on the condition of the great bell of the new Palace of Westminster* (1860). It contained reports by Tyndall, Astronomer Royal George Biddell Airy (1801–92), and Professor of Metallurgy at the Royal School of Mines John Percy (1817–89).
3. *the Great Bell of Westminster:* the Elizabeth Tower's bell in London, also commonly known as 'Big Ben'.
4. *PRELIMINARY OBSERVATIONS.:* Tyndall worked on this topic throughout November and December 1859, and occasionally recorded on his progress in his journal.
5. *soundbow:* the soundbow, also spelled sound–bow or sound bow, refers to the thickest part of a bell (*OED*).

From Arnold Escher von der Linth 20 December 1859 1670

Charpentier Essai sur les glaciers Lausanne 1841, p.79 attribue la direction oblique des crevasses à  la différence de la progression laquelle dans son hypothèse est la plus rapide le long des bords du glacier.

Agassiz Études sur les glaciers Neuchâtel 1840, p.86. Les grandes crevasses ont en général une direction perpendiculaire à celle du glacier. Mais comme le massif de glace chemine ordinairement plus vite près des bords qu'au centre etc.

Agassiz Système glaciaire—révoque l'hypothèse que le glacier marche plus vite près des bords qu'au centre.

Escher établit au mois de Juillet 1841 au glacier d'Aletsch, à peu près ¾ de lieue en amont du Lac de *[Mariel]* dans la région dans laquelle des crevasses commencent (en aval d'une région étendue très peu inclinée et qui n'en montrait pas *[)]* le système suivant de pieux.

700 mètres

Rocher Rocher

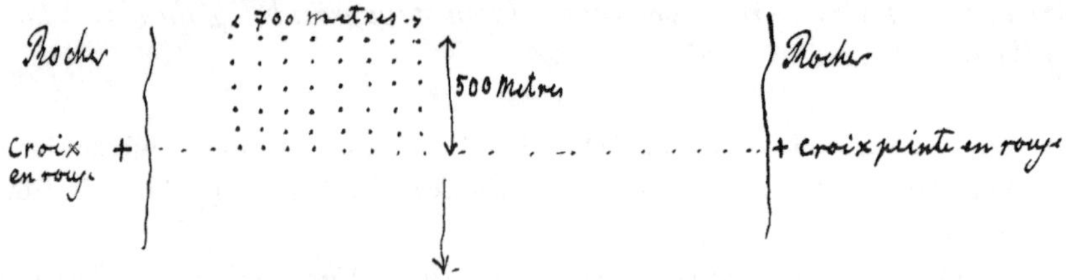

Croix en + + croix peinte en rouge

rouge

La position rectangulaire des pieux était déterminée au moyen d'un petit instrument d'Amici.

Les pieux sont éloignés l'un de l'autre 100 mètres dans la direction transversale.

id id id longitudinale.

id se trouvent enfoncés dans la glace à peu près 3 pieds le 8 juillet.

" ne montrent pas un changement de place bien évident et incontestable——- le 16 juillet.

(Les distances étaient mesurées au moyen d'une ficelle vernie de *circa* 150 mètres de longueur. Le vernis ne la préservait cependant pas tout à fait de l'influence de l'humidité etc. *[)]*

Escher départ le 17 juillet

" se trouve avec M. *[Studer]* le 10 août soir à l'hospice du Grimsel où arrivent M. Agassiz, Forbes et Heath, chassés du glacier de l'Aar par le mauvais temps. On parle beaucoup de bandes bleues et blanches.

" avec Studer trouvent le———-14 août presque tous les pieux sur le glacier d'Aletsch tombés; même des trous il n'existait plus de traces chez la plupart. L'ablation était donc au moins 3 pieds en 29 jours. Le Föhn avait régné d'une manière puissante.

" stupéfait de cette expérience et voyant qu'il faudrait employer un autre procédé pour obtenir des observations suivies sur la manière de progression, continue avec M. Studer le voyage en Valais.

" rencontre le 29 août à Viesch M. Agassiz et Desor qui reviennent justement de *[*l'ascion de*] [*1 word illeg*]*31 août jusqu'au 5 sept. 1841 séjour avec Agassiz et Desor au glacier de l'Aar, où l'on fixe la position d'un certain nombre de blocs et dont si je ne me trompe plusieurs se trouvaient sur une ligne à peu près transversale du glacier.

C'est très probablement pendant ces journées, qu'Agassiz se résolut de faire exécuter par M. Wild (maintenant Professeur de Géodésie et Topographie au Polytechnieum à Zurich) le levé des glaciers de l'Aar.

Voilà, M. Tyndall! Les notices que je puis donner, mais veuillez me permettre que je vous prie, de ne pas les employer pour une guerre de priorité, guerres, qui toujours me paraissent être fâcheuses et ne servant pas à l'avancement de la science. Je termine en félicitant vous et nous autres de vos brillantes découvertes sur la cémentation de la glace et de la véritable origine des bandes bleues et blanches et en vous remerciant bien sincèrement du mémoire si instructif que vous bien voulu me donner.

Votre très dévoué ami Escher | Zurich 20 Dec. 1859.

Charpentier, Essay on the glaciers, Lausanne, 1841 p. 79,[1] attributes the oblique direction of the  crevasses to the difference in the progression, which, he hypothesises, is fastest along the edges of the glacier.

Agassiz, Studies on the glaciers, Neuchâtel, 1840 pg. 86.[2] The large crevasses are generally perpendicular to the direction of the glacier. But as the ice block usually advances faster on its edges than in its centre, etc.

Agassiz Glacial System[3]—dismisses the hypothesis that the glacier moves faster on its edges than in its centre:

Escher, in the month of July 1841 at the Altesch glacier,[4] at a distance of approximately ¾ of a league upstream from Lake *[*Märjelen*]*[5] in the region where crevasses are appearing (downstream in a very large zone ever so slightly inclined and where they were not visible before*[*)*]* set up the following system of stakes:

700 meters

Rock Rock

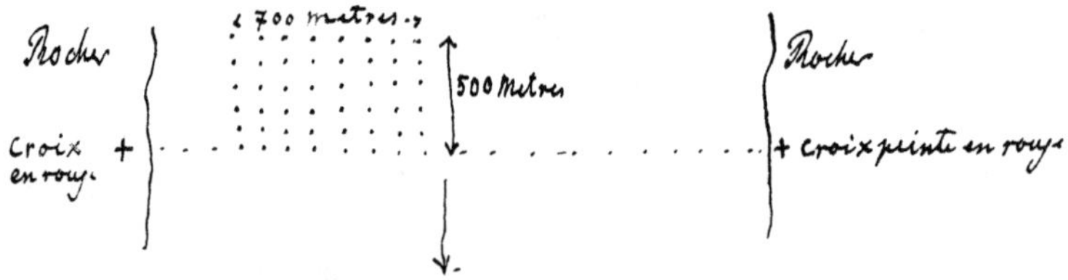

A red cross + + a cross painted in red

The rectangular position of the stakes was determined using a small Amici device.[6]

The *[stakes]* have been spaced at a distance of 100 meters in the transversal direction.

“ “ in the longitudinal direction.

“ are buried in the ice at approximately 3 feet on July 8th.

“ do not indicate any obvious and indisputable change of place .. . . . on July 16th.

(The distances were measured using a varnished string of circa 150 meters in length. However, the varnish did not protect it entirely from the effects of humidity, etc.*[)]*

Escher separates from the group on July 17th.

“ is found with Mr Studer[7] on the evening of August 10th, at the Grimsel hospice where Mr Agassiz, Forbes, and Heath[8] arrive, chased away from the Aar glacier by the bad weather. The blue and white stripes are discussed extensively.

“ with Studer finds———on August 14th that almost all of the stakes on the Aletsch Glacier have fallen; for most of them even their holes had disappeared. The erosion was thus of at least 3 feet in 29 days. The Föhn wind[9] had reigned powerfully.

“ stunned by this experience, and seeing that a different method should be used to obtain consistent observations on the ways of progression, continues the trip in Valais[10] with Mr Studer.

“ meets on August 29th at Fiesch[11] Mr Agassiz and Desor,[12] who just returned from *[the Ascion]*.

August 31st to September 5th 1841, stay with Agassiz and Desor at the Aar Glaciers, where the position of a certain number of blocks is set, and if I am not mistaken, many were on a line roughly transversal to the glacier.

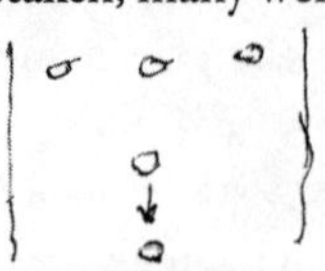

It is most likely during those days, that Agassiz decided to have Mr Wild[13] (now Professor of Geodesy and Topography at the Polytechnic in Zurich) carry out the topographic study of the Aar Glaciers.[14]

There you have it, Mr Tyndall! The notes that I can give you, but please allow me to beg that you not use them in war of priority,[15] wars, which always seem to me to be disagreeable and a disservice to the advancement of science. I will conclude by congratulating you and the rest of us with your brilliant discoveries on the cementation of ice and the veritable origin of the blue and white stripes. And I sincerely thank you for the very informative memoire that you have given me.[16]

Your very devoted friend Escher
Zurich, December 20th 1859.

RI MS JT/1/E/108

---

1. *Charpentier, Essay on the glaciers, Lausanne, 1841:* J. Charpentier, *Essai sur les Glaciers et sur le terrain erratique du Bassin du Rhône,* Lausanne: Imprimerie et Librairie de Marc Ducloux, (1841).
2. *Agassiz, Studies on the glaciers, Neuchâtel, 1840:* L. Agassiz, *Etudes sur les Glaciers* (Neuchatel: Jent et Gassman, 1840).
3. *Agassiz Glacial System:* Agassiz undertook extensive research on glaciers, see letter 1664, n. 1.
4. *Escher, in the month of July 1841 at the Altesch glacier:* Agassiz and Escher undertook research across the Altesch glacier and the Aar glacier, see also letter 1664, dated 7 December, where Tyndall requested Clausius for further details on their work.
5. *Lake [Märjelen]:* conjectured translation from the French *[*Lac de Marial*]*, likely referring to Lac Märjelen, the Märjelensee, on the eastern side of the Aletsch Glacier.
6. *The rectangular position of the stakes was determined using a small Amici device:* i.e Amici Prism.
7. *Mr Studer:* the Swiss geologist Bernhard Studer (1794–1887).
8. *Heath:* likely refers to Josiah Marshall Heath (d.1851), a metallurgist and ornithologist.
9. *The Föhn wind*: a type of warm and dry wind on the side of a mountain. The wind can be very powerful.
10. *Valais:* a canton in Switzerland, see also letter 1604, n. 3.
11. *Fiesch*: a village in the canton of Valais.
12. *Desor:* the German-Swiss geologist Pierre Jean Édouard Desor (1811–22).
13. *Mr. Wild:* see letter 1664, n. 2.
14. *It is most likely during those days, that Agassiz decided to have Mr Wild . . . carry out the topographic study of the Aar glaciers:* Tyndall requested this information in a letter Clausius (see letter 1664).

15. *but please allow me to beg that you not use them in war of priority:* this was in fact what Tyndall was building evidence for. Tyndall was engaged in a dispute with Forbes over glacial motion where he accused Forbes of plagiarising the work of Rendu, and also of misrepresenting Agassiz' work.
16. *I sincerely thank you for the very informative memoire that you have given me:* not identified.

To Sarah Faraday[1] 24 December 1859 1671

Geneva | Christmas eve, 1859

My dear Mrs. Faraday,

I have just dined, feel exceedingly well, and cannot I think adopt a better plan to convert the whole dinner into nutriment than the plan of writing to you. I take the opinion of the celebrated Hufeland[2] on this point, and he affirms that there is nothing better than the company of pleasant friends for promoting health and digestion. I am here at Geneva—I succeeded in crossing the channel on Thursday night—the commotion had almost entirely subsided and nobody that I saw was ill. From Calais half way to Paris we had a good carriage, we then changed, and I found myself with five others, and three footwarmers[3] in the same small compartment. The smell was like the cabin of a badly ventilated ship: but the travellers like so many toads seemed to revel in the noisome atmosphere. They would not open a chink nor let a breath of fresh air into the carriage, and so we were compelled to partake of each others carbonic acid again and again. Having arrived at Paris I drove straight to the Lyons station, and being determined to get rid of the diabolical footwarmers at least I took a second class to Macon.[4] Further than this it was not possible to go. So I had to content myself with the space from London to Macon unbroken by a night's delay. At Macon I slept—and so soundly! The boots almost shook my door off its hinges before he could wake me this morning. The streets of Paris were in a dreadful state of puddle from the melted snow—for 50 miles beyond Paris there was but little snow upon the ground—here and there along the fringes of the river the ice was piled in long barriers, but the country generally was free from it. At Macon however the whole land lay in the stern clasp of winter; between Macon and Geneva for some short distance the covering of snow again lightened but near Geneva and around the town all is thickly covered. The chain of the Jura[5] shining white is quite majestic. I have paid my fare in the diligence to Sallenches,[6] but I was informed[7] to my comfort that there was no communication with Chamouni,[8] and that it was very doubtful whether the snow would permit my reaching it. My informant did not know what a good pair of legs can do when all other modes of convayance fail. I journeyed part of the way with a Frenchman, or a Swiss whose language was French. He had a theory of England which did not please me—England was sinking must sink in accordance with the law of things—she had developed all that she could develop, and must now go down hill. France was now to rise, but when I pushed him as to the powers of development which France possessed and we did not possess he was a little put out. These old Continental shavers imagine that England is like a gouty old man, rich and comfortable, but with a vitality which is fast dying out. I should very much like to have my antagonist with me for half a day upon the glaciers, just to give him a taste of the vanishing vitality of England.

I have seen nobody at Geneva, but to night I purpose calling at the house of Mr. De la Rive. I hope Mr. Faraday is better; previous to leaving I saw Dr. Bence Jones and his opinion was that Mr. Faraday would soon be well—I sincerely trust that nature may do her work of renovation even quicker than Dr. Jones anticipates.

Tomorrow morning (Xmas morning) I start for Sallenches; and shall try and get a sledge from there to Chamouni—There my trouble really commences: but if I can only see across the Mer de Glace, and if the guides at Chamouni be neither cowardly or lazy I hope to get my work done—At all events I shall try and having done my part should I even fail. I shall return content.

A merry Christmas to you and very many of them—I shall make myself as merry as it is possible to be in the interior of a diligence. —good bye.

Ever Yours sincerely | John Tyndall

RI MS JT/1/TYP/12/4093
Faraday Correspondence 5: 3699

---

1. *Sarah Faraday:* (née Barnard, 1800–79), daughter of Mary Boosey and Edward Barnard, a successful London silversmith. She was Sandemanian in her religion and probably met Michael Faraday at the Sandemanian meeting house in Paul's Alley, London. After they married in 1821 she moved into his rooms at the RI. Tyndall frequently dined and socialized with Sarah and Michael Faraday and their niece Jane Barnard.
2. *Hufeland:* likely refers to Christoph Wilhelm Hufeland, a German physician.
3. *Footwarmers:* small stoves designed to keep feet warm (*OED*).
4. *Macon:* Mâcon or Mascon, a small city in central France.
5. *The chain of the Jura:* mountain range by the French-Swiss border.
6. *Sallenches:* a town near Geneva.
7. *I was informed:* may refer to the letter 1668, from Balmat.
8. *Chamouni:* see letter 1633, n. 2.

To Michael Faraday [26 December 1859] 1672

My dear Mr. Faraday,

I wrote to Mrs. Faraday from Geneva,[1] and will now report to you a portion of my subsequent proceedings. I started from Geneva at 8 A.M. on Christmas morning and reached Sallenches[2] a little after 4 P.M. The time at my disposal being so extremely short my aim was to reach Chamouni[3] on the night of Christmas day, in order to commence, weather permitting, my observations on Monday morning.[4] I therefore hired a vehicle at Sallenches and started thence at 5 P.M. Not liking to run a risk, the exact nature of which I did not see, I soon forsook the carriage and placed myself beside the postillion.[5] He was young, strong and resolute, had good eyes, and needed them extremely at times. We had no moon, no lamps, and a dense gray cloud canopy over head cut away from us the feeble light of the stars. As far as Servoz[6] our journey was pleasant enough but afterwards the road was difficult and our progress slow. Deep snow flanked us right and left, which, in the middle of the road, was more or less beaten down, filling the space however with knobs and hummocks. Trotting was here quite out of the question. Having passed the pont Pelissier, a little above Servoz, we alighted to lessen the strain upon the horses, and I went on in advance. Bare, brown, and motionless the trees stood right and left, while the cliffs and precipices, mottled with the snow which clung to their ledges took any form which the imagination choosed to give them. To my left the Arne rushed along a deep ravine, and sent upwards through darkness a sad and broken murmur. Sometimes on coming behind an eminence the sound was suddenly and utterly cut away, and the consequent stillness was solemn in the extreme. It was the silence of a churchyard; and the huge black pines which threw their gloom upon the road seemed like the hearseplumes of a dead world.[7] I walked on till I reached a wooden hut where batons crystals and eaudevie[8] are offered to the passing travellers in summer; leaning against the door I enjoyed for a time the intense sternness of the scene. The sound of the river was here audible, but dim, distant, and melancholy, yet still the voice of life. I was far in advance of my conveyance, and the intermittent tinkle of the horses bells informed me of the progress and the pauses of the vehicle. At the summit of the incline we again mounted and proceeded slowly towards lesouches.[9] We passed several houses, all dark and dismal, loaded with snow and without light or sound of merriment to denote that it was the pleasant Christmas time. The horses seemed affected by a kind of torpor, and leaned listlessly against each other as they crawled along. Vainly the postillion tried with tongue and whip to kindle their enthusiasm: once or twice he succeeded in urging them to a trot down the steeper slopes, but they quickly subsided into the previous monotonous pull. As we approached Chamouni the drowsiness of nature began to operate upon me, and I quitted the carriage within what I supposed to be a quarter of an hour's walk to the village. Previous to this however the wind from the mountains had met us in strong and hostile gusts, as if Mont Blanc wished to warn us from his territory. At the top of the last inclination of the road the surface became more exposed to the action of the wind, and on passing the summit the road disappeared. The wall to the left was quite covered, and a few isolated stones which rose above the snow at intervals to the right were the sole land marks of the right hand wall. The snow, drifted by the storm, had ranged itself in oblique ridges across the road; they could not be seen; there was no light to shew them, but I staggered over four or five of them in succession, sinking kneedeep, until finally on stepping promptly forward to regain a lost equilibrium, I found myself immersed to the waist. The idea occurred to me that I had missed the road, and I vainly looked round in search of some object by which I might check my position. I retraced my steps: the carriage, when I reached it, had got among the ridges, and one of the horses was down; his haunches were immersed in the snow, one of his legs sunk to the shoulder, and the other thrown forward upon the surface. My companion, as I have said, was strong and resolute, and he needed both qualities. In extreme cases I have known guides to give way to sullenness and ill humour, but this man shewed nothing of the kind. He succeeded in getting his horse upon his legs; but being, like myself, doubtful as to whether we had not quitted the road, I went back exploring, and thus satisfied myself that we were right. Afterwards I walked on in front, choosing the harder portions of the snow, while he slowly followed, holding his horses by their heads. After half an hour's struggling we found ourselves in the streets of Chamouni. All seemed dead here also—no sound, no light; the thawing snow splashed as it fell from the loaded eaves, the fountain made a melancholy gurgle; here and there a loosened window shutter, swung creaking in the wind, and banged against the object which limited its oscillations. All was desolation. The Hotel de l'Union so gay and full of life in summer, was nailed up and deserted. We rang the bell at the Royal Hotel, but the deep bark of a watch dog, the resonance of which proclaimed the hollowness of the house was long our only answer. Ringing seemed powerless, so I tried the energy of my bootheel against the door, and by perseverance roused the sleepers. In ten minutes I was seated beside a stove in a comfortable room, listening to the wild harmonies of the storm as it roared through the chimney funnel.

This morning, Monday, the snow falls heavily: tomorrow I shall make an attempt on the Mer de Glace. It is absolutely necessary that I should be able to see across the glacier. If this be not possible I return defeated, but still with the consolation that I have done

"The best my circumstances allows"
Believe me | Ever Yours | John Tyndall

RI MS JT/1/TYP/12/4096
Faraday Correspondence 5:3700

1. *I wrote to Mrs. Faraday from Geneva:* see letter 1671.
2. *Sallenches:* see letter 1671, n. 5.
3. *Chamouni:* see letter 1633, n. 2.
4. *on Monday morning:* 26 December 1859 (*Faraday Correspondence* 5:3700).
5. *postillion:* also spelled postilion, a coach driver.
6. *Servoz:* commune in south-eastern France.
7. *seemed like the hearseplumes of a dead world:* Tyndall used the same phrase in *Glaciers of the Alps* (p. 197).
8. eaudevie: also spelled eau-de-vie, brandy (*OED*).
9. lesouches: presumably Les Houches, which would then have been a small village by the Avre between Sallanches and Chamonix.

To Michael Faraday 27 December 1859 1673

Montanvert, Chamouni[1] | 27 Dec 1859

My dear Mr. Faraday,

At 3 o clock to day just as you were commencing your lecture in Albemarle Street, I drank your health at the Montanvert, and wished you vigour to accomplish your work to your own satisfaction and to the delight of "the juveniles".[2] The wine in which I drank it had to be dug out of the earth where it had been buried to keep it from the hands of an adventurous robber, and also from the sterner hand of the frost. I left Chamouni this morning under the most unpromising auspices it had snowed heavily during the night, and the clouds when I arose gathered all around the mountains obscuring them from top to base. It was also spitting snow. My men came to me at 7 o clock and we agreed to wait till 8, and then decide as to our future movements. The idea of coming all the way to Chamouni and quitting it without at least making an attempt to accomplish my work was extremely repugnant to me, so that when my men came a second time, and appeared rather in favour of the expedition than otherwise, I at once resolved to start. We did so at a quarter before 9 A.M. Two of my men had tried to force their way from the source of the Arverron straight up the mountain to the Montanvert on Thursday last,[3] but found the effort so painful that today they decided on following as close as possible the usual route. For the sake of security I have engaged four men who are to be tied together while ranging out my lines; the crevasses are covered with snow, but some of them only lightly, so the man who fixes my stations must have three others to take care of him. Two additional men were hired to carry our provisions to the Montanvert. Our work, for some time, was comparatively easy: the men however halted after the first three quarters of an hour, attached pattens to their feet.[4] These were pieces of wood, about sixteen inches long and ten wide, which presented a large surface to the snow and prevented the wearers from sinking too deeply. I tried the pattens, but not being accustomed to them rejected them after a brief essay. I had reason to repent of this afterwards. The pine trees were quite majestic: the snow of the preceeding night had fallen through a perfectly still atmosphere and loaded the branches, forming in some cases, a layer at least ten inches deep. The branches drooped under the weight and presented the appearances of arms terminated by vast claws turned downwards; some of the smaller pines with closer branches were quite covered, and presented most fantastic appearances; one in particular was exactly like a huge white parrot with folded wings and drooping head. The average depth of the snow was between 3 and 4 feet; but in some places it exceeded five feet. The men took the lead in turns; the snow was like flour, and the first man laboured through it up to his knees and often to his hips he and the last man viz myself suffered most: he, because he had to break fresh ground, and I because the steps which my precursors made were not always sufficiently rigid to sustain my weight, and I sometimes found myself in a shaft waist deep, from which it required a considerable effort to extricate me. Two or three times one of the men in advance of me sank suddenly, leaving absolutely nothing but his head above the surface of the snow. On the steeper slopes the work was very difficult, for here a certain component of the slope added itself to the true vertical height of the snow. The couloirs were encumbered with the debris of avalanches, and on one open space we worked upwards for three quarters of an hour over the ruins of an avalanche that had fallen the night before. We choose it, because the snow of the avalanche was more compact than the fresh snow which it had swept away in its downward rush. At several different places my attention was excited by a dull booming sound like the report of a distant gun: it was caused by the breaking asunder of the layer of snow, which had been for some time previous in a state of strain, the forward portion being held back by its attachment to the portion behind. We actually in several instances saw the crevasse formed at the moment that the sound was heard. The weather improved as we ascended, and some of the cloud effects were wonderful: language is a mere lumbering vehicle to describe the magnificence of their arrangement. They shifted incessantly: sometimes leaving the grandest summits drawn against the deep blue heaven, sometimes obscuring all above and casting an angry indigo-blue light upon the snow and pines. The Aiguille du Dru was noble in the extreme—half its mighty cone strongly illuminated, and half in shade; while the light snow powder scattered upon its ledges was caught by the wind and unfurled itself as a banner which streamed far southward from the mountain head. We reached the Montanvert at 2 P.M. after five hours and a quarter of extremely hard work; I in a ferocious state of hunger as I had not taken any food during the journey. The Mer de Glace was quite glorious, not as in summer wasted and dirty, but pure and white with its frozen billows steep, high, and sharply crested. It was all in shadow save the portion opposite to the tributary which descends from the Col de geant; this tributary being 14 degrees from the meridian, at 2 o clock the sun beams gushed straight down the corridor of le geant, ruled the Mer de Glace with a transverse beam and flooded the Aiguille de Morne with radience from base to summit. Having swept the snow from the floor of our domicile a large pine fire was made beside which I now sit. A thermometer at 8 feet distance from the fire still shews a temperature of half a degree centigrade under zero. I have just placed a minimum thermometer on a chair which I have fastened in the snow, and tomorrow morning I shall be able to state the lowest temperature of the night. The stars are shining, and the north wind an hour ago shook all the windows and made some of the panes perfectly musical. It has now subsided and I have strong hopes of being able to accomplish what I intended. I have given you but a feeble idea of the interest of this day's experience. The frozen figures on the window panes of the Montanvert Auberge are the most beautiful things imaginable, and when I place my fingers against the glass and melt the frozen layer the recrystallization of the liquid water, looked at through a pocket lens, is perfectly exquisite. It gladdens one's very heart to see those beauties: Some artist proudly remarked on viewing a fine picture "I too am a painter"[5]—Is it presumption if when looking at those wondrous productions, and feeling a certain affinity with them one should bless the gods for being a natural philosopher?

Good bye | Ever Yours | John Tyndall

RI MS JT/1/TYP/12/4100
Faraday Correspondence 5: 3701

---

1. *Chamouni*: see letter 1633, n. 2.
2. *accomplish your work to your own satisfaction and to the delight of "the juveniles".*: this refers to the RI Christmas lecture, which had been postponed until 31 December 1859. Published as, *A course of six lectures on the various forces of Matter, and their relations to each other,* London and Glasgow: Richard Griffin and Company, (1860) (*Faraday Correspondence, 5:3701*).
3. *on Thursday last*: 22 December 1859 (*Faraday Correspondence, 5:3701*).

4. *attached pattens to their feet*: a type of overshoe (*OED*), this particular type appears to be similar to a snowshoe.
5. *"I too am a painter"*: a saying attributed to the Italian painter Antonio Allegri Corregio (*c.* 1494–1534), upon seeing Raphael's (1483–1520) painting 'The Ecstasy of St. Cecilia'.

From Rudolf Clausius 28 December 1859 1674

Zürich, 28/12/59.

Lieber Tyndall

Ich habe die Erkundigungen welche Sie wünschen eingezogen, und ich muss Sie nur um Entschuldigung bitten, dass ich Ihnen nicht schon früher geschrieben habe; aber es machte einige Umstände, die genauern Data von den 3 Herrn zu erhalten.

Der erste Versuch einer Messung der verschiedenen Bewegung ist von Escher allein, welcher im Juli 1841 Pfähle in den Aletschgletscher eiinschlagne liess bis auf die Tiefe von etwa 1 m., um sie im August wieder zu beobachten. Er fand sie aber bei seinr Rückkunft umgefallen. Im September desselben Jahres 1841 bestimmte er mit Agassiz die Lage von einer Anzahl von Blöcken (N.B. Sehe am Schluss) auf dem Aare-gletscher. Im Frühlung 1842 forderte Agassiz Herrn Wild auf, die nunmehrige Lage der Blöcken zu bestimmen, und eine Aufnahme des Gletschers zu machen. Im Juli 1842 begab sich Herr Wild mit Agassiz, Descor und anderen am Ort und Stelle un machte die Messungen.

Die Ansicht dass die Gletscher sich am Rande schneller bewegen als inder Mitte, ist vor Agassiz von Charpentier ausgesprochen.

Herr Escher hat Ihnen eine kurze Zusammenstellung der fraglichen Untersuchungen mit dem Zeitangaben gemacht,welche ich beilege.

Die Betrachtungen von Mayer uber d. mech. Wärmetheorie finden sich, so mir bekannt, im folgenden Orten.

(1) Eine Abhandlung in den Ann. de Chem. u. Pharm. von Liebig u. Wähler B.XLII seite 233–40 1842)
(2) ein eigene Schrift: die organsiche Bewegung in ihrem Zusammenhange mit der chemischen Wirkung. Heilbronn 1845.
(3) eine Note in den Pariser Compt. rend. T.XXVII p. 385. 1848.
(4) eine eigene Schrift: Bemerkungen über das mech. Aequivalent der Wärme. Heilbronn 1851.

Ich habe mich im November verheirathet. Meine Frau verbindet ihren Dank mit dem meinigen für Ihre freundliche Gratulation zu unserer Verlobung. Sie wünscht sehr, dass sich bald eine Gelegenheit finden möge, Sie kenne zu lernen.

Dass ich unter diesen Umständen gegenwürtig nicht viel wissenschaftlich arbeite, können Sie Sich wohl denken. Ich hoffe aber mit der Zeit wieder im Zug zu kommen.

Indem ich Ihnen noch für Ihr letzte schäne Abhandlung über die Gletscher sehr danke, verbleibe ich freundschaftlich

Ihr | Clausius.

N.B. Soeben erfahre ich von Hrn Escher auf meine Auffrage, dass am Anfange des September 1841 auf dem Aare-Gletscher nicht blos die Lage von Blöcken bestimmt ist, sondern auch Reihen von Pfählen quer über den Gletschern eingesenkt sind, 10 Fuss tief, um im nachsten Jahre die Krummerung der Linien, in welchen die Pfähle standen zu beobachten.

Zürich, 28/12/59.

Dear Tyndall

I made the enquiries you[1] requested,[2] and I must apologize for not writing to you sooner; but it was somewhat troublesome to obtain the exact data from the 3 gentlemen.[3]

The first attempt to measure the different movements was done by Escher alone. In July 1841 he hammered posts into the Aletsch glacier to the depth of ca. 1 m, with the intention of monitoring them again in August. On returning, however, he found that they had fallen over. In September of the same year 1841 he and Agassiz determined the position of a number of blocks on the Aare glacier (n.b. see end of letter). In the spring of 1842, Agassiz asked Mr. Wild[4] to determine the eventual position of the blocks and to take a picture of the glacier. In July of 1842, Mr. Wild went with Agassiz, Descor[5] and others to take measurements on the spot.[6]

It was Charpentier who expressed the opinion[7] that glaciers move more quickly on the edges than in the middle before Agassiz did.

I have included a brief list of the relevant studies with time data that Mr. Escher compiled for you.

As far as I know, Mayer's reflections on the Mechanical Theory of Heat can be found in the following places:

(1) a study in the Ann. de Chem. u. Pharm from Liebeig + Wöhler Vol. XLII page 233–40 1842)[8]

(2) a separate paper: die organische Bewegung in ihrem Zusammenhange mit der chemischen Wirkung, Heilbronn 1845.[9]

(3) a note in the Pariser Compt. rend. Vol. XXVII p. 385. 1848.[10]

(4) a separate paper: Bemerkungen über das mech. Aequivalent der Wärme. Heilbronn 1851.[11]

I married in November. My wife[12] and I both thank you for your friendly congratulation to our engagement. She hopes very much that there will soon be an opportunity to become acquainted.

I am sure that you can think for yourself that I am not doing very much scientific work in this situation. I hope however to come back to it again in time.

I thank you very much for your last, wonderful paper on the glaciers[13] and send my fond regards, | Your | Clausius

N.B. I have just heard from Mr. Escher in regard to my query that at the beginning of September 1841 not only the position of blocks on the Aare glacier was determined, but that a row of posts was also sunk across the glacier, 10 feet deep, in order to enable an observation of the curvature of the lines in which the posts had stood in the next year.

RI MS JT/1/TYP/7/2236–37
LT Typescript Only[14]

---

1. Clausius used the formal version of 'you' (Sie).
2. *I made the enquiries you requested:* letter 1664, n. 1.
3. *the 3 gentlemen:* Tyndall requested information from Arnold Escher von der Linth, Johannes Wild, and Albert Mousson, see letter 1664.
4. *Mr. Wild:* see letter 1664, n. 2.
5. *Descor:* Desor, see letter 1670, n. 12
6. *to take measurements on the spot:* in the spring of 1842, Agassiz asked Mr. Wild to determine the eventual position of the blocks and to take a picture of the glacier. In July of 1842, Mr. Wild went with Agassiz, Désor and others to take measurements on the spot, see letter 1664.
7. *It was Charpentier who expressed the opinion:* see J. G. F de. Charpentier, *Essai sur les glaciers et sur le terrain erratique du bassin Rhône* (Lausanne: 1841).
8. *a study in the Ann. de Chem. u. Pharm from Liebeig + Wöhler Vol. XLII page 233–40 1842:* refers to J. R. Mayer, 'Bemerkungen über die Kräfte der unbelebten Natur', *Annalen der Chemie und Pharmacie,* 42 (1842), pp. 233–40.
9. *die organische Bewegung in ihrem Zusammenhange mit der chemischen Wirkung, Heilbronn 1845:* refers to J. R. Mayer, *Die organische Bewegung in ihrem Zusammenhang mit dem Stoffwechsel. Ein Beitrag zur Naturkunde,* (Heilbronn: 1845).
10. *a note in the Pariser Compt. rend. Vol. XXVII p. 385. 1848.:* J. R. Mayer, 'Sur la transformation de la force vive en chaleur et réciproquement', *Comptes Rendus,* 27 (1848), pp. 385–87.
11. *Bemerkungen über das mech. Aequivalent der Wärme. Heilbronn 1851:* refers to J. R. Mayer, *Bemerkungen über das mechanische Äquivalent der Wärme,* (Heilbronn: 1851).
12. *My wife:* Clausius married Adelheid Rimpham on 13 November 1859, see also letters 1628, n. 3, and 1802, n. 5.

13. *I thank you very much for your last, wonderful paper on the glaciers:* Tyndall published three papers on glaciers in 1859; J. Tyndall, 'On the Physical Phenomena of Glaciers—Part I. Observations on the Mer de Glace', *Phil. Trans.*, 149, (1859a): 261–78; J. Tyndall, 'Remarks on Ice and Glaciers', *Phil. Mag.* 17, (1859b): 91–96; and J. Tyndall, 'On the Veined Structure of Glaciers; with Observations upon White Ice-Seams, Air-Bubbles and Dirt-Bands, and Remarks upon Glacier Theories', *Phil. Trans.*, 149, (1859c): p. 279–307.
14. *LT Typescript Only:* the handwritten corrections to the typescript appear to be in John Tyndall's hand, and it is likely that he edited parts of German transcription.

1860

To John Murray [6 January] 1860[1] 1675

Royal Institution | Friday 2½ P.M.

Dear Mr. Murray

Your note[2] has just fallen into my hands, and I regret to say that I was bound before it reached me.

I have just made a winter expedition to the Mer de Glace,[3] and in the face of no small difficulty have succeeded in determining the winter motion along two transverse lines, including twenty-one different points. —The winter observations are thus raised to the same degree of accuracy as that already proposed by the summer ones.

The aspect of the glacier and its surroundings was grand and singular— Indeed irrespective of science it was worth while going all the way to the Montanvert to see it.

I have recommenced writing for Mr Clowes[4] & expect have the whole book in his hands by the middle of March.[5]

Very sincerely yours | John Tyndall.

National Library of Scotland, John Murray Archive, Ms. 41214

---

1. *[6 January]* 1860: see n. 3, which implies a date in early January 1860. The first Friday in January 1860 was the 6th.
2. *Your note*: note missing.
3. *I have just made a winter expedition to the Mer de Glace*: Tyndall's visit was a quick one to fit between his other commitments. He left London on 22 December 1859 and was back on New Year's Day 1860.
4. *Mr Clowes*: William Clowes & Sons, founded in 1803, were John Murray's printers. See also letter 1695.
5. *expect have the whole book in his hand by the middle of March*: the manuscript of *Glaciers of the Alps,* see also letter 1595, n. 4.

To Mrs Buxton[1] [25] January 1860[2] 1676

My dear Mrs Buxton,

You must I think have noticed my hesitation last night[3] when you asked me to take Miss D.[4] to dinner. Possibly my taciturnity during dinner may also have attracted your notice[5]—Will you allow me to write two words in explanation of this?

For a considerable time I was honoured by the intimacy of Mrs D. but it is now almost a year since that intimacy ceased, and it is perfectly understood between Mrs D. and myself that it is never to be renewed.

You could not have known this when your kindness prompted you to make an arrangement which under other circumstances would have been so pleasant to me; but which, as it was, placed me in a position of no small difficulty.

If I dared I would ask you to regard this as a confidential note: it is indeed many months since the name of Mrs D.[6] has escaped either my lips or pen, and it should not do so now were I not extremely anxious to prevent the recurrence of an incident which must have been the reverse of agreeable to those whom I should be sorry to annoy.

Believe me dear Madam | most truly yours | J.T.

RI MS JT/2/13c/1184

---

1. *Buxton*: likely Emily Mary Buxton (née Holland, 1824–1908), wife of Charles Buxton (1822–71).
2. *[25] January 1860*: this letter was recorded in Tyndall's journal for Saturday 28 January 1860 together with letter 1638 (Journal, RI MS JT/2/13c/1184–85). In the Saturday 28 January journal entry, Tyndall recorded his activities for Monday, Tuesday, Wednesday, Thursday and Friday. The facsimile of these two letters were included between Tuesday and Wednesday. Tyndall wrote that, 'On Tuesday I wrote my epitome and experimented with Gassiot. I dined at Buxton's'. It is therefore likely that this letter was written on Wednesday 25 January 1860, and letter 1677 was written on Thursday 26 January.
3. *last night*: Tyndall attended a dinner at the Buxton's, which was also attended by Mary Drummond. Tyndall courted Mary for a period in 1858 (see for example letters 1564 and 1576 to Hirst in Volume 6), but their relationship ended badly (see letter 1604).
4. *Miss D.*: Mary Drummond.
5. *Possibly my taciturnity during dinner may also have attracted your notice*: Tyndall recorded this letter and his thoughts about the dinner in his journal entry for Saturday 28 January 1860. He noted that 'Though I did not think that Mrs Buxton at all knew the true state of things: I thought she might have known a little and that the attempt was kindly

made to heal old wounds. Not this mode of healing however suited me'. (Journal, RI MS JT/2/13c/1184).

6. *Mrs D*: Maria Drummond, mother of Mary.

To Mrs Buxton[1] [26] January 1860[2] 1677

My dear Mrs B.

I am very much obliged to you for your note[3]—will you allow me to say one word in addition to what I have already said?[4] Do not let my note[5] cause you to think unkindly either of Mrs D.[6] or me. The difference between us[7] is the result of a series of unfortunate mistakes on both sides. I do not entertain the slightest hostile feeling towards her—I would do any thing to render her a service, and if I were in need of her good offices I believe she would not refuse them. You would do me a great favour to burn these two notes, and I feel perfectly at rest as to your keeping their contents entirely to yourself.[8]

J.T.

RI MS JT/2/13c/1185

1. *Buxton*: likely Emily Mary Buxton (née Holland, 1824–1908), wife of Charles Buxton (1822–71), see also letter 1676.
2. *[26] January 1860*: see letter 1676, n. 1.
3. *your note*: the note is missing, but Tyndall recorded in his journal for Saturday 28 1860, that 'A very kind note from Mrs Buxton on the following day assured me that she was perfectly innocent of any knowledge of our relations'. (Journal, RI MS JT/2/13c/1184).
4. *what I have already said*: refers to his summary of events in letter 1676.
5. *my note*: letter 1676.
6. *Mrs D*: Maria Drummond, mother of Mary.
7. *The difference between us*: see letter 1604, n. 7.
8. *keeping their contents entirely to yourself*: Tyndall recorded in his journal for Saturday 28 January 1860 that he had discussed the matter with Faraday (Journal, RI MS JT/2/13c/1185).

To Juliet Pollock [late January or early February 1860][1] 1678

My dear M[rs]. Pollock

Do you think that boy Walter[2] would care anything about the axe with which I ascended Monte Rosa?[3] If so I would present it to him as a Valentine.

Ever yours | J. Tyndall

RI MS JT/1/T/1229
RI MS JT/1/TYP/6/1966

1. *[late January or early February 1860]*: the year is suggested by LT, the month by the context of Valentine's Day.
2. *boy Walter*: Walter Herries Pollock. His tenth birthday was on 21 February 1860. See letter 1744.
3. *the axe with which I ascended Monte Rosa*: Tyndall ascended Monte Rosa twice, a week apart, in 1858. On the second occasion, making the first solo ascent, he dropped his ice axe at the summit and only realised on the descent what a disaster it would have been had it slid out of reach. See *Glaciers of the Alps,* p. 160.

From Edward Frankland 05 February 1860 1679

42 St. John's Park Road | Haverstock Hill N.W. | Feb. 5/60.

My dear Tyndall

Herewith I enclose[1] the promised experimental results. When my paper on the Air of Mont Blanc is printed,[2] I will send you a copy: you ought to have come in to the Chemical[3] & heard it, instead of listening for nearly 2 hours to the anything but sweet or honeyed liver-sugar controversy.[4]

In a recent experiment I found a candle to burn actually faster with the barometer at 9 inches than under atmospheric pressure. I am having an apparatus constructed to burn candles in air at 5 atmospheres.

Sincerely yours | E. Frankland.

RI MS JT/1/F/48
RI MS JT/1/TYP/12/3959

1. *Herewith I enclose*: not identified. LT added the note, 'Sending experimental results on air'.
2. *When my paper on the Air of Mont Blanc is printed*: printed as E. Frankland 'On the Composition of Air from Month Blanc', *Quarterly Journal of the Chemical Society of London,* 13 (1861), pp. 22–30.
3. *the Chemical*: the Chemical Society of London.
4. *For nearly 2 hours to the anything but sweet or honeyed liver-sugar controversy*: likely referring to the ongoing debate on the effect of sugar on the liver, and the manifestation of diabetes. It is uncertain what specific talk Frankland was referring to.

From Charles Frédéric Martins[1] 15 February 1860 1680

Montpellier le 15 février 1860

Monsieur,

Je vous remercie de l'envoi de votre mémoire sur les phénomènes physique des glaciers. Il m'a vivement intéressé. La théorie de la plasticité des glaciers de M. Forbes m'avait toujours parue insoutenable comme, je crois, à tous ceux qui ont vécu sur les glaciers. Vos belles expériences sur la glace ont résolu une partie du problème. J'ai été heureux de voir que vous avez refait l'expérience pour voir si le fond du glacier marche moins vite à cause du frottement en enfonçant des piquets à diverses hauters comme je l'avais fait sur le petit glacier du Grünberg au dessus de celui de l'Unteraar et non en employant la méthode compliquée et sujette à correction du Professeur Forbes.

Votre explication des bandes bleues me parait très admissible. Agassiz l'avait pressenti en 1847 (Nouvelles Etudes et expériences sur les glaciers actuels p. 301 quand il dit qu'elles se multiplient surtout dans les endroits où la pression est à son maximum. Mais je ne pense pas que toutes les bandes bleues se forment de cette manière.

J'ai l'honneur de vous envoyer quelques brochures parmi lesquelles se trouve un mémoire intitulé Nouvelles Observations sur le glacier du Faulhorn qui a paru dans le bulletin de la Société géologique de France en 1845. J'ose vous prier de lire de la page 15 à la page 27. Cet exemplaire, le dernier que je possède, n'a point de planche ; celle-ci se trouve bulletin de la Société géologique 2è série 8 II p. 223. En expérimentant sur le glacier du Faulhorn j'étudiais pour ainsi dire l'Embryologie des glaciers, tous les phénomènes sont plus simples, plus visibles, plus rapprochés je vous *[*engage*]* beaucoup à les visiter l'année prochaine.

Le jour où vous faisiez l'ascension du mont blanc j'étais avec deux demoiselles anglaises sur l'Oldenhorn montagne du Canton de Berne elevée de 3120 mètres. Je suppose que le vent vous a empêché de séjourner au sommet. Nous avions ce projet en 1844 avec mes amis *[*Bravais*]* et *[*Lepibus*]* les froids terrible de la nuit nous a empêché de l'exécuter nous avons seulement séjourné trois jours au grand plateau. N'avez-vous aucune trace des piquets de notre *[1 word illeg]* que nous y avions laissés ils doivent être maintenant dans le glacier de Bossons et de Taconnay. Vous avez veux-je penser un exemplaire de mon mémoire sur le Froid thermométrique où je résume nos expériences du Mont Blanc. Je vous envoie les séries originales qui pourraient vous intéresser. Je serai heureux d'apprendre l'année prochaine que vous aurez rénové ; tâchez de faire l'ascension en juillet et de trouver des jours calmes. Je fais des vœux pour votre succès.

Veuillez agréer je vous prie Monsieur l'assurance de ma considération la plus distinguée.

Ch. Martins

Professeur d'histoire naturelle à la Faculté de Médecine de Montpellier.

Montpellier February 15th 1860

Dear Sir,

I thank you for sending me your memoir on the physical phenomena of glaciers. I was greatly interested in it. Mr Forbes' theory on the plasticity of glaciers always seemed untenable to me as to, I think, all those who lived on the glaciers. Your great experiments on the ice have solved a part of the problem.[2] I was pleased to see that you have done the experiment again in order to see if the bottom of glaciers moves less quickly because of the friction by putting poles at different heights like I did on the little Grünberg glacier above that of the Unteraar and not using Professor Forbes's complicated method of correction.

Your explanation of blue stripes seems very likely to me. Agassiz gave it in 1847 (Nouvelles Etudes et expériences sur les glaciers actuels[3] p.301 when he says that they increase especially in places where the pressure is at a maximum. However I do not think all the blue stripes are formed this way.

I have the honour of sending you some brochures among which is a memoir called Nouvelles Observations sur le glacier du Faulhorn[4] which appeared in the newspaper of the geological Society of France in 1845. I venture to suggest that you read from page 15 to page 27. This copy, the last one I have, has no plate; which can be found in the newspaper of the geological Society 2nd series 8 II p. 223. By experimenting on the Faulhorn glacier I was studying the embryology of glaciers, all phenomena are more simple, more visible, closer, and I encourage you to visit them next year.

The day you were climbing the Mont Blanc I was with two young English ladies[5] on the Oldenhorn mountain of the Canton of Berne at a height of 3120 metres. I suppose the wind prevented you staying on the summit. We had that plan in 1844 with my friends *[*Bravais[6]*]* and *[*Lepibus[7]*]* the terrible cold during the night prevented us from achieving it and we only stayed for three days on the Grand Plateau. Did not you see any marks of the poles of our *[1 word illeg]* that we had left there, they must be in the Bossons or Taconnay glaciers now. You have, I think, a copy of my memoir on thermometric cold[8] in which I summarize our experiments on the Mont Blanc. I send you the original series which might interest you. I would be happy to hear next year that you would have tried again; try to climb in July and during quiet days. I wish you success.

Please, Sir, accept the assurance of my highest consideration.
Ch. Martins
Professor of natural history at the Faculty of Medicine in Montpellier

RI MS JT/1/M/50

1. *Charles Frédéric Martins:* (1805–89), a French botanist and geologist at the University of Montpellier. The letterhead listed Martins' address as Plants Garden of Montpellier (Jardin des Plantes de Montpellier). See also letter 1687.
2. *Your great experiments on the ice have solved a part of the problem:* refers to Tyndall's explanation for glacier motion, see letter 1594, n. 3.
3. *Nouvelles Etudes et expériences sur les glaciers actuels:* refers to Agassiz' book entitled *Nouvelles études et expériences sur les glaciers actuels: leur structure, leur progression et leur action physique sur le sol* (Paris: Victor Masson, 1847). Translated from French the title means 'New studies and experiments on present-day glaciers'.
4. *Nouvelles Observations sur le glacier du Faulhorn*: Refers to Martin's article 'Nouvelles Observations sur le glacier du Faulhorn', *Bulletin de la Société Géologique de France,* 2 (1845), pp. 223–57. Translated from French the title means 'New Observations on the Faulhorn glacier'.
5. *two young English ladies:* not identified.
6. *Bravais:* not identified.
7. *Lepibus:* not identified.
8. *my memoir on thermometric cold*: likely refers to Martin's *Du Froid Thermométrique, et de ses relations avec le froid physiologique dans les plaines et sur les montagnes* (Montpellier: Boehm, 1859).

From Edward Sabine 20 February 1860 1681

My dear Tyndall,

It was in the summer of 1841, at the Lower Grindelwald Glacier,[1] that I first saw, and was greatly impressed and interested by examining and endeavouring to understand (in which I did not succeed), the veined structure of the ice. I do not remember when I mentioned it to Forbes, but it must be before 1843, because it is noticed in his book,[2] p. 29. I had never observed it in the glaciers of Spitzbergen[3] or Baffin's Bay,[4] or in the icebergs of the shores and straits of Davis[5] or Barrow.[6] I feel the more confident of this, because, when I first saw the veined structure in Switzerland, my arctic experience was more fresh in my recollection, and I recollected nothing like it.

Veins are indeed not uncommon in icebergs, but they quite resemble veins in rocks, and are formed by water filling fissures and freezing into blue ice, finely contrasted with the white granular substance of the berg.

The ice of the Grindelwald Glacier (where I examined the veined structure) was broken up into very large masses, which by pressure had been upturned, so that a very poor judgment would be formed of the direction of the veins as they existed in the glacier before it had broken up.

Sincerely yours, | Edward Sabine. | Feb. 20, 1860.

Glaciers of the Alps, p. 378

---

1. *Lower Grindelwald Glacier*: see letter 1612, n. 2.
2. *his book*: refers to Forbes's book *Travels through the Alps of Savoy* (see letter 1594, n. 13).
3. *Spitzbergen*: island in the Arctic Ocean.
4. *Baffin's Bay*: Sea between Baffin Island and Greenland.
5. *straits of Davis*: strait located west of Greenland.
6. *Barrow*: Strait in Northern Canada, connected to Davis Strait through Lancaster Sound.

To Heinrich Debus 23 February 1860 1682

23rd Feb. | 1860

My dear Heins,

I want you to tell me whether it was not Berthelot[1] who first produced organic bodies out of inorganic.[2] You remember the speculations which we indulged in on the banks of the Dinan in Brittany[3]—I am about introducing some similar thoughts into my book,[4] referring of course to the relationship of organic & inorganic, so I want to know whether it was not Berthelot who first broke down the division wall between both.

I think you & Tom[5] & myself ought to have a rum somewhere during the Easter holidays[6]—I should like much to take a race through North Wales and to climb Snowdon & other mountains—would that be too far for you?—It would do us all three a great deal of good.

Seven of my lectures[7] are over today—five only are now to come. I am working very hard at my book[8] and I hope when its ready I shall have no occasion to be ashamed of it.

Simpson[9] is here at present and purposes working in some of the laboratories for some time—in which I do not know.

Now goodbye Heins: I wish you success in all your undertakings—the longer I live however the more I feel what a grand thing patience is. A man with brains and patience can conquer any thing—It is really what a wonderful power we obtain over a subject by continuing to look at it.

Goodbye Heins | & believe me | Yours most Sincerely | John Tyndall

RI MS JT/1/T/259
RI MS JT/1/TYP/2/366

1. *Berthelot:* Pierre Eugène Marcellin Berthelot (1827–1907), a French chemist best known for his work in synthetic organic chemistry.
2. *who first produced organic bodies out of inorganic:* Friedrich Wöhler is typically credited with the first instance of producing an organic compound from an inorganic. Wöhler synthesised urea from the inorganic materials ammonium cyanate in 1828, a process which is known as the Wöhler synthesis. However, Wöhler's synthesis was neither a definitive blow to vitalism, nor did it unify organic and inorganic chemistry. For a brief overview of Wöhler's synthesis and its role in the development of organic chemistry, see Peter Ramberg 'Myth 7: That Friederich Wöhler's synthsis of urea in 1828 destroyed vitalism and gave rise to organic chemistry' in Ronald Numbers and Kostas Kampourakis (eds), *Newton's Apple and Other Myths about Science,* (Cambridge MA: Harvard University Press, 2015) pp 59–66. In the late 1850s, Berthelot formulated an ambitious program to synthesize all organic compounds, and published *Chimie organique fondée sur la synthèse,* 2 vols (Paris: Mallet-Bachelier, 1860).
3. *Dinan in Brittany:* Dinan is a small town in Brittany, France.
4. *I am about introducing some similar thoughts into my book:* May refer to what later became the book, *Heat as a Mode of Motion* (1863), where Tyndall addressed the topic of organic structures.
5. *Tom:* Thomas Archer Hirst.
6. *to have a rum somewhere during the Easter holidays:* according to Hirst's journal, Hirst was very sick over the Easter period and Tyndall was also unwell. (*Hirst Journal*, RI MS JT/2/32c/1531–33).
7. *my lectures:* Tyndall lectured on 'Light' at the RI in 1860.
8. *my book:* likely refers to *Glaciers of the Alps* published in 1860.
9. *Simpson:* likely James Simpson (1811–70); see letter 1648, n. 3.

To Charles Babbage[1] 28 February 1860 1683

The following are among the names already sent in[2]:—

John Stuart Mill, Esq.
George Grote, Esq.
Right Hon. Lord Stanley, M.P.
Charles Darwin, Esq., F.R.S., F.L.S., F.G.S.
Prof. Huxley, F.R.S., F.L.S., F.G.S.
Neil Arnott, Esq., M.D., F.R.S.
Sir Charles Lyell, F.R.S., F.L.S., F.G.S.
R.G. Latham, Esq., M.D., F.R.S.
J.D. Hooker. Esq., M.D., F.R.S., F.L.S., F.G.S.
Prof. Tyndall, Ph D., F.R.S.
Sir John Trelawney, Bart., M.P.
Prof. Busk, F.R.S., F.G.S., F.L.S.

Dear Mr Babbage

M[r] Spencer has worked much to fill this page with names of weight, now he has asked me to write a line to ask you whether you could add yours:—You may trust him for thorough liberality of thought.

Ever yours | J. Tyndall | 28th Feb. 1860

BL, Add MS 37197, ff. 504b

---

1. *Charles Babbage*: the mathematician and computer pioneer Charles Babbage (1791–1871). Babbage held the Lucasian professorship in mathematics at Cambridge from 1828 to 1839, and was both a highly influential and controversial figure.
2. *The following are among the names already sent in*: several prominent figures signed their names to a letter dated March 29, 'THE BRITISH MUSEUM', *The Times* (12 April 1860) p. 10 (letter 1690) arguing against a restructure of the British Museum.

To Juliet Pollock [February 1860][1] 1684

Saturday

My dear M[rs]. Pollock

I thank you very much for permitting me to send you the proofs:[2] what I want you to do for me is, not to correct the proofs, or to give yourself any trouble about verbal inaccuracy; which I shall ferret out myself, but to give me your advice about the style. My aim is to keep my eye closely upon the things which I attempt to describe; still I may be tedious, or bald, or turgid, and if I am any of these I want you to correct me.

I send you two pieces—the Introduction, and a little of the early part of the expedition of 1857.[3] I will afterwards send you the narratives of the ascents and a little of the more purely scientific portion—this will enable you to judge of the temper of the thing, and I shall be extremely obliged to you for the 'free & fearless utterance of your opinion'—do not spare me in any way: I can bear a terrible amount of plucking.

I have an idea of your present residence—five years ago I met a friend on the Railway platform[4] at Conway[5] who wished me to go to Llandudno[6] to dine with him. But I didn't go—next day I went up the 'foaming flood' of the Conway,[7] and had a choice dinner of trout at Llanrwst.[8] What a glorious country it is!—I will make myself acquainted with every nook corner & summit in Great Britain and Ireland before I go to the Alps again.

I have lectured four times this week. my audience is far less terrible than that of the R.I.[9] 20 or 30 students—who I am happy to see feel an interest in what I tell them

—Thank M[r] Pollock for his kind & pleasant letter to me.[10] Good bye most truly yours | J. Tyndall

I have not yet broken up the book into sections—intending to wait until it is all printed.

RI MS JT/1/T/1214
RI MS JT/1/TYP/6/1975

1. This letter has been dated to February 1860, based on when Tyndall gave the School of Mines lectures and mentions of his proofs, probably *Glaciers of the Alps.*
2. *permitting me to send you the proofs:* Tyndall asked Juliet Pollock to read parts of *Glaciers of the Alps;* see letter 1642.
3. *and a little of the early part of the expedition of 1857:* likely refers to what would become chapter 4 of *Glaciers of the Alps.*
4. *five years ago I met a friend on the Railway platform:* Tyndall recorded in his diary for 27 September 1854 that he met a friend named Latham at the railway station in Conway (Journal, RI MS JT/2/13b/692). This was likely the philologist and physician Robert Gordon Latham (1812–88).
5. *Conway:* also known as Conwy, a town in Wales.
6. *Llandudno:* town by the sea in Wales.
7. *'foaming flood' of the Conway:* a reference to a poem 'The Bard' by Thomas Gray (1716–71).
8. *Llanrwst:* town in the Conwy Valley.
9. *I have lectured four times this week. my audience is far less terrible than that of the R.I.:* Tyndall took up teaching at the School of Mines in 1859.
10. *for his kind & pleasant letter to me:* letter missing.

To Heinrich Debus [late February to early March 1860][1] 1685

Monday[2]

My dear Heinrich,

You are not 'schuldig'[3]—it was the want of time that prevented me from writing—I have been compelled to make a hermit of myself, and not to go out anywhere for several weeks; and until the 20th of March I am determined to adhere to this rule. I am compelled to do this order to get my book done, and I hope to have it very nearly accomplished by the time that I have mentioned.

Meanwhile I am devoting to it every hour that I can command—when finished I think there will be some parts of it at least which will meet your approbation.

The death of Mrs Hofmann[4] is one of the saddest things that I have heard of for some time—it affected me very deeply—as you say there is something bewildering in the thought of such youth, beauty & apparent strength being so suddenly cut away.

I am rejoiced to hear such good news from Queenswood.[5] I hope it will continue to prosper.

Tom & I had a long walk yesterday:[6] he is very well & busy with his mathematics.

A bookseller here wishes him to edit a French work on prizms—regarding which however he has not yet made up his mind.[7]

If I can I will join you for a race at Easter—to the Isle of Wight or elsewhere—before that as you say, my leaving London is quite out of the question

Believe me dear Heins | Ever yours | John Tyndall

RI MS JT/1/T/330

---

1. *[late February to early March 1860]*: this a conjectured date based on the discussion of Mrs Hofmann, who died 30 January 1860, and Tyndall's decision to 'make a hermit of myself' in the several weeks leading up to 20 March.
2. The letter included a marginal note summary of the letter, possibly added by LT: 'Shall be a hermit till 20th March. My book *[compels]* this *[talk]* of *[Mrs]* Hofmann. Tom & I *[took a]* walk yesterday—*[good humoured and pleasing]* | 1859'.
3. *schuldig*: German, translates to 'guilty'.
4. *The death of Mrs Hofmann*: Rosamond Hofmann (née Wilson), born 11 February 1838 died 30 January 1860. Rosamond was August Hofmann's second wife.
5. *I am rejoiced to hear such good news from Queenwood*: the letter with news from Debus is missing.
6. *yesterday*: a Sunday, uncertain date. Tyndall and Hirst spent much time together in this period.
7. *A bookseller here*: not identified.

To Juliet Pollock [late February to early March 1860][1] 1686

My dear Mrs Pollock

I am sorry to confess to <u>you</u> the existence of the word 'impossible' in my vocabulary. But some time ago the pressure of my work was so heavy, the completion of my book[2] so doubtful that in my extreme need I swore by Zeus that excepting on Sundays, until the 20th of March I would not dine out. I am now bound by this chain. and it must be borne until the performance of my vow enables me to shake it off.

Ever yours | <u>John Tyndall</u>

<u>Monday night</u>

My vindication of Tennyson[3] is the most triumphant piece of physical poetical demonstration hitherto produced. I have smitten all his critics into everlasting dumbness—their pestilent voices will annoy him no more

RI MS JT/1/T/1219
RI MS JT/1/TYP/6/2015

---

1. *[late February to early March 1860]:* dated with reference to the comment about not going out until 20th March in letter 1685.
2. *completion of my book:* Tyndall was completing the final stages of *Glaciers of the Alps.*
3. *My vindication of Tennyson:* Tyndall mentions Tennyson in 'Physics and Metaphysics' published in the *Saturday Review,* 10 (4 August 1860), p. 141: 'Tennyson . . . is full not only of the imagery derived from strict study, but also of intuitive applications of scientific truth', and had perhaps shown a draft to Juliet earlier. This piece was to have been a chapter in *Glaciers of the Alps* until he was dissuaded from publishing it there. The article, though published anonymously, is the first substantive public statement of his 'materialist' philosophy.

From Charles Frédéric Martins[1] 1 March 1860 1687

Montpellier 1er Mars 1860

Mon cher Monsieur,

Les dirt bands du glacier du Faulhorn paraissent vous intéresser. Je vous envoie un petit dessin de ce glacier qui vous montrera qu'il est tout à fait simple et que la progression n'a peut être pas une part aussi forte que la fusion dans la formation de ces lignes paraboliques même sur les grands glaciers. J'ai mis dans le petit dessin ci-joint des lettres qui le mettent en rapport avec le mémoire. Je me réjouis de lire le récit de vos excursions dans les Alpes.

Votre bien dévoué serviteur | Ch. Martins

Fig. 1.

S. Sommet du glacier

E. Escarpement terminal ou intérieur a b c d dirt bands

Fig. 3

Coupe verticale suivant la ligne SE de la Fig. 1 a b c d f g h i k. dirt bands.

Fig. 5

Bandes bleues.

Petit Glacier du Faulhorn en 1844

Fig. 1

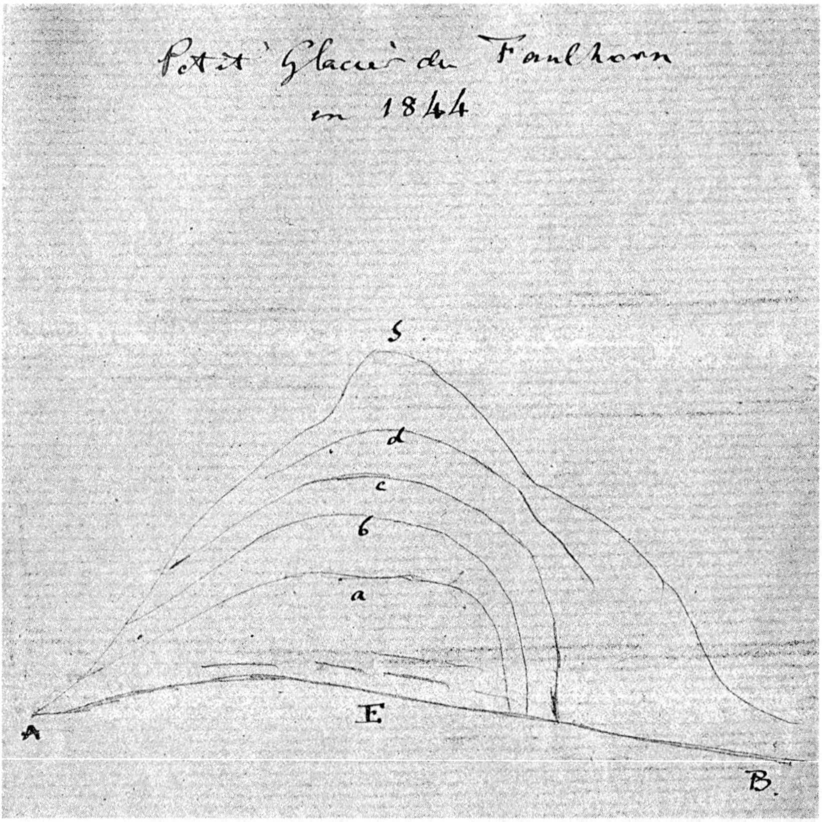
Petit Glacier du Faulhorn
en 1844
S
d
c
b
a
E
A
B.

Fig. 3

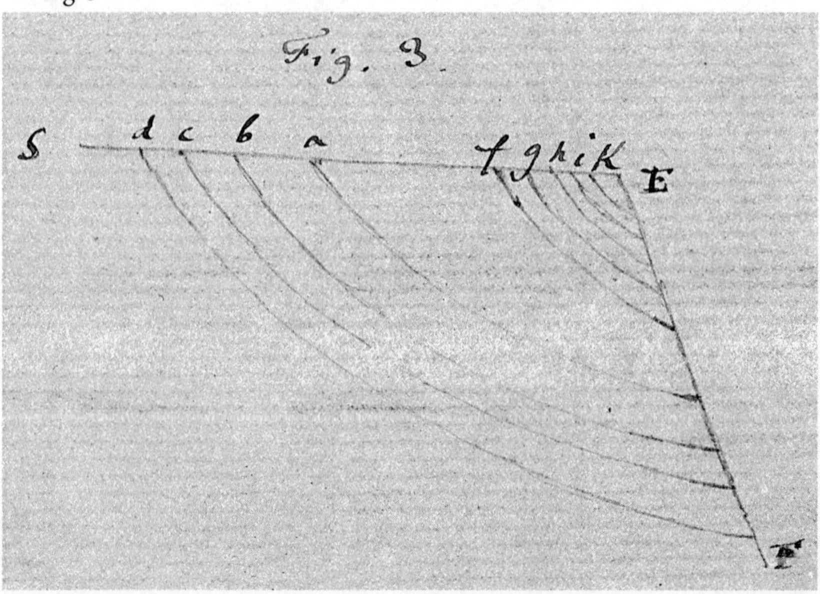
Fig. 3.
S
d c b a
f g h i k
E
F

Fig. 5

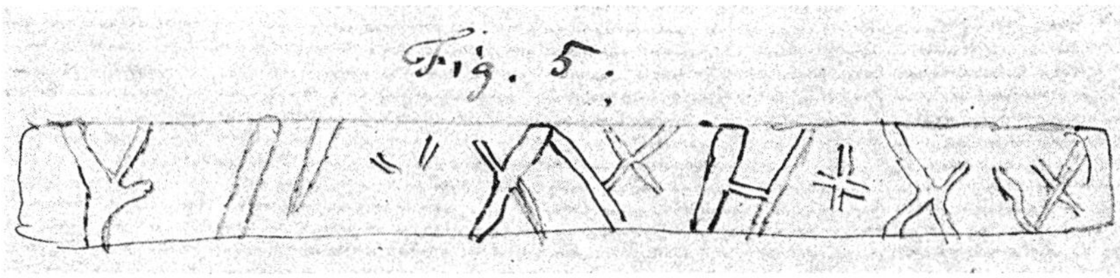

Montpellier March 1st 1860

My dear Sir,

The dirt bands of the Faulhorn glacier[2] seem to be of interest to you. I am sending you a small drawing of that glacier which will show you that it is quite simple, and that perhaps the advance does not play such an important role as fusion for the formation of those parabolic lines even on large glaciers. In the small enclosed drawing, I have placed the letters that relate to the memoir.[3] I cannot wait to read the account of your excursions in the Alps.[4]

Your most devoted servant | Ch. Martins

Fig. 1

S. Summit of the glacier.

E. Terminal Escarpment or the interior a b c d dirt bands.

Fig. 3

Vertical section along the line SE of Fig. 1 a b c d f g h i k dirt bands.

Fig. 5

Blue bands.

Small Faulhorn Glacier in 1844

Fig. 1

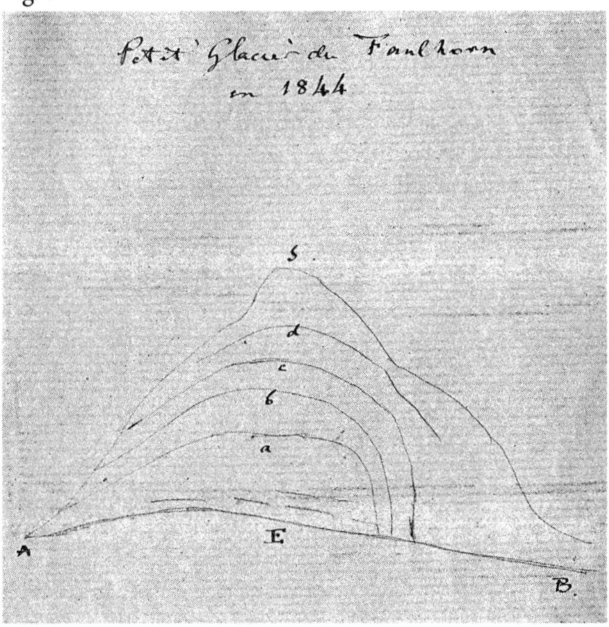

Fig. 3

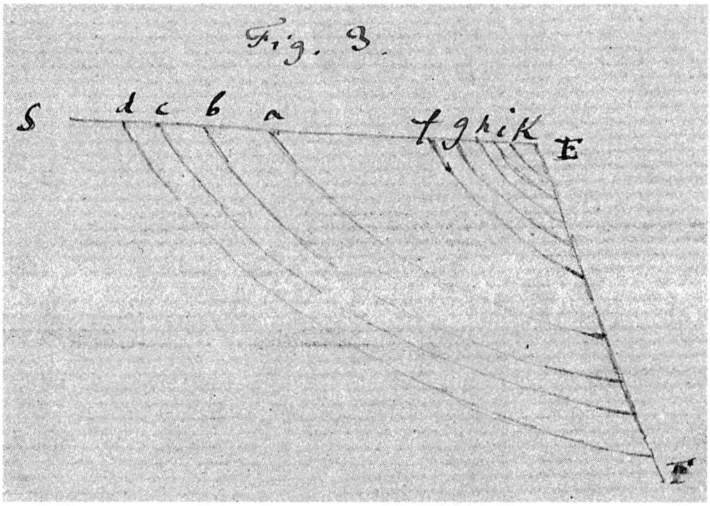

Fig. 5

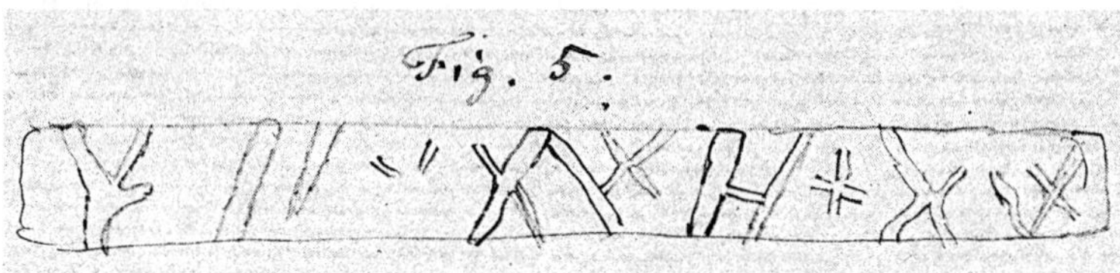

RI MS JT/1/M/51

1. *Charles Frédéric Martins:* (1805–89), a French botanist and geologist at the University of Montpellier.
2. *Faulhorn glacier:* the Faulhorn mountain is above Grindelwald on the opposite side of the valley and some distance from the Eiger and its glaciers.
3. *the memoir:* it is uncertain which of Martin's publications this refers to. Given the reference to Faulhorn, it is likely the book *Remarques et Expériences sur les Glaciers sans Névé de la Chaîne du Faulhorn* (Paris: Fain et Thunot, 1842).
4. *the account of your excursions in the Alps:* refers to Tyndall's *Glaciers of the Alps* published in 1860.

To Lady Colvile[1] 2 March [1860][2] 1688

2nd March | Royal Institution

My dear Lady Colvile

I will call down & explain to you how it came that I had to make a vow more than a week ago not to dine out for 6 weeks.[3] It arises from a piece of work which I have now on hand, of the most perplexing & difficult character, which requires all my time and all my strength, and will not permit me for the time mentioned the slightest social indulgence.

I fear I have made some of my friends angry with me through this resolve. But they do not know how I am driven—& I hope I may calculate that you will not */range/* yourself among the angry ones.

Yours always | John Tyndall

BL, Add MS 60634 ff. 119–35

---

1. *Lady Colvile*: Frances Elinor (née Grant), Lady Colvile (1838–1919), married to Sir James William Colvile (1810–80).
2. *2 March [1860]:* the letter is dated to 1860 based on the reference to his vow to stay home for an extended period of time, see n. 3.
3. *I had to make a vow more than a week ago not to dine out for 6 weeks:* Tyndall decided to stay in until 20 March 1860 to complete writing *Glaciers of the Alps.* He also referred to this decision in a letter to Debus (letter 1685) and to Pollock (letter 1686).

From Edward Sabine 12 March 1860 1689

Dear Tyndall,

You will find in page 506 et seq. an acc[oun]t of the experiments in Jamaica on Solar & Terrestrial Radiation at different heights in the Atmosphere,[1] which I named to you.[2] They were very decisive in character, and might be repeated with ease & with great advantage by the travellers of the Alpine Club,[3] either in the Alps or elsewhere. The same fact, 'the increased effect of radiation'[4] is found in the high latitudes, when compared as favorably as it may be with radiation in the Tropics: & probably due t[o] the same cause, the smaller portion of atmosphere thro' which the rays have to pass, the atmosphere acts as a fog.

That the height of the atmosphere is considerably greater at the equator than at the Poles is probable from the effect of the rotation of the earth. The same cause which produces the flattening of the one at the Poles, will produce a greater flattening of the other.

Having a few copies of the volume left, I will ask you to keep the one I send.

Sincerely yours | Edward Sabine

March 12th, 1860

RI MS JT/1/S/18
RI MS JT/1/TYP/4/1313

1. *You will find in page 506 et seq. an acc[oun]t of the experiments in Jamaica on Solar & Terrestrial Radiation at different heights in the Atmosphere*: refers to Sabine's book *An Account of Experiments to Determine the Figure of the Earth* (London: John Murray, 1825).
2. *which I named to you*: not identified.
3. *the Alpine Club*: mountaineering club founded in 1857. Tyndall had joined in 1858.
4. *'the increased effect of radiation'*: the source of this exact reference is unknown, but Sabine was likely referring to Tyndall's experimentation on the effect of radiation on water vapour and carbon dioxide.

To Editor of *the Times* 29 March 1860 1690

THE BRITISH MUSEUM. | A FEW REASONS WHY THE BRITISH MUSEUM SHOULD BE PRESERVED IN ITS ANCIENT INTEGRITY, AS BY LAW ESTABLISHED.[1]

From the time of the foundation of the British Museum,[2] more than a century ago, it has been a depository of objects of natural history and of antiquities of various kinds, united with books and manuscripts. As such it was instituted, and the Act of Parliament[3] (26th year of George II.) which gave it vitality declared that "the several collections, additions, and library, so received into the said general laboratory, shall remain and be preserved therein, for public use to all posterity."[4] In order to ensure this permanence and to control the progress of the great national institution thus founded, trustees were appointed by the Act; some representing the chief public offices of the State in the person of its high functionaries, and these, with the exception of the Archbishop of Canterbury, were necessarily, therefore, trustees only for the period during which they held office; others were appointed as the representatives of liberal benefactors, to whose munificence the British Museum, in fact, mainly owed its origin.

By a fresh enactment, these trustees were to be the electors of 15 more to be added to their number, and one in addition was to be nominated by the Crown.

So unwieldy a body eventually had to be curtailed for the purposes of regular business, to transact which an improved Act, resulting from Royal Commission, directed the nomination by the trustees of 15 of their number, who were to act as a standing committee. These, in addition to the three principal trustees—viz., the Archbishop of Canterbury, the Lord-Chancellor, and the Speaker of the House of Commons, were to undertake in fact the real administration of the Museum.

The century during which the Museum has existed has been one unparalleled in respect of the gigantic progress which scientific research has given to natural history; while the increased wealth of the nation, and the advance in the cultivation of public taste, have during that period communicated a great stimulus to the collection in the national Museum of objects connected with the purposes for which it was instituted. The natural history collections have increased to what are in fact several departments, each vying with and many surpassing the finest collections of their kind in other parts of the world. The books have expanded to a library highly worthy of this country, and the manuscripts, the prints, and the antiquities are of equal importance.

The trustees have watched this rapid progress of the collections under their care with a constant anxiety to meet the fresh requirements for space which that progress demanded by gradual but continual accessions of ground contiguous to the Museum.

Until recently the enlargements thus contemplated steadily progressed. A check, however, was given to them by the promise of space within the Museum walls which was entertained at least by those trustees who represent natural history, on the occasion of Mr. Panizzi[5] proposing his admirable scheme for converting the quadrangle of the Museum into what should be at once a noble reading-room and a most comprehensive library. By the carrying out of this scheme, which has been so eminently successful, a space was, it was hoped, to be created for the accession to the library for the next half-century, and also that some of the large rooms now occupied by the old library were to have been devoted to the reception of other collections. This hope has not, however, been fulfilled.

About the same time that the new library and reading room were in the course of erection, the question was raised, whether the collections of natural objects should not be separated from those devoted to the works of man? Had this proposition merely contemplated the segregation of the Museum into distinct parts, or into buildings contiguous to one another and to it, no naturalist could have opposed the suggestion. Indeed, those of the trustees who especially take an interest in its natural history departments have always desired so to extend the building northwards as to give to natural history a separate part of the Museum, though at the same time one in juxtaposition with the rest. But for the last four years no steps whatever have been taken to carry out the recommendation of this committee; it having been understood that some great change was impending, and that nothing could be done in the way of accommodation, adaptation, or improvement, until the decision of the Government and the Parliament was obtained as to this intended operation. The management of the different departments of the Museum is assigned to distinct committees of the trustees; and it is an important fact that the Committee for Natural History has always been unanimous in the opinion that it is of the utmost importance, with a view to preserve the popular regard for, no less than to facilitate the study of, the collections of natural history, that those collections should not be removed from Bloomsbury. That they should be separated from the halls and galleries devoted to art need never necessitate their detachment from the great national library. It was under the influence of a similar conviction that in the year 1858 a very considerable body of men of science addressed to the Chancellor of the Exchequer the following document,[6] embodying the opinions which had been unanimously advocated by the Natural History Committee of the Trustees:—

"MEMORIAL OF THE PROMOTERS AND CULTIVATORS OF SCIENCE ON THE SUBJECT OF THE PROPOSED SEVERANCE FROM THE BRITISH MUSEUM OF ITS NATURAL HISTORY COLLECTIONS; ADDRESSED TO HER MAJESTY'S GOVERNMENT.

"The Report of the Royal Commission appointed to inquire into the best site for a National Gallery, and recent discussions in Parliament, having led to the contemplation of breaking up the British Museum by severing from it the Natural History Collections, we, the undersigned, promoters and cultivators of natural knowledge, beg to record our strong objections to such removal, and for the following reasons:—

"I. The British Museum, when established by Act of Parliament in 1755, was essentially a Natural History Collection, the enlightened views of its founder, Sir Hans Sloane,[7] being that it should 'be rendered as useful as possible, as well towards satisfying the desire of the curious as for the improvement, knowledge, and information of all persons.'[8]

"II. The object of Sir Hans Sloane has been so satisfactorily carried out that, according to the Report of the last Royal Commission which inquired into the whole state of the Museum (1849), 'the evidence of men of the highest authority in science was referred to with great satisfaction to show that the Natural History Collections were, as a whole, equal, it not superior to any in the world.'[9]

"III. While we are aware that much greater space is required to provide for the reception of antiquities and ancient sculpture (chiefly on the ground floor), it has been ascertained by the trustees that, when additional buildings shall be called for, they can be extended northwards in halls requiring little embellishment, and, according to a plan laid before the trustees by Mr. Smirke,[10] involving a considerably less expenditure than that which must be incurred by a transference of those collections to any other spot, and the consequent erection of an entirely new edifice. On this point we beg to quote the following resolutions adopted by the trustees of the British

Museum, composing the standing committee, as printed in a return to the House of Commons dated the 4th February, 1858:—

" 'The committee having had under their consideration the report of the principal librarian,[11] dated 10th of November, 1857, resolved. —

"'1. That it appears from such report that there is a great deficiency of space at present for the proper exhibition of the different collections in the different departments of the Museum, and that there is no vacant space now belonging to the trustees which will be sufficient to provide for such deficiency.

"'2. That in providing an adequate space for that purpose it is very desirable to contemplate the future and progressive, as well as the actual and immediate, requirements of the British Museum.

"'3. That it appears to the trustees that the best mode of providing for such present and future requirements will be by adopting the plan submitted by Mr. Smirke for the purchase of land to the north of the Museum, as contained in the report of the principal librarian.

"'4. That in the opinion of the trustees, even if the increase of the collections which are under their care should at some future time make is necessary to transfer any of those collections to some other place, the land of which the purchase is now recommended must always be of great and peculiar value to the Museum.

"'5. That such plan, together with the present resolutions, shall be laid before Her Majesty's Ministers, with the view of requesting their concurrence therein, and their recommendation thereof to the consideration of Parliament."

"IV. Presuming that few persons will be found to advocate the removal of the grand masses of ancient art from their present position, so it is manifest that, if all the Natural History Collections be taken away, their transference will afford no place for the classical works which now temporarily encumber the principle façade, or crowd the crypts below; for, as such massive objects much be placed on the ground floor, so an extension of the basement is inevitable if the antiquities remain part of the Museum, and all that the Natural History Collections can require for their future development will be an allotment of space above such extension of the ground floor.

"We would also observe that the prolongation of the present building northwards, on the abovementioned plan, besides being much less costly than the formation of an entirely new building, will put a stop to all controversies respecting the appropriate site; and the style of architecture to be applied to a new building.

"V. In reference to other suggestions that have been vaguely thrown out of the breaking up of the Natural History Collections of the nation into several parts, by transferring, *e.g.,* the minerals, to the Government School of Mines; the stuffed animals to the Zoological Society; the insects and shells to the Linnaean Society, &c., we have first to observe that not any of the above institutions, two of which are only voluntary associations of individuals, possesses the space or means for the reception and display of such constituent parts of the great national series of illustrations of nature; and, further, that as the chief end and aim of natural history is to demonstrated the harmony which pervades the whole, and the unity of principle which bespeaks the unity of the creative cause, it is essential that the different classes of natural objects should be preserved in juxtaposition under the roof of one great building.

"VI. We further strongly object to the proposed transference, because those engaged in the study of natural history have in the British Museum the paramount advantage of consulting every work which can aid their researches; while a removal of the collections would either involve a conjoint transference of a very large portion of the national library, or necessitate a very expensive purchase of a special natural history library.

"VII. While such are among the prominent reasons against the removal of the Natural History Collections from the site where they have been established for upwards of a century in the centre of London, we beg to add the expression of our opinion, that such removal, particularly if to any situation distant from that centre, would be viewed by the mass of the inhabitants with extreme disfavor, it being a well-known fact that by far the greater number of visitors to the Museum consists of those who frequent the halls containing the Natural History Collections; while it is obvious that many of those persons who come from the densely peopled districts of the eastern, northern, and southern parts of London would feel it very inconvenient to resort to any distant locality.

"For these reasons, as based on scientific advantages, the convenience and instruction of the people and the saving of a large sum to the nation, we earnestly hope that the Natural History Collections may not be interfered with, but be allowed to remain associated with the many other branches of human knowledge which are so admirably represented in this great national establishment.

"Her Majesty's Government, we trust, will never yield to the argument that, because in some countries the products of nature and art are exhibited in distinct establishments, therefore the like separation should be copied here. Let us, on the contrary, rejoice in the fact that we have realized what no other kingdom can boast of, and that such vast and harmoniously-related accumulation of knowledge is gathered together around a library, illustrating each department of this noble museum.

"*Wrottesley, President Royal Society; *Rod. L. Murchison F.R.S., D.C.L., President Geographical Society; *P.M. Grey Egerton, M.P., F.R.S.; J.E. Portlock, Major-General R.E., F.R.S., &c; *Rosse, F.R.S.; C. Wheatstone, F.R.S.; G.G. Stokes, Secretary Royal Society; Thomas H. Huxley, F.R.S.; James Paget, F.R.S.; W. Sharpey, M.D., Secretary Royal Society; John P. Gassiott, F.R.S.; W. Bowman, F.R.S.; Jos. D. Hooker, F.R.S.; William B. Carpenter, F.R.S.; W. R. Grove, V.P.R.S.; Leonard Horner, V.P.R.S. and V.P.G.S.; Thomas Bell, F.R.S., President of the Linnean Society; Wm. J. Burchell, F.L.S.; S. James A. Salter, F.L.S., F.G.S.; Wilfred D. Speer, F.L.S.; John Lindley, F.R.S.; Richard Owen, V.P.R.S., Superintendent Natural History Departments, British Museum; John Barlow, F.R.S., Vice-President and Secretary Royal Institution; Robert M'Andrew, F.R.S., F.L.S.; George Rennie, F.R.S., F.G.S., &c.: William Hopkins, F.R.S., F.G.S., and M.A.; J. G. Bowerbank, F.R.S., F.L.S., &c.; Thomas Graham, F.R.S., Master of the Mint; John Marshall, F.G.S.; *B. C. Brodle, F.R.S., D.C.L.; A. W. Hofmann, LL.D., F.R.S.; *T. Mayo , F.R.S., President of the Royal College of Physicians; Charles Lyell, F.R.S., D.C.L., F.G.S.,; Andrew C. Ramsay, F.R.S., F.G.S.; William Henry Fitton, F.R.S., F.G.S.; H. Falconer, M.D., F.R.S., L.S., and G.S.; J. Crawford, F.R.S., F.G.S., John Forbes, M.D., F.R.S.; John Tyndall, F.R.S., Professor, &c; John Phillips, M.A., LL.D., President of the Geological Society, F.R.S.; John Prestwich, F.R.S., F.G.S.; Thomas Davidson, F.R.S., G.S.; Robert G. Austen, F.R.S., G.S.; Alfred Tylor, F.G.S.; Robert William Mylne, F.G.S.; Warrington W.

Smyth, Secretary Geological Society, F.R.S.; Thomas Sopwith, M.A., F.R.S., F.G.S.; Selkirk, M.A., F.R.S., F.G.S.; J. J. Bigsby, M.D., F.G.S.; Robert Kane, Knight, M.D., F.R.S., Director of the Museum of Irish Industry; *Cawdor, F.R.S., F.G.S.; Baden Powell, F.R.S., Professor of Geometry Oxford; Charles Frederick White, M.M.S.; James Glaisher, F.R.S., &c.; M. Marshall, F.Z.S., Bank of England; Robert Warington, F.C.S., M.M.S.; J. George Appold, F.R.S.; John Birkett, F.R.E.G., England, F.L.S., &c.; John Quekett, F.L.S., M.R.C.S.E., &c.; Conrad Loddiges, F.H.S.; Henry Christy, F.S.S.; J. Claude Webster, F.S.A., Middle Temple; E. W. Cooke, A.R.A., F.L.S.; Ducie, F.R.S., F.G.S.; John Morris, F.G.S., and Professor of Geology, University College, London; Edward Sabine, Treasurer and Vice-President of the Royal Society; William Whewell, D.D., F.R.S., Master of Trinity College; G. W. Featherstonhaugh, F.R.S., F.G.S.; Robert Hunt, F.R.S.; Enniskillen, F.R.S., &c.; R. G. Latham, M.D., F.R.S.; W. J. Broderip, F.R.S., F.G.S., P.L.S.; J. E. Gray, F.R.S., V.P.Z.S., President of the Entomological and Botanical Societies of London, &c., Keeper of Zoology, British Museum; George R. Waterhouse, Keeper of Geology, British Museum; Nevil Story Maskelyne, F.G.S., Keeper of Mineralogy, British Museum; J. Gwyn Jeffreys, F.R.S.; John Miers, F.R.S., F.L.S.; John Hogg, F.R.S., F.L.S., &c.; Samuel Stevens, F.L.S., Treasurer of Entomological Society; John S. Gaskoin, F.L.S.; D. W. Mitchell, Secretary to the Zoological Society of London, F.L.S., &c.; Philip Lutley Sclater, F.L.S., M.A. Fellow of Christ Church, Oxford; George Busk, F.R.S., F.R.C.S.E., &c.; W. Macdonald, M.D.; E. W. H. Holdsworth, F.L.S., F.Z.S.; John J. Bennett, F.R.S.L.S.; John Percy, M.D., F.R.S.; M. Henderson, M.D., F.H.S.; John Carrick Moore, M.A., F.R.S.; John Gould, F.R.S.; Robert Stephenson, M.P., F.R.S.; George Peacock, Dean of Ely., F.R.S.; Charles Darwin, F.R.S., &c.; J. F. W. Herschel, M.A., D.C.L., F.R.S., London and Edinburgh; Adam Sedgwick, M.A., F.R.S., Woodwardian Professor, Cambridge; W. H. Miller, M.A., F.R.S., Professor of Mineralogy in the University of Cambridge; Charles C. Babington, M.A., F.R.S., Cambridge; G. D. Living, M.A., Lecturer on Natural Sciences, St. John's College, Cambridge; William Clark, M.A., F.R.S., Professor of Anatomy, Cambridge; Charles Daubeny, M.D., F.R.S., Professor of Botany, Oxford; Henry W. Acland, M.D., F.R.S., Regius Professor of Medicine, Oxford; John R. Mummery, F.L.S.; Nathaniel Ward, F.R.S., F.L.S.; Stephen H. Ward, M.D., London; Richard King, M.D.; Bedford Pym, R.N., F.R.G.S.; Francis Walker, F.G.S. and L.S.; Edward Newman, F.L.S.; John A. Power, M.D., F.R.G.S.; John Douglas, F.L.S.; H. T. Staunton, F.L.S., &c.

"This document, signed by so many distinguished men of science, including the Presidents of the chief scientific societies, was, indeed, in the course of the autumn of 1858, followed by another, signed by eight leading naturalists, three of whom had given their names to the previous memorial. They asked for a reorganization of the Natural History Collections, and with that a distinct administration of them. But, though they adverted to the then prevalent report, that the collections were to be transferred to 'another locality,' the prayer of their memorial might in all essential respects be fully complied with if the collections were to remain where they are.

"Recently a special general meeting of the trustees was called to receive a report prepared by a sub-committee appointed to ascertain the relative expense of erecting such buildings as would embrace all that the eminent Superintendent of the Natural History Department considered requisite for the different collections under his supervision. The comparison was instituted between the expense of adding to the present Museum to the extent of five acres, and the cost of erecting a new museum of similar extent on the only easily obtained ground, which is situated on the southwestern extremity of the land at Kensington-gore, purchased by the Royal Commissioners of the Exhibition of 1851.

"It has been recently stated in the House of Commons that on that occasion, when 17 trustees were present, the vote in favour of the removal of the Natural History Collections in Kensington was carried by a majority of one only; and, as it was further stated that six Cabinet Ministers voted in that majority, it follows that the trustees, who are really charged with the managements of the Museum, and who may be asserted to have the interests of the collections most vividly present to their minds, constituted the minority. It was, in fact, made a question of expenditure as balanced against the true interests of science.

"Members of the Government probably may think more of the relative importance to the country which 150,0001., or even 200,0001., may represent than of the removal to a suburban district of the departments of Natural History. But the trustees who voted in the minority certainly represented the feelings of men of science, and also, we believe, of a large metropolitan public, when they considered that a central position in London, propinquity to the great national library, and the maintenance in the eye of the public in the middle, as it were, of their daily paths, of our noble collections of natural history, was of more importance for that public than the sum of money to which the interests of science were to be postponed.

"But even the argument drawn from economy, which is the main argument that has been ostensibly put forward, loses much of its importance when critically investigated.

"It is doubtless true that the purchase of land covered with houses at Bloomsbury must involve a large outlay than the acquisition of a similar extent of grass land at Kensington; but, on the other hand, it is to be borne in mind that the purchase of the latter and the cost of the new building must both be paid for at once or in the ensuing three or four years, whereas the extension hitherto contemplated by the trustees around the Museum involves only a gradual acquisition of the ground, and consequently an expenditure extending over many years. The design for such an extension and for the buildings contemplated in it would, indeed, be formed on a complete and comprehensive plan, but all that would be necessary for the present towards carrying that plan out would consist in the securing the ultimate reversion of the land, and in the purchase for immediate requirements of a comparatively small number of leases. The fortunate circumstance that the land required is all the property of one nobleman would doubtless render the ultimate acquisition of the reversion of it by the British Museum a simple negotiation.

"Furthermore, the expense of removal and the risk of injury to the collections, no less than the years that must ensue during which no relief can be afforded to the existing establishment-years during which the façade must remain encumbered, as it now is, with the antiquities of Halicarnassus[12],—are no light arguments against the project of breaking up the establishment. When to these considerations is added the additional expense involved in the formation of a scientific library—to say nothing of the costliness of the façade of a new building,—assuredly the gradual though systematic extension of the British Museum northwards for the natural history, and westward for the antiquities, the uninterrupted exhibition to the public of the collections even while that extension is going forward, and the profound conviction on the part of so many men of science that the welfare of science is involved in this extension, are arguments to which an economy of a few tens of thousands is no answer.

"Undoubtedly the multitudes who inhabit the more densely populated regions of the metropolis—so large a fraction of the British nation—would share in the disappointment of scientific men at the removal of the Natural History Collections from a spot so near to the centre of the great metropolitan area, to the western extremity of the Kensington grounds, where the entrance to the new museum would be nearly half a mile beyond the existing South Kensington[13] or Brompton Museum.

"The complaint that the British Museum presents now a confused medley of heterogeneous collections, even if it were just, would be a very different thing from declaring the advisability of separating those collections by the whole length of the radius of London. But the public, whose interests are in fact most deeply concerned, will probably prefer the middle course, and will concur with the sentiments above expressed in favour of separation without removal,—of a distinct administrative separation of the three great series of collections, the Natural History, the Antiquarian, and the Bibliographical, so that each may form a department entire in itself, but contiguous to the others, the great Reading-room and Library forming the nucleus and centre of union for the whole. The British Museum will then still remain the rich storehouse of knowledge, to be read in books and to be illustrated by collections, which has hitherto made it an unique institution, of which every Englishman feels he may be proud, and to which every foreign traveler bends his steps, as giving him at once a general view of the intelligence, enterprise, and practical sense of this nation.

"March 29."

John Tyndall et. al, 'THE BRITISH MUSEUM', *The Times* (12 April 1860) p. 10.

---

1. While the letter was dated March 29, it did not appear in print until 12 April, as 'THE BRITISH MUSEUM', *The Times* (12 April 1860), p. 10.
2. *the foundation of the British Museum*: the British Museum was founded in 1753, and opened in Montagu House in Bloomsbury in 1759.
3. *Act of Parliament*: the British Museum Act 1753 (26 Geo. 2, c 22).
4. *"the several collections, additions, and library, so received into the said general laboratory, shall remain and be preserved therein, for public use to all posterity."*: passage of the British Museum Act 1753 (26 Geo. 2, c 22). (*Acts and votes of Parliament relating to the British Museum with the Statutes and rules thereof, and the succession of trustees and officers* (London, 1805) pp. 20–30.)
5. *Panizzi*: Anthony Panizzi (1797–1879), an Italian exile in Britain, he was a librarian with the British Museum (*ODNB*).
6. *It was under the influence of a similar conviction that in the year 1858 a very considerable*

body of men of science addressed to the Chancellor of the Exchequer the following document: this was a petition to the government on the subject of the relocation and organization of the natural history museum. The petition against removing the natural history collections away from the British Museum was initiated by Roderick Murchison. For more details, see Frederick Burkhardt and Sydney Smith (eds), *Darwin Correspondence Volume 7* (Cambridge: Cambridge University Press, 1992), Appendix VI; Richard Drayton, *Nature's Government: Science, Imperial Britain, and the 'Improvement' of the World* (New Haven and London: Yale University Press, 2000) pp. 198–217.

7. *the enlightened views of its founder, Sir Hans Sloane*: the physician Hans Sloane (1660–1753) donated his personal collection with the intention to found a museum (*ODNB*).
8. *be rendered as useful as possible, as well towards satisfying the desire of the curious as for the improvement, knowledge, and information of all persons*: recorded in the document *Authentic Copies of the Codicils Belonging to the Last Will and Testament of Sir Hans Sloane, Bart. deceased, Which relate to his Collection of Books and Curiosities*, (1753) p. 19.
9. *'the evidence of men of the highest authority in science was referred to with great satisfaction to show that the Natural History Collections were, as a whole, equal, it not superior to any in the world'*: the passage appeared in a slightly different version in the *Report of the Commissioners Appointed to Inquire into the Constitution and Government of the British Museum with Minutes of Evidence* (London: 1850), p. 44, as 'Lastly, we refer with great satisfaction to the evidence of men of the highest authority in science, showing that the Natural History collections of the Museum are, as a whole, equal, if not superior, to any in the world'.
10. *Mr. Smirke*: Robert Smirke (1780–1867) was the architect of the British Museum.
11. *of the principal librarian*: that is, Panizzi (see n. 3), who was principal librarian of the British Museum between 1856 and 1866.
12. *Halicarnassus*: ancient Greek city.
13. *South Kensington or Brompton Museum*: originally founded in 1852 with specimens from the Great Exhibition in 1851, the museum moved to the South Kensington location and was named the South Kensington Museum in 1857 (Lara Kriegel, *Grand Designs: Labor, Empire, and the Museum in Victorian Culture* (Durham and London: Duke University Press, 2007), pp. 164–67).

To Heinrich Debus [March 1860][1] 1691

Saturday

My dear Heinrich.

Until now I have not had time to answer your last letter.[2] With regard to the Royal Society I should like to have the names of all the considerable Chemists in England attached to your certificate[3]—

Now to do this you have allowed very little time: but still I am of opinion that if you should wish to become a Fellow your claims are so high that there would be every chance of your election.

With regard to Easter I do not yet know what our plans will be—we must get off somewhere, and I have no objection to the Forest[4] if you think Wales too far. But it would never do to leave poor Tom[5] at home—he will need a run also and we shall be all three very jolly together—I do not know what his plans are—He has recently been engaged at the University College teaching mathematics,[6] and this may keep him in London during the Easter holidays.[7] If this be the case you shall have me all to yourself to pet me and fondle and feed me & above all to teach me metaphysics. I want to argue some mechanico-metaphysical points with you—

I am very tired today and long to have a run in the country: but on the whole I have not enjoyed such working power for many years as I have this year.

Goodbye dear Heins | Ever yours | John Tyndall

RI MS JT/1/T/352

1. *[March 1860]:* Easter Sunday in 1860 was 8 April. Given Hirst's appointment to University College teaching mathematics which he was offered on 28 February 1860, (see *Hirst Journal*, RI MS JT/2/32c/1528), this letter dates to a Saturday in March 1860.
2. *your last letter:* not identified.
3. *your certificate:* for proposal as a FRS. Debus was elected on 6 June 1861.
4. *the Forest:* the New Forest, in southern England.
5. *poor Tom:* Thomas Hirst.
6. *He has recently been engaged at the University College teaching mathematics:* Hirst was offered the position on 28 February 1860, *Hirst Journal*, RI MS JT/2/32c/1528.
7. *may keep him in London during the Easter holidays:* in the event, Hirst became ill and stayed in London.

To John Murray [March 1860][1] 1692

Queenwood | near Stockbridge | Hants | <u>Sunday</u>.

My dear M[r] Murray,

I left the introductory portion of my manuscript[2] at the Institution[3] with directions to have it sent to you. Would you kindly have it put in type for me—I want to ask a friend[4] how it reads—When you are done with it I shall have more ready for you.

I intend to give a sketch of each of my expeditions making use of the sketch to elucidate the minor scientific points of the journey—and afterwards to deal with the question of glaciers relieved from all lumber.

Ever yours | <u>John Tyndall</u>

John Murray Esq[ui][r[e]] | &c. &c. &c.[5]

National Library of Scotland, John Murray Archive, Ms. 41214

---

1. This letter was likely sent in March 1860, as Tyndall was near to submitting the full draft of *Glaciers of the Alps*. It may be that the friend (n. 4) was Juliet Pollock; see letters 1642 and 1684. It is also possible that it was Huxley, who 'urged him' to finish, see letter 1695.
2. *my manuscript*: refers to *Glaciers of the Alps*, see also letter 1595, n. 4.
3. *Institution*: the RI.
4. *a friend*: not identified, but may have been Juliet Pollock or Huxley, see n. 1.
5. Verso included the note '1859 Tyndall J n.d.'

From Jules Duboscq[1] 05 April 1860 1693

Paris, 5 Avril 1860

Monsieur

J'ai reçu ce matin le bon de 44£ 8/4 pour solde de ma facture remise en mars denier. Je m'occupe de votre commande vous la recevrez à la fin de la semaine de Pâques.

Vous me faites dans votre dernière lettre une observation sur les prix de la lanterne & du régulateur. Ceux que je vous ai fournis pour Royal Institution sont si je me rappelle bien les instruments que j'avais emportés lors de mon voyage à Liverpool. J'aurai lors de l'expédition de la *[caisse]*, l'honneur de vous informer de son départ.

Veuillez, Monsieur, recevoir je vous prie, l'assurance des sentiments avec lesquels je suis votre obéissant serviteur

J. Duboscq | Pour mon Père

Mr Tyndall London

Paris, April 5th 1860

Sir,

This morning I received the sum of 44£ 8/4 in payment of my invoice which was sent last March. I am taking care of your order;[2] you will receive it at the end of the Easter week.

In your last letter[3] you make an observation on the price of the lantern & of the regulator. Those that I provided to you for the Royal Institution are, if I remember correctly, the instruments that I carried on my trip to Liverpool.[4] Once the parcel is sent, I will have the honour of informing you of its departure.

Please, Sir, I beg you to accept the assurance of the warmest wishes with which I am your obedient servant

J. Duboscq | For my Father[5]

Mr Tyndall, London

RI MS JT/1/D/181

1. Letter signed 'J. Duboscq', Louis Jules Duboscq (1817–86), a highly esteemed French scientific instrument maker. The letterhead included the following information, 'J. Duboscq. | Student and successor of SOLEIL | 21, Odéon Street. | at the back of the courtyard (J. Duboscq. | Eleve & Succ*[esseu]*r de SOLEIL. | 21, Rue de l'Odéon. | au fond de la Cour)'. Duboscq and the optician Jean Baptiste François Soleil (see n. 5) were both based at that address.
2. *your order*: Duboscq specialized in optical and photographic equipment. (John Hannavy, *Encyclopedia of nineteenth-century photography*: *A*-I, 1 (Routledge: 2008), p. 445). It is uncertain what Tyndall had ordered.
3. *In your last letter*: letter missing.
4. *on my trip to Liverpool*: Duboscq exhibited instruments at the annual meeting for the BAAS, which in 1854 was held in Liverpool.
5. *For my Father*: it is possible that this a reference not to Duboscq's own father, a cobbler, but to his father-in-law, Jean Baptiste François Soleil (1798–1878). Duboscq trained under Soleil (John Hannavy, *Encyclopedia of nineteenth-century photography*: *A*-I, 1 (Routledge, 2008), p. 445), and his stamp named himself as student and successor of Soleil.

To Thomas Archer Hirst [16 April 1860][1] 1694

Royal Institution | Monday

My dear Tom.

I wish I was near you—by Jove I would stir you up. I can answer that question for you:[2] it <u>is</u> bodily weakness & it is <u>not</u> indolence—My dear fellow I could shew you half a dozen entries in my journal where I mourn over my lost mental vigour—and blame myself as the cause of it, calling myself unfaithful & so forth, when at the same time it was my bodily mechanism that did the whole mischief. Stay away as long as you can—you will be all the better for it—Do <u>not</u> let the women[3] nurse you—ride as much as you can & never mind what Mrs Hirst[4] says about the walking (don't shew her this) I walked 20 miles a day while on the Isle of Wight,[5] and after 4 days just felt myself becoming myself. I wish I had stayed two days more away, but Debus's time was up. I am hard at work at the proofs[6] but it is terrible work. I will never write a book again.

am now preparing for my lecture this ev[enin]g.[7] and have half a dozen other letters to answer;—good bye dear Tom: walk, ride, jump & <u>be patient</u>—all will come right in the end

<u>Affec[tionaltel]y John</u>

RI MS JT/1/T/949

1. The date is suggested based on a visit by Hirst to relatives in Yorkshire, to recover from an attack of blindness, and by the dates of Tyndall's evening lectures at the Royal School of Mines (see n. 7). Hirst left London on 5 April and wrote to Tyndall in the week ending 15 April (*Hirst Journal*, RI MS JT/2/32c/1534). His journal says he was 'conscious of a great want of energy' (*Hirst Journal*, RI MS JT/2/32c/1534). He noted receipt of a letter from Tyndall in the week ending 22 April (*Hirst Journal*, RI MS JT/2/32c/1535), which may be this one. We do not know Tyndall's exact movements in the weeks prior to this, though if the date is correct, he must have made a trip with Debus to the Isle of Wight.
2. *I can answer that question for you:* not identified. The letter from Hirst written in the week ending 15 April is missing.
3. *the women:* Hirst's journal mentions that he is with Rachel and Emily. Rachel was the wife of William Hirst, Thomas Hirst's younger brother, who was a farmer, and Emily was their daughter.
4. *Mrs Hirst:* Rachel (see n. 3).
5. *while on the Isle of Wight:* see n. 1.
6. *I am hard at work at the proofs: Glaciers of the Alps* published on 6 July 1860.
7. *my lecture this evg.:* on 16 April, Tyndall was due to give one of the Lectures to Working Men at the Royal School of Mines. These were scheduled for Mondays 12th, 19th, 26th March; and 2nd, 16th, 23rd April, though a note minuted in the meeting of the Council of the Royal School of Mines on 18 February 1860 relates that Tyndall felt unable to give all the lectures that he was scheduled to do, because of pressure of work (Anne Barrett, personal communication).

To John Murray 30 April [1860][1] 1695

Monday | 30th April

Dear Mr. Murray.

Huxley has 'urged me'[2] and in consequence I can now state to you that the whole of the work is now in the hands of Mr. Clowes,[3] and I undertake to have it ready for printing off in two days after I receive the revises. It has been very strongly represented to me by my friend Prof. Wiedemann[4] that a German translation[5] of the book ought to appear as closely after its appearance in England as possible—He says an arrangement ought to be made with Vieweg[6], with whom he (Wiedemann) is in communication; and from whose staff he would select some excellent translators. Wiedemann says that it is certain to be translated, and if not guarded in the way he recommends runs the chance of being spoiled—I told him that I would mention the matter to you.

The index is the only thing remaining, but this cannot be completed until the book is finally /bound/. I write this morning to Mr. Clowes[7] 'urging' him to send me the Revises.

Very sincerely yours | John Tyndall

National Library of Scotland, John Murray Archive, Ms. 41214

---

1. *30 April [1860]*: Verso gave the date 'April 30 1860 | Tyndall John'. Letter also included a note, 'Huxley has urged him to complete his work'. Both additions were presumably written by the John Murray publishers.
2. *Huxley has 'urged me'*: not identified where and when this took place.
3. *I can now state to you that the whole of the work is now in the hands of Mr. Clowes*: *Glaciers of the Alps* was printed by William Clowes and Sons (Stamford Street and Charing Cross) (see also letter 1675, n. 4 and 5).
4. *Prof. Wiedemann*: Gustav Heinrich Wiedemann (1826–99) was a German physicist and a founder of the Berlin Physical Society. Together with his wife, the translator Clara Wiedemann, he published German translations of Tyndall's later work. (Rudolf Reiger, 'Wiedemann, Gustav Heinrich', *Allgemeine Deutsche Biographie,* 55 (1910), pp. 67–70.
5. *a German translation*: the plans for a German translation did not come to fruition.
6. *Vieweg*: the publishing house Vieweg and Son.
7. *I write this morning to Mr. Clowes*: letter missing.

To Heinrich Debus [April 1860][1] 1696

My dear Heins,

If a situation were open for which you wished to become a candidate there is no doubt that the fact of your being a fellow of the Royal Society[2] would be of service to you. —This therefore might be a means of improving your Weltliche Stellung.[3]

Perhaps it would be as well if you allowed the matter to remain over for another year. —It is now rather late to commence operations.

I do not entertain a doubt as to your election.

If you see any prospect of applying for a situation during the coming year then I would say come forward at once: Still even should it occur; testimonials would do & I should myself in a few words settle the question as to you not being a Fellow of the Royal Society.

Ever yours | J. Tyndall

RI MS JT/1/T/329

1. This letter has been dated to April 1860 based on the discussion regarding the possibility of Debus becoming Fellow of the RS 'during the coming year'. Debus was elected at the Annual Meeting for the Election of Fellows held on 6 June 1861. As Tyndall noted that it would be 'now rather late to commence operations' for the current year's election, and the list of nominees for 1860 was decided in May 1860, this letter was likely written around April 1860. LT added a note 'Monday | 1859 or 1860'. See also letter 1779.
2. *being a fellow of the Royal Society*: Debus was elected FRS in 1861.
3. *Weltliche Stellung*: worldly position.

To Thomas Henry Huxley [April 1860][1] 1697

Royal Institution. | Thursday.

My dear Huxley

I have just received the Westminster[2] and looked simply at the last paragraph of your article.[3] My head, which is awfully out of order, would permit me to do no more; but I cannot help expressing to you the great admiration which the passage[4] excited within me. —It is really magnificent.

Ever Yours | John Tyndall

IC HP 8.282
RI MS JT/1/TYP/9/3157

1. *[April 1860]*: the IC HP dated this letter to April 1860, in part because the type-written letter included a handwritten note '?April 1860'. The date fits with the publication of Huxley's review of Charles Darwin's *Origin of Species*, which appeared in the *Westminster Review* in April 1860.
2. *the Westminster*: the *Westminster Review* was a radical quarterly.
3. *your article*: presumably T. H. Huxley 'Art.VIII.-Darwin on the Origin of Species', *Westminster Review*, 17 (April 1860), pp. 541–70.
4. *the passage*: the last sentence of Huxley's review reads as follows: 'And viewed as a whole, we do not believe that, since the publication of Von Baer's "Researches on Development," thirty years ago, any work has appeared calculated to exert so large an influence, not only on the future of Biology, but in extending the domination of Science over regions of thought into which she has, as yet, hardly penetrated'. T. H. Huxley. 'Art.VIII. —Darwin on the Origin of Species', *Westminster Review*, 17 (April 1860), p. 570.

From Henrietta Huxley [April 1860][1] 1698

14 Waverley Place | April 1860

My dear Uncle John,

If you are not engaged to dine with your cousin[2] on Sunday—will you dine with us at six o'clock? Mr and Mrs Lukes[3] are coming and some other item of humanity whom I do not know

We will have a good substantial disagreement which you like so much. I am quite prepared to oppose you in anything and everything

Yours sincerely | Henrietta Huxley

2?th of April 1860[4]

RI MS JT/1/TYP/9/2900
LT Typescript Only

1. *[April 1860]:* this is an uncertain dating based on a note by LT, see n. 4.
2. *dine with your cousin*: not identified, it may refer to Tyndall's American cousin Hector Tyndale, but it is not certain if he was in England during this period.
3. *Mr and Mrs Lukes:* not identified.
4. *2?th of April 1860:* the handwritten letter is missing, and this date is likely a later addition by LT.

To George Gabriel Stokes 10 May 1860 1699

10th May 1860

My dear Sir.

For several days the state of my health has been such as to render it dangerous for me to dwell much upon scientific matters; but I had read a good portion of Prof. Maxwell's papers previously and have looked over the remaining portion. The part to which I have devoted closest attention shows so much ability and accuracy, that I believe I cannot err in thoroughly recommending the publication of the paper in the Philosophical Transactions.[1]

Believe me | very sincerely yours | John Tyndall

Prof. Stokes | Sec. R.S.[2]

RS RR/4/181

1. *publication of the paper in the Philosophical Transactions*: the paper was published as J. C. Maxwell, 'On the Theory of Compound Colours and Relation of the Colours of the Spectrum', *Roy. Soc. Proc.*, 10 (1860), pp. 404–9, 484–86.

2. At the end of the letter, a note was added which appears to be in Stokes' handwriting: '1860 | Maxwell | by | Tyndall'.

From Robert Bunsen 12 May 1860 1700

Theuerster Freund

Ich wende mich mit einer Bitte an Sie, durch deren Erfüllung Sie einen meiner genausten Freunde und dadurch auch mich zu grossem Danke verpflichten würden. Sie haben gewiss die schöne Arbeit Kirchhoffs über Emission und Absorption gelesen. Dieselbe ist bis jetzt nur durch Notizen in den englischen Zeitschriften bekannt geworden und hat zu verschiedenen Bemerkungen Veranlassung gegeben, die Kirchhoff zu einer Erörterung zwingen, die ihm dadurch fast unmöglich wird, dass die eigentliche Arbeit, welche seine wichtige Entdeckung enthalt, Ihren Landsleuten gar nicht vorgelegen hat. Ich wende mich daher an Sie, mein theuerster Freund, mit der Bitte die beifolgende Abhandlung Kirchhoffs nicht im Auszuge sondern vollständig sammt der im Manuscripte beiliegenden Nachschrift im Philosophic. Magazine in Uebersetzung abdrucken zu lassen. Verursacht die Uebersetzung Kosten, so würde ich deren Berichtigung gern besorgen. Ihre herrlichen Gletscher Untersuchungen habe ich mit der innigsten Freude verfolgt und Sie nicht wenig um Ihre merkwürdigen Montblanc Besteigung beneidet!

In aufrichtiger und herzlicher Freundschaft | Ihr | R. Bunsen

Heidelberg den 12 Mai 1860.

Dearest friend

I turn to you with a request, the fulfilment of which will make one of my closest friends, and therefore also myself, greatly indebted to you. You have surely read Kirchhoff's beautiful work on emission and absorption.[1] Until now, his work has only become known in the English journals through short notices and has elicited various comments that force Kirchhoff to an explanation, which is almost impossible for him to do as the actual work containing his important discovery is not available to your fellow countrymen. I therefore turn to you, my dearest friend, with the request to have Kirchhoff's enclosed paper translated and published in the Philosophic. Magazine—not as an excerpt but in its entirety[2]—together with the enclosed handwritten postscript.[3] If there are expenses associated with the translation[4], I will gladly arrange for their settlement. I have followed your wonderful investigations of glaciers with the deepest pleasure and envy you not a little for your remarkable climb of Mont Blanc.

In sincere and cordial friendship | Your | R. Bunsen

Heidelberg, 12 May 1860

RI MS JT/1/B/147

---

1. *read Kirchhoff's beautiful work on emission and absorption:* G. Kirchhoff, 'Ueber das Verhältniss zwischen dem Emissionsvermögen und dem Absorptionsvermögen der Körper für Wärme und Licht', *Poggend. Annal.,* 109 (1860), pp. 275–301.
2. *not as an excerpt but in its entirety:* the article was published as G. Kirchhoff, 'I. On the relation between the radiating and absorbing powers of different bodies for light and heat', *Phil. Mag.,* 20 (1860), pp. 1–21.
3. *together with the enclosed handwritten postscript:* the postscript was included in the article as requested.
4. *translation:* the article was translated by the physicist and chemist Frederick Guthrie (1833–86). Guthrie later collaborated with Tyndall, and took over Tyndall's post at the Royal School of Mines in 1869 on Tyndall's recommendation (ODNB).

From Michael Faraday 16 May 1860 1701

Royal Institution | 16 May 1860

My dear Tyndall,

Just a line to say I am very sorry you grumble at your head. I was in hopes the weather and country had refreshed you as it has done me. Thomson is here at work for Friday[1] and inquiring earnestly after you. De la Rive is busy diplomatically.[2] Barry is gone from us;[3] it strikes me the more because I have had occasion to be at the houses of parliament lately and it was only on Saturday week we had a long conversation at the Royal Academy Dinner. Health and long life to you is the sincere wish of your friend

RI MS JT/1/TYP/12/4151
Faraday Correspondence 5: 3781

---

1. *Thomson is here at work for Friday*: William Thomson gave the Friday Evening Discourse of 18 May 1860 on 'Atmospheric Electricity', *Roy. Soc. Proc.,* 3 (1862), pp. 262, 277–90.
2. *De la Rive is busy diplomatically*: following the annexation of Savoy by France, De la Rive was appointed to be a Swiss special ambassador to England. (*Faraday Correspondence* 5: 3781; *The Encyclopædia Britannica: A Dictionary of Arts, Sciences, Literature and General Information,* 7 (1910) p. 944).
3. *Barry is gone from us*: the architect Charles Barry (1795–1860) died 12 May 1860.

From Lyon Playfair 16 May 1860 1702

Edinburgh | 16 May /60

My dear Tyndall

I have been reading your own & others views on Regelation[1] & it has struck me that a good deal too much stress is put on Water & Ice being the same Chemical Body. I do not believe this to be so. Let me start with my assumption, Water according to my view is constituted on the type of Peroxide of Hydrogen

$$\left.\begin{matrix} O \\ O \end{matrix}\right\} H \text{ like } \left.\begin{matrix} O \\ HO \end{matrix}\right\} H \text{ Water,}$$

The reason for this I need not enter into at present. Ice according to my fancy is

$$\left.\begin{matrix} O \\ HO \end{matrix}\right\} H + 9HO$$

or a solid hydrate of the chemical body Water represented in its exact analogue Sulphuretted Hydrogen, peroxide of Barium, Hydrates of Chlorine, Sulphurous Acid & e.

Hydrate of Chlorine,Bra$^{e?}$

$$\left.\begin{matrix} O \\ Cl \end{matrix}\right\} H + 9HO$$

Wöhlers Hydrate of HS. (probably).

$$\left.\begin{matrix} HO \\ S \end{matrix}\right\} H + 9HO$$

Hydrate of Sulphurous Acid

$$\left.\begin{matrix} O \\ SO_2 \end{matrix}\right\} H + 9HO$$

Hydrate of Peroxide Barium

$$\left.\begin{matrix} O \\ O \end{matrix}\right\} Ba + 6HO$$

or Hydrate Barytes

$$\propto^{e} \left.\begin{matrix} O \\ HO \end{matrix}\right\} Ba^{2} + 9HO$$

Now this assumed Hydrate of Water constituted on its analogous solid hydrates, I further assume begins to be formed at the point of maximum density of Water, or rather just below it, but that it is held in solution by the Water which has not passed into that state.

Hence Water or rather the solution of Ice between 39.1 & 32°[2] expand because Ice

$$\left.\begin{matrix} O \\ HO \end{matrix}\right\} H + 9HO \quad \text{is less dense than Water} \quad \left.\begin{matrix} O \\ HO \end{matrix}\right\} H$$

Hence also its refraction angle increases as though it contracted, because the hydrate ice has a greater refractive angle than the body Water.

Hence also the solution of Ice may be cooled below 32° without freezing, just as a supersaturated solution of Sulphate of Soda can introduce a crystal of the Salt into the latter & of ice into the former & the crystals shoot out

$$\left.\begin{matrix} O \\ SO_3 \end{matrix}\right\} Na + 10HO \quad + \quad \left.\begin{matrix} O \\ HO \end{matrix}\right\} H + 9HO$$

If then there be any foundation in the assumption that Water between 39° & 32° is not simply Water but a solution of its solid Hydrate Ice, you will readily see an application to regelation, for anything that favours the crystallization of this dissolved Hydrate may produce junction at a temperature which according to the favourable conditions presented may be anywhere between 39° & 32°.

The first part of the letter involves a theory on which you could consult Frankland (to whom I have explained it) viz. that neutral salts & hydrates are constituted on the only neutral oxide which we know

$$\left.\begin{matrix} O \\ O \end{matrix}\right\} H \quad \text{or} \quad \left.\begin{matrix} O \\ O \end{matrix}\right\} Mn$$

& not on the highly basic protoxides

$$\left.\begin{matrix} H \\ H \end{matrix}\right\} O_2$$

the favourite type of the Young School.[3] But I send the suggestion to you merely as one worth thinking of—as a speculation not a theory

Yours truly | Lyon Playfair

RI MS JT/1/P/122
RI MS JT/1/TYP/3/992–93

---

1. *Regelation*: see letter 1594, n. 3.

2. *39.1 & 32°*: that is, Playfair was examining the nature of water and ice at the melting point, 32° Fahrenheit, or 0° Celsius.
3. *Young School*: Playfair was likely referring to the work of the physicist Thomas Young (see letter 1662, n. 22) in determining the size of molecules.

From William Hopkins 19 May 1860 1703

Cambridge | May 19 '60

My dear Sir,

I have received certain portions of your work on Glaciers,[1] but have not yet had time to make much way into it, which I shall be able to do however in a few days. You speak (p. 286) of having made an experiment on a vertical precipice to ascertain the rates of motion at different depths.[2] I infer that this precipice extended down to the bed of the glacier. This is an important point and ought not I think be left to inference. A word or two would correct it.

You ask me whether I would undertake to review it.[3] I cannot promise but shall be much disposed to do so, if any good periodical would insert it. I don't recollect any thing of the kind to have appeared in the Quarterly. If they would take it I would write it; or perhaps if Parker would insert it in Fraser's Mag. I would not write it for any inferior periodical. Have you any idea whether the Quarterly would accept it. There wants a general review of the whole subject. All that I have seen, full of Forbes and nothing else, have been wretched one-sided things.

I am sorry to hear that your health is banishing you from town. Take care of yourself in time.

Yours truly | W. Hopkins

I have just finished a Review of Darwin for Fraser.[4] Tell your friend Huxley he mustn't quarrel with me for dealing severely with his friend's theory.[5] I like scientific freedom, with personal courtesy, but I hate scientific quarrelling.

RI MS JT/1/TYP/2/615
LT Transcript Only

---

1. *your work on glaciers:* the manuscript or proof pages of *Glaciers of the Alps,* see also letter 1595, n. 4.
2. *rates of motion at different depths:* these measurements are described on pp. 289–93 of *Glaciers of the Alps.* Tyndall included the sentence 'Our next care must be to compare the motion of the surface of a glacier with the motion of those parts which lie near its bed' (p. 289).
3. *You ask me whether I would undertake to review it:* letter missing.
4. *I have just finished a Review of Darwin for Fraser:* W. Hopkins, 'Physical theories of the

Phenomena of Life', *Fraser's Magazine,* 61 (June 1860), pp. 739–52, and W. Hopkins, 'Physical theories of the Phenomena of Life. Part II', *Fraser's Magazine,* 62 (July 1860), pp. 74–90.

5. *Tell your friend Huxley he mustn't quarrel with me for dealing severely with his friend's theory:* Hopkins's review, while asserting the right of the naturalist to regard the phenomena of life as physical processes, argues regarding the theories of Lamarck and Darwin, that they do not show any 'very essential difference in the fundamental hypotheses on which the two theories rest as physical theories'. He states his admiration for Darwin's marshalling of facts, but expresses the view that Lamarck's theory is preferable to the 'less aristocratic process' of Darwin's theory of continuous variation and natural selection, based on 'the accident of birth'. In the second part he argues in particular the lack of experimental evidence for the theory, the lack of intermediate forms in the fossil record, and our lack of understanding of the physical causes, i.e. of the 'molecular forces which govern the combinations of animate matter'.

To [George Gabriel Stokes][1] 24 May 1860 1704

24th May 1860

My dear Sir,

A cursory inspection of Mr. Maxwell's 'postscript' which is all the state of my health permits me to bestow upon it, leaves no doubt upon my mind as to the desirableness of its publication in the Philosophical Transactions. I would strongly recommend its publication.[2]

Very sincerely yours | John Tyndall

RS RR/4/180

---

1. *[George Gabriel Stokes]*: Stokes is the assumed recipient of this letter, based on the reference to Maxwell's paper, discussed in letter 1699.
2. *I would strongly recommend its publication*: see letter 1699, n. 1.

To Sir John [unidentified][1] 25 May [1860][2] 1705

Royal Institution | 25th May

Dear Sir John

It would give me exceeding pleasure to accept your invitation were I a free agent. But the fact is that I have recently hurt my brain very seriously and am under strict orders to keep myself quiet. On the 4th of June moreover I expect to be in the /Frolic/ Isle.[3]

Most truly yours | John Tyndall.[4]

Wellcome Library MS. 7777/03

1. *[unidentified]:* recipient unknown, possibly John Frederick William Herschel.
2. *1860:* dated from the writing on the reverse (see n. 4).
3. *Frolic:* a possible reading of the text, perhaps a jokey reference to the Isle of Wight, which he often visited for relaxation, by comparison to 'Hogarth's Frolic' on the Isle of Sheppey. Conceivably it could be a visit to the Isle of Sheppey.
4. The following is handwritten on the reverse of the letter: '1860 Sunday | Tyndall can't come on the 4th June'. It might have been added by the recipient on Sunday 27 May. 4 June was a Monday.

To [George Gabriel Stokes][1] 26 May 1860 1706

26th May 1860

My dear Sir,

I consider the paper of Messrs Matthiessen & Holzmann a very important one, and most heartily recommend it for publication in the Philosophical Transactions.[2]

Yours very since[rel]y | John Tyndall[3]

RS RR/4/174

1. *[George Gabriel Stokes]*: the recipient is assumed to be Stokes as the letter concerns the referee process for the physical section of the *Phil. Trans.*.
2. *and most heartily recommend it for publication in the Philosophical Transactions*: the paper was published as A. Matthiessen and M. Holzmann, 'On the Effect of the presence of Metals and Metalloids upon the Electric Conducting Power of Pure Copper', *Phil. Trans.*, 150 (1860), pp. 85–92.
3. At the end of the letter a note which appears to be in Stokes's handwriting was added: '1860 | Matthiessen | & Holzmann | Tyndall'.

To John William Parker 26 May [1860][1] 1707

Royal Institution | Private Saturday 26th

Dear Mr Parker

I have just received a note[2] in which your name is mentioned and which induces me to write to you.

Mr W[illia]m Hopkins of Cambridge, who has just completed for you a Review of Darwin's work,[3] is the author of a series of most valuable papers upon glaciers. I am not acquainted with anybody in England more competent to write upon the subject: indeed his high character and scientific eminence are in themselves a sufficient guarantee that any thing he might write would be both able and impartial. He has never had a suitable opportunity of laying before the general public what he has himself done in connexion with glaciers, and I for one should like to see the opportunity granted to him. In a recent note to me he says 'There needs a general review of the whole subject—all that I have seen have been wretched one sided things'. Now such a review he is disposed to write, for some first rate periodical,—such for example as Frazer's magazine—Would you feel disposed to accept it?[4]

Very sincerely yours | John Tyndall

Wellcome Library MS. 7777/80–80b

1. *26 May [1860]:* see n. 2.
2. *I have just received a note:* letter 1703.
3. *Review of Darwin's work:* W. Hopkins, 'Physical theories of the Phenomena of Life', *Fraser's Magazine,* 61 (June 1860), pp. 739–52.
4. *Would you feel disposed to accept it?*: Hopkins did write a review for *Fraser's Magazine.* See letter 1731, n. 11.

From William Hopkins 29 May 1860 1708

Cambridge | May 29. 60

My dear Sir,

I was just finishing the last pages I have received of your 'Glaciers'[1] when your note[2] reached me this morning. I return them without further delay, with a few annotations, of very minor importance, which I have made. Any other sheets you may send for my inspection, should you wish it, address to me St. Peter's Coll. Camb., if sent on or before Thursday. If later, my address will be 3 Belvoir Terrace, | Scarboro'.

The style of your book is I think very happy for popular reading, and the work will be far superior to any other I know, for giving at once a correct and popular acquaintance with glaciers and their phenomena. Some of the discussions on the theories of sound, light, and heat may hardly appear perhaps to have a sufficiently obvious relation to 'Glaciers', and may appear, I think, to some readers rather discrepant, unless you could include them, as it were, in your table; so as to intimate (if it could be done with sufficient brevity) that the work would contain some explanations of physical principles applicable to Glacier Phenomena. Or possibly this might be much better done explicitly in your introduction. But you will best judge of this.[3]

Yours will be the first work in which any thing like justice has been done to Rendu.[4] If Forbes knew his writings, (as I doubt not he did), he has acted most unworthily towards him.[5] But it is the same want of candour which he showed towards Agassiz.

Yours truly | W. Hopkins.

RI MS JT/1/TYP/2/616
LT Transcript Only

---

1. *your 'Glaciers':* the manuscript or proof pages of *Glaciers of the Alps,* see also letter 1595, n. 4.
2. *your note:* missing.
3. *But you will best judge of this:* Tyndall explained his rationale in the preface, *Glaciers of the Alps,* p. v.
4. *Rendu:* see letter 1619, n. 13.
5. *he has acted most unworthily towards him:* the question of whether or not Forbes had properly acknowledged Rendu was disputed. Forbes and his friends believed he had; Tyndall and his friends believed the opposite. See e.g. N. K. L. Kaalund, 'A frosty disagreement: John Tyndall, James David Forbes, and the early formation of the X-Club', *Annals of Science,* 74 (2017), pp. 282–98, on p. 289.

From William Hopkins 4 June 1860 1709

Scarboro' | 3 Belvoir Terrace | June 4. 60

My dear Sir,

I brought with me here the concluding sheets of your work,[1] and found in them much pleasant reading during my journey. I have made no correction, I think, except the substitution of the word 'oval' for 'ellipse', a change of very minor importance; but though every ellipse is an oval, very few ovals, you know, are ellipses. I think your general statements respecting the structure of glaciers, and of the evidence on which your views rest, extremely clearly put and perfectly comprehensible to any reader of ordinary intelligence. Forbes's theory of the veined structure[2] is unquestionably wrong, and it is impossible for him to either wiggle out of it, or to twist it into any admissible form. He can have no right however to make the slightest complaint of any thing you have said. My impression is in fact that, throughout the whole of your paper on glaciers, you have dealt out the soft solder rather too freely to him. But if an error at all it was assuredly on the right side.

On thinking over your whole work, two omissions occur to me, one of which I ought to have mentioned when I wrote last,[3] if I am correct in supposing that the point I allude to is omitted. According to Forbes's observations on the extremity of one of Chamouni[4] glaciers, it appeared that the sliding motion of the surface of the glacier was about one half of the whole motion of the surface of the mass; and a similar result is deducible (or pretty nearly so) from your own observations on the flank of the Mer de Glace. It is just as important therefore to explain how such uniform sliding takes place over the inclined bed of the glacier, as to explain that part of the motion which may be due to the structure and change of form of the glacial mass, or any other cause. Forbes in the first edition of his 'Travels'[5] laid great stress (and some others did at the time) on the difficulty of explaining a uniform, instead of an accelerated, motion down the incline plane of the glacial valley. I believe no one has ventured to express a doubt of my own explanation[6] being entirely correct. But in the discussions of glacial motion scarcely any one has ever alluded to it. You have I know distinctly recognised the fact of all glaciers sliding, but I don't recollect your mention of the explanation I alluded to. However I will not abuse you for the omission, though, should I ever review your book, I must of course correct it.

The other omission I allude to is the explanation which Prof. William Thomson[7] gave of the formation of the aqueous strata of liquefied ice, due to the action of pressure in a direction perpendicular to those strata. I recollect its appearing in the Proceedings of the Royal Society,[8] and I suppose the explanation due to him, though you have not I think mentioned his name in connection with it. But you will know better than myself what claim he has to such mention. Still I would recommend your giving him whatever he may have just claim to, because of circumstances and associations will naturally tend to place him more or less in the faction opposed to you.

I mentioned to you before that I thought you had said more about light, heat &c than your subject called for. But I object to your 'Conclusion'[9] far more decidedly, as altogether irrelevant to the general purpose of your book. It is not that I object to your views as expressed in it, for you may see in the 'Fraser' of this month[10] I have expressed very similar views, and the subject is one of great interest and importance at this particular moment; but what, it will be asked, has it do with 'The Glaciers of the Alps'? You must recollect that your book will have its strong opponents, and you may depend upon it they will not fail to designate your 'Conclusion' as altogether mal apropos and introduced for a bit of display, an accusation which, as an earnest and truth-seeking philosopher, I know you would heartily detest. If you wish to express your views publically on the extent to which the applicability of physical research may be legitimate, why not do it independently, and in some way in which you might give it more of that development which so grave a subject demands? But I have said enough, and would not have said half so much, had I not felt sure that the criticism will be taken in good part. Still don't, I pray you, depend solely upon my judgment, I may be over fastidious on such points. Consult some other friend on the subject, and then judge for yourself.

Believe me | Yours very truly | W. Hopkins.

I send the Review[11] by the same post as this letter.

RI MS JT/1/TYP/2/617–19
LT Transcript Only

---

1. *your work: Glaciers of the Alps,* see also letter 1595, n. 4.
2. *Forbes's theory of the veined structure:* Forbes had originally proposed that the veins were small fissures formed in the ice by its motion, that these were filled with the water of the melted ice in summer, which froze in winter so as to form the blue veins (J. D. Forbes, *Travels through the Alps of Savoy,* p. 377 (see letter 1594, n. 13). He retracted this theory in his 'Thirteenth Letter' (J. D. Forbes, *Occasional Papers on the Theory of Glaciers,* pp. 187–205. This letter was originally published in the *Edinburgh New Philosophical Journal,* January 1847 as J. D. Forbes, 'Thirteenth Letter on Glaciers; addressed to Professor Jameson', *Edinburgh New Philosophical Journal,* 42 (1847), pp. 136–54; see letter 1620, n. 4). Tyndall was unaware of it when he wrote his first paper on glaciers with Huxley, but believed Forbes had substituted an equally incorrect explanation: see *Glaciers of the Alps,* p. 380.
3. *when I wrote last:* letter 1708.
4. *Chamouni:* see letter 1633, n. 2.
5. *'Travels':* see n. 2.
6. *my own explanation:* Hopkins gave his explanation of the importance and mechanism of the sliding of glaciers, and of glacier properties overall, in a later extensive paper: W. Hopkins, 'On the Theory of the Motion of Glaciers', *Phil. Trans.,* 152 (1862), pp. 677–745.
7. *Prof. William Thomson:* Hopkins appears to be mistaken, as the paper was by William Thomson's brother James Thomson (1822–92), Professor of Engineering at Queen's College, Belfast. See n. 8.
8. *in the Proceedings of the Royal Society:* J. Thomson, 'On Recent Theories and Experiments Regarding Ice at or near Its Melting-Point', *Roy. Soc. Proc.,* 10 (1859), pp. 151–60.
9. *your 'Conclusion':* Tyndall had originally intended to include in *Glaciers of the Alps* a section stating his 'materialist' position that the visible phenomena of life could be reduced to mechanical attractions and repulsions; i.e. that the processes of life (though not its meaning and purpose) could be explained solely in terms of natural philosophy. This would have been seen by many as highly controversial, and an assault on religion and theology. Tyndall published his views, anonymously, in the *Saturday Review* just as he left for the Alps: 'Physics and Metaphysics', *Saturday Review,* (4 August 1860), pp. 140–41.
10. *the 'Fraser' of this month:* W. Hopkins, 'Physical theories of the Phenomena of Life. Part I', *Fraser's Magazine,* (June 1860), pp. 739–52. See also letter 1703, n. 4 and n. 5.
11. *the Review:* see n. 10.

From Jane Benson[1] 5 June 1860 1710

1 Albert Terrace | Moss Side[2] | June 5th 1860

Dear Dr. Tyndall,

My pen had almost written the old address, 'Dear brother John' for the account from Queenwood of your weary head and worn down strength awakened the old desire to nurse and comfort you which originated the title. I hope the change and quiet have been of use to you; this 'interesting contribution to Alpine literature' which I see Mr Murray promises,[3] must have been a heavy addition to the ordinary demands on your time and thought, for a book, to be worth reading cannot be hastily written. I should have thanked both you and Dr. Hirst,[4] long ago, for your kind expressions of good wishes, which I valued much; but 'some how or other' one does not settle down all at once in the new life, and letter writing is too often set aside. One's habits are as much broken up as the particles of glacier ice but do not regelate so quickly as I think you say they do.[5] Do you remember Mr Benson in those early unsettled days at Queenwood?[6] We both hope so, and that the acquaintance may be renewed at some not very distant time. I should have felt considerable contempt for any prophet who had told me the last time you and I said good bye, that, he—my old friend—would prove the strong man 'with power to rule my will' for whom I was waiting: yet so it was; we met after ten years' separation, shook hands, and behold! 'the little wilful sister'[7] finds herself managed beautifully—and happily, or she would not be so ready to confess it. We were delighted to notice in your note,[8] the little 'perhaps' inserted as an afterthought in your assertion that 'it was good for man in your case to be alone'; it implied a doubt after all, in which we quite agreed, most sincerely wishing that some loving little heart had the right to tend and cheer you. You would only be a 'terrible plague' to a wife who had not full faith in you. We will not despair of you yet. Next summer is a long time to look forward to, but my good husband wishes me to remind you that the British Association is to be held here then,[9] and to assure you how much pleasure it would give us to welcome you to our little home. Small as it is, we have two spare rooms for our friends, and if we cannot offer them the freedom of country doings out of doors, they may depend on full liberty to do just as they please within the house; cocoa or mutton chops—puddings or oatmeal porridge &c. &c. being obtainable here as well as at Queenwood. So—we will say nothing of the honour we should feel, but tell you simply that at any time if you wander in this direction there will be a warm welcome awaiting you at No. 1 Albert Terrace. At No. 2. lives Mr F. Crace Calvert,[10] he removed from this house when we took it, preferring a front room for his study, which he could not arrange here.

His nerves must be strong, to enable him to write amid the rumble of the carriages in the front; but he does not look as if he were troubled with nerves, as he walks up from the omnibus with portly frame and elevated chin, at peace with himself apparently, if not all the world. I surely ought to apologise for so long a scribble, but you know of old I am a rambling letter writer when once my pen is at work.

I will not tax your memory, but if it should occur to you when you see Dr. Hirst or Dr. and Mrs Frankland[11] will you kindly remember me to them. I am looking forward with much pleasure to spending a few weeks at Queenwood doing the summer holidays, the dear old place will be looking beautiful soon.

United by Mr Benson in kindest regards | Believe me | Yours sincerely | Jane Benson

RI MS JT/5/14/33–34
LT Typescript Only

---

1. *Jane Benson*: Jane Benson, née Edmondson (1823–1906), was the only daughter of George Edmondson, headmaster of Queenwood College, and his wife Anne. She had been born in St Petersburg, and married a Lancashire business man, David Benson (1820–92) in 1859.
2. *Moss Side*: in Manchester. 'Manchester' is handwritten on the typescript.
3. *'interesting contribution to Alpine literature' which I see Mr Murray promises*: *Glaciers of the Alps,* published by Murray on 6 July 1860.
4. *Dr. Hirst*: Thomas Hirst.
5. *One's habits are as much broken up as the particles of glacier ice but do not regelate so quickly as I think you say they do*: an amusing reference to Tyndall's theory of fracture and regelation.
6. *Mr Benson in those early unsettled days at Queenwood*: Tyndall had joined Queenwood College in 1847 as it was being established under George Edmondson. 'Unsettled' may refer to the inevitable challenges of early days and the tensions that existed between some of the teachers, including Tyndall, and the Edmondsons. See volume 2 for letters of the period.
7. *'the little wilful sister'*: see *Tyndall Correspondence,* vol. 4, letter 0890, where Tyndall had addressed Jane as 'my wilful sister'.
8. *your note*: missing.
9. *the British Association is to be held here then*: the Annual Meeting of the British Association was held in Manchester from 4–11 September 1861. Tyndall did not attend, as he was still in the Alps.
10. *Mr F. Crace Calvert*: Frederick Crace Calvert (1819–73), industrial and analytical chemist (FRS 1859). Tyndall reviewed at this time a paper of which Calvert was a coauthor (see letter 1716).
11. *Dr. and Mrs Frankland*: Edward and Sophie Frankland.

From Henry Bowie[1] 6 June 1860 1711

Philosophical Institution | Edinburgh 6 June 1860

Dear Sir

I am again instructed by the Directors of the Philosophical Institution[2] to convey to you their very Cordial and unanimous invitation to deliver, some time in the Course of next Session, two Lectures to the Members, and they all will be much gratified to learn that you can accept the invitation. The subject which they beg to propose to you for treatment is the Phenomena of the Glaciers.

Our platform is happily a broad one, and from which men holding different views on Scientific and Literary subjects can, with the utmost freedom and good feeling, announce their views. In this spirit you would be welcomed to the Institution—heartily welcomed; and permit me also to assure you that not only to the Members generally, but to not a few men of Science here, and several of these friends of your own, your visit would, likewise, be peculiarly gratifying.

Our session begins in November 1860, and ends about the middle of March 1861:—the lecture days being the Tuesday and Friday of each week. You could thus, if necessary, leave London on a Monday night and be back on Saturday. The Directors would take it very kind could you, if you can do no more, give this short space to a visit here; and they will, with pleasure, endeavor to accommodate you with the exact date in the Session most suitable to you.

I formerly mentioned[3] that the fee for the Two Lectures is £21—

I send by this post a Copy of the Report of the affairs of the Institution[4] for the year lately closed—and I beg to direct your attention, in particular, to the portion of the Report bearing on our lecture department.

My Directors will be glad to have a reply at your early convenience.[5]

I am Dear Sir | Yours faithfully | Henry Bowie; Sec[retar]y

Professor Tyndall

RI MS JT/1/B/121

---

1. *Henry Bowie:* Henry Bowie (1812–85), secretary and treasurer of the Edinburgh Philosophical Institution.
2. *the Philosophical Institution:* the Edinburgh Philosophical Institution, formed in 1846, was a reformed version of the earlier Edinburgh Philosophical Association. Shapin argues that it was 'less scientific' but 'more socially elevated' than its predecessor, with lawyers being a much larger share of the membership: S. Shapin, '"Nibbling at the teats of science':

Edinburgh and the diffusion of science in the 1830s', in I. Inkster and J. Morrell (eds), *Metropolis and Province* (London: Routledge, 1983), p. 177. For a detailed analysis of the Philosophical Institution, see D. A. Finnegan, 'Placing science in an age of oratory: spaces of scientific speech in mid-Victorian Edinburgh', in D. N. Livingstone and C. W. J. Withers (eds), *Geographies of Nineteenth-Century Science* (Chicago: University of Chicago Press, 2011), pp. 153–77.

3. *I formerly mentioned:* letter missing.
4. *Report of the affairs of the Institution:* missing.
5. *a reply at your early convenience:* reply is missing. It does not appear that Tyndall lectured on this occasion. He did lecture at the Edinburgh Philosophical Institution in 1866, on sonorous vibrations: 'Philosophical Institution—Professor Tyndall on Sonorous Vibrations', *Caledonian Mercury,* 15 December 1866.

From Elizabeth Steuart [9 June 1860][1] 1712

June 9, 60.

My dear John,

I lose no time in writing to offer you my best thanks for your kind and most welcome present,[2] received this morning, and which I shall highly value, as the gift of one in whose welfare I have felt such warm interest for so many years.

I trust your late excursion to this country[3] has been attended with benefit to your health, though your stay was so short, still your native air would enliven and refresh you. I shall anxiously look for a few lines from the Alps, which I truly hope you may reach and <u>leave</u> in perfect safety.

And now, my dear John, accept the kindest and best | wishes of your sincere friend | E.D. Steuart

RI MS JT/1/TYP/10/3341
LT Typescript Only

1. *[9 June 1860]*: only the typescript exists, and the date may have been inferred and added by LT. Tyndall was in Ireland from 14–26 June 1860, where Elizabeth Steuart lived, so if the typescript is correct in giving 1860 as the year, the letter may have been written a few weeks later. Tyndall left for the Alps on 2 August 1860.
2. *your kind and most welcome present*: unidentified.
3. *your late excursion to this country*: see n. 1.

To Miss Moore[1] 12 June 1860 1713

Royal Institution | 12th June 1860

My dear Miss Moore

I thought this morning as I lay resting my weary head an hour longer than usual that I would write to you today. Indeed my first impulse when I heard of your recent loss[2] was to write to you. I used to trust my impulses, but not quite so much now as formerly; and on reflection I thought it might be more agreeable to your feelings if I refrained from expressing mine. But I cannot allow this to pass without saying to you that I heard with deep sorrow the intelligence of your father's death. I do not forget how kindly and cordially he used to welcome me, and with what great interest I have listened to the recital of his wonderful memories. One of my earliest meetings with him was at Wimbledon,[3] and as I left the house and bade him goodbye, he squeezed my hand and hoped that his son,[4] and his son's son[5] & myself should be good friends. Well he is gone—gone at an age which mortals rarely attain;[6] still, however the mind assents to this, as the inevitable law of nature, well I know the hearts by which he was surrounded will not so easily reconcile themselves to his loss. But that which was pain at first will melt into a gentler feeling by and by, in which pain will scarcely form an element.

On Monday next I hope to be able to commence my excursion to Ireland.[7] My book will be published during my absence[8] & I shall leave directions to have a copy sent to you. I should rather have liked to write my name in it, but should not like to keep you waiting till my return. Will you present my kindest remembrances to your mother & sister; to Mr John Carrick Moore; & to his son—good bye

Most sincerely yours | J. Tyndall

RI MS JT/1/T/1073
RI MS JT/1/TYP/3/867

---

1. *Miss Moore:* Harriet Moore or Julia Moore.
2. *your recent loss:* her father James Carrick-Moore (1762–1860) had died on 1 June 1860, at 9 Clarges Street.
3. *One of my earliest meetings with him was at Wimbledon:* on 12 July 1856, Tyndall recalled a visit to the Moores in Wimbledon, commenting that the old Mr Moore's brother was the Sir John Moore who fell at Corunna (Journal, RI MS JT/2/13c/861–62).
4. *his son:* John Carrick Moore (1805–98). He was a Manager of the Royal Institution 1857–58.
5. *his son's son:* John Graham Carrick Moore (1845–90).

6. *an age which mortals rarely attain:* he died aged 97.
7. *my excursion to Ireland:* According to Hirst, Tyndall was in Ireland for a 'short holiday' from 14–26 June (*Hirst Journal*, RI MS JT/2/32c/1540–42). On 18 June he was in Killarney (see letter 1665).
8. *My book will be published during my absence:* in the event, *Glaciers of the Alps* was published on 6 July 1860, after his return. See also letter 1595, n. 4.

To Juliet Pollock 16 June [1860][1] 1714

Killarney | 16th June.

My dear Mrs Pollock.

I thought of calling to see you before quitting London[2] but couldn't. I wanted to beg your pardon for not joining you at the cricket ground—but I was so entangled, and so worn down even if I were not entangled that I found myself wholly unfit to go. I reached this place yesterday evening—throughout the day heavy showers fell, some on the train, but others on the flanking mountains, dense and black while we sped on in sunshine. You would love this place—Despite the rain it is delightful. Last evening I took a boat to the Island of Innisfallen. I was quite alone. The island is cumbered with ruins: covered with clustering foliage with patches of the richest verdure between. The lake waters lapped audibly on the rocks all round, giving tone to the stillness of the spot. It was quite a fairy land. As I returned the western heaven was walled with black clouds. I reached some eminences which commanded a view of the lake. Round about me were crumbling ruins, all old, very old. The sun once managed to look through a tunnel in the clouds, and hit Innisfallen and a series of other islands with his last beams; all right & left being black with gloom. The foliage glittered splendidly in the beam which at its end struck the breast of a cloud on which it painted as far as it reached a vivid rainbow. This hotel overlooks 'the lower lake'. There are three in all & today I walked through the gap of Dunloe, reached the head of the 1st lake, where a boat had been sent to meet me, and rowed through all three of them. I have never seen anything in England, Scotland, Wales or Switzerland so truly lovely, lovely is the word. I have seen grander & larger lakes; but lovelier, never. I wish you were here in fine weather. It would enter into your very soul.

The mountains however are horrible condensers, and the rain fell down upon me today as if some fury were pouring it out of a watering pot. I am getting strong; more and more springy every hour; and if I can get only a few hours fair weather I shall clamber up Carrantual[3] and drink your health there. I have already done on the highest spot in England[4]—and the highest in Europe,[5] and I don't see why I should not do it on the highest in Ireland.

Now excuse all this rambling—I wanted to fill an hour before dinner which would otherwise be a vacuum and lo! here is the result

Ever most truly yours | John Tyndall.

RI MS JT/1/T/1167
RI MS JT/1/TYP/6/2039

1. *[1860]*: Tyndall visited Ireland, including Killarney, from 14–26 June 1860 (*Hirst Journal*, RI MS JT/2/32c/1540 and 1542). It was this visit that inspired him to revisit the Alps, after he had thought in 1859 that he would not return.
2. *before quitting London*: Tyndall left on 14 June.
3. *Carrantual*: Carrauntoohil, the highest mountain in Ireland (1,038m or 3,404ft).
4. *the highest spot in England*: Scafell Pike, in the Lake District. In April 1859, according to Edward Frankland, he and Tyndall climbed Scafell (by which he probably meant Scafell Pike, 978m or 3,209ft, which is slightly higher than Scafell, 964m or 3,164ft). See Sketch of the Ascent of Scafell, Papers of Sir Edward Frankland, ex Joan Bucknell, Box 2, John Rylands Library, University of Manchester. Tyndall did not mention this ascent in his journal.
5. *the highest in Europe*: Tyndall meant Mont Blanc (4,810m or 15,780ft), which he climbed three times. Today, Mount Elbrus (5,642m or 18,510ft) in the Caucasus is considered the highest mountain in Europe.

From Gustav Magnus 16 June 1860 1715

Berlin 16 Juni 60

Mein theurer Tyndall

Ich habe jetzt eine neue Ausgabe meiner Schrift über die Abweichung der Geschoße drucken lassen, die einiges Neue enthält. Ich werde ihnen durch Asher ein Exemplar zugehen lassen, ebenso denke ich an General Sabine ein Exemplar zu senden, wenn ich aber nicht irre, so interessirte sich damals noch General Yorke für die Sache; soll ich diesem auch ein Exemplar senden oder wem sonst, oder wollen Sie es übernehmen diese Exemplare zu vertheilen? Gern möchte ich dem Prinzen Albert auch eine Aufmerksamkeit erweisen, da er sich der Humboldt Stiftung so thätig angenommen hat; aber ich möchte nichts mit diplomatischen oder *[treuhänderischen]* Verhandlungen zu thun haben. Ich denke mir die Sache interessirt gerade jetzt die Militaire.

Ihre Abhandlung, betreffend das Vorrücken der Gletscher, habe ich mit großem Interesse gelesen, und bin gespannt auf das größere Werk das Sie über die Alpen zu veröffentlichen gedenken. Ich habe mich viel mit Versuchen aus dem Gebiete der Wärme beschäftigt und einige Facta gefunden, die zwar keine besondere Bedeutung haben, aber mir doch ganz interessant erschienen sind, so bald ich fertig theile ich Ihnen dieselben mit. Lassen Sie mir recht bald ihre Antwort zukommen, denn ich schicke die Exemplare nicht eher ab als bis ich Ihren Brief habe. Soll ich für Faraday ein Exemplar beifügen? Sie wissen wie hoch ihn schätze! Wenn Sie glauben daß er die Aufmerksamkeit, denn weiter ist es nichts, gern sieht, so macht es mir die größte Freude sie ihm zu erweisen.

Von Herzen | Ihr | GMagnus.

Prof. Tyndall.

Berlin 16 June 60

My dear Tyndall

I have now had a new edition of my paper on the deflection of projectiles printed,[1] which contains quite a few new things. I will send you a copy via Asher;[2] I am also thinking to send a copy to General Sabine[3]—if I am not mistaken, General Yorke[4] was also interested in this subject back then; shall I also send him a copy or to whom else, or do you want to assume the distribution of these copies? I would like to show attention to Prince Albert, as he has so actively looked after the Humboldt Foundation,[5] but I would not like to have to do with any diplomatic or *[fiduciary]* negotiations. I think that the matter is of interest to the military especially now.

I have read your paper concerning the movement of glaciers[6] with great interest, and am curious to see your larger work on the Alps[7] that you are thinking about publishing. I have been very busy with experiments in the field of heat[8] and found some facts that admittedly do not have any special significance, but seemed to me quite interesting; as soon as I am finished I will tell you about them. Let me have your reply[9] quite soon, as I will not send the copies before I have your letter. Shall I add a copy for Faraday? You know how much I admire him! If you think that he will appreciate this small attention, for it really is nothing more, it will bring me the greatest joy to give it to him.

Cordially | Your | GMagnus.

RI MS JT/1/M/23

---

1. *a new edition on the deflection of projectiles:* Magnus was referring to his 'Ueber die Abweichung der Geschosse', *Abhandlungen der Königlichen Akademie der Wissenschaften in Berlin,* (1852), pp. 1–23; and his paper 'Ueber die Abweichung der Geschosse, und: Ueber eine auffallende Erscheinung bei rotirenden Körpern', (translated as 'On the deviation of projectiles, and: On a Sinking Phenomenon among rotating bodies'), which was published in *Poggend. Annal.,* 164 (1853), pp. 1–29. The paper was reprinted in 1860 by Ferdinand Dümmler's publishing house in Berlin.

2. *Asher:* Adolphus Asher.
3. *General Sabine:* Edward Sabine was a Colonel at the time, and did not become a General until 1870 (*ODNB*).
4. *General Yorke:* Probably Philip James Yorke (1799–1874), elected FRS in 1849. He retired from the army in 1847 as a Lieutenant-Colonel, and was Colonel for the Herefordshire Militia from 1852–59 (ODNB). Magnus appears to have over-promoted both Sabine and Yorke.
5. *Humboldt Foundation:* Prince Albert supported the establishment of the proposed Humboldt Foundation for Physical Science and Travels. At his inaugural address as President of the BAAS in Aberdeen in 1859, he declared his support for the proposed Foundation, providing £100 and heading the subscription list, 'Address by His Royal Highness the Prince Consort', Report of the Twenty-Ninth Meeting (Sept 1859), *Brit. Assoc. Rep.*, 29 (London: John Murray, 1860), pp. lviii–lxix.
6. *your paper concerning the movement of glaciers:* the most recent paper was 'On the Veined Structure of Glaciers; with Observations upon White Ice-Seams, Air-Bubbles and Dirt-Bands, and Remarks upon Glacier Theories', *Phil. Trans.*, 149 (1859), pp. 279–307.
7. *your larger work on the Alps:* presumably *Glaciers of the Alps*, published on 6 July 1860 (see also letter 1595, n. 4). Magnus may not have been aware how close it was to publication.
8. *experiments in the field of heat:* these experiments would give rise to a priority dispute between Magnus and Tyndall, and differences of interpretation over many years. See the Introduction.
9. *your reply:* no reply is extant. This volume contains only one letter from Tyndall to Magnus; the next is letter 1773.

To George Gabriel Stokes 18 June [1860][1] 1716

My dear Stokes.

I am very sorry that you have been put the trouble of writing to me again[2]—but had I had any hope of being able to report in favour of Prof. Calvert's paper[3] I should have strained a point to get it done I have looked at it a little and heard it read before the society[4] and my opinion is not in favour of its insertion in the Phil. Trans. But perhaps a reference of the paper to some other person would be safest.

I write by this post to my hostess[5] directing her where to find the paper, and to send it to you. If you like to wait till the 26th I shall be happy to look more carefully over it and give you a more decided opinion.[6]

Yours very sincerely | John Tyndall

Killarney| 18th June[7]

RS RR/4/32

1. *18 June [1860]*: dated by context. See n. 3.
2. *writing to me again*: letter missing.
3. *Prof. Calvert's paper*: F. G. Calvert and G. C. Lowe, 'On the Expansion of Metals and Alloys', *Roy. Soc. Proc.*, 10 (1859), pp. 315–18.
4. *read before the society*: the paper was read before the Royal Society on 16 February 1860.
5. *my hostess*: presumably his landlady.
6. *a more decided opinion*: there is no evidence that Stokes asked again. The paper was not published in the *Philosophical Transactions.*
7. '1860 Calvert & Lowe by Tyndall' is written upside down below this date, apparently in Stokes's hand.

To Henry Bence Jones [22 June 1860][1] 1717

Victoria Hotel, Killarney. | Friday

My dear Dr Jones

I must now report progress and am happy to be able to send a good report. I reached this place on Friday evening last; next day I walked through the Gap of Dunloe;[2] had a boat to meet me at the head of the Upper Lake and rowed through all three lakes home. It rained all the time. Next day I went to the summit of Mangerton:[3] the next rowed on the lake and rambled through Muckross Abbey and Demesne.[4] I have climbed the Eagle Rock mounted the Purple mountain,[5] and done sundry other things. My health strengthened from the very first, and if I could only remain I should absolutely get stout. I never experienced so great and quick a change for I was very much worn down. The exercise I went through yesterday was very trying, but this morning I feel as springy as a greyhound. My sleep is good and my appetite enormous. I can drink beer, wine brandy and Cork whiskey; and can also smoke with impunity. Had I not some examination papers to prepare I would forego Oxford[6] and remain here. These Lakes are exquisite—the loveliest things I have ever seen. Yesterday the scene from the summit of the Purple Mountains was indescribably beautiful. I shall be at home on the 26th, good bye

Ever Yours | John Tyndall

RI MS JT/1/TYP/3/793b
LT Transcript Only

1. *[22 June 1860]*: according to Thomas Hirst, Tyndall left for Ireland on 14 June and returned to England on 26 June (*Hirst Journal*, RI MS JT/2/32/c/1540–42). It is therefore likely that he arrived in Killarney on Friday 15 June, and that this letter was written a week later on Friday 22 June.

2. *Gap of Dunloe*: a mountain pass between the MacGillicuddy Reeks and the Purple Mountain remains a famous tourist excursion in a beautiful landscape of glacial origin, containing several lakes.
3. *summit of Mangerton*: Mangerton Mountain, in County Kerry, has a height of 839m or 2,753ft.
4. *Muckross Abbey and Demesne*: the Franciscan friary of Irrelagh, known as Muckross Abbey and now a ruin, was founded for the Observatine Franciscans about 1448 by Daniel McCarthy Mor. The Muckross Demesne is the estate around Muckross House, built in 1843.
5. *climbed the Eagle Rock mounted the Purple mountain*: the Eagle's Rock is a formation overlooking one of the lakes; the Purple Mountain has a height of 832m (2,730ft).
6. *I would forego Oxford*: Tyndall did attend the Annual Meeting of the BAAS in Oxford, arriving on 28 June. He was present at the now famous encounter between Thomas Huxley and Bishop Wilberforce over Darwin's recent *On the Origin of Species.*

To William Frederick Pollock [June 1860][1] 1718

Royal Institution | Friday

My dear Pollock

I am sorry to say that I shall not be here at 4 o'c[lock]. My head frightens me, and I run to the country to day to rest it. I shall be here on Monday—let us then talk the matter over.

Nothing has passed between Murray[2] & me since he promised to send me the book.[3]

What passed between us was this—I said to him that I thought you would undertake the Review, and that some months ago you spoke of doing so. He said something about your great literary powers, I know not what exactly, and at once said that he would send you an early copy of the work.

My feeling is & was & shall be, that there are no hands into which it would give me greater pleasure to see my book fall. But I w[d] sooner bury it as deep as Prospero's[4] than endeavor in any way to sway my friend towards a task to which he might feel disinclined. —I was probably wrong in mentioning the thing to Murray at all—if so it was an error of judgment which I know you will forgive.

Ever yours | John Tyndall.

RI MS JT/1/T/1256
RI MS JT/1/TYP/6/2199

1. *[June 1860]:* this letter indicates that Tyndall envisaged Pollock writing a review of *Glaciers of the Alps.* It gives a likely date in mid-1860.

2. *Murray:* John Murray, publisher of *Glaciers of the Alps.*
3. *the book: Glaciers of the Alps.*
4. *as deep as Prospero's:* a reference to Act 5, Scene 1 of *The Tempest* by William Shakespeare: 'Some heavenly music, which even now I do, | To work mine end upon their senses that | This airy charm is for, I'll break my staff, | Bury it certain fathoms in the earth, | And deeper than did ever plummet sound | I'll drown my book'.

To Lyon Playfair June 1860 1719

Edin. June /60

My dear Tyndall

My difficulty about publishing my letter to you[1] is that I have not made up my mind whether my molecule Ice[2] is

$$\left.\begin{matrix} O \\ HO \end{matrix}\right\} H + 8\,HO \quad \text{or} \quad \left.\begin{matrix} O \\ HO \end{matrix}\right\} H + 9\,HO$$

Recent analyses of the Hydrates of Barytes & Strontia by Bloxam[3] would lead to the former view.

Hydrate of Barytes (readily got by boiling Cryst[d] $BaO_1\ NO_5$ with a solution of 2Eq. NaO) has remarkably the power of keeping in solution a supersaturated quantity of

$$\left.\begin{matrix} BaO \\ HO \end{matrix}\right\} 8\,HO \qquad \text{corresponding to Ice—}$$

Would two crystals of

$$\left.\begin{matrix} BaO \\ HO \end{matrix}\right\} 8\,HO \qquad \underline{\text{regelate}} \text{ in such a solution?}$$

Yours Sin[cere]y | Lyon Playfair

P.S. If you accept the proposal[4] of the Phil. Inst. to lecture here in winter, I shall have a room at your service while you are in Edinburgh

RI MS JT/1/P/127
RI MS JT/1/TYP/3/994

1. *my letter to you*: letter 1702. Tyndall was asking Playfair about the properties of ice because of his research on glaciers, his idea of 'regelation' and therefore the importance of the behaviour of ice near its freezing point.
2. *my molecule Ice*: Playfair assumed a chemical distinction between ice and water to account for the behaviour of water as it reaches its freezing point. As water cools, its density increases, until at 4°C it starts to decrease. At the freezing point, unusually for a substance, ice is less dense than water, so it floats. Playfair argued that the formation and dissolution

of a separate particle 'ice' in the water sample at low temperatures would account for the expansion below 4°C. The 'freezing' of water would be nothing more than 'the separation of ice from a saturated solution of ice'. See Frederic Guthrie, 'On Cryohydrates and Water of Crystallization', in *Conferences held in Connection with the Special Loan Collection of Scientific Apparatus,* ed. Various, (London: Chapman & Hall, 1876), p. 99. See also letter 1702.

3. *Bloxam*: see Charles L. Bloxam, 'On the crystallized hydrates of baryta and strontia', *Quarterly Journal of the Chemical Society of London,* 13 (1861), pp. 48–50.
4. *if you accept the proposal*: see letter 1711.

From Sarah Faraday[1] 11 July 1860 1720

Hampton Court | 11 Jul 1860

Dear Dr. Tyndall,

I was in hopes of thanking you personally last Saturday for the flattering mark of your kind feeling which I received in the present of your book,[2] but as I did not see you, I must send a few words to tell you how highly I esteem it; I shall not attempt to speak of its merits for which I am quite incompetent, but I may say that in the little I have read so far, I see there is a great deal to interest even me. Your short visit was very tantalizing,[3] for we hope it will be repeated before long when we shall have more time.

Believe me dear Dr. Tyndall | with warm interest in your welfare | Most sincerely yours | S. Faraday

RI MS JT/1/TYP/12/4168
Faraday Correspondence, 5:3801

1. *Sarah Faraday*: see letter 1671, n. 1.
2. *your book*: *Glaciers of the Alps,* published 6 July 1860. See also letter 1595, n. 4.
3. *Your short visit was very tantalizing*: not identified.

From Jules Jamin[1] 14 July 1860 1721

Paris 14 Juillet 1860

Cher Monsieur,

Je serai très heureux de présenter votre livre à la Société philomathique, veuillez donc me l'envoyer et je tâcherai, après en avoir pris connaissance, ce que m'intéressera beaucoup, de donner sur son contenu toutes les explications que je pourrai.

Ruhmkorff qui vous a envoyé dernièrement un appareil devait se charger de vous faire parvenir le second volume de mon traité de physique. Il l'a oublié mais je le tiens à votre disposition si vous avez une occasion pour le faire prendre. Si non je vous l'enverrai par la poste.

Veuillez me croire cher Monsieur. | votre bien dévoué | J. Jamin

Paris, July 14th 1860

Dear Sir,

I would be very glad to present your book[2] to the Philomatic Society,[3] please send it to me and I will try, after reading it, which would be of great interest to me, to give on its contents all the explanations I can.

Ruhmkorff, who recently sent you a device, should have forwarded you the second volume of my treatise on physics.[4] He forgot it but I am keeping it at your disposal, if you have an opportunity to take it. If not, I will send it to you by mail.

Please believe me, dear Sir. | Your much devoted | J. Jamin

RI MS JT/1/J/7

---

1. *Jules Jamin*: Jules Célestin Jamin (1818–86), French physicist. He was professor of physics at the École Polytechnique from 1852 to 1881 and received the Rumford Medal in 1858 for his work on light. Tyndall corresponded with Jamin, whom he met in Paris in 1855 and became a good friend, over many years.
2. *your book*: *Glaciers of the Alps*, see also letter 1595, n. 4. Tyndall's letter, presumably offering this gift, is missing.
3. *Philomatic Society*: the Société Philomathique de Paris, a prestigious scientific and philosophical society, was founded on 10 December 1788 by Augustin-François de Silvestre. Members have included Lavoisier, Laplace, Cuvier, Gay-Lussac, Ampère, Cauchy, Fresnel, and Pasteur.
4. *second volume of my treatise on physics*: J. Jamin, *Cours de physique de l'École polytechnique, Tome I* (Paris: Mallet-Bachelier, 1859). Jamin's classic textbook of physics was published in four volumes from 1858, and revised in several editions.

From Auguste Balmat 20 July 1860 1722[1]

Chamonix le 20 Juillet 1860

Monsieur le Professeur

Vous me pardonnez le retard que j'ai mis à vous accuser reception mais pour contre je me trouve bien aise de pouvoir vous annoncer aujourd'hui qu'une ascension au Mont Blanc vient d'avoir lieu et qu'on à vue les trois thermomêtres tout à fait intact sur le sommet mais on a rien aperçu au corridor au Grand Plateau ni aux Grands Mulets.

Je joins ici les observations que j'ai faites depuis votre depart j'y ajoute celles faites par M^r L'Abbé Veillet

En attendant le plaisir de vous revoir vueillez me croire | votre tout devoué | seviteur | Auguste Balmat

Chamonix 20 July 1860

Dear Professor,

You will forgive me the delay of confirming reception but for which I am glad to be able to inform you today that an ascent of Mont Blanc has just been made and that the three thermometers were seen intact on the summit but none on the Grand Plateau nor at the Grands Mulets.[2]

I enclose here the observations I have made since your departures and add those made by M^r L'Abbé Veuillet.[3]

In expectation of the pleasure of seeing you again believe me | your devoted | servant | Auguste Balmat

1859
x^bre Deg Centigrade

31	0	
1860 janvier	Deg	
1er	1	
2	3	
3	3	½
4	0	

janvier Deg

5	0	
6	0	
7	3	¼
8	12	
9	7	
10	3	

<table continues separately on a second page>

janvier	Deg	
11	2	
12	3	
13	6	
14	6	½
15	4	
16	6	
17	4	½
18	1	
19	1	[illeg] zéro

Fevrier	Deg	
8	11	
9	7	
10	6	
11	17	
12	15	½
13	12	¾
14	17	
15	16	
16	17	

Mars	Deg	
1	2	
2	7	½
3	5	
4	10	
5	4	
6		
7	14	¾
8	5	
9	15	

20	0	
21	3	
22	3	¼
23	5	
24	11	
25	4	½
26	4	½
27	4	
28	1	½
29	9	
30	1	½
31	2	
Fevrier	Deg	
1	5	
2	12	
3	17	½
4	16	½
5	15	½
6	15	
7	9	¾

17	11	¼
18	13	¾
19	14	¾
20	13	¼
21	12	
22	16	½
23	15	½
24	15	
25	7	½
26	9	
27	2	
28	3	½
29	1	½

10	19	
11	18	
12	15	
13	14	¾
14	3	
15	3	½
16	8	¾
17	12	
18	9	
19	6	
20	4	½
21	5	
22	4	
23	8	¾
24	7	
25	6	½
26	4	½
27	5	
28	7	¾
29	2	½
30	3	
31	4	

RI MS JT/1/V/6

---

1. This letter is written on the same piece of paper as letter 1723.
2. Balmat acted as guide, accompanying James Forbes, Alfred Wills and Tyndall on various expeditions. In 1858, having had the idea of burying a thermometer on the summit of Mont Blanc to register the winter temperature, Tyndall secured £15 from the Royal Society and went with Balmat, Wills, and several porters to place the instrument. It was sought in August 1859 but not found. See: *Glaciers of the Alps,* pp. 189–91; 'Particulars of an Ascent of Mont Blanc', Report of the Twenty-Eighth Meeting (1858), *Brit. Assoc. Rep.,* 28 (London: John Murray, 1859), pp. 39–40. In 1859, Tyndall arranged a more substantial expedition to establish thermometers at 6 stations, including the summit of Mont Blanc (see ref. in letter 1723, n. 1), in a party including Edward Frankland. They slept overnight on the summit, the first time it had been done. In bad weather in 1860, Tyndall was not able to access the thermometers, and those in the lower stations were destroyed. By 1861 the thermometers on the summit were also broken. The locations of the thermometers for these winter readings by Balmat and Veuillet is not clear.
3. *add those made by Mr L'Abbé Veuillet*: see letter 1723.

From Abbé Veuillet[1] 20 July 1860 1723

Monsieur le Professeur,

Vous trouverez une certaine différence entre les observations de Mr Balmat[2] et les miennes. La différence provient de ce que le thermomètre de Mr Balmat était exposé au Sud, tandis que le mien était à l'Est, d'où il s'en est suivi que celui du Sud marquait plus de froid en temps humide que celui de l'Est; par contre celui de l'Est en temps sec donnait au moins trois degrés de plus que celui du Sud. J'ai fait des observations le matin à 7 heures et le soir à neuf heures. Si ces petites observations peuvent vous être agréables, je vous les envoie avec le plus grand plaisir et quand je pourrais, Monsieur le Professeur, vous obliger, je le ferai toujours avec un nouveau plaisir. En attendant d'avoir le bonheur et l'avantage de vous posséder à Chamonix, je suis toujours avec dévouement,

Votre très humble serviteur

Dear Professor,

You will find a certain difference between the observations of Mr Balmat and mine. The difference is explained by the fact that Mr Balmat's thermometer was exposed to the South, whereas mine was exposed to the East, hence the South one showed more cold and a more humid weather than the East one; however, the East one a dry weather showed at least three degrees more than the South. I made observations at 7 o'clock in the morning and at 9 o'clock at night. If these observations are agreeable to you, I shall send them to you with the greatest pleasure and when I can, dear Professor, oblige you, I will do it with a new pleasure. While awaiting the pleasure and the advantage of having you in Chamonix, I am always your very humble servant | [Veuillet]

7 heure du matin
x^{bre}

31	0
janvier	Deg
1	5
2	6
3	9
4	½
5	½

7 heure du matin
janvier | Deg

6	0
7	½
8	12
9	11
10	¾
11	2
12	7

9 heure du soir
x^{bre} Deg

31	1	
Janvier 1.	4	
2	7	½
3	0/6	
4	0/1	
5	0/1	
6	1	

janvier 9 h. soir
Deg

7	6
8	14
9	5
10	3
11	3
12	9

<table continues separately on a second page>

Janvier 7 heure du matin	Deg		9 heure du soir			7 heure du matin			9 heure du soir		7 heure du matin			9 h soir		
			13	10		Fevrier										
13	11		14	5		16	18	½	16		21	8	½	21	0/2	
14	12		15	10		17	14		17		22	0/3		22	3	
15	5		16	6		18	14	¼	18		23	10		23	3	
16	4		17	0/2		19	20	½	19		24	0/4		24	0/2	
17	5		18	0/1		20	12		20		25	4		25	4	
18	0		19	2		21	13		21		26	3		26	3	½
19	0/2		20	2		22	20		22		27	2		27	0/2	
20	1		21	1		23	20		23		28	3		28	0/½	
21	4		22	4		24	16		24		29	4		29	0/1	
22	2	½	23	8		25	14		25		30	0/3		30	0/2	
23	3		24	4		26	12		26		31	5		31	0/4	
24	7		25	3		27	0/2		27							
25	4		26	3	¾	28	4		28							
26	3	½	27	1		29	0		29							
27	1		28	3		Mars	7 du soir		9 h du soir							
28	1		29	2		1er	1		1er	1						
29	6		30	0/2		2	11		2	3						
30		½	31	4		3	7		3	4						
31	1	½				4	13		4	6						
Fevrier 7 heure du matin	Deg		9 h du matin			5	4		5	4						
1er	4		1er	12		6	5		6	11						
2	13		2	16	½	7	14		7	4						
3	21		3	19		8	5		8	8						
4	22		4	18		9	15		9	11						
5	18		5	15		10	20	½	10	15						
6	14		6	5		11	21		11	18						
7	7		7	10		12	17		12	3						
8	14		8	10		13	13		13	½						
9	6		9	2		14	1		14	7						
10	5		10	9		15	2		15	3						
11	19		11	11		16	8		16	9						
12	15		12	10		17	15		17	8						
13	12		13	11	½	18	11		18	5						
14	16	¾	14	19	¾	19	4		19	5						
15	19	¼	15	12		20	9		20	5						

RI MS JT/1/V/6

1. *Abbé Veuillet:* the Abbé Veuillet of Chamonix. Tyndall describes the involvement of Veuillet in 'IV. Thermometric Station on Mont Blanc', *Hours of Exercise in the Alps* (London: Longman, Green, & Co., 1871), pp. 53–58. For Chamonix, see letter 1633, n. 2.
2. *Mr Balmat:* Auguste Balmat. See letter 1722, which was written on the same piece of paper as this letter.

To Thomas Henry Huxley [Late July 1860][1] 1724

Tuesday

Dear Huxley,

Would you send this to Spencer for me?[2] I read that excellent article of yours in the Saturday.[3] Francis[4] is in ecstasy with it and wants to import a portion of it into the Annals.[5] It was very nice of you to say those handsome words near the commencement.[6]

Always yours | J. Tyndall

RI MS JT/TYP/9/2853
IC HP 8.285

1. *[Late July 1860]*: see n. 3. If this assumption is correct, possible dates are Tuesday 24 or 31 July.
2. *Would you send this to Spencer for me?*: there was presumably an enclosure, forwarded on to Herbert Spencer, which is missing.
3. *that excellent article of yours in the Saturday*: the evidence points to: T. H. Huxley, 'The Glaciers of the Alps', *Saturday Review,* 10 (1860), pp. 81–83. Huxley opens the review article, published on 21 July, by describing the work as the 'mightiest evangel' of the 'sect of muscular philosophers' who made up the Alpine Club, hence indicating Tyndall (without naming him directly) as the leading member.
4. *Francis*: William Francis.
5. . . . *the Annals*: *Annals and Magazine of Natural History*: the 'import' has not been found.
6. *those handsome words near the commencement*: see n. 3

From Thomas Archer Hirst 6 August 1860 1725

1 Brougham Place | Wellington Road | Dublin | Aug^st^ 6th /60

My dear John.

I arrived here very early this morning after a very pleasant journey which I broke into two parts at Bangor where I descended for a day's exercise in the neighbourhood of the Menai Straits. Before leaving town I dispatched to your address the Saturday review containing two excellently written articles. They entertained me greatly on my journey. The first[1] is full of original and deep thought. On a first perusal it is the peculiarity of the speculations which most strikes the reader but the concluding portion will rank amongst the boldest of well founded speculations of the day. You have directed your mental telescope furnished with intellectual lenses of high power under good arrangement to the nebulae of metaphysics and thus given definition and shape to what was before mere nebulosity. Your article must I think command attention from the best thinkers and help to give you the place amongst them to which you are entitled. To me such thoughtful piercing glances into the depths of philosophy possess a peculiar fascination and I muttered audibly Bravo John! as I read. The article on Faraday[2] pleased me almost as much, it will please Faraday too I am sure. For certainly it has never before fallen to his lot to be so thoroughly and nobly appreciated. Again I cried Bravo John, and thrust the Saturday Review into my pocket to be carefully preserved amongst my other valuables.

The distribution of prizes at University College passed off well on Friday last. In reply to the answer I sent to Key[3] (to the effect that an offer must proceed from the council) he informed me that the portion of work which I proposed to resign amounted to one fourth of the whole and that the salary if diminished accordingly would amount to £180 he was however prepared to recommend 180 guineas. I replied briefly saying that if 180 guineas were offered to me I should accept the appointment. Nothing more passed between us up to the day of the distribution when without any apparent reason they handed a scrap of paper to me on which were written these words

> "I intend to recommend to the council a salary of £200, instead of the amount I last stated"

I thanked him and replied that after the numerous proofs he had given me of his willingness to meet my needs and consult my wishes I was quite prepared to leave the matter of salary entirely in his own hands.

I am expecting daily to hear of the final settlement of the question.[4]

Having only just arrived I cannot of course say when I shall reach you. You may rest assured however that I shall not delay unnecessarily what I look forward to with so much pleasure. I hope you are proceeding satisfactorily with what you wish to do and are daily gaining strength to be afterwards expended upon work into which you will throw your whole force.

Wishing you every success and blessing I remain

Dear John | Yours as ever | T. A. Hirst

RI MS JT/1/H/245

1. *The first*: 'Physics and Metaphysics', *Saturday Review,* 10: 249 (4 August 1860), pp. 140–41.
2. *The article on Faraday*: 'Faraday's Juvenile Lectures', *Saturday Review,* 10: 249 (4 August 1860), pp. 149–50.
3. *Key*: Thomas Hewitt Key (1799–1875), a classicist and mathematician, was headmaster of University College School from 1832–75. Hirst taught mathematics at the school from 1860–64.
4. *final settlement of the question*: Hirst's journal also notes that Key initially proposed a salary of 180 guineas and then, without any particular comment from Hirst, recommended 200 guineas to the Council (*Hirst Journal,* RI MS JT/2/32c/1547), though Hirst does not note if that was the final figure.

To Juliet Pollock 17 August 1860 1726

17th Aug. 1860.

My dear Mrs Pollock,

I reached this place[1] late yesterday evening on my way to Zermatt, and this morning the air is crammed with fog which empties itself in thick rain upon the earth—while waiting for it to clear the thought of writing to you occurred to me—more indeed for my own amusement than for yours, for I have but little to write about. At the station of Troyes in France a Cambridge man named Hawkins[2] whom I knew to be both a good mountaineer & an able man came into the carriage which I occupied, and his plans being indefinite we agreed to work together for a time. He has been my companion ever since. We first went to Grindelwald and measured the motion of the glacier: we should have climbed the Wetterhorn had the weather been favourable. From Lauterbrunnen our desire was to get over to the Canton de Valais in a day, which could only be done by scaling the mountain wall which divides the said Canton from that of Berne, the salient points of the wall being the Junfrau,[3] the Breithorn, the Mittagshorn[4] and other formidable mountains which linked together by lofty ridges which had never been crossed. Between 3 & 4 A.M. on Thursday the 9th we quitted Lauterbrunnen and a journey of 4½ hours brought us to the base of the wall. We chose a kind of couloir[5] against which some snow was firmly stuck, being deeply furrowed by the avalanches which had been discharged from the summit. We had two strong men with us as guides;[6] but before we had climbed half the declivity they showed signs of giving way, manifestly thinking the thing impracticable. They saw the snow whirled by the wind at the summit, and said that this alone would defeat us. Once I quitted the couloir and sought to get up among the rocks. They were very steep, and I soon found myself at some height above my companions. I had at length to get round a promontory which overhung, and which afforded me but half a grasp for my hands, while underneath me was an edge of snow. I edged round as far as I could but had at length to submit and ignominiously call to my guide for a helping hand. He slowly worked his way up to me and helped me off my perch. For seven hours we tugged up this pass—near its summit we were sometimes on ice, and some of the rocks over which we clambered had all their crevices & depressions filled with ice. The snow-dust moreover was often bewildering, and we suffered from the cold. The men showed no great spirit—they were weary & daunted, and would have given way at any moment had they received the slightest encouragement to do so. But seven hours work placed us on the top, and the slope on the other side led us gently down to the great Aletsch glacier, which we quitted as night began to fall. We made our way over the Alps towards the hotel of the AEggischhorn, but we were all tired and our progress was slow—one of the men became ill, and finally darkness came and completely hid the route. The men sat down upon the Alp and sounded the jodel,[7] and after some time it was answered by a cry from a distance, and we were fortunate enough to find a senner[8] who undertook to conduct us. But he also lost his way, and finally the jodel again sounded brought a light toward us from the hotel, which we reached between 9 and 10 P.M.

We remained in the neighbourhood of the Aletsch glacier for a week. The Barnetts[9] were there during a part of the time, and we made some excursions together. Hawkins and myself were also engaged in determining the motion of this grand ice-stream. Some of the views to be obtained from it are of exceeding magnificence, and all the more precious inasmuch as the fair intervals which permitted us to see them were exceedingly rare. Yesterday morning we quitted the AEggischhorn & rode down the Rhone valley to Visp, walking afterwards to St. Nicholas,[10] on our way to Zermatt. At St. Niklaus we now remain, perfectly hemmed in by the weather—the rain is atrocious but I look forward to better times. A north wind creeps up the valley—while the motion overhead seems from the south—at the common boundary of these two currents the rain is formed, and I hope the northern element will finally prevail, and purify the atmosphere.

To-morrow I hope we shall be able to cross the Theodule Pass[11] to Breul,[12] where we intend to remain for some days—If the weather be only propitious there might be a chance of my being able to write you a more interesting letter after we have finished at Breul:[13] But no plans are safe at present.

I was very sorry not to be able to help you as regards the German hymn: the friend to whom I purposed to write is a profound man, but not a poetic one, and hence on reflection I did not write to him. I shall have my pleasure in following you through Frazer,[14] and in cutting you up if you give me any opportunity. Good bye. Love to the boys

Always yours | J. Tyndall.

I hope to be at Chamouni[15] in ten days from the present time.

British Library, Add MS 63092, ff. 205–6
RI MS JT/1/TYP/6/1968–69

---

1. *this place:* St Niklaus (St Nicholas).
2. *a Cambridge man named Hawkins:* Francis Vaughan Hawkins (1833–1908), lawyer and mountaineer.
3. *Junfrau:* Jungfrau.
4. *Mittagshorn:* Mittaghorn.
5. *a kind of couloir:* a steep, narrow gully, often filled with ice or snow. Tyndall published an account of the ascent: 'From Lauterbrunnen to the Aeggischorn by the Lauwinen-Thor', in Galton, F. (ed) *Vacation Tourists and Notes of Travel in 1860* (Cambridge: Macmillan, 1861), pp. 305–17, and later as the opening chapter of *Hours of Exercise in the Alps* (London: Longmans, Green and Co., 1871), pp. 1–17, illustrated with the first illustration after the frontispiece, the 'Ascent of the Lauwinen-Thor', p. 11.
6. *two strong men with us as guides:* Christian Lauener and Ulrich Kaufmann, the brother of Christian Kaufmann.
7. *sounded the jodel:* a distinctive lilting use of the voice, to carry in the mountains.
8. *senner:* an Alpine herdsman and dairyman, a maker of cheese.
9. *The Barnetts:* probably the family of Henry Barnett (1815–96), of Glympton Park, Oxford. He married Emily Ann Stratton in 1838 (Westminster Marriage Record) and they had a large family. His father, George Henry Barnett, who had inherited the estate in 1846, died in 1871.
10. *St Nicholas:* a village in the valley leading to Zermatt from Visp, now St Niklaus.
11. *Theodule Pass:* the Théodule Pass between Zermatt in Switzerland and Breuil in Italy.
12. *Breul:* Breuil, the village below the Matterhorn on the Italian side, a base for climbing the mountain.
13. *there might be a chance of my being able to write you a more interesting letter after we have finished at Breul:* Tyndall was planning an attack on the Matterhorn. The weather was bad in 1860, but Tyndall, Hawkins, Bennen and Jean-Antoine Carrel reached the highest point yet attained. See F. Vaughan Hawkins, 'Partial Ascent of the Matterhorn', in Galton, F. (ed) *Vacation Tourists and Notes of Travel in 1860* (Cambridge: Macmillan, 1861), pp. 282–304.
14. *following you through Frazer:* the Wellesley Index identifies forty-seven articles by Juliet Pollock in the periodical press, but none at this time. Her novel *Hanworth* was serialized, anonymously, in *Fraser's Magazine* in 1858 ('Hanworth', *Fraser's Magazine,* 58 (1858), pp. 1–15, 173–89, 271–84, 437–54, 554–69, and 675–88). Tyndall is referring here to a subsequent serialized novel by her, mistakenly attributed to J. M. Capes in the Wellesley Index. See letter 1729, n. 6.
15. *Chamouni:* see letter 1633, n. 2.

To Editor of *The Times* 7 September 1860[1] 1727

The Accident on the Col-Du-Géant.

Sir,—I am induced to lay before you the following remarks on the recent fatal occurrence on the Col-Du-Géant,[2] leaving it to you to decide whether they are of sufficient interest to render their publication desirable.

On the 18th of August, while staying with a friend at Breuil, rumours of the disaster reached us; but the accounts at first were so contradictory as to inspire the hope that they might be altogether false. Clearer intelligence, however, soon arrived, and before I quitted Breuil it was perfectly certain that three Englishmen, named Fuller,[3] Rochester,[4] and Vavasour,[5] with a guide named Tairraz[6] had perished on the Col. On the 21st I saw the brother of Tairraz at Aosta, and learned from him all he knew of the occurrence. The more I heard of it, however, the stronger became my desire to visit the scene of the catastrophe myself, and obtain through actual contact with the facts information more reliable than could possibly be conveyed to me by description. On the afternoon of the 22nd I accordingly reached Courmayeur, and, being informed that M. Curie,[7] the resident French pastor, had visited the place and made an accurate sketch of it, I at once called upon him. With the most obliging promptness he placed the sketches in my hands, furnished me with an account of the occurrence drawn up by himself, and volunteered to accompany me to the spot, so that we might inspect it together. I gladly accepted this offer, and early on the morning of Thursday, the 24th of August, we reached the Pavilion,[8] which it had been the aim of the travellers to reach on the day of their death. Wishing to make myself acquainted with every inch of the ground over which, from the commencement of their glissade,[9] the unfortunate men had passed, I walked straight up from the Pavilion to the base of the rocky couloir along which they had been precipitated. This couloir was described as being so dangerous that a chamois hunter had declined ascending it some days before; but through the kindness of M. Curie I secured at Cormayeur the services of an intrepid man, named Baireux,[10] who had once made the ascent, and whom it was now my intention to follow. We commenced our climb at the very bottom of the rocks, while the pastor made a détour and joined us on the spot where the body of the guide Tairraz had been found, and where some fragments of his skull—an extremely thick one—still furnished sad evidence of his fate. From this point upwards M. Curie shared the dangers of the ascent—strongly, I confess, against my will—until we reached the place where the rocks ended, and the fatal snow slope commenced. Here we parted company, he deeming it more prudent to resort to a strong arête to the right, than to trust himself upon the snow. I was urged by M. Curie to content myself with an inspection of the place, but no inspection, however close, could have given me the information I desired. I asked my guide whether he feared the slope, and, his reply being negative, we entered upon the snow, and ascended it along the course of the fatal glissade, the traces of which had not been entirely obliterated. Among the rocks below we had frequent and often melancholy occasion to assure ourselves that we were on the proper track, being guided by pieces of broken bottles, of wood, lucifer matches, fragments of leather, of crania, and other objects not necessary to mention. We also found a penknife and a small magnetic compass but the most important discovery was made upon the snow, being that of a fine new ice-axe, the property of one of the guides, who had relinquished it in his descent. This implement accompanied me yesterday across the channel,[11] and the discovery of it helps to throw a most needful light upon the disaster. Thus, then, from the beginning to the end of this fatal track I made myself acquainted with its true character, and as I stood upon the summit of the incline and scanned the ground over which I had passed a feeling of augmented sadness took possession of me. There was no sufficient reason for this terrible catastrophe. With ordinary precaution the glissade made,[12] in the first instance, have been avoided, and with average capacity to cope with such an accident the motion might, I am persuaded, have been arrested after it had been commenced.

Bounding the snow slope to the left, looking downwards, is a stony arête which forms the track by which travellers to Cormayeur usually descend. It is rough, but absolutely without danger. Any lady of ordinary walking powers might be conducted down it with perfect safety. But the men were tired when they reached this place, and to avoid the roughness of the arête they betook themselves to the snow. In connexion with their first entrance upon the snow a statement has been made which needs correction. On the morning of the fatal day a party, including two ladies, had actually descended the Col to Cormayeur, and M. Curie was one of this party. They also had trodden upon the snow, and it is alleged that their trace was followed by the party which attempted to descend in the evening. This is not true. M. Curie's party kept close to the arête, whereas the others were several yards to the right of it. The traces of both parties were perfectly plain on the day of my visit. An instrument for measuring the inclination of the snow was not, I am sorry to say, in my possession, but I should estimate its upper portion as forming an angle of 30 degrees, or thereabouts, with the horizon. It steepened considerably lower down, but its steepest portion I should estimate at less than 45 degrees of inclination. At all events, practised mountaineers are accustomed to trust themselves on steeper slopes without fear, and many of them, I imagine, might be found who would offer themselves to be tossed from a carpet on the slope in question, with perfect reliance on their power to arrest their downwards motion.

There are two other points which must be disposed of if we would have an unclouded view of this deplorable occurrence. It is alleged that when the party was approaching the summit of the Col on the Chamounix side, the guides proposed to return, but the Englishmen persisted in going forward. One condition of things alone could justify the proposition of the guides, namely, a fog as thick as to prevent them from reaching the summit of the Col at the proper point, and to render it necessary for them to return by the traces which they had already stamped in the snow. The only portion of the Col hitherto regarded as dangerous had been passed, and hence, unless in the case assumed, it would have been simply absurd to propose recrossing the really perilous portion of the glacier, instead of proceeding to Cormayeur. Such a fog, the guides affirm, did exist, but they, nevertheless, reached the summit of the Col, and the weather cleared after the summit had been attained. The subsequent portion of the journey must be discussed purely on its own merits, and whether the Englishmen refused to return or not upon the Montanvert side ought in no way to influence our judgment of what occurred on the Cormayeur side. Indeed, the alleged refusal seems to be justified by the facts; the guides acted upon it, the Col was surmounted, the descent at the other side was commenced, and here the weather, which prompted the proposal to go back, ceased to be blamable. The second point to which I would refer is this:—A statement is current which conveys the impression that the unfortunate men were carried down by an avalanche. In connexion with this point M. Curie writes to me thus:—"Il parait qu'à Chamounix on repand le bruit que c'est une avalanche qui a fait périr les voyageurs. C'est là une fausseté que le premier vous saurez démentir sur les lieux."[13] I have no hesitation is subscribing to this opinion of M. Curie, that a considerable quantity of snow was brought down during the glissade is exceedingly probable; but an avalanche, properly so called, there was not, and it simply leads to misconception to introduce the term avalanche at all.

We are now prepared to discuss the descent. The travellers, it is alleged, reached the summit of the Col in a state of great exhaustion; and this I can readily believe. It is also certain that such a condition would deprive them, to a great extent of that firmness of tread which is necessary in perilous places. But every guide and every experienced traveller is acquainted with this effect of fatigue, and this knowledge ought, one would think, to prevent them from entering upon the snow slope at all, when the rocky arête was at hand. Still I would excuse them here; for who would anticipate serious danger on a slope of 30 degrees? We are informed, however, that even on this slope the foot of one of the men repeatedly gave way, and that it was necessary continually to exhort him to fix his heels in the snow. Thus warned, why did the guides not quit the snow and resort to the arête? They must have had full confidence in their powers to arrest the glissade, which seemed so imminent, or else they were perfectly reckless of the lives committed to their charge. We must assume the former, though, unfortunately, when this confidence was put to the test it proved delusive. At some distance down the slope the heaviest of the three gentlemen slipped, and was shot downwards; he jerked his next neighbour off his feet, he again the next, until, finally, the whole of the men were in swift motion. What efforts were made to check this fearful descent we know not. At what point of the glissade the two guides relinquished the rope we are not accurately informed; which of them gave way first we do not know, though all this might be, and ought to be, known. All that we really know is, that the foremost and the hindmost guides did let go the rope, and escaped, while the three Englishmen and Tairraz went to destruction; and, like Englishmen, they met their doom without a word of exclamation.[14] At the bottom of the slope a ridge of rocks rose slightly above the level of the snow, thus forming a little barrier of which practised men might possibly have availed themselves at the last moment; as it was, however, all four were tilted over this barrier, and immediately afterwards descended a precipice, at the base of which they were certainly placed beyond all feeling. I do not think a single second's suffering could have been endured. During the wild rush downwards the men were too bewildered to experience fear, and at the base of the precipice life and pain ended suddenly together. A steep slope of rocks connected the base of this first precipice with the brow of a second, but lower one, at the bottom of which the body of Mr Fuller was found. Another slope ran from this point to the summit of another ledge, where the body of Mr Rochester was arrested, while attached to the latter by a rope, and quite overhanging the ledge, was the body of Mr Vavasour. The body of the guide Tairraz was precipitated much further, and was correspondingly mutilated.

The question has been raised whether it was right under the circumstances to tie the men together. In my opinion, it was perfectly right, if only properly done. But the actual arrangement was this:—the three Englishmen were attached together by a rope which was tied firmly around the waist of each of them; one end of this rope was held in the hand of the guide who led the party, the other end was similarly held by the hindmost guide, while Tairraz laid hold of the rope near the middle. Against this mode of attachment I would enter my most solemn protest; in no case ought it to be permitted. In all probability it was this which led to the destruction of the unfortunate Russian gentleman on the Findelen glacier last year,[15] and to it I believe is to be ascribed in a great measure the present fearful disaster. Whatever the attachment may be which is necessary for the tourist, that attachment the guide ought to share—their destinies ought to be inseparable. Let me now show cause for this emphatic protest, basing myself as I do so on the actual facts. At a little depth below the surface the snow upon the slope in question is firm and consolidated; but upon this firmer substratum a superficial layer partly fresh, and partly disintegrated by the weather, lay to a depth of eight or ten inches, or perhaps a foot, on the day of the disaster. By a proper action of the feet upon such snow its granules may be made to cement themselves together so as to afford a secure footing; but when the foot slips and the body is thrown prostrate on a slope covered with such snow the granules act like friction wheels, and the body shoots downwards almost without resistance. Suppose a smooth oilcloth to be placed upon a polished table, and the latter tilted up to form an angle of 45 degrees with the with the horizon; imagine a boy laid flat upon the oilcloth, the latter would slide over the wood, and the boy along with it, as the loose snow slid over the firm snow in the case before us. But suppose the boy to be furnished with a stick armed with a spike of iron at its end, what might he to do to check his descent? He ought to drive his spike with energy through the oilcloth, and seek to anchor it firmly in the wood underneath. And a precisely similar action ought to have been resorted to on the slope of the Col-du-Géant. Each man as he fell ought to have turned promptly on his face, pierced with his bâton the superficial layer of soft snow, and pressed with both arms the spike of his staff in to the consolidated mass underneath. A break would thus have been applied sufficient not only to bring a single man to rest, but if well done, sufficient, I believe to enable any single man to stop himself and another. I do not lightly express this opinion; it is founded on my own experience upon slopes at least as steep as that here under consideration.

Now, look at the bearing of the mode of attachment above described upon this mode of rescue. When the rope is fastened round the guide's waist both his arms are free, and both are needed to drive the baton into the snow and hold it there by pressure. But, as it was, one arm of each guide was virtually powerless; on it the strain of the falling body in advance was thrown, and, instead of being free, it was fettered. But this is not all; when the attached arm perceives the jerk the guide instinctively grasps the rope with the second hand; but in doing so he relinquishes his bâton, and thus loses the sheet anchor of salvation. Such was the case of the foremost and of the hindmost guide on the fatal day now referred to. The one lost his bâton, the other lost his axe; thus they became powerless to render any effectual resistance to others, and probably had to make expert use of their legs and arms to save even themselves. Tairraz was in the midst of the party; whether it was in his power to rescue himself or not, whether he was caught in the coil of the rope or laid hold of by one of his companions, we can never know. Let us believe that he clung to them loyally, and went with them to death sooner than desert the post of duty. And such fidelity the guides of the Alps must be made to regard as their bounden duty, and that no member of their fraternity is fit for his vocation who is unwilling to share the fate of these whose lives are committed to his charge.

A catastrophe of this kind ought not to be suffered to be skimmed over by a mere judicial acquittal. What can an ordinary Government official know of the power or the duty of a guide under such circumstances? Cast such an official upon such a slope; he will infallibly go to smash, and he naturally concludes that all others must do so in similar circumstances. But little weight is to be attached to the opinion of any man unpractised in such matters, and who derives his information from the very persons whose conduct is the subject of investigation. Instead, therefore, of imagining that the question has been set at rest by the publication of a verdict[16] whereby all concerned are so lightly relieved of blame, the guides of Chamouni ought to regard this terrible disaster as a stain upon their order which it will require years of service faithfully and wisely rendered to wipe away.

I am, Sir, your obedient servant, | (s[igne]d) John Tyndall. | Royal Institution, Albemarle-street, Sept. 7.

RI MS JT/6/5/1/9–17
'The Accident on the Col-du-Geant', *The Times,* (8 September 1860), p. 8.

---

1. Letter was written on 7 September and published on 8 September. The handwritten version is used here, with notes of any changes in the published version. For a detailed account of this accident see C. Gos, 'Disaster at the Col du Géant', *Alpine Tragedy* (London: George Allen & Unwin Ltd, 1948), pp. 12–17.
2. *Col-Du-Géant*: a high glacier pass between Chamonix and Courmayeur.
3. *Fuller*: Benjamin Fuller (*c.* 1827–60).
4. *Rochester*: John M. Rochester (*c.* 1833–60).
5. *Vavasour*: Frederick Vavasseur (*c.* 1834–60).
6. *Tairraz*: Frédéric Tairraz.
7. *M. Curie*: the pastor of the French parish of the commune.
8. *the Pavilion*: a location on the route from Courmayeur to the Col du Géant, at a height of about 2,300m or 7,546ft. The summit of the Col is at 3,356m or 11,010ft.
9. underlined words were given in italics in the published version.
10. *Baireaux*: not identified.
11. *channel*: 'Channel' in the published version.
12. *made*: 'might' in the published version.
13. *Il parait qu'à Chamounix on repand le bruit que c'est une avalanche qui a fait périr les voyageurs. C'est là une fausseté que le premier vous saurez démentir sur les lieux*: It appears that the news is spread at Chamonix that an avalanche killed the travellers. This is a falsehood that you will be the first to be able to deny on the spot.
14. *met their doom without a word of exclamation*: though in *Hours of Exercise in the Alps* Tyndall wrote: 'Tairraz screamed, but, like Englishmen, the others met their doom without a word of exclamation', *Hours of Exercise,* p. 23.

15. *unfortunate Russian gentleman on the Findelen glacier last year*: Edoard de Grotte had tried to cross the Findelen Glacier with his two guides on 13 August 1859. Whilst his body was fastened to a rope, the local guides had the rope fastened to their left arms. He had also refused to bring an alpenstock with him. Passing over a large patch of snow, De Grotte sank through it into a hidden crevasse. He was pulled up after 5 hours, but had frozen to death. The initial report in the *Times* suggested that part of the reason the man might have fallen was because the rope snapped, possibly due to it being cut by the guides: J. J. C., 'Fatal Accident In Switzerland', *The Times,* (25 August 1859), p. 8. A later letter argued that it was more likely to be due to a shortness of the rope, with the guides simply not being able to hold on to it: 'A MEMBER OF THE ALPINE CLUB. The Accident To The Russian Gentleman At Zermatt', *The Times,* (14 September 1859), p. 12.
16. *publication of a verdict*: not identified.

From Charles Mackay[1] 17 September 1860 1728

The London Review[2] | Conducted by Charles Mackay |
11 Southampton Street | Strand W.C. | Sep. 17th 1860.

My dear Dr. Tyndall,

Though I have not seen you—or heard from you, for a long time,[3] I have not ceased to take an interest in your career; and have watched your progress upwards with great satisfaction.

Should you ever be in this neighbourhood, I shall be delighted to receive a call. We wish particularly to include a sketch of Professor Faraday, among our 'Men of Mark' commenced in last week's London Review; of which I directed a copy to be sent to you. Is it hopeless to expect to number you among our contributors?[4]

Ever Yours truly | Chas. Mackay
Professor Tyndall.

RI MS JT/1/TYP/2/782
LT Typescript Only

---

1. *Charles Mackay:* Charles Mackay (1814–89), Scottish author, journalist and editor, best known for his books entitled *Songs and Poems* (London: Cochrane and McCrone, 1834) and *Extraordinary Popular Delusions and the Madness of Crowds* (1841). He began working for the *Illustrated London News* in 1848 and became its editor in 1852. Mackay had written encouragingly to Tyndall in 1848, after Tyndall thanked him for his writings and told him of his own ambitions (see *Tyndall Correspondence,* vol. 2, letter 0362). Tyndall admired Mackay's poetry, which he read frequently. He had met him while teaching at Queenwood, as Mackay had a son there.

2. *The London Review:* The London review and weekly journal of politics, literature, art, and society. The journal was published weekly from 7 July 1860 to 27 March 1869. Intended to be non-partisan, it was under the sole direction of Charles Mackay.
3. *for a long time:* perhaps not since 3 October 1850, when Tyndall noted visiting him in Camden Town (Journal, RI MS JT/2/13b/513). Mackay had written on 25 June to invite him to visit (see *Tyndall Correspondence,* vol. 3, letter 409).
4. *Is it hopeless to expect to number you among our contributors?:* 'Men of Mark. —No. 5. Faraday' was published in October: *The London Review,* 15 (13 October 1860), pp. 344–45. The style does not appear to be Tyndall's.

To Juliet Pollock [25 September 1860][1] 1729

Tuesday evening

My dear M[rs] Pollock.

I have just 20 minutes at my disposal and shall fill them by writing a random scroll to you. Last Sunday week I was with Sir Benjamin & Lady Brodie[2] at Broome Park: His eyesight is much damaged[3] but still not so bad as I expected—he is able to walk about the house & grounds though he cannot read or write. On my return home on Monday evening I was stunned by the intelligence that my poor friend Huxley had lost his eldest child & only boy[4]—a fine little fellow nearly 4 years old. I knew the hopes that were centered in that boy—I knew that both his father & mother doated on him, and was bewildered by the intelligence of their bereavement. We buried the boy on Tuesday at Finchley Huxley, his brother & myself being the only persons present. It is sad very sad.

I suppose this atrocious weather stretches to the Lakes[5]—Last night the air howled as if agitated by a winter storm—wild, dashing rain, which might be fine were it not accompanied by the thought of the crops that are still exposed, and the still sadder reflection as to the lives that were exposed at sea.

I have my eye upon 'Ida',[6] certainly there are materials in these two first chapters for many interesting entanglements—I shall follow the threads with interest as you cast them, assured that you will work them all into a web of beauty at the end.

Ever yours | J. Tyndall

Prof. Forbes has written a somewhat bitter pamphlet against me[7]—but I do not think I shall notice it. At all events not at present. I could 'flutter' him if I liked.

RI MS JT/1/T/1194
RI MS JT/1/TYP/6/1908

1. *[25 September 1860]:* Noel Huxley died on Saturday 15 September 1860 and was buried on Tuesday 18 September: see L. Huxley, *Life and Letters of Thomas Henry Huxley* (2 vols) (London: Macmillan, 1900), vol. 1, p. 213. Tyndall was the only close friend at the funeral. It seems likely that this letter is from the following Tuesday, not the day of the funeral.
2. *Sir Benjamin & Lady Brodie:* Sir Benjamin Collins Brodie (1783–1862), physiologist and surgeon, was President of the Royal Society 1858–61. He married Anne Sellon in 1816. Tyndall was a welcome visitor at their country house. The family was later keen for him to take a house close to them at Box Hill: see Tyndall to Elizabeth Steuart, 20 October 1868, RI MS JT/1/TYP/10/3383.
3. *His eyesight is much damaged:* this prompted his resignation the following year as President of the Royal Society (RS Minutes of Council (1858–69) vol. 3, p. 89–90), and he was succeeded by Edward Sabine in November 1861.
4. *my poor friend Huxley had lost his eldest child & only boy:* Noel Huxley, who died of scarlet fever, age 4, see n. 1.
5. *the Lakes:* The Lake District, where the Pollocks regularly went on holiday.
6. *'Ida':* Juliet Pollock's novel *Ida Conway* was serialized in twelve monthly instalments in *Fraser's Magazine* between September 1860 and August 1861. The byline is 'By J. M. C.', presumably Juliet M. Creed (her maiden name). The novel is incorrectly attributed in the Welleseley Index to J. M. (John Moore) Capes. The first two chapters were published in September 1860 (*Fraser's Magazine,* 62 (1860), pp. 285–98) and the final chapters in August 1861 (*Fraser's Magazine,* 64 (1861), pp. 219–33).
7. *Prof. Forbes has written a somewhat bitter pamphlet against me:* see letter 1731, n. 4.

From William Wright[1] 16 October 1860 1730

Springfield Clack | Co. Ohio. | Oct. 16/60
John Tyndall Esq. | London

Sir,

Permit me to introduce to you my friend C. A. How Esq.[2] who is on a short visit to your city and I embrace the opportunity to recommend him to the favourable attention of my oldest friend in Great Britain, hoping that during his stay in London you would have time and opportunity to accompany him to some of the objects of interest with which your city abounds. Any favours extended to Mr How will be an especial gratification to

Yours truly | William Wright.

RI MS JT/1/TYP/5/1832
LT Transcript Only

1. *William Wright:* William Wright (1818/19–92), from Leighlin Bridge, was a close childhood friend of Tyndall, with whom he subsequently maintained a long-term corre-

spondence. He married Sarah Maria Neale on 7 February 1841 at St Nicholas Parish Within, Dublin (Irish Church Records). Possibly due to opposition from her family, the couple rapidly emigrated to America, sailing from Liverpool in early March 1841. They settled in Cincinnati, Ohio, where William worked in the dry goods and grocery trades.

2. *C. A. How Esq.:* not identified.

From William Hopkins 18 October 1860 1731

H. Thomas Esq.[1] | Llwyn Madoc[2] | Builth Wells | S. Wales. | Oct 18. 60

My dear Sir,

Possibly you may and probably you may not have heard of the very serious accident I have recently met with. While staying with my friend Mr Thomas at this place, I slipped in descending a steep bank down which I fell with such violence as to dislocate the joint of the left knee which was only reduced after three painful efforts. It is now upwards of three weeks ago, and ever since I have been tied by this unhappy leg to the bed in which I am now lying, and where I must probably lie for three weeks longer. However the knee is progressing I believe to the satisfaction of the surgeon, though its progress seems to me at least very slow. I am with friends too who have ample wills as well as ample means of affording me every comfort and I have my wife and daughter with me, so that there is much to counteract all grumbling tendencies. Since the first few days also I have not suffered so much pain as to prevent my being able to think sufficiently to do some work which of course has been my best resource. I have chiefly occupied myself with the review of your 'Alpine Glaciers',[3] but I am unfortunately without the slightest means of reference on the subject, having nothing but your book and the pamphlet which Forbes has just put forth.[4] Of course I discuss the questions between you not as a personal question, but in reference to Rendu's claims.[5] I don't think however that you have said a word in the whole book to give Forbes any ground of personal complaint. One question is, how far Forbes gave as much of Rendu's views as the public had any right to expect of him in his 'Travels';[6] I don't know that we have a right to hold him responsible for not giving more, and you have not said that he is so. But the real question is, as regards justice to Rendu, whether the scientific public did or did not possess sufficient means of duly comparing his claims with Forbes's. Now it is manifest that Rendu's views were far too little known for this purpose, and you have only rendered him an act of common justice in the more ample view you have given of them. But I must see Rendu's original memoir.[7] Can't you send it to me here by post? Forbes's insulated extracts evidently give a very imperfect notion of it. Can you also send me a copy of Forbes's memoir in the 'Transactions' of the Royal Society?[8] I can't get here at my own. I suspect that the notion he has put forth there, of the internal pressure of a glacier arising from the hydrostatic pressure of the water contained in the capillary tubes of the ice, to involve an absolute hydrostatical blunder. In fact I believe that all his statics, hydrostatics and dynamics are nothing but a demonstrable series of blunders. Send me those two memoirs as soon as you can. I like your book exceedingly.

My review will form, I believe, at least an attempt at a far more complete discussion of glacier theories than has yet appeared, and I hope too a more impartial one. I confess to you that I rather regret not having reserved the first offer of it to the Quarterly, in which I don't recollect any decent article on the subject. Do you think Parker[9] would like its being offered to the Quarterly? If not, I should of course not think of it. It would thus constitute a higher notice of your book, and would probably be of more use to it than any notice in Fraser.[10] Is this a question which you can suggest to Parker and discuss with him. He is said to be rather peppery and it may require some delicacy. However I will leave it in your hands. I should have no objection, either, to write a shorter article for Fraser, expressly as a review (comparatively short and popular) of your book, and reserve a more general Essay to offer to the Quarterly. I could have the former ready for Fraser Dec. number.[11] The Essay would make a long article for Fraser, but not a long one for the Quarterly. Let me know your thoughts about all this.

I have no means here of hearing any scientific gossip. What do they say of Forbes's pamphlet? I think it was utterly uncalled for; but people always prick up their ears at the prospect of little fightings, whether it be Sayers and Heenan[12] or Forbes and Tyndall.

Yours truly | W. Hopkins.

Further thoughts—my more general Essay will be, I find, too long for Fraser. It would have to be divided which is a great evil. Whatever may become of it; I must write a shorter review for Fraser.

RI MS JT/1/TYP/2/620–22
LT Typescript Only

---

1. *H. Thomas Esq.:* Henry Thomas (1808–63), landowner and magistrate (from Burke's Landed Gentry 1846, vol. 2, p. 1386).
2. *Llywn Madoc*: Llywn Madoc was an estate belonging to Henry Thomas, near Builth Wells.
3. *your 'Alpine Glaciers': Glaciers of the Alps,* see also letter 1595, n. 4.
4. *the pamphlet which Forbes has just put forth:* Following Tyndall's section 'Rendu's Theory' in *Glaciers of the Alps,* pp. 299–308, Forbes published a separate pamphlet to supplement the 'Prefatory Note' in Forbes (1860), pp. xiii–xxix: J. D. Forbes, *Reply to Professor Tyndall's Remarks, in His Work "On the Glaciers of the Alps," Relating to Rendu's "Théorie Des Glaciers"* (Edinburgh: A. and C. Black, 1860).

5. *Rendu's claims:* see letters 1619, n. 13 and 1708, n. 5.
6. *his 'Travels':* see letter 1594, n. 13.
7. *Rendu's original memoir:* see letter 1619, n. 13.
8. *Forbes's memoir in the 'Transactions' of the Royal Society:* In 1846, Forbes had published his memoir in three parts: 'Illustrations of the Viscous Theory of Glacier Motion', *Phil. Trans.,* 136 (1846), pp. 143–210.
9. *Parker:* see letter 1707.
10. *Fraser: Fraser's Magazine,* edited by John Parker.
11. *ready for Fraser Dec. number:* Hopkins produced this article (anonymously): 'The Glaciers of the Alps', *Fraser's Magazine,* 62 (December 1860), pp. 793–809. Much later he produced, again anonymously, an article for the *Quarterly Review:* 'Voyages dans les Alpes . . .', *Quarterly Review,* 114 (1863), pp. 76–125.
12. *Sayers and Heenan:* on 17th April 1860, the all-American boxing champion John Carmel Heenan came to England to fight 'Brighton Titch' Tom Sayers, the champion of England, in an illegal, bare-knuckle fight, perhaps the first 'world-title' contest. It was declared a draw after 42 rounds, when the fight was broken up by the police, and the contestants shared the £400 purse.

To Heinrich Debus [21 October 1860][1] 1732

Sunday.

My dear Heins

It is long since I wrote to you—far too long. I have however for many weeks been fighting with great difficulties in my investigations and had no great heart to write to any body. I am happy to say that all are now substantially overcome, and that the progress of my work is in the highest degree satisfactory. The more I think and work on the subject the more I see its great extent and importance. If I mistake not before the year closes I shall be able to furnish a memoir which will exceed in interest all that has been published on the subject of radiant heat for many years. In a chemical point of view the results also promise to be of great interest and importance.

Among other things I believe I have discovered an allotropic condition of oxygen.[2] I find that ozone presents most striking characteristics: but I wish you were near me so that we might talk over these matters—Will you kindly inform me of some of the ways by which ozone may be produced; and also by which it may be reduced to the condition of ordinary oxygen?

What is your belief regarding the substance? My experiments harmonise with the idea that it is a peroxide of hydrogen; but they do not yet prove it!

My health since I returned from the Alps has been excellent. I work hard every day and do not find my strength sensibly diminished. I am glad that the glaciers are off my hands,[3] for if I mistake not I shall soon shew the world that I can produce something better. Forbes has written against me[4] as I suppose you are aware, but he shall not make me swerve from my work at present. Besides his pamphlet has produced no effect in London. When a second edition of my book is published[5] I shall make a few remarks upon his pamphlet, but not before.

I think you know a good deal about Brodie's ideas of grouped elements.[6] With regard to the nitrogen the element which I obtain from the nitrite of potassium, and which chemists state to be perfectly pure, is a totally different body from the nitrogen of the atmosphere—in all chemical tests they agree, but my physical test shews an enormous difference between them.

Does any impurity occur to your mind likely to mix itself with nitrogen obtained from the nitrite of potassium? I have prepared it on several occasions, but with the same unvarying result.

What amount of gaseous impurity do you think ordinary chemical means could detect in the atmosphere? I can readily measure the effect 0·008 per cent of certain gases mixed with air. I can also accurately determine the absorption of a gas possessing no more than 1/15th of a millimetre of tension.

Do not mention these things to any body Faraday has warned me against speaking of the subject at all:[7] I would again express the wish that you were nearer to me so that I might have the benefit of your suggestions.

Believe me dear Heins | Most truly Yours | John Tyndall

RI MS JT/5/11/1365–66
LT Transcript Only

1. *[21 October 1860]*: letters 1733 and 1734 reply to this letter, making the likely date Sunday 21 October 1860.
2. *allotropic condition of oxygen*: the typescript text reads 'nitrogen', but it can be assumed that Tyndall wrote, and certainly meant, 'oxygen'. Though Tyndall incorrectly concluded at this point that oxygen was monatomic, he was correct to identify ozone as a form of oxygen, which was uncertain at the time. See also letter 1734, n. 2.
3. *I am glad that the glaciers are off my hands*: *Glaciers of the Alps* had been published on 6 July.
4. *Forbes has written against me*: see letter 1731, n. 4.
5. *When a second edition of my book is published*: a second edition was not published until 1896, after Tyndall's death. Though the first edition was published by Murray, the second edition was published by Longmans: *Glaciers of the Alps* (London: Longmans, Green, & Co., 1896).
6. *Brodie's ideas of grouped elements*: Benjamin Collins Brodie (1817–80), chemist. In the 1860s, Brodie rejected an atomic interpretation of chemistry, and proposed a new set of symbols for chemistry which took into account the laws of chemical combination. He

gave a seminal exposition of his theory to the Chemical Society in 1867. See W. H. Brock (ed.), *The Atomic Debates* (Leicester: Leicester University Press, 1967), pp. 13–15.

7. *Faraday has warned me against speaking of the subject at all*: this indicates the importance of establishing Tyndall's priority for the absorption of heat by gases, and indeed that Faraday was not unconcerned at such earthly reputation.

From Heinrich Debus 23 October 1860 1733

Queenwood College | Stockbridge | Oct[o]b[e]r 23rd 60.

My dear John,

I was very much pleased with the letter I received from you this morning.[1] As I have a little time to spare I will answer to day one of your questions.

Nitrite of ammonia or a mixture of nitrite of potash and chloride of ammonium undergoes according to condition of experiment different decompositions. Dissolved in water and heated, it is asserted that perfectly pure nitrogen and water are formed. If heated in the dry state nitrogen, nitric oxide (NO_2), nitrous oxide (NO), ammonia and water are obtained. Now I can discover no where a proof that the gas obtained from the solution is really pure nitrogen.

Knowing how much assertions of such a nature are worth I should not be content by assuming that the nitrogen prepared by boiling a solution of nitrite of potash and chloride of ammonium was pure, but experiment in the following manner.

Prepare the nitrogen by boiling a solution of nitrite of potash and chloride of ammonium. Pass this nitrogen through a tube containing pumic-stone[2] (Bimsstein) moistened with a solution of caustic potash, after this through a tube containing pumic-stone moistened with strong sulphuric acid, and after this through a tube 2 feet long containing red hot copper turnings.

Only after the gas has been treated thus you may be sure to have pure nitrogen.

The copper turnings are prepared for this process in the following manner: heat the commercial turnings in crucible to redness for some time; transfer the same to a glass tube of bohemian glass (of the size of a combustion tube), heat to redness and pass hydrogen over the red hot copper turnings. Thus you obtain pure metallic copper. If you like I will set Flight[3] to prepare you some turnings.

As to the nitrogen of our atmosphere, pray how did you prepare it? Besides carbonic acid, ammonia and water and oxygen the air contains organic substances of a nature utterly unknown to us. All these must be removed. It may be well for chemical experiments simply to abstract the ammonia, water, oxygen and carbonic acid but for such a delicate reagent as radiant heat only a perfectly pure material will do.

Pure nitrogen may be obtained from air as follows:

Pass the air through 2  tubes, one containing pumic-stone and potash and the other pumic-stone and sulphuric acid. Here water carbonic acid and ammonia are withdrawn from it. Thus purified pass the air through a red hot tube containing bright copper turnings. The copper removes the oxygen and the high temperature destroys the organic matter. Or what is better let there be where the gas enters the tube a layer of oxide of copper of a few inches in length and after this the metallic copper. The oxide of copper burns the organic matter to carbonic acid and water. At the other end were the gas leaves the tube you have nitrogen and this water and carbonic acid. Therefore, after the passage of the gas over the oxide of copper and metallic copper let it proceed through two tubes containing potash and sulphuric acid, in order to purify it.

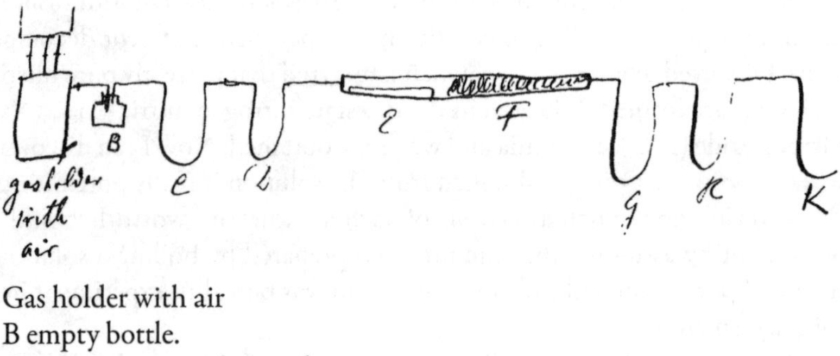

Gas holder with air
B empty bottle.
C pumic-stone with potash
D pumic-stone with SO_3HO[4]
E oxide of copper 6 inches long } both red hot.
F metallic copper 18 inches long }
G pumic-stone with potash
H. pumic-stone with sulphuric acid
K. gas conducting tube.

The air must pass very slowly through this apparatus in order that the different reagents have sufficient time to act. It is even advisable in order to be quit sure to double the number of the tubes and the length of the layer of copper shavings.

Phosphorous does not remove the organic matter and other methods, except the one described here, to not furnish a pure nitrogen from air.

In September I bought a bottle of pale brandy; it is still corked and I suppose I must drink it myself.[5] More in my next letter. If I can prepare anything for you I shall do so with pleasure.

Always your | H. Debus

RI MS JT/1/D/25

1. *the letter I received from you this morning:* letter 1732.
2. *pumic-stone:* pumice stone, for which the German word is Bimmstein.
3. *Flight:* Walter Flight (1841–85). He was educated at Queenwood College and then studied chemistry in Germany. He returned to England in 1867 and became a mineralogist specialising in meteorites (ODNB).
4. *SO3HO:* sulphuric acid.
5. *I suppose I must drink it myself:* it was a frequent complaint of Debus that Tyndall did not visit him enough.

From Heinrich Debus 30 October 1860 1734

Queenwood College | Stockbridge | Hants | Oct[o]b[e]r 30th 60.

My dear John,

I suppose my last letter[1] came to your hands on Wednesday last.

There appear to be, without doubt, two bodies of similar properties, which are called by the same name—ozone.[2] If the fact is true, that perfectly pure oxygen can by means of electric sparks be converted into what is called ozone, and if it is true, that the oxygen obtained from water by passing a electric current through the same, contains a body of the composition HO_3 and comporting itself like the electrified oxygene, then it follows that we have

a) modified oxygene
b) a peroxide of hydrogen

both in properties very much alike and going by the same name.

About Baumerts and Bunsens experiments[3] I have no doubt and scarcely any with regard to the electrified oxygen. De la Rive[4] and Marignac,[5] Becquerel[6] and Fremy,[7] Berzelius,[8] Marchand[9] &c &c all profess to have experimented with dry oxygen;

Have you not examined perfectly pure and dry oxygen enclosed in a glass tube and treated with electric sparks? If the presence of nitrogen is no objection, then the best mode of obtaining ozone would be by leaving a piece of phosphorous in moist and confined air. If the presence of nitrogen is objected to then pass pure oxygen over phosphorous, 2/3 of which is immersed in water, at a temperature of 360C. Part of the oxygen is converted into ozone and it is only necessary to pass the ozonised oxygen over anhydrous chloride of calcium or better phosphoric acid in order to obtain it dry.

According to my knowledge the best way to reduce ozone to the common oxygen is, to pass the said ozone through a red hot tube; also the powder of charcoal, if shaken with ozone, converts the latter in common oxygen, or oxygen and water, perhaps a small quantity of carbonic acid is formed; at all events all the ozone is destroyed.

We can determine 1/50,000 part of ammonium and I dare say less than 1/100 000 part of chlorine or any other gas of decided chemical properties, in our atmosphere; of course such numbers depend entirely on the nature of the gas.

I am happy to hear you are so strong in body and mind and so successful in your investigation. You may depend, I shall not divulge or mention the subject of your labours to any one.[10]

Always yours | H. Debus

RI MS JT/1/D/26

1. *my last letter:* letter 1733.
2. *ozone:* the chemical nature of ozone was not clear at the time. Ozone was considered either to be a form of oxygen or a compound with hydrogen. Though Tyndall thought at the time from his absorption of heat experiments that oxygen was atomic, not molecular, he correctly deduced that ozone was a form of oxygen. We now know that oxygen molecules contain two atoms, O_2, and ozone molecules three, O_3.
3. *Baumerts and Bunsens experiments:* Friedrich Moritz Baumert (1818–65), see M. Baumert, 'Ueber eine neue Oxydationsstufe des Wasserstoffs und Ihr verhältnis zu Ozon', *Poggend. Annal.*, 89 (1853), pp. 38–55; 'Zur Ozonfrage', *Poggend. Annal.*, 99 (1856), pp. 88–94; 'Ueber das elektrolytisch entwickelte Ozon', *Ann. Chem.* 101 (1857), pp. 88–90. Robert Bunsen (1811–99), see R. Bunsen, *Gasometrische Methoden* (Braunschweig: Friedrich Vieweg und Sohn, 1857). Specific publications by Bunsen on ozone have not been identified, but his key work in developing techniques to separate, measure, and identify chemical compounds was picked up on by others and resulted in new findings on ozone. See, for example, M. Soret, 'XXIX. Experiments on Ozone. In a Letter to Professor Tyndall', *Phil. Mag.*, 25 (1863), pp. 208–10. In this letter, Soret describes how he used one of Bunsen's methods of gas analysis, as well as Baumert's experiment using electrified oxygen in his investigations on ozone.
4. *De la Rive:* Auguste-Arthur de la Rive. He published on his work with Marignac in the postscript of: A. de la Rive, 'Sur les mouvements vibratoires que déterminent dans les corps, soit la transmission des courants électriques, soit leur influence extérieure', *C. R. Hebd. Seances Acad. Sci., Ser. C,* 20 (1845), pp. 1287–91.
5. *Marignac:* Jean Charles Marignac (1817–94). J. C. Marignac, 'Sur la nature et la production de l'ozone', *Archives de électricités. (Supp. a la Bibliothéque Universelle de Genève),* 5 (1845), pp. 5–11. A similar paper in *Comptes Rendus* omitted the part about the dry oxygen.
6. *Becquerel:* Alexandre Edmond Becquerel (1820–91). E. Fremy and E. Becquerel, 'Recherches électrochimiques sur les proprietés des corps électrisés', *Annal. Chim. et Phys., 3rd series,* 35 (1852), pp. 62–105.

7. *Fremy:* Edmond Frémy (1814–94). See n. 6.
8. *Berzelius:* Jöns Jacob Berzelius (1779–1848). J. J. Berzelius, 'Ueber das Ozon', *J. Prakt. Chem.*, 40 (1847), pp. 242–46.
9. *Marchand:* Richard Felix Marchand (1813–50). R. F. Marchand, 'Ueber Ozon', *Poggend. Annal.*, 67 (1846), pp. 143–44.
10. *I shall not divulge or mention the subject of your labours to any one:* see letter 1732.

To John Chapman[1] [October 1860][2] 1735

Tuesday

My dear Chapman,

I regret to say that I am pledged to the 'Thatched House'[3] for Thursday next.

I have been reading that excellent article on Neo-Christianity.[4] Wont it open the eyes of those Oxford men! It is an extremely able production—If I can get away sufficiently early from St. James's Street[5] I will join you, but that I can hardly hope for.

[sincerely] yours | John Tyndall

Huntington Library Harr 1/111.66

---

1. *John Chapman*: John Chapman (1821–94) British publisher and physician. In 1851 he became proprietor and editor of the *Westminster Review,* which was a quarterly liberal publication, initially founded by Jeremy Bentham.
2. *October 1860*: October 1860 is written in another hand at the top of the letter. See also n. 4, which implies an October date.
3. *"Thatched House"*: this probably refers to Thatched House in St. James's Street, which was the meeting place of a number of clubs until the building was demolished in 1863. Thatched House was the meeting place for 'The Literary Club', but Tyndall was not elected until 1871. See N. A. Rupke, *Richard Owen: Biology without Darwin, a Revised Edition* (Chicago: University of Chicago Press, 2009), p. 62.
4. *that excellent article on Neo-Christianity*: 'Art. I—Neo-Christianity', *Westminster Review,* 18 (October 1860), pp. 293–332. It is a review of the second edition of *Essays and Reviews,* by the historian and positivist Frederic Harrison (1831–1923).
5. *St. James's Street*: see n. 3.

To Hector Tyndale [late October or early November][1] 1860 1736

POLYTECHNIC INSTITUTION,[2] | (LIMITED), |
309 REGENT STREET.

The Premises of this Institution having been materially Altered and Improved, rendered thoroughly Secure for the Use of the Public, Re-Decorated, and Supplied with Works of Art, Models of Machinery, Specimens of Manufacturers of the latest character, and new Apparatus, will be

RE-OPENED to the PUBLIC on MONDAY, the 12th of NOVEMBER, 1860.

AN EDUCATIONAL DEPARTMENT HAS BEEN FORMED,

with suitable class-rooms,

FOR THE STUDY OF ART, SCIENCE, AND LITERATURE,

For both male and female pupils.

Dear Hector

Would you and Mrs T.[3] care about going to this?

J. T.

PROGRAMME—PRIVATE VIEW

Saturday Evening, November 10th, 1860.

Doors open at half-past seven o'clock.

Eight o'clock—OPENING ADDRESSES.

Half-past Eight—REVELATIONS of IMMENSITY, with a Particular Reference to EARL ROSSE'S DISCOVERIES. With fac-similies of Telescopic Views, executed on a new principle. By Mr Mackintosh.[4]—In the Large Theatre.

Quarter to Nine—DESCRIPTIONS of COLLIN'S PATENT ELECTRO-PRINTING BLOCKS for ENLARGING and DIMINISHING at Pleasure, to any Extent, all kinds of DRAWINGS AND ENGRAVINGS. By Mr E.V. Gardiner,[5] Professor of Chemistry—In the Front Theatre.

Quarter-past Nine—EXPLANATION of MODELS of NEW INVENTIONS, and DESCRIPTION of the ILLUMINATED CASCADE, Exhibited by DUBOSCO'S ELECTRIC LIGHT APPARATUS. by Mr. King.[6]—In the Gallery.

Half-past Nine—EXPERIMENTS with RHUMKORF'S[7] ELECTRIC COIL APPARTUS, Exhibiting GASSIOT'S CASCADE, *[8]TYNDALL'S GAS TUBES, &c &c. by Mr. E.V. Gardiner. —In the Front Theatre

Quarter to Ten—DESCENT of the DIVING BELL, and EXPERIMENTS under Water, with HEINKE'S PATENT DIVING DRESS and HELMET. Description by Mr King.

Ten o'clock—EXHIBITION of a NEW SERIES OF DISSOLVING VIEWS, by a much IMPROVED OPTICAL COMBINATION: SCENES in SICILY and ITALY, painted by Messrs Prout and Hine:[9] the DESCRIPTIVE LECTURE written by the Rev. C.M. DAVIES,[10] M.A. To be followed by a very beautiful Series of PHOTOGRAPHS of Greek and other SCULPTURE, supplied by Messrs Negretti and Kambra.[11]

The band, under the direction of Mr. V. Waud,[12] of the Theatre Royal, Covent Garden, will play favourite selections during the evening.

RI MS JT/1/TYP/5/1630–31
LT Transcript Only

---

1. *[late October or early November]:* dated with reference to the meeting on 10 November.
2. *POLYTECHNIC INSTITUTION:* Established in 1838 by the engineer Sir George Cayley (1773–1857), the Polytechnic Institution was created to instruct and entertain the public through popular science lectures, and demonstrations of new scientific inventions and technologies. The Institution provided public classes in science, music, history, geography, and languages. Its public attractions included science lectures, rides in a diving bell, photography demonstrations, and exhibitions of models and machines. In 1841 Prince Albert became Patron of the Institution, and the name was changed to the Royal Polytechnic Institution. Under new management in 1860, the Institution was reopened to the public following a brief closure.
3. *Mrs T.*: Julia Tyndale. In 1842 Hector had married Julia Nowlen (1823–97), the daughter of Carlow immigrants (see *Tyndall Correspondence,* vol. 5, letter 1052, n. 18). There is no evidence that Tyndall or the Tyndales went to this event.
4. *Mr Mackintosh:* John Mackintosh provided science lectures for the Polytechnic Institution, such as his 'Properties of Matter' (1860). He also presented at the BAAS in 1858 on 'The Application of Combustible Compounds to be used in War' and on 'Constructing and Laying Telegraph Cables'.
5. *Mr E.V. Gardiner:* Professor E.V. Gardner was the Professor of Chemistry at the Polytechnic Institution during this period, and directed the experimental science department at the Polytechnic College in connection to the Institution. He was also a professor at the College of Experimental, Military, and Naval Sciences at Berners College of Chemistry, where he taught private pupils preparing for examinations. (*Chemical News,* (Sept 18, 1868), p. 139; (Oct 24, 1873), p. 217).
6. *Mr. King:* J. L. King of the Polytechnic Institution was one of the main lecturers on popular topics including astronomy, such as his 'Starwatching, and The Transit of Venus' (1860) and 'The Worlds Above' (1873). He also lectured in zoology, such as his 'Muscle

and Muscular Power' (1871), and on the oxyhydrogen microscope which was one of the main exhibits at the Institution.

7. *RHUMKORF'S:* Heinrich Daniel Rhumkorff.
8. *: the asterisk was added by hand to the printed material, presumably by Tyndall.
9. *Messrs Prout and Hine:* Henry George Hine (1811–95), landscape painter, comic artist, and wood-engraver (DNB); John Skinner Prout (1806–76), watercolour painter (*DNB*).
10. *Rev. C.M. DAVIES:* Rev. Charles Maurice Davies (1828–1910), Anglican clergyman, journalist and author. Ordained deacon in 1851, and priest in 1852. He was also headmaster of West London Collegiate School (1861–68) (*DNB*).
11. *Messrs Negretti and Kambra:* The Anglo-Italians Henry Angelo Ludovico Negretti (1818–79) and Joseph Warren Zambra (1822–97) formed a partnership in 1850 as successful makers of scientific instruments. They were instrument makers to the queen, Greenwich Observatory, and the British Meteorological Society (*DNB*).
12. *Mr. V. Waud:* Mr. W. Waud of the Royal Italian Opera, renamed the Royal Opera House in 1892. He provided the musical arrangements for the Polytechnic Institution.

From Hector Tyndale 2 November 1860 1737

Friday Morning (Nov 2/60)

My dear John

Bent upon sight-seeing with Julia—Mrs Tyndale—I could not reply to your note of yesterday[1] until this moment.

I fear I can give you no information, satisfactory as to the matter of our name and family in their connection with the Reformer Wil^m^ Tyndale.[2] For that which is entirely satisfactory to me and to all our family—viz: the internal evidence of our direct and true descent from the same stock and branch as William, will not amount to "information" for others. Therefor I never cite it to others, although I will do so to you, telling you, however, "an oft told tale".

There has been in our family so long as we can remember or know, an unbroken and unquestioned tradition of our direct connection by blood with the "Martyr Tyndale". This had the more belief from us because of the known family idiosyncrasies, in which, I think, you largely participate, tending to Theological investigation (with no especial dislike to discussion!) indeed to all manner of investigation; to a religiousness of character making itself manifest, very curiously, in reverent analysis.

A character the more reverent and the more religious that it takes heed of forms but secondarily, and only as they are useful to human or humane relations; and that it subjects tradition to conviction. Knowing well, even the untaught amongst them, that God is always out-flowering into new forms—that is, there is a recognition of the living God within—they are enthusiastic.

Then, too, there are temperamental and moral qualities in common—as the same quick temper, the same quick regret for it's showing, the same hardly given but always given apology for that showing, if unfair; and the same disposition to urge the right even to the bitterest end. All which seem to us to have been characteristic of William Tyndale. Another quality, nearly universal in the family, is that their courage is based mostly, in some cases wholly, upon Conviction or Conscience, wherein they are strong in Courage—while otherwise, tho' not deficient, they are not particularly or noticeably so—as was with William Tyndale, who, as you remember, fled from place to place from his blood-seekers, as if having a dread of death, or, rather, a care for life, even when life could give no rewards peculiar to itself. Unlike Luther he did not love the fight, tho' he could die or live (a harder matter in his case) without any disturbance from the Devil, or any shaking of his faith. Again, our family names are (or were until my own Father gave a commemorative set to his children), mostly the same as those of the family of William Tyndale—viz: Matthew, William & John.

But there is another matter, the one you refer to in your note, which gives a more plausible and objective reason than the internal and family-satisfying evidences I have named. It is that of my Brother,[3] who died many years ago leaving us without any portrait or likeness of himself, except the very strong and ever-reminding one of William Tyndale an engraving of which we had in a copy of the "Neue Testament", Tyndale's translation, a fac simile Old English print published here in London about 25 years ago. My Brother's loss was very notable to his friends, for he was made of all Goodness, Sweetness, Courage, Manhood. He was entirely faithful. He was, besides all, of fine and even noble personal presence, and exceedingly well gifted. The Martyr's portrait gave great comfort to us all, for it was as like my Brother as though drawn from his living face—even with all the difference in their years of life, and between the bearded Man and the unbearded youth of twenty one years.

I should mention as coincidental with this resemblance that one of the Progenitors of the Reformer married a Miss Hytchins[4] here in London and that one of my Mother's Grand Parents bore the same name and was the child of one of the followers of William Penn to America.

I fear I have bored you with a long and, as you may have reason to think, irrelative reply to your short note, but having time I have 'found it in my heart to bestow all my tedium upon you'.

Y[ou]rs Affectionately | Hector Tyndale

What has become of Hirst and yourself or, rather, what of your evenings?

I return to you the circular.[5] Can you procure me one or two, <u>or more</u> copies of it for sending to America?

RI MS JT/1/T/50
RI MS JT/1/TYP/5/1627–29

1. *your note of yesterday*: possibly a reply to letter 1736. The response was probably also inspired by the Earl of Ducie asking Tyndall if he was descended from William Tyndale (Journal, RI MS JT/2/13c/1187). See letter 1740.
2. *Reformer Wilm Tyndale*: William Tyndale.
3. *my Brother*: Troilus Tyndale (1817–38). His eldest brother Sharon Tyndale (1816–71), was secretary of state for Illinois 1865–69.
4. *Miss Hytchins*: not identified.
5. *I return to you the circular*: probably the London Institution notice (see letter 1736).

To Heinrich Debus 4 November 1860 1738

Sunday | 4th, Nov 1860.

My dear Heinrich

I received both your letters[1] and was very thankful to you for them. They were and will be of very great use to me. Indeed the simple addition of the pumice stone and caustic potash to the apparatus, caused the nitrogen formed from the nitrite of potash to descend to the level of ordinary nitrogen, but what can it be that produced the effect? Caustic potash would have no influence on nitrous oxide and I do not think nitric oxide could have been present for there was not the slightest trace of red fumes. The passage of the gas through pumice stone & sulphuric acid seemed to have no influence on the result.

I am still at work every day,—I get to the Institution as early as I can & work while I have light. The subject requires the greatest care and delicacy, and for more than six weeks I did little else than find out the conditions of accurate experiment. I have now reached an almost miraculous command over my apparatus, and shall devote the remainder of the time at my disposal this year to the determination of quantitative results. When you see these I think they will excite your surprise.

Of course one cannot work at a subject of the kind without being led into speculation—I think I see molecular qualities, and relations between the molecules of ordinary matter & the surrounding ether more clearly than ever they have been apprehended before. But while I give free reign to my imagination in these matters,[2] I drag her sternly down at last to the touchstone of experiment.

I shall be very grateful for any suggestions which you may be kind enough to make to me. I am much in want of some volatile liquids, as I am now engaged on the subject of vapours—have you any that you could give me? A thimble full of each would be sufficient for my purpose.

I hope you will visit us in London this year oftener than you used to do. Hector[3] is still here, and often speaks of you. The more I see of him the more I am struck by his remarkable powers: it is a thousand pities that such a man has been tied down to a vocation with which he has no sympathy, and which is unable to develop the extraordinary faculties which he possesses.

I have been lately not a little interested in looking into the life of an old fellow named William Tyndale[4] who quitted Northumberland in the times of the wars of the Roses and settled in Gloucestershire. He translated the bible, wrote against the Church of Rome; had to run for his life from place to place, and finally was caught, strangled and burnt near Brussels. When a boy this man's example was often set before me, and it was affirmed that I belonged to his stock. I always regarded the thing as a myth. Still the tradition is a very old one. But it interests and amuses me to see the striking similarity between the temper of this old reformer and those of Hector & myself. Curious also it is to notice the tendency towards theological discussion in us. I now know all about the Romish controversy so well that I should not fear to hold my ground against Cardinal Wiseman[5] himself. I owe this knowledge to my father, he to his father & so on. It certainly is curious and amusing.

But I am boring you. When you come to town I must show you my methods of experimenting. Goodbye dear Heinrich.

Always yours, | John Tyndall

RI MS JT/1/T/260
RI MS JT/1/TYP/7/2367

1. *both your letters*: letters 1733 and 1734.
2. *give free reign to my imagination in these matters*: Tyndall's belief in the power of creative scientific 'imagination' to give explanations for observed phenomena is best captured in his 'Scientific Use of the Imagination', delivered before the British Association at Liverpool, 16th September 1870. J. Tyndall, *Essays on the Use and Limit of the Imagination in Science* (London: Longmans, 1870).
3. *Hector*: Hector Tyndale.
4. *William Tyndale*: see letters 1737 and 1740.
5. *Cardinal Wiseman*: Nicholas Patrick Stephen Wiseman (1802–65), created first Catholic Archbishop of Westminster in 1850 by Pope Pius IX.

From [William Fullarton] Cumming[1] 4 November 1860 1739

Kinellan[2] | Nov. 4th. 1860

My dear Tyndall

If you only knew how my conscience has been smiting me for my scandalous behaviour to you, I am sure you would forgive me for not sooner acknowledging your most kind letter[3]—It reached me ages ago, when located at my snuggery in Mull; and droll enough, when put into my hand, the young Marquis of Lorne[4] was seated at the breakfast table with me!—so after reading, I passed it to him, and he was greatly amused, and not a little pleased at your description of his rubicund locks &c. &c. I wish you had had longer acquaintance with the young Marquis, for I am sure you would have liked him much.

But what excuse am I to make for my long neglect in writing? Oh Professor, may you never know the miseries of a paralytic limb, and all the "inertia" of body and mind it brings along with it! Day after day I have been resolving and re-resolving, yet never have acted. And I now almost wonder that I should have the courage to take pen in hand—! I am literally good for nothing. A mere cumberer of the ground—a vegetating fungus rather than a living and sentient being. But I am not going to fill a paper with groans and lamentations. This would neither be fair to you nor just to myself, for though sorely smitten, life has still some enjoyments for me, and I desire rather to be thankful for the mercies yet vouchsafed to me than to murmur over those that have been taken away.

True it is that in <u>this</u> world, I have nothing to look forward to but ευθαναϭια,[5] but fortunately for us all, there is <u>another</u> world, to which my thoughts are often pointing. But you will say when is all this twaddle to have an end? Not until I have made a proposal which I hope you will entertain favourably.

You know well the importance to an over-worked brain of rest from all mental toil in a picturesque and healthy locality.

Now, some years ago, Argyle,[6] at my suggestion, built a cottage on the shore of a beautiful little loch on an estate of his in Mull as a fishing box for himself and me—Thither I repair every autumn, and pass the months of July, August, September, enjoying immensely the fine air, fine scenery, and—though last, not least—the sport of catching some 6, or 8, or 12 fine sea-trout daily. Even if I can get no fish, I am always overpaid by living on the loch all day in my boat.

Now, why should not you come and take a spell of 'Tusculum'[7] next summer?

I cannot promise you 'Proud frowning rocks of sublimity'[8] like Mont Blanc but we have a mountain—'Benmon' to wit—of some 4000 ft. high,[9] which even an Alpen climber like you might find pleasure in exploring. Argyle brings the Duchess and several of the children for 2 or 3 weeks, and then our cottage is full, but at all other times it shall be wide open to receive you. Now Professor, I hold you engaged next summer, (if I live so long) to come to my Tusculum when I shall do every thing in my power to make your stay agreeable, besides indoctrinating you into the "gentle art" of angling.

I do not bind you to a promise now, but when the time approaches, I shall write and lay hands upon you.

I find my poor sister[10] in great depression on account of her eldest daughter Mrs Whately[11] who is now here, and in a state of health so bad as to leave little or no hope of recovery. By the way I sent to you 2 or 3 days ago, a copy of a great work! you spoke of in your letter. You will find it sad twaddle, but never mind, it will at least, light your fire, or your pipe—and it is all I have to offer in token of the great regard I bear to you. Of course I expected no reply to this letter—only bear in mind I hope to see you in Mull next autumn. Meantime believe me

Ever my dear Tyndall | most truly Yours | [W. F.] Cumming | Professor Tyndall

RI MS JT/1/TYP/1/279

---

1. *[William Fullarton] Cumming:* a handwritten note on the typescript reads [George Cumming | Nov. 19. 60]. However it is likely that this letter is from Mary Coxe's brother, William Fullarton Cumming.
2. *Kinellan:* home of James and Mary Coxe in/near Edinburgh. It was occupied by William Fullarton Cumming after their deaths.
3. *your most kind letter:* letter missing.
4. *the young Marquis of Lorne:* John Campbell (1845–1914), who succeeded his father as 9th Duke of Argyll in 1900.
5. ευθαvασια: euthanasia.
6. *Argyle:* George Campbell, 8th Duke of Argyll (1823–1900). Politician, and writer on science and religion. A regular correspondent with Tyndall.
7. *Tusculum:* Tusculum was a Roman city in the Alban Hills, famous for its luxurious villas, yet accessible to Rome.
8. *Proud frowning rocks of sublimity:* this may be a loose reference to the poem 'Modern Greece' by Felicia Hemans (1793–1835). She is best known for her opening line 'The boy stood on the burning deck' from her 1826 poem 'Casabianca'.
9. *'Benmon' to wit—of some 4000 ft. high:* Ben More, the highest mountain on Mull, is in fact 3,178ft or 966m, in height.
10. *my poor sister:* Mary Coxe.
11. *her eldest daughter Mrs Whately:* not identified.

To Hector Tyndale [5 November 1860][1] 1740

Royal Institution | Monday m[ornin]g

My dear Hector.

Would you make an addition to your kindness of looking up for me the works of Tyndale,[2] by looking up some photographer who will either reproduce that stereoscopic miniature of yourself, or take one as nearly equal to it as his art or the expression of your countenance in London will admit of. I would undertake this labour myself were I not so held by my experiments.

Fill your soul with a good thought, let it irradiate your features, and then let the photographer do his work—I do not at all object to that dash of sorrow which mingles with the general expression of the face.

Yours | John

I must read the "Martyr"[3] through and through. I must pierce him and fuse him, reduce him to his component atoms and deduce his actions from their very spring—What I have read interests me amazingly by the reflected light which it throws upon the writer.

RI MS JT/1/T/1473
RI MS JT/1/TYP/5/1625

1. *[5 November 1860]:* LT dated this letter 25 July 1860. Tyndale was in London at this time, but 25 July was a Wednesday. It is more likely that this letter dates to early November, around the time of Tyndale's letter of 2 November (letter 1737), when Tyndale was also in London, and it may be a response to that letter.
2. *the works of Tyndale:* see letter 1737.
3. *the "Martyr":* William Tyndale. See letter 1737.

From William Hopkins 5 November 1860[1] 1741

Llwyn Madoc | Builth, S. Wales | Nov. 3 /60.

Dear Tyndall,

I have finished the review of your book[2] with the exception of a single passage in which I wish to announce the fact of my having read Rendu's memoir,[3] which I have never been able to lay my hands upon.[4] In writing to Parker[5] the other day, it occurred to me that the volume of Transactions containing it[6] might be obtained perhaps from the Royal Society or the Geol. Sec., should you not get back your single copy in time. I want much to see it though; I doubt not, my impression will be the same as that from your section respecting it.[7] I am glad you do not think of replying directly to Forbes's pamphlet,[8] more especially as I have spoken of it myself; and I think your immediate justification (if you can be said to require any at all, which you do not) will come better from me than from yourself. I have insisted on Rendu's claim to a share of such honour as the first conception of the molecular mobility of glacial ice ought to confer, allowing Forbes all the merit of his persistent endeavours to establish the viscous theory,[9] but not of being its first originator. I have pointed out also that we have really little concern with the question as a personal one—whether Forbes did justice to Rendu or not in his 'Travels'. The real question is a much broader one; whether, when the scientific public of this country assigned the whole merit of the Viscous Theory to Forbes, had or had not (thro' Forbes or anybody else) sufficient means before them of weighing Rendu's merits with his? We had really nothing of the kind. Forbes's insulated quotations gave no real idea even of what your section gives one. But I want much to bear testimony myself to the character of the real memoir.

Forbes is a selfish, uncandid and ungenerous minded man. If I ever did swear (which I never do) I should say that his d—-d Scotch prudence keeps him just on the safe line of propriety beyond which he never tumbles over on the generous side. Such was singularly the case in his dispute with that kindly open-hearted fellow Agassiz, and such has been exactly his conduct towards Rendu. His extracts from Rendu's memoir are such as might perhaps gain him a verdict against a charge of injustice in a court of scientific law, but in a court of scientific equity and generosity he would be utterly lost. While he makes bald and insulated quotations from Rendu's memoir he usually tries to bespatter him with more or less of contempt, and has encouraged in a remarkable degree a most unjustly deprecatory tone towards him on the part of his friends and reviewers. He has endeavoured to smother my own views and arguments by ignoring them. He will not find that ruse successful. In some form or other I intend to publish a complete statement of my own views[10] as founded on my own earlier investigations and your recent researches, in which I shall criticize without reserve Forbes's theoretical reasonings which I regard as contemptible from beginning to end. His attempt to merge your experiments in his theory, in that Jesuitical paragraph[11] in his introduction to his Occas. Papers offends me excessively. If his tearing and bruising should be your only process of continued cracking breaking and regelation! If indeed it should be so! The whole passage is a kind of fac simile of the man.

I hope my review[12] won't make you conceited. It is complimentary. I shall be here till the 12th at least.[13] Don't forget the memoir. The review is long one for one in Fraser. It will occupy a dozen pages.[14]

Yours truly | W. Hopkins.

My knee mends but rather slowly I think I am just mounting my crutches and become a sort of quadruped an instance of a mutation of species perhaps. I shall be glad to resume biped character again, but that will require time. I know not when I shall be scientifically in town again.

Nov. 4.

I could not get my letter off on Saturday[15] and to-day we had no post. I add a few lines more. As you are about to write an article for Fraser,[16] if it be on glaciers and you are likely to write it soon, had you not better see my review before it is printed; unless it appear in the Dec. number, in which case you would gain little by seeing it before hand. As it was impossible for want of space to enter fully into the analysis of glacier theories, this part of the review is but imperfect. I have made some observations on the sliding theory. There is not a fact connected with glacial motion more certain than that glaciers do slide, and that one part of the motion of their surfaces is due to that cause. Your observation makes it about 1/2 of the whole motion; Forbes makes it 5/8. But the curious thing is that, while nobody doubts this, everybody forgets it, apparently because it is independent of the internal structure of the mass. It must however, be recognized in any general view of the motion of a glacier. Till my experiment exemplifying the slow and uniform motion of a block of ice down a plane of small inclination, Forbes used to maintain that if the glacier slided at all, it would descend with an accelerated motion and be discharged into the valley beneath. When he dared no longer to use that argument, he has tried as much as possible to ignore the subject. In one of my letters to the Phil. Mag.[17] I said that both the sliding and viscous theories were founded on verae causae, but that the relative amounts of motion produced by them could only be determined by observation, and I insisted on the determination of the velocities at the surface and bottom of a *[1–2 words missing]* glacier as the only one which could decide the relative claims of the two theories. I anticipated that a portion of the whole motion would be due to each cause, but my impression was so far erroneous that I expected more would be due to the molecular mobility of the mass. Forbes's observation at the end of the Glacier des Bois was in fact made at my instigation. Having found the two motions in the ratio of 5/3 in favour of sliding, he has the impudence (I can call it nothing else) to say that the result leaves us no alternative but the adoption of the viscous theory, and, as he evidently means, to the exclusion of the sliding theory which he certainly does thoroughly exclude, for I don't think he ever mentions it again with reference to my experiment, and only in his notice of De Saussure's theory[18]—the most absurd and illiberal piece of criticism that ever man was guilty of. However the learned professor must be taught that his decision is not final in such matters. Your theory of regelation requires a large crushing force. This I am satisfied is derived principally at any point of the valley where the motion is obstructed from the sliding of the mass in the higher and more unobstructed parts of the valley. An enormous crushing force on an obstructed part of the glacier will thus be produced a tergo,[19] and there comes in so beautifully your regelation, after the change of form permitted by the molecular mobility of the crushed mass and required by the conditions of the glacial valley, has been effected. Where the valley is unobstructed the quicker motion of the centre of the glacier is constantly both by tension and pressure, producing both this infinitesimal cracking (as well as finite fissures) and consequent molecular mobility. I very much doubt whether a glacier would move at all if it were every where totally incapable of sliding. Forbes has felt the want of a force of this kind to bring his viscosity into play and his endeavour to find it in his "enormous hydrostatic pressure" the most blundering bosh that ever was written. I alluded to the passage you have quoted from the third Part of the Royal Society Memoir.[20] Well may you say you don't perfectly understand it. But all this requires careful and accurate development. I have not yet said any thing about it in the review. I may add a paragraph about it.[21] I wish I could see the review of your book in the New Edin. Ph. Jour.[22] I don't wish of course to make my review an answer to any ignorant fellow who may choose to abuse you in that periodical, but if there be any thing to which I could indirectly reply I would do so, if you think it worth while to get me a copy thro' Parker.

Your smashing of Forbes's absurd ripple theory[23] is complete. What will become of his differential theory of the veined structure without his ripples?

Your elucidation of the internal pressures and tensions of a glacier, due to the more rapid motion of the centre, will be very valuable. Forbes's friends in the plentitude of their ignorance, may ignore a mathematical investigation which neither they nor their readers can understand, but they cannot so easily ignore the conversion of your circle into an oval, and the obvious inferences therefrom. It only recently occurred to me how simply your diagram would determine also the line of greatest differential motion.

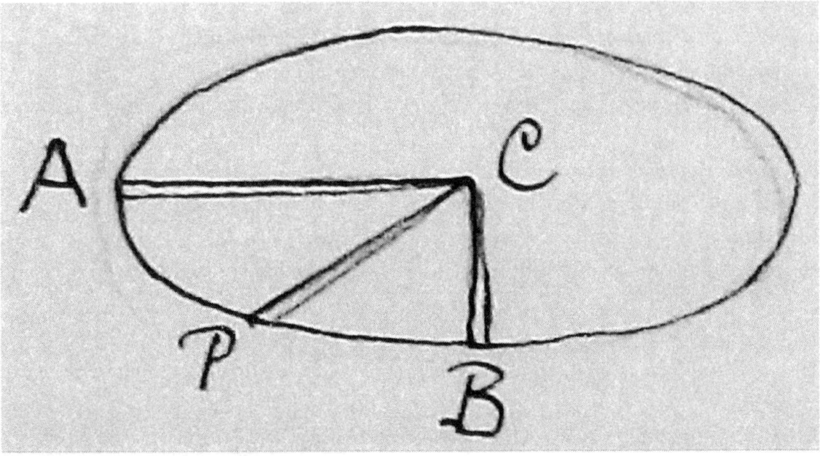

It is manifestly that in which the difference between two consecutive radii drawn from the centre to the circumference of the oval, is greatest along C A and C B the directions of greatest tension and pressure it is zero, and a max. along C P which (the oval being but little elongated) will manifestly be inclined at 45° to each of the above lines. Would it not be well to add this in your next edition?[24] It comes so directly to the point. That Forbes should have mistaken C A for C B is nothing to be surprised at; it is in accordance with all his other mechanical blunders and obscurities; but that Whewell[25] should have copied him only shows the blundering hurry in which he does all such things.

Nov. 5th

I got my letters unexpectedly yesterday. I find my review must be in the hands of the editor of Fraser by the 15th.

Don't be afraid of Forbes and his sychophants. I'll support you. With the sense of having truth on our side we can fight with better temper than they can, and it will be strange if men of science, experimental and mathematical, can't beat 'Authorities of the press.'

I am mending though slowly. I have no pain but the knee remains stiff and swelled and the leg of course useless to me at present.[26]

Yours truly | W. Hopkins.

RI MS JT/1/TYP/2/623–29
LT Transcript Only

---

1. This letter was started on or before 3 November, with additions dated 4 and 5 November.
2. *your book: Glaciers of the Alps,* see also letter 1595, n. 4.
3. *Rendu's memoir:* see letter 1619, n. 13, and letter 1731.
4. *which I have never been able to lay my hands upon:* see letter 1619, n. 13, and letter 1731.
5. *Parker:* see letter 1707.
6. *volume of Transactions containing it:* Rendu's memoir 'Théorie des Glaciers de Savoie', was published in the *Memoirs of the Royal Academy of Savoy* in 1841.
7. *your section respecting it:* in *Glaciers of the Alps,* Tyndall had a section (14) entitled 'Rendu's Theory'.
8. *Forbes's pamphlet:* see letter 1731, n. 4.
9. *the viscous theory:* the idea that a glacier flows as a viscous substance.
10. *publish a complete statement of my own views:* see letter 1731, n. 11.
11. *that Jesuitical paragraph:* Forbes had effectively claimed that Tyndall's results were confirmatory of his own theory: Forbes, *Occasional Papers* (1860), pp. xvi–xvii.
12. *my review:* see letter 1731, n. 11.
13. *I shall be here till the 12th at least:* Hopkins was recovering from a knee injury. See letter 1731.
14. *It will occupy a dozen pages:* in the event seventeen pages.

15. *on Saturday:* 3 November 1860.
16. *As you are about to write an article for Fraser:* it does not appear that Tyndall published an article in *Fraser's Magazine* at this time.
17. *In one of my letters to the Phil. Mag.:* Hopkins had published three letters in the *Philosophical Magazine* in 1845: 'Mr. Hopkins on the Motion of Glaciers', *Phil. Mag.,* 26 (1845), pp. 1–16; 'Mr. Hopkins on the Mechanism of Glacial Motion', *Phil. Mag.,* 26 (1845), pp. 146–69; 'Mr. Hopkins on the Mechanism of Glacial Motion. Third Letter', *Phil. Mag.,* 26 (1845), pp. 237–51. The suggestions referred to appear on p. 250.
18. *De Saussure's theory:* Horace Bénédict De Saussure is seen as the originator of the idea that glaciers move primarily by sliding, see letter 1666, n. 3.
19. *a tergo:* from behind
20. *the Royal Society Memoir:* see letter 1731, n. 8.
21. *I may add a paragraph about it:* Hopkins did not specifically address this point in his article, which concentrated on demolishing other aspects of Forbes's theories.
22. *the review of your book in the New Edin. Ph. Jour.:* 'Reviews and Notices of Books: the Glaciers of the Alps', *Edinburgh New Philosophical Journal* 12 (1860), pp. 249–65.
23. *Forbes's absurd ripple theory:* Forbes set out his ripple theory in his 'Seventh Letter on Glaciers,—on the Veined Structure of the Ice', *Edinburgh New Philosophical Journal* (1844), pp. 244–49. He proposed that differential motion of parts of the glacier could produce the marginal structure, as ripple-marks are produced on the surface of a flowing river. Tyndall challenged this argument in a section of his *Glaciers of the Alps,* pp. 398–403. Hopkins likewise considered Forbes's arguments incorrect: see 'The Glaciers of the Alps', *Fraser's Magazine,* 62 (December 1860), pp. 808–9.
24. *your next edition:* the second edition was not produced until 1896, after Tyndall's death.
25. *Whewell:* William Whewell (1794–1866), with whom Forbes exchanged many letters about how to handle his dispute with Tyndall.
26. *I have no pain but the knee remains stiff and swelled and the leg of course useless to me at present:* Hopkins had badly injured his knee and was recuperating at Llwyn Madoc near Builth Wells, the estate of Henry Thomas. See also letter 1731.

From Heinrich Debus — 7 November 1860 — 1742

Queenwood College | Nov[em]b[e]r 7th 60.

My dear John,

Your kind letter[1] was duly received yesterday morning, and read by me with great pleasure.

It is difficult to say what kind of substance was absorbet[2] by the caustic potash, without knowing something about the way how you prepared the nitrite of potash or making some experiment on the subject. Carbonic acid, laughing gas (NO) ammonia, nitrous acid or nitric oxide (NO_2) might have been present. If sulphuric acid had, as you say, no effect on the gas so as to alter your results, ammonia, or nitric oxide (NO_2) can not have been present, because both are absorbed by $HOSO_3$.[3] Nitrous acid is of an intense red colour and I have no doubt, even a very small quantity might be perceived. Laughing gas is a little absorbed by caustic potash.

Now I suppose you prepared the nitrite of potash by passing nitrous acid into solution of potash and afterwards adding chloride of ammonium to this liquid. The question here occurs, was this liquid acid or alkaline? If alkaline then the nitrogen would most probably be contaminated with carbonic acid and the action of the caustic potash would be comprehensible. However I cannot be certain about this point without knowing all the conditions under which the N[4] was prepared.

as for volatile liquids I give you a bit of those I possess and shall be most happy to let you have as much of them as is at my disposal. Or if you have your eye on one I will have it prepared for you.

Methylic alcohol—boils at about 600C

Amylene—boils at about 300

Bisulfide of carbon

Bromide of ethyle boils at 440C

Chloride of acetyle, boils at about 550 but is affected very much by moisture and metals.

Bromide of ethylene boils at 1290C

Sulfocarbonate of ethyle boils at 2000C

acetate of glycol boils at 1800C

chloroacetate of glycol boils at about 1400

chloride of Benzoyl boils at 2000C

nitrite of ethyle boils at about 170C

I am greatly delighted with the splendid results you have lately obtained and shall be right glad to listen to you, and have them explained to me. Very likely I shall come to town either on Thursday next week or on the Thursday a fortnight later.

I beg to return the little paper concerning William Tyndale's monument.[5] If the man belongs to your family you may be proud of him, and if he does not, then you may at least look up to him with high reverence and be pleased that the name of Tyndale has had already so worthy a representative. It is a curious fact, and I believe one to little understood and examined into, the fact that peculiarities and tendencies of the father are inherited by the son and very often are even met with for several generations. How is this to be accounted for? It is perfectly true I have often noticed in you a tendency to examine into religious questions and you certainly know much more about them than I do. —

Miss Tribe[6] has been staying at Broughton during September and part of October and I went sometimes down to spend the evening there. On such an occasion I met the Miss Nicol and Miss Betsy Nicol[7] and they inquired most kindly how Dr Tyndall was, and seemed to remember him very well. Mr Tribe[8] is engaged to Miss Lander[9] and as soon as he has obtained a more lucrative position he is going to marry her. I only saw the lady from some distance at church on Sunday evening last, but from all I hear she is likely to make an excellent wife to Tribe. With my kind regards to Tom.

Always your | H. Debus.

RI MS JT/1/D/27

---

1. *Your kind letter:* letter 1738.
2. *absorbet:* absorbed. Debus betraying his German heritage.
3. *HOSO3*: sulphuric acid (now written H_2SO_4).
4. *N:* nitrogen.
5. *I beg to return the little paper concerning William Tyndale's monument:* missing. The paper is not mentioned in letter 1738. A monument to the martyr William Tyndale, Tyndall's supposed ancestor, was erected at Nibley Knoll, north of Bristol, in 1866. It is a 100ft tower, still standing today overlooking the valley across to the River Severn.
6. *Miss Tribe:* not identified. Tyndall's journal mentions meeting Tribe's sister (see n. 8) on 6 September 1857 (Journal, RI MS JT/2/13c/1032). At the time of the 1841 Census, Tribe may have had three sisters, Jane, Elizabeth and Ann.
7. *Miss Nicol and Miss Betsy Nicol:* not identified. According to Tyndall's journal, Nicol was a clerical friend of Tribe (see n. 8) and these may have been his sisters (Journal, RI MS JT/2/13c/1032).
8. *Mr Tribe:* Walter Harry Tribe (1832–1909) had been appointed to teach classics at Queenwood in July 1855. He took holy orders in 1857 and was Rector of Stockbridge (1860–67). He served on the north-west frontier of India (1867–85) and was Archdeacon of Lahore (1885–92). He ended his days as Vicar of Awliscombe in Devon. His daughter Mary (1865–1937) became the 11th Duchess of Bedford. See *Tyndall Correspondence,* vol. 3, letter 1245, n. 7.
9. *Miss Lander:* Sophy/Sophia/Sophie Lander married Walter Tribe in 1861. Tribe does not seem to have waited for a 'more lucrative position'.

From John Douglas Cook[1] 9 November 1860 1743

My dear Tyndall

I am told you have a lot to say about the Matterhorn.[2] Pray write every word of it, and then send it to me, like a very good Boy, which you really would be if you would do more for the S.R.[3] and

Yours very truly | John D. Cook | Friday Novbr 9/60

RI MS JT/1/C/48
RI MS JT/1/TYP/1/227

1. *John Douglas Cook:* John Douglas Cook (*c.* 1808–68). Editor of the *Morning Chronicle* until its sale in 1854, then founding editor of the *Saturday Review.*
2. *I am told you have a lot to say about the Matterhorn:* see letter 1726, n. 13.
3. *for the S.R.:* for the *Saturday Review.* There is no article from Tyndall on the Matterhorn, but he did write one on his winter ascent of Snowdon: 'Snowdon in Winter', *The Saturday Review,* (5 January 1861), pp. 11–12.

From Juliet Pollock 10 November 1860 1744

59 Montague Square | Nov. 10th 1860

My dear Mr Tyndall

We have lost our friend Mr. Parker.[1] It is a great loss & your warm, kind heart will readily conceive how much it pains us. —he was such a zealous friend and there was an unconventionality about him, not amounting to eccentricity, that was very endearing. —I have written to my boys not to think any more about the charade party. —It was so associated with him that the attempt at pleasure would be too painful. —I believe my sons will sympathise in this feeling—and I will try to amuse them in some other way.

You will be sorry to hear that John[2] is too ill to remain in service and that we are looking out for another servant.

Good friends and good servants are more easily lost than replaced—such is life. —You had better dine with us on Saturday the 17th: at ½ past 6,—that we may set off together.[3] I have taken a place for Walter.[4]

Yours most truly | Juliet Pollock

RI MS JT/1/P/170
RI MS JT/1/TYP/6/1971

1. *Mr. Parker*: John Parker.
2. *John*: unidentified.
3. *that we may set off together*: after dinner Tyndall went with Juliet Pollock, Walter (her son) and Miss Creed (a relative, as Juliet's maiden name was Creed) to the opera Robin Hood (Journal, RI MS JT/2/13c/1191). Robin Hood, an opera in three acts with libretto by John Oxenford and music by George Macfarren, had opened at Her Majesty's Theatre on 11 October 1860.
4. *I have taken a place for Walter*: Walter Herries Pollock.

To Juliet Pollock 10 November 1860 1745

10th Nov. 1860

My dear Mrs Pollock

You judged aright when you considered that I should feel grieved by the intelligence of John Parker's death.[1] The good dear fellow so warm and so upright, it fills me with sorrow to think that he is gone. Every one who knew him must share this feeling. I had hoped for his growing intimacy and friendship; for half by instinct or sympathy I had seen in him an Israelite indeed in whom there was no guile.[2] And now he is silent—no joyous jumping any more on your little stage—Well it is sad.

I will gladly come to you on Saturday.

good bye Ever yours | J Tyndall

RI MS JT/1/T/1134
RI MS JT/1/TYP/6/1970

---

1. *John Parker's death:* see letter 1744.
2. *an Israelite indeed in whom there was no guile:* John 1:47, 'Jesus saw Nathanael coming to him, and saith of him, Behold an Israelite indeed, in whom is no guile!'.

From William Hopkins 12 November 1860 1746

Llwyn Madoc | Nov. 12 /60

Dear Tyndall,

I merely write to say that I leave this place on Tuesday (tomorrow) morning. Should Rendu's memoir[1] not have started when you receive this, address it to me J. Martins Esq.[2] | Whatten House | Loughbro' | Leicestershire. | Should you have sent it off I shall either get it this afternoon or it will be immediately forwarded to me. I must send the review to the Editor[3] on the 15 or 16th. If I can see Rendu previously tant mieux[4], if not I must speak on your report of it. I am quite sure my own impression will be the same, but having seen it one could speak with rather more authority.

Thank you for the anonymous review.[5] It is by reviews of this kind, written by wretched advocates without the smallest amount of exact knowledge on the subject, that so much harm is done in retarding the sound advance of science. Why is Forbes to take up the subject as a matter of personal dispute with every one who may put forward any thing opposed to his own particular views. This is not the conduct of a lover of scientific truth, but of a jealous lover of his own reputation. I know of no other instance at all like it. I wonder he was not ashamed of quoting such an 'authority of the press' as the review you sent me. I knew it would come to this, that all your experiments on regelation and on the formation of the veined structure would be claimed to feed his voracious theory. You will see what I have said about your section on Rendu. That your very right desire to do justice to him should be interpreted into a personal attack on Forbes is intolerable.

My leg is improving,[6] but it still, and will for many weeks hence, keep me very helpless.

Yours truly | W. Hopkins.

RI MS JT/1/TYP/2/630–31
LT Transcript Only

1. *Rendu's memoir:* see letter 1619, n. 13.
2. *J. Martins Esq.:* John Martin (b. 1815). He married Augusta-Louisa, second daughter of William Hopkins, in 1852.
3. *to the Editor:* John Parker, of *Fraser's Magazine.* Hopkins had presumably not heard of his recent death (see letter 1744).
4. *tant mieux*: so much the better.
5. *the anonymous review:* not identified.
6. *My leg is improving:* see letter 1731.

From Robert Bunsen 13 November 1860 1747

Theuerster Freund

Ich erhalte eben zu meiner größten Ueberraschung von Miller die Nachricht, daß mir die Royal Society die Coplay Medal hat zu Theil werden laßen. Je weniger ich auf eine Auszeichnung von solcher Bedeutung Anspruch machen kann, um so dankbarer erkenne ich darin einen neuen Ausdruck der freundlichen Gesinnung, welche Sie mein theuerster Tyndall gleich meinen vielen übrigen Englischen Freunden so oft mir bethätigt haben. Ich bitte Sie Allen die sich so freundlich für mich interessirt haben meinen herzlichsten Dank zu sagen.

Bei aller Freude über die mir so unverdiente Weise gewordene Ehrenbezeugung drückt mich doch einige Besorgniß, über die Sie mich beruhigen müßen. Sollten die letzter von mir und Kirchhoff gemeinschaftlich publicirten Spectralarbeiten für den Entscheid des Councils maßgebend gewesen sein? Wäre dieß der Fall, so würde ich der Auszeichnung nicht froh werden können, da sie dann vielmehr Kirchhoff als mir gebührte, denn mein Antheil an den wichtigen Untersuchungen meines Collegen Kirchhoff ist nur ein untergeordneter. Was davon auf mich kommt, sind nur Abfälle seiner großen und herrlichen Entdeckung der den nicht atmospharischen Fraunhoferschen Linien zum Grunde liegenden Ursache. Ich bitte Sie liebster Freund, schreiben Sie mir ganz aufrichtig, wie weit meine Besorgniß gegründet ist.

Ihre herrliche Alpenarbeit, für deren freundliche Uebersendung ich immer noch nicht gedankt habe, macht jetzt, nachdem ich sie mit dem höchsten Intereße und Vergnügen gelesen, die Runde bei meinen hiesigen Freunden. Bitte mein theuerster Freund schreiben Sie mir bald ein paar Worte der Beruhigung im Betreff meiner oben ausgesprochenen Besorgniß.

In treuer Freundschaft | Ihr | RWBunsen.

Heidelberg den 13^{n} Nov. 1860.

Dearest friend

To my greatest surprise, I have just received from Miller[1] the news that I have been awarded the Coplay Medal[2] by the Royal Society. The less I can lay claim to awards of such significance, the more gratefully do I perceive therein a new expression of the friendly disposition that you, my dearest Tyndall, like my many other English friends, have so frequently shown towards me. I ask you to communicate my most heartfelt thanks to all who have shown such kind interest for me.

With all delight at the honour that has so undeservedly been granted to me, you must ease me of some concerns that are weighing me down. Were the works on spectra that I published in collaboration with Kirchhoff[3] a determining factor in the council's decision? If this was the case, I could derive no pleasure from the award as Kirchhoff would be more deserving of it, since my part in my colleague Kirchhoff's important investigations was only secondary. Whatever I contributed to his great and magnificent discovery of the causes of non-atmospheric Fraunhofer lines[4] is only rubbish. I beg you, dearest friend, write me most candidly how far my worries are justified.

Your splendid work on the Alps,[5] for the kind sending of which I still have not thanked you, is now making the rounds among my local friends after I read it with the greatest interest and pleasure. Please, my dear friend, write me soon some words of reassurance concerning the worries that I expressed above.

In sincere friendship | Your | RWBunsen

Heidelberg, 13 Nov. 1860.

RI MS JT/1/B/148

---

1. *Miller*: probably William Hallowes Miller (1801–80), Foreign Secretary of the Royal Society 1856–73.

2. *Coplay Medal*: Copley Medal, the Royal Society's premier medal, awarded at the Anniversary Meeting each year on 30 November.
3. *works on spectra that I published in collaboration with Kirchhoff*: Bunsen was in fact awarded the medal 'for his researches on Cacodyl, Gaseous Analysis, the Voltaic Phenomena of Iceland, and other researches'. See *Royal Society Minutes of Council, 1858–1869,* 3 (8 November 1860), p. 72. The paper by Kirchhoff and Bunsen 'Chemical Analysis by Spectrum-Observation', was published in *Phil. Mag.,* 20 (1860), pp. 89–109.
4. *Fraunhofer lines*: see letter 1595, n. 10.
5. *Your splendid work on the Alps*: *Glaciers of the Alps,* see letter 1595, n. 4.

To Henry Enfield Roscoe[1] 23 November 1860 1748

Royal Institution of Great Britain | 23rd Nov. 1860

My dear Roscoe,

From the combustion of metals in the arc you get quite magnificent effects with the electric lamp. The mode of proceeding is this: place a pillar of carbon on your lamp with the top hollowed out so as to form a crucible; place on this a fragment of metal—say of copper or silver or zinc, or if you wish for an alloy use brass & bring your upper point down on it. Make your slit narrow and you get with each metal a most glorious & characteristic system of bright lines.

Send the light of a second lamp thro' the arc thus formed & I doubt not you will get the effects of Kirchhoff & Bunsen.[2]

I have desired Anderson[3] to send you a carbon pillar—this will save you a little trouble.

The bright bands are obtained with care, but I have not myself made the experiment with the second lamp; but nothing can be easier than to try it.

Yours very since[re]ly | John Tyndall

The Royal Society of Chemistry—Ex Coll H.E. Roscoe, 1915

---

1. *Henry Enfield Roscoe:* Henry Enfield Roscoe (1833–1915), chemist, who had worked with Bunsen at Heidelberg. He returned to England in 1857, as Professor of Chemistry at Owens College, Manchester.
2. *the effects of Kirchhoff & Bunsen:* Roscoe gave a Friday Evening Discourse 'On Bunsen and Kirchoff's Spectrum Observations' on 1 March 1861; *Roy. Inst. Proc.,* 3 (1861), pp. 323–28.
3. *Anderson:* Charles Anderson (1790–1866), lecture and laboratory assistant at the Royal Institution 1832–66.

From John Ball[1] 28 November 1860[2] 1749

Dear Tyndall,

Many thanks for your kind note.[3] I merely cease to be President of [the] Alpine Club by virtue of law—having held it for 3 years.[4]

Though I knew it was doubtful whether you could give the little time required my wish was that you might be induced to take my place. I had not spoken on the subject but lately ascertain that several of the active men in the Club were looking the same way.

Meantime, however, in addition to the doubt whether you would accept the Glacier controversy has waxed hotter (be hanged to it) and though no one who has spoken to me is in the least influenced one way or the other—it was feared that in some quarter or other that might interfere with the thorough cordiality with which the proposition would otherwise have been received by every individual in the Club.

This being the case I am sure that I acted as you might wish your friend to act by advising that the choice should fall on a completely neutral man who will have time to attend whenever wanted. The worthy Kennedy[5] does not care, so long as he has glaciers and peaks to climb, whether the ice moves up hill or down and at the present moment he is the best man we can have.

I trust you will not think of leaving the Club.[6] I find that many men (originally mere climbers) are beginning to take an interest in the physical questions that have to be decided in the Alps. In a few years there will be an organization which you will be able to wield with effect.

Don't forget that you are to send me *Rendu.*[7]

I hope you are all right. If you don't otherwise desist from overwork you must be provided with a wife to control you.

Ever yours, | J. Ball

A. S. Eve and C. H. Creasey, *Life and Work*, p. 387.

---

1. *John Ball*: John Ball (1818–89), an Irish politician, naturalist, and alpinist. He had been Under-Secretary of State for the Colonies from February 1855 to June 1857, and was the first President of the Alpine Club, elected in 1857. Tyndall climbed in the Alps with Ball in 1864.
2. *28 November 1860*: date according to *Life and Work,* p. 387.
3. *your kind note*: letter missing.
4. *having held it for 3 years*: Ball was succeeded as President of the Alpine Club by Edward Shirley Kennedy (1817–98), mountaineer, and editor of the second volume of *Peaks, Passes and Glaciers* (1862).

5. *Kennedy*: E. S. Kennedy (see n. 4).
6. *I trust you will not think of leaving the Club*: Tyndall resigned in January 1862 shortly after being elected vice president (see letters 1835, 1849, and 1848). The reasons are discussed in *The Ascent of John Tyndall*, p. 225.
7. *Rendu*: see letter 1619, n. 13.

From John Ball[1] 3 December 1860[2] 1750

Dear Tyndall,

I have been in the country or should have written[3]. I quite understand that the Alpine Club and its affairs are of infinitesimal importance to you, but as you have joined the body and contributed to its volume[4] and as there is in it the promise of future work of a kind that you may turn to real good account, I still think it a pity that you should separate from it. You should not misconceive the feelings of the men in it towards yourself. Several, and those the men whose minds run beyond mere climbing, have spoken or written their wish to see you President both as an honour to the Club and as giving a prospect of future good guidance. Others look on you simply as the boldest and most successful mountaineer and would on that score welcome you as their chief. If you would have taken the place there would certainly have been no opposition but I was led to fear that among a very few who have been affected by recent controversies[5] there might be a want of cordiality instead of the unanimous acclamation that I desired. Perhaps I was wrong.

Be this as it may I am still desirous that you should not leave the Club and especially at a moment when your doing so will be misunderstood.

If you are fixed on so doing, let me suggest that it would be a graceful thing for you—if not too inconvenient—to come to the dinner on the 13th and propose Kennedy's health.[6] If you can do so I think it would be well whether you remain or leave the club especially in the latter case.

Pray send me *Rendu.*[7] I will return it as soon as you please.

Ever yours, | J. Ball

A. S. Eve and C. H. Creasey, *Life and Work*, pp. 387–88

---

1. *John Ball*: see letter 1749, n. 1.
2. *3 December 1860*: date according to *Life and Work,* p. 387.
3. *or should have written*: see letter 1749. Tyndall's response to this letter, in which he seems to have signalled his intention to resign from the Alpine Club, is missing.
4. *contributed to its volume*: 'A Day among the Séracs of the Glacier du Géant', *Peaks, Passes, and Glaciers: A Series of Excursions by members of the Alpine Club* (London: Longman, Green, Longman, and Roberts, 1859), pp. 39–57.

5. *recent controversies;* the dispute with Forbes. See letters 1703, 1708, 1709, 1731, 1741, and 1746. See also *The Ascent of John Tyndall,* pp. 143–44, 148–51, and N. K. L. Kaalund, 'A frosty disagreement: John Tyndall, James David Forbes, and the early formation of the X-Club', *Annals of Science,* 74 (2017), pp. 282–98.
6. *come to the dinner on the 13th and propose Kennedy's health*: Tyndall did attend the dinner, and proposed the health of E. S. Kennedy, the incoming President (Journal, RI MS JT/2/13c/1194). He resigned a year later: see letter 1749, n. 6.
7. *Rendu*: see letter 1619, n. 13.

From William Hopkins [6 December 1860][1] 1751

23 Hans Place | Sloane Street

Dear Tyndall,

Your letter[2] has reached me here this morning. I learn from Rendu's memoir[3] together with the other publications you were kind enough to send me to Llwyn Madoc.[4] Mind you don't lose Rendu's memoir on any account. I am now staying with a friend of ours Miss Marston,[5] whose address I have given above. She desires me to say how glad she should be to see you if you would give her the pleasure of your company at dinner either to-morrow (Friday) or Saturday at 6 o'clock. There will be no party besides Mrs Hopkins[6] and myself and a Miss Haworth[7] who is residing with Miss Marston. You will have time to let us know which day we may expect you. Do come if you can as I should like to have a little talk with you exceedingly. If you cannot come I will write to you again very soon. I stay here till Monday, on my way to Hastings.

Yours truly | W. Hopkins.

RI MS JT/1/TYP/2/632
LT Transcript Only

1. *[6 December 1860]:* probably Thursday 6 December, as Tyndall dined with Hopkins at Miss Marston's on 8 December (Journal, RI MS/JT/2/13c/1194).
2. *Your letter:* letter missing.
3. *Rendu's memoir:* see letter 1619, n. 13.
4. *Llwyn Madoc:* the residence of Henry Thomas, where Hopkins stayed until mid-November while recovering from a leg injury. See letters 1731 and 1746.
5. *Miss Marston:* Ann Marston (1808–1901), anti-vivisectionist and animal welfare advocate, who conducted Christian missionary work in India. She was interested in mathematics, and frequently attended meetings of the British Association and other scientific societies.
6. *Mrs Hopkins:* Caroline Frances Hopkins (1799–1881). He had married Anne Braithwaite in 1818 but she died c. 1821 and he married Caroline Boys in 1827.

7. *Miss Haworth:* Miss Haworth (1807–88), spelt Howarth in the magazine article referenced here, seems to have lived with Ann Marston (see n. 5) for fifty-two years, and the two were very close. See 'Ann Marston', *East and West,* Vol. 1, No. 8 (June 1902), pp. 839–44.

From Jack Tidmarsh[1] 12 December 1860 1752

Adelaide, South Australia. | Dec. 12th 1860.

Dear Tyndall,

It is now twelve years since you and I met last,[2] 12 years!—a little age in itself but big enough to look back in wonderment at the various changes reverses and turnings up of matters sublunary. And as far as our own particulars of incident play. I have no doubt that (at least on your part) many sheets of paper would be interestingly covered. I would have written to you before this did not the vain idea take possession of my imagination of returning to Ireland in decent time to play there before you an apology, but things were ordered otherwise—Matters and men changing spoiled my plans. So it was with no little chagrin that I had to wipe away the fair picture that was then before me.

The thoughts of returning to one's native clime from any place are of course rich with a thousand convictions, but in a distant region such as this—brown, barren, and hot wind-dried, with no green spots for the eye to rest on—no lake to reflect at "<*1 word omitted*> eve" the figure of the stag—no—but enough, it is not Ireland consequently full of imperfections. In such a region then living but not loving, must not the feelings be doubly keen—the memory minutely faithful, that scroll has all delineated on it vividly and fresh back to childhood. Memory! how much has been said and sung of that ethereal part of our composition—how pleasing to look back upon its line, and how much more pleasing and satisfactory must it be when the whole can be traversed free from blemish or dark spot, 'tis one of the many gifts to man, and one of the many proofs of an all-pervading Deity. I am not aware of your religious views now—once indeed I could readily measure the exact amount of this ingredient enclosed within you, 'twas easy work, for like my own the space was waste—void of the verdure that ever throws a warmth and freshness o'er one's life. I however am a little changed in these matters—that is I think more seriously of late than I had been in the habit of doing. Mistake me not Tyndall for always a strange animal, but man of the present day is wonderfully so. It would be rashness and folly to suppose that he is now engaged in a game of chance—the very exertion of proving it such is now more difficult than ever, and few there are who dare exert. But look at the ease, simplicity, and grace of the argument on the other side. This present day as one ascends and gathers strength and force, the other falls, still gathering <*1 word omitted*> as it rolls along in its dark career. One carries with it civilization and happiness—the other chaos and death. One leaves the spirits buoyant with a hope of a glorious eternity in which friend will again meet friend—the other covers the spirit like a pall fostering therein a miserable misgiving, a weight growing heavier under uncertainty till bowed down earth takes her own, the talent never left it. The philosopher is indeed a true one that has and cherishes the creed of the Christian, and who can plant his standard of hope firmly beyond the abyss of all doubt. Well I shall conclude this strain by assuring you again, in case I may stamp a false impression, that I am no Saint. So now for a word or two about the Australian Tidmarshs'. I arrived here as one Tidmarsh only in July 1849 a few years afterwards, and by a process not known to you in chemistry I "<*1 word omitted*> and resolved" myself into two, which anon and with the assistance of time has become five, and this is at present the family circle, but should any part of the circumference be <*1 word omitted*> disconnected, no doubt the link or links will in due course take their places. Now in order to find bread and butter for the above, I follow the highly respectable calling of a soap and candle maker, and which I shall very probably pursue for some years to come, or at least till I find myself possessed of sufficient means to ship the circle centre and all, and be once more an Irishman,[3]—and then I shall give you the very particulars of my 'strange eventful' life out here. I write now for the purpose of renewing our correspondence, say a letter every six months this I should be satisfied with provided you send me in the interim a few papers. You will perceive by this that I have not been much accustomed to letter writing, at the same time assure you that I am still Jack. I don't know how it may be with you, whether you have grown a phlegmatical, dogmatical, critical sort of a fellow, abounding in pomposity &c. &c. or that you are the same as when we met in that 'sweet village'[4] where dwelt a doctor's daughter.[5]—The same as when we swam its harbour—climbed its sandy cliffs. —The same as when at midnight Ginty's[6] tongue went fast and hotter grew "discussion". The same as when at eve, and in that long room, and before the presence of old Youghal's very elite so dark and fair so sweet and pie-ous everything was settled,—but perhaps I had better shift the scene, taking you by the Sibyl of the raven plume,[7] and leaving you pierced[8] in the fortress of Ferguson Street.[9]

In conclusion I must ask you to give me information of Hurst,[10] and any of our friends that you may know the whereabouts of—*enclosed is the facsimile of your ancient friend. Send me yours in return, that I may see what a good looking old man you have made.

Good bye, Tyndall and believe me | Yours very sincerely | J. Tidmarsh.

Address—| John Tidmarsh & Co. | Adelaide.
*This was intended—now accompanies

RI MS JT/1/TYP/11/3874–76
LT Transcript Only

---

1. *Jack Tidmarsh:* John Francis Tidmarsh (nicknames Jack or Tid) (1824–1906). He joined the Irish Ordnance Survey on 11 May 1840, working in the 5th Division, C District under the command of Lieutenant George Wynne, and worked closely with Tyndall in Youghal, where he lodged with him and William Ginty, and again in Kinsale and Cork, where Tyndall enjoyed the hospitality of Tidmarsh's family. He was separated from Tyndall after being transferred to the English Survey on 20 May 1842, where he worked in the 5th Division and undertook field work in Lancashire, Yorkshire and Westmoreland. In the summer of 1843 he contemplated emigrating to the Cape of Good Hope to take up a job with relatives, but instead remained in England where, by 1845, he was employed, alongside Tyndall, making surveys for the West Riding Union Railway. At this time, he and Tyndall were once more lodging together in Halifax. In March 1849 Tidmarsh emigrated to Australia, settling in Adelaide, where he initially worked as a surveyor and valuator for the city's first building society and later prospected during the Australian gold rush in 1852. He established a successful candle and soap manufacturing company, although overwork impaired his health. ('Death of Mr. John Tidmarsh', *Register* (Adelaide), 12 November 1912, p. 6).
2. *It is now twelve years since you and I met last:* when they were both surveyors, based in Halifax.
3. *and be once more an Irishman:* it does not appear that Tidmarsh ever returned from Australia.
4. *that 'sweet village':* possibly Youghal, in southern Ireland, where they were stationed as assistants on the Irish Survey.
5. *a doctor's daughter:* not identified.
6. *Ginty:* William Ginty (*c.* 1820–66), one of their surveying colleagues and a close friend of Tyndall.
7. *the Sibyl of the raven plume:* Ellen (the 'lady of the raven plume'), a love interest and romantic ideal of both Tyndall and William Ginty, whom they presumably met while stationed in Kinsale in County Cork in the summer of 1841 (see *Tyndall Correspondence,* vol. 1).
8. *pierced:* doubtless a reference to the two beauteous Piercy sisters in Halifax. On one day in early 1847 Tyndall recorded the attractions of Miss Piercy—'that type of beauty which the young Mahommedan expects to meet in heaven'—and Miss Fanny Piercy—'an excellent piece of physiological workmanship', Journal, RI JT/2/13a/173.
9. *Ferguson Street:* the lodging house in Halifax in which Tyndall and many of his friends stayed in the 1840s (see *Tyndall Correspondence,* vol. 2).
10. *Hurst:* Thomas Hirst.

To Richard Dawes 17 December [1860][1] 1753

Royal Institution | 17th December

My dear Mr Dean,[2]

I am almost ashamed to ask your acceptance of this book[3] so long after you ought to have received it. But do take it from me as a Xmas box.

Ever yours | John Tyndall.

Would you have the kindness to hand the enclosed to your schoolmaster[4] for me—His name escapes my memory.

Private Collection, via Frank James
RI MS JT/1/TYP/1/320

1. *[1860]:* dated with reference to letter 1754.
2. *Mr Dean:* Richard Dawes was Dean of Hereford from 1850–67.
3. *this book: Glaciers of the Alps,* see letter 1595, n. 4.
4. *your school-master:* George Henry With.

From Richard Dawes 18 December 1860 1754

Deanery, Hereford. | Dec. 18th 1860.

My dear Tyndall

You must not suppose that your Glaciers of the Alps have not been read by me; I got the book on its first coming out and I don't know that I ever read anything with greater interest or which gave me more pleasure, in fact you are both a philosopher and a poet, and I shall only make you too vain if I tell you the way in which I hear many of my friends speaking of Tyndall on Glaciers.

Your sending me a copy[1] has given me very great pleasure, and I shall value it as a memorial of your friendship. I cannot resist sending you a few school-boy answers to questions set by Mr With the schoolmaster, you recollect him, to the senior boys in the school, and which have arisen out of reading the Glaciers. The lark soaring in the air and the volume of air put in motion when she sings. He, (With) has been giving a few lectures on light and he gave me the enclosed[2] yesterday to look at, being their first attempts at answering in writing two or three questions which he had set them; these are selected out of a dozen or 15 which are much the same, but for boys so young not a bad attempt. Just cast your eye over them and then throw them into your waste-paper basket.

We have just now adopted some of Constable's reading books in two of our schools here, and seeing your name as a writer in one called the Advanced Reading Book,[3] led to our ordering and the lessons on Light suggested this subject as within their reach.

With, the master,[4] is a very excellent fellow and has a great deal of science in him. I gave him your note this morning with which he was much pleased and I told him I should send a few of his boys' papers to shew you that we take a hint and show you that the throwing stones into a pool of water, or seeing a bird rise on its wings in the air were becoming with us more than mere matters of amusement.

And now, my dear Tyndall, let me tell that I have for the last two or three months almost every day been threatening to write both to yourself and to Frankland,[5] to ask what you were about and if there was any chance of our being able to persuade you both to pay us a visit here. You are both of you such busy men that I fear it is difficult to persuade you to leave London. I can only say that whenever you can come and spend a week or ten days with us, at whatever time of year suits you best, we should be most glad to see you, and I shall write to Frankland in a few days and tell him we hope to have Mrs Frankland with him.

I have few friends that I should be more glad to see than Frankland and yourself, or in whom I feel a greater interest, and Mrs Dawes begs me to add you must come although you did not send your kind regards to her today, which she missed much, but begs me to give hers to you.

It is post-time so I must conclude, and with our kindest regards,

Believe me, my dear Tyndall, | Always sincerely yours | R. Dawes.

It was with much regret that I heard of our friend Barlow's resignation[6] of his post at the Royal Institution, and still more so to hear his health is not good.

RI MS JT/1/TYP/1/320A–20B
LT Transcript Only

1. *Your sending me a copy:* see letter 1753.
2. *the enclosed:* enclosure missing.
3. *the Advanced Reading Book: Constable's Educational Series, Advanced Reading Book, Literary and Scientific (1860)* (Edinburgh: Thomas Constable and Co., 1860). The section on Light is on pp. 51–60. Tyndall also contributed sections on Sound, Heat, and Electricity.
4. *the master:* George Henry With.
5. *Frankland:* Edward Frankland.
6. *our friend Barlow's resignation:* The Reverend John Barlow (1798–1869), Secretary of the Royal Institution 1843–60. He was succeeded by Henry Bence Jones.

From Charles William Siemens[1] December 1860 1755

IX. *On a new Resistance Thermometer.* | *To Professor John Tyndall, F. R. S. &c., Royal Institution.*

3 Great George Street, Westminster, S. W. | December, 1860.

My DEAR SIR,

YOU will probably be interested to hear about a very direct application of physical science to a purpose of considerable practical importance, which I had lately occasion to make. Having charge, for the British Government, of the Rangoon and Singapore telegraph cable,[2] in so far as its electrical conditions are concerned, I was desirous to know the precise temperature of the coil of cable on board ship at different points throughout its mass, having been led by previous observations to apprehend spontaneous generation of heat. As it would have been impossible to introduce mercury thermometers into the interior of the mass, I thought of having recourse to an instrument based upon the well-ascertained fact that the conductivity of a copper wire increases in a simple ratio inversely with its temperature. The instrument consists of a rod or tube of metal about 18 inches long, upon which silk-covered copper wire is wound in several layers so as to produce a total resistance of, say 1000 (Siemens) units[3] at the freezing temperature of water. The wire is covered for protection with sheet india-rubber, inserted into a tube and hermetically sealed. The two ends of the coil of wire are brought, by means of insulated conducting wires, into the observatory, where they are connected to measuring apparatus, consisting of a battery, galvanometer, and variable resistance coil. The galvanometer employed has two sets of coils, traversed in opposite directions by the current of the battery. One circuit is completed by the insulated thermometer coil, and the other by a variable resistance coil of German silver wire. Instead of the differential galvanometer, a regular Wheatstone's bridge arrangement may be employed.

You will readily perceive that if the thermometer coil before described were placed in snow and water, and the variable resistance coil were stoppered so as to present 1000 units of resistance, the currents passing through both coils of the differential galvanometer would equal one another, and produce, therefore, no deflection of the needle. If, however, the temperature of the water should rise, say 1° Fahr., its resistance would undergo an increase of 1000 x ·0021 = 2.1 units of resistance, necessitating an addition of 2·1 units to the variable resistance coil in order to re-establish the equilibrium of the needle.

The ratio of increase of resistance of copper wire with increase of temperature may be regarded as perfectly constant within the ordinary limits of temperature; and being able to appreciate the tenth part of a unit in the variable resistance coil employed, I have the means of determining with great accuracy the temperature of the locality where the thermometer resistance coil is placed. Such thermometer resistance coils I caused to be placed between the layers of the cable at regular intervals, connecting all of them with the same measuring apparatus in the cabin.

After the cable had been about ten days on board (having left a wet tank on the contractors' works), very marked effects of heat resulted from the indications of the thermometer coils inserted into the interior of the mass of the cable, although the coils nearer the top and bottom surfaces did not show yet any remarkable excess over the temperature of the ship's hold, which was at 60° Fahr. The increase of heat in the interior progressed steadily at the rate of about 3° Fahr. per day, and having reached 86° Fahr., the cable would have been inevitably destroyed in the course of a few days, if the generation of heat had been allowed to continue unchecked.

Considering the comparatively low temperature of the surface of the cable, much incredulity was expressed by looker-on respecting the trustworthiness of these results; but all doubts speedily vanished when large quantities of cold water of 42° temperature were pumped upon the cable, and found to issue 72° Fahr. at the bottom.

Resistance thermometers of this description might, I think, be used with advantage in a variety of scientific observations,—for instance, to determine the temperature of the ground at various depths throughout the year, or of the sea at various depths, &c. &c. In the construction of this instrument, care has to be taken that no sensible amount of heat is generated by the galvanic currents in any of the resistances employed.

By substituting an open coil of platinum wire for the insulated copper coil, this instrument would be found useful also as a pyrometer.

But, finding this letter already exceeds its intended limits, I shall not enlarge upon these applications, which, no doubt, are quite obvious to you.

I am, dear Sir, | Yours very truly, | C. Wm. Siemens.

C. W. Siemens, 'On a new Resistance Thermometer', *Phil. Mag.,* 21 (1861), pp. 73–74.

---

1. *Charles William Siemens*: Charles William Siemens (1823–83). Born Carl Wilhelm Siemens, he was an engineer and businessman who worked in Britain for much of his life and became a British subject. He was elected a FRS in 1862.
2. *the Rangoon and Singapore telegraph cable*: in the autumn of 1860, the Siemens Company was contracted to superintend the making and laying of the cable between Rangoon and Singapore: McDonald, Donald & Hunt, Leslie B. *A History of Platinum and its Allied*

Metals. (London: John Matthey, 1982), pp. 354–55. First intended by the British Government for a telegraph connection from Falmouth to Gibraltar, the cable was temporarily reassigned for a connection between Rangoon to Singapore, before finally being used for the Malta to Alexandria Telegraph in May of 1861. It was produced by the Gutta-Percha Company: John Timbs, *The Year-Book of Facts in Science and Art* (London: Lockwood and Co., 1862), p. 177.

3. *(Siemens) units*: fearing that the current resistance of metal wires was amendable by their transport, use and varying impurities of the metal, Siemens believed that the resistance of such wires could not reliably be used to define a Standard Measure of current resistance. Instead, he proposed an independent Standard Unit of resistance based on the resistance of mercury, as this metal was easily procurable, rendered pure and its resistance was more independent of temperature than alternative metals: M. W. Siemens, 'Proposal for a new reproducible Standard Measure of Resistance to Galvanic Currents', *The London, Edinburgh, and Dublin Philosophical Magazine and Journal of Science, 4th series,* 21 (1861), pp. 25–38. It is not to be confused with the Siemens Unit used today, which was named in honour of Siemens, but is otherwise unrelated to his previous Standard.

1861

To William Frederick Pollock [7 January 1861][1] 1756

Monday

My dear Pollock.

Would you kindly cast your eye over this for me? You will see that after the establishment of the principle of reciprocity in radiation & absorption with regard to gases I try to unite together Radiation, absorption & conduction. What I chiefly want you to look at is whether the mechanical conception of the molecules & ether & their mutual relations is fairly & clearly carried out. The gist of the thing lies as you will see in the union of the atomic and undulatory theories, and the reduction by this means of Radiation absorption & conduction to the same principle.

Ever yours | John Tyndall

If you could manage to glance over it before Wednesday night I should be really obliged: I want to hand it in to the Royal Society on Thursday.[2]

I should rather like to have been "Ernest" in the last chapter of Ida:[3] but poor Warburton: what is to be done with him? If he could make experiments on Radiant Heat he would be all right. The revelation of the fact to him[4] is to my mind extremely well done.

RI MS JT/1/T/1248

---

1. Tyndall is asking for comments on the draft of his paper of the Bakerian lecture, which he produced for publication in the *Phil. Trans.* It was submitted to the Royal Society on Thursday 10 January 1861. See letter 1757 for Pollock's reply.
2. *I want to hand it in to the Royal Society on Thursday*: see n. 1.
3. *"Ernest" in the last chapter of Ida*: Juliet Pollock's novel *Ida Conway* was being serialised in *Frasers's Magazine* (see letter 1729, n. 6). Chapters 10 and 11 were published in the January 1861 issue. In the latter, Ida, who is secretly enamoured of Count Ernest, instigates a horse race with Captain Richard Warburton, who is secretly enamoured of Ida. Ida's horse bolts, and she swoons into Ernest's arms after he brings her horse under control.

4. *The revelation of the fact to him*: as Ida regains consciousness in the presence of Warburton she calls for Ernest, revealing to the disappointed Warburton where her affections lie. At the end of the novel she marries Count Ernest.

From William Frederick Pollock 7 January [1861][1] 1757

59 Montague Square | Monday night | 7th Jan

My dear Tyndall,

I have been exceedingly gratified by the perusal of your paper.[2] It surpassed the expectations raised by your note which accompanied it.[3] You have most lucidly, & effectually brought forward your point—and I am surprised that you should have been able to pack so much into so limited a space. I cannot conceive that any one will have the slightest difficulty in following your chain of reasoning. To myself it is most beautiful and satisfactory, and I felicitate you very much upon having made so great an advance in the physical conception of the probable reactions between the ultimate particles of matter & the all-pervading ether. All such theories must, I take it, be accepted provisionally, but with the atomic theory of chemistry on the one hand, & the undulatory theory of physical optics on the other,[4] you are indulging in no vain speculation, but, in the highest spirit of genius, utilizing & adopting what it finds, you show the harmony of both, and carry them onwards in a new combination, which confirms the probability of their individual truth; while it seems to project a bridge into the future history of science, of which it is impossible to predict, where the ultimate abutments on the other side will some day (& I think in your day) be found to rest.

Ever yours truly | W. F. Pollock

RI MS JT/1/P/251

1. This letter replies to 1756, and was written later the same day.
2. *your paper*: the draft of Tyndall (1861).
3. *your note which accompanied it*: letter 1756.
4. *the atomic theory of chemistry on the one hand, & the undulatory theory of physical optics on the other*: Pollock is drawing a comparison here between a particle and a wave theory for understanding physical processes. These Victorians would have had no problem with quantum mechanics and wave/particle duality.

To Mary Helen Dawes 9 January 1861 1758

Royal Institution | 9th Jan. 1861

My dear Mrs Dawes,

The Dean's last note[1] to me contained a reproach the sting of which I still feel. But indeed indeed, it was to you I intended to send the book[2] & not to him. Only the day previous to sending it I had some conversation about him, which brought him so prominently before my mind's eye that on the spur of the moment I sent the book off to him.

I wish you ten times over a happy new year! I wish sincerely I could respond to the Dean's kind invitation and escape to you for a week; but strong chains at present surround me. I have just emerged from the entanglements of a difficult original investigation and find myself in the entanglements of lecturing—I have two courses at the School of Mines this year, one at the Royal Institution; am examiner for four or five distinct classes of military candidates, and also examiner for the University of London—Thus you see my hands are filled & my head too, heaven knows. Still I shall not always be bound thus, & at some future day I hope to join Frankland again at your hospitable board.

Will you say to the Dean that I looked over the examination papers which he sent me, and the effect they had upon me was to excite the desire to be of some service to your schoolmaster Mr With,[3] if I could. I know the people at Kensington, could I do anything for him there. The papers are exceedingly sound, and I should be glad to have an opportunity of stating this; if the statement would do any good. Do be good enough to ask the Dean to pardon my tardiness in responding to his last kind note.

Goodbye | & believe me | always yours most truly | John Tyndall.

Private Collection, via Frank James

RI MS JT/1/TYP/1/1321

1. *The Dean's last note*: letter 1754.
2. *the book*: *Glaciers of the Alps,* see also letter 1595, n. 4.
3. *Mr With*: George Henry With.

From Edward Sabine 11 January 1861 1759

13 Ashley Place S. W. | Jan[uar]y. 11th 1861

Dear Tyndall.

If I understand you right you wish to show your class[1] the way in which the motions of the magnet are recorded in our observatories by eye-observation.

And you purpose to exhibit to them first the motions of the magnet themselves, by reflecting a Dubosc lamp[2] on a skreen[3] from a mirror attached to the magnet—so far so well.

Your subject will then lead you into two divisions—First you will show them how the image of a scale which is fixed to the Telescope, & illuminated, is reflected by the mirror onto the field of the Telescope & so read off. This is Gauss's[4] original plan, and is used by us whenever the disturbances occur of so large amount as to make the substitution of a collimator & lens affixed to the magnet <u>instead of a mirror</u> an inadequate apparatus by reason of the collimator scale going beyond the limits of the field of the telescope, as it does sometimes even in England. You should therefore exhibit also a magnet furnished with a collimator scale & lens,—one of the 12 or 15 inch magnets will exhibit this very well, and will be quite comprehensible, by your explanation, without looking into the telescope—You will find the whole process by collimator minutely described by Lloyd in the Report of the Com. of Physics of the R.S. 1840[5]—and the whole process of Gauss's mirror method in the Sci Mem vol. 2. p. 37.[6]

I fear that I am forming a very uncertain notion of what your precise purposes are. But you must be more particular in defining your own purposes to enable me to help you more effectually, which I should delight in doing.

I think you will find a spare 12 or 15 inch magnet fitted with collimator & lens, & a telescope at Kew—& you will also find, a small magnet and a mirror, & a telescope furnished with the scale. Brooke[7] has also (I doubt not) both if you do not find them at Kew.

Sincerely yours | Edward Sabine

RI MS JT/1/S/19
RI MS JT/1/TYP/4/1314

1. *your class*: probably Tyndall's students on his course of general physics at the School of Mines.
2. *Dubosc lamp*: Louis Jules Duboscq (1817–86), instrument maker and inventor of the Duboscq lamp, an arc lamp. Tyndall bought these for demonstrations in his lectures. See, for example, letter 1646.

3. *skreen*: screen.
4. *Gauss's*: Carl Friedrich Gauss (1777–1855), German mathematician and astronomer. Gauss was the first director of the Göttingen observatory in Germany, and was one of the principal figures in the international programme to systematically study the Earth's magnetic field. Gauss developed a mathematical theory which correlated the data collected onto a model of the Earth's internal magnetic structure. His theory rejected the notion, which had been previously held by Sabine, that the Earth's magnetism could be explained by two magnetic axes, both in motion. Gauss also invented what he called the magnometer, an early type of magnetometer which measured the direction and strength of a magnetic field.
5. *described by Lloyd in the Report of the Com. of Physics of the R.S. 1840*: in the section entitled 'Terrestrial Magnetism' in the Royal Society's *Report of the Committee of Physics, including Meteorology on the objects of Scientific Inquiry in those Sciences* (London, 1840), pp. 1–37.
6. *Gauss's mirror method in the Sci Mem vol. 2. p. 37:* R. Taylor (ed), *Scientific Memoirs* (London: Richard and John E. Taylor, 1841), vol. II, p. 37.
7. *Brooke*: probably Charles Brooke (1804–79), surgeon and inventor of measuring instruments. He was a FRS from 1847, and President of both the Meteorological Society (1865–67) and the Royal Microscopical Society (1863–64 and 1873–74). Brooke was best known for his inventions of self-recording instruments, such as barometers, thermometers, psychrometers, and magnetometers which were adopted at the observatories at Greenwich, Paris, Toronto, and Cambridge, Massachusetts. These instruments would automatically register variations by means of photography (*ODNB*). See, for example, C. Brooke, 'On the Automatic Registration of Magnetometers, and Other Meteorological Instruments, by Photography', *Phil. Trans.*, 137 (1847), pp. 59–77.

From Edward Sabine 18 January 1861 1760

13 Ashley Place S. W. | Jan[uar]y 18. 1861

Dear Tyndall.

Perhaps you will like to see in operation the next stage to that which you saw at Kew. The photograms are taken to Woolwich, where tables are formed from them such as I enclose a specimen.[1] The direction of the magnet at every solar hour is measured from the Photogram, & entered each hour in the monthly table is then separately summed up & a mean taken. These are the first normals—Then every entry which differs $\overset{m}{0}$. 150 from its first normal is marked with a ring in pencil. The first normals are then corrected by the omission of those pencil ringed directions, which gives second normals. The process is then repeated till the means at each hour in a month are the true means of all the directions, excluding those which differ more than $\overset{m}{0}$. 150

from these, the final normals—The larger disturbances, so indicated, are then marked with a ring in ink, as you see them—and the sheet comes up to me to have marked in small figures, (as you will see done in September $^{D}_{0}$, $^{D}_{1}$, *and* $^{D}_{2}$) the lunar hour to which each direction most nearly corresponds.

I then return the sheets to the office, then Lunar Monthly Tables are written from them (omitting all the directions which have an ink ring). Then commences the process of distributing the larger disturbances into the hours, months & years of their occurrence & comparing these, which I do not go into, as it is a stage beyond the table which I now send.

Sincerely yours | Edward Sabine

RI MS JT/1/S/20
RI MS JT/1/TYP/4/1315

---

1. *such as I enclose a specimen*: enclosure missing. From 1858–64, staff at Kew Observatory were involved in calculating the lunar-diurnal variation of the magnetic elements shown from magnetograms. In order to determine the lunar-diurnal variation, photograms were used to record the hourly changes in position, from a permanent zero, of a magnet whose motion was limited to a horizontal plane, and the amount of horizontal or vertical elements of force acting on a freely suspended magnet. An instrument was devised at Kew to tabulate the values, and their accuracy was checked by repetition of the process. However, there were not enough staff at Kew to undertake these tabulations, and so Sabine had the positions tabulated at the Magnetic Office at Woolwich. See letter 1764 for further details.

To Thomas Hirst [22 January 1861][1] 1761

Tuesday

My dear Tom

I hope you have not forgotten the examination.[2] Just hand the enclosed[3] to the superintending officer[4] & it will make all clear,

alw[a]ys | John

Turn over[5]

Important.

When the Examination is over be good enough to forward the papers at once to Frankland[6] 42 St. John's Park Road Haverstock Hill.

If you thought you could look over all the answers by Friday you might keep them yourself, but if not send all to Frankland

RI MS JT/1/T/956

1. *[22 January 1861]:* the journals of both Tyndall and Hirst mention this episode (Journal, RI MS JT/2/13c/1198; *Hirst Journal*, RI MS JT/2/32c/1567).
2. *the examination:* see letter 1762, n. 6.
3. *the enclosed:* letter 1762.
4. *the superintending officer:* the addressee of letter 1762, or his representative.
5. *Turn over:* the remainder of the note is written on the other side.
6. *Frankland:* Hirst noted in his journal that Frankland was to correct the papers (*Hirst Journal*, RI MS JT/2/32c/1567), and Tyndall noted that he had the papers conveyed to Frankland (Journal, RI MS JT/2/13c/1198). The load proved such that Frankland's marking was late. When Tyndall received a 'very tart letter' (letter missing) from Captain Greentree in his response to his explanation (letter missing), he sent in his resignation (letter missing). It was ignored, and a further polite exchange followed (letters missing). See Journal, RI MS JT/2/13c/1198 and 1202.

To [Major-general Joseph Ellison Portlock][1] [22 January 1861][2] 1762

My dear General[3]

Two public departments[4] clash today in my poor person. Yours, and that of the School of Mines.[5] I would go to you if I could get a friend to lecture for me.

I therefore have asked my friend Mr Hirst (who is intimately aquainted with physics & is head mathematical master in University College) to take my place at Chelsea[6] today.

Most truly yours | John Tyndall

RI MS JT/1/T/955

1. *[Major-general Joseph Ellison Portlock]:* this note is enclosed with letter 1761 to Hirst. The addressee is possibly Major-general Joseph Ellison Portlock (1784–1864), whom Tyndall knew well, consistent with a familiar mode of address. Portlock was a member of the Council of Military Education, and had been Inspector of Studies at Woolwich until 1856. It is also possible that the letter is to Major-general Randal Rumley, Vice-President of the Council.
2. *[22 January 1861]:* see letter 1761.
3. *General:* see n. 1.
4. *Two public departments:* Tyndall's lecturing at the School of Mines and examining for the Council of Military Education. In the evening of 22 January, Tyndall gave a lecture at the School of Mines on electricity, one of a weekly series of ten lectures for Working Men that had started on 8 January. Hirst managed to return in time to attend it (*Hirst Journal*, RI MS JT/2/32c/1567).

5. *School of Mines:* Government School of Mines in Jermyn Street.
6. *Chelsea:* Chelsea Hospital, the location for the examination of cadets for entry to Woolwich.

To [Benjamin Ward Richardson][1] 26 January 1861 1763

Royal Institution | 26th Jan. 1861

Dear Sir,

I have been so extremely busy of late that I had no time to reply to your note[2]—I am much obliged to you for the offer of the old machine,[3] which I shall be glad to refer to in one of my lectures.

Very truly yours | John Tyndall

Science Museum, London, T/1876–39

---

1. *[Benjamin Ward Richardson]:* probably Benjamin Ward Richardson (1828–96). This note was found at the Science Museum, London, attached to an electrical machine purportedly belonging to Benjamin Franklin, donated by 'Dr Richardson FRS' in 1876. From January to April 1861, Tyndall gave 12 lectures on Electricity at the Royal Institution.
2. *your note:* letter missing.
3. *the old machine:* see n. 1.

From Edward Sabine 29 January 1861 1764

13 Ashley Place S. W. | Jan. 29. 1861

Dear Tyndall

I send for your examination[1] (as you desire) a similar sheet to the one you had before.[2] This sheet is advanced <u>two stages</u> beyond the preceding. It is marked in small figures with the <u>lunar</u> hour to which each observation most nearly belongs. (By these Lunar Tables have been formed) and 2[d] The "daily means omiting disturbances" are calculated. This enables me to collect the daily means into weekly means, of which there will be 52 in each year; always commencing with Jan[uar][y] 1st—and throwing Feb. 29 (if there are 29 days in that month) and Dec. 31 with the mean of the nearest 7 days. —(Feb. 29 with the mean of the 1st March, and Dec 31 with that of Dec 27).

Having then a series of 52 weekly declinations in each year, I can investigate the mean amount of secular change *<1 word illeg>* 7 Declin, diminishing) the systematic regularity or irregularity of the secular change in aliquod parts of the year. The possible <u>annual</u> variation; season or otherwise—Systematic influences on the Declination of the Suns revolution on its axis, the Moon's declination, her phases, &c. &c.

I shall very shortly have 3 years Kew Ob[servatio]^{n} ready for this examination—and have 5 years at Hobarton[3] of similar investigations ready.

I had a letter from Becher[4] today from Giessen. He too is engaged in experiments on Radiant Heat.

Sincerely y[ou]^{rs} | Edward Sabine

RI MS JT/1/S/21
RI MS JT/1/TYP/4/1316

1. *I send for your examination*: enclosure missing.
2. *a similar sheet to the one you had before*: see letter 1760. Following on from letter 1760, this letter describes the next stage in the work at Kew Observatory to create lunar tables from the data collected. The records of the positions were then entered into monthly lunar tables, according to the lunar hours each position corresponded to the most. From these the mean variation in each lunar hour in each year was calculated. Every hourly position which differed from the mean of the same hour in the same month was then regarded as evidence of the presence of a magnetic disturbance. See E. Sabine, 'Results of the Magnetic Observations at the Kew Observatory. No. III', *Phil. Trans.,* 156 (1866), pp. 441–51 (on pp. 441–42).
3. *Hobarton*: the observatory in Tasmania.
4. *Becher*: possibly Dr Emil Becher (d. 1884), a German physician. He was appointed Acting Assistant-Surgeon in the Medical Services of the British Army in 1855, and promoted to staff surgeon in 1868. A paper by Becher was presented at the Royal Society of London on 26 February 1863, with Sabine as President in the Chair: E. Becher, 'On the Effect of Temperature on the Secretion of Urea, as Observed on a Voyage to China, and at Hong Kong', *Roy. Soc. Proc.,* 12 (1863), pp. 440–41. In the paper Becher reported a series of observations during his voyage in China, including meteorological observations, on the relation between extreme variations of temperature and his bodily secretions of urine, urea, and sodium chloride during each day.

From William Hopkins 29 January–3 February 1861 1765

22 Grand Parade, | Brighton. | Jan. 29/61

Dear Tyndall,

I have just been reading the review of "Forbes & Tyndall" in the Edinburgh[1]—a holy alliance I scarcely expected to see cemented in the North. The whole tone is marvelously modified from the former denunciations which have hitherto proceeded from that quarter against every opponent of

Viscosity. The viscous theory[2] is in fact departed. The claim for a presentiment on the part of Forbes of the property of ice made out by Faraday and yourself is simply ludicrous. The reviewer, too, effectively gives up Forbes's theory of the veined structure, though not venturing to state the real reason why it cannot possibly be maintained. How Forbes will like this I don't know because it is quite certain that it was the part of his theory on which he prided himself most. I suppose I have the honor of the reviewer's notice as 'an ingenious critic' in the note on the sliding of glaciers (p. 323) where he makes the absurd assertion that the surface of a glacier would move very nearly in the same way whether it should adhere to its bed or not. Even Forbes never dared to make this assertion. How singularly unjust those northern reviewers are to Agassiz.[3] The 'Nouvelles Etudes'[4] of the latter are but just noticed, and yet for details of glacial phenomena there is nothing at all comparable to it. I have no idea who the reviewer is. Can you tell me? He seems to me to be standing forth as a sort of mediator with the olive branch in his hand, and I should rather infer that Forbes himself may intend to withdraw from further controversy within the dignity of his new station at St. Andrews.[5]

But we must not suppose that because certain of Forbes's theories may be passing away, the theory of glacier motion is complete. There are one or two points to which I wish to draw your attention.

There is a fallacy or misrepresentation which has pervaded all that Forbes and others have written on the points to which it relates. When glacial ice conforms itself to the contracted form of the glacial valley, it must of course do so by virtue of a certain degree of plasticity, which it displays in consequence of the enormous compressing forces both a tergo[6] and laterally to which it is necessarily subjected in such cases. This is the kind of plasticity which is common in a greater or less degree to all solid bodies. Now it has been always tacitly assumed that admitting this degree of plasticity in glacial ice, a glacier if placed on a smooth level plane would be crushed, or squeezed out, by its own weight so as to produce in it a lateral motion; or in other words, its component molecules instead of remaining fixed with respect to each other would move inter se, and the cohesive power be entirely overcome. Now this would certainly not be necessarily the case for that degree of plasticity which would allow a glacier to be squeezed into a narrower portion of its valley; and one obvious reason is this—that we may conceive the force à tergo in this squeezing process, to be much greater than the crushing weight of the superincumbent mass in the case in which the lateral motion is produced. This last case might require a much greater degree of plasticity, or a nearer approach to viscidity than the other. Now it is asserted (as in the note above alluded to in the review) that there could be no sliding without plasticity, by which it is here (as it is usually) meant, without that degree of plasticity which would under the ordinary conditions of a glacier, be an independent cause of continuous motion. Forbes obviously never admitted the notion of a solid body with a certain amount of compressibility; but glacial ice, according to his views, was either rigid, or possessed a degree of plasticity which implied a molecular cohesion sufficiently feeble to be overcome entirely by the crushing superincumbent weight of the upper portions of the glacial mass, and this not only in the deeper portions of the mass, cut in nearly all portions of it generally. Now I maintain, on the contrary, that there may be, and probably is a wide interval between that degree of plasticity (compressibility) which may admit of a glacier being squeezed into a smaller volume, and that degree of plasticity (approximate viscosity) which, under the pressure to which a glacier at rest and without any sliding motion, would be commonly subjected, would be an independent cause of continuous progressive motion. But let us consider the different forces which exert a tendency to crack and break the glacial mass, with the view of ascertaining which are the most important.

The external force producing an active tendency to dislocate the mass is that of gravity or the weight of the mass. Resistance arising from friction on projecting obstacles being considered to act as it were, passively. Let us take as the approximate type and for the simplicity of explanation, a glacier descending a canal shaped glacier at a constant inclination (α) to the horizon, α being small. We may resolve the weight (W) of the glacier into W sin α parallel to the bottom of the valley and W cos α perpendicular to it. The tendency of the latter force will be to crush the mass, or rather its lower parts, by the superincumbent weight of the upper portion. I much doubt, however, whether this tendency can ever be very considerable. For suppose we could, for instance, remove Mont Blanc, and form for it a new pedestal by a horizontal stratum of crystalline limestone, and replace the mountain upon it. The limestone I imagine would be crushed and its crystalline structure destroyed. But if a bed of limestone had been originally found there and gradually hardened and crystallized under its mountain load, that same load would not suffice to crush it again. And so with glacier ice. In all those parts of a glacier on which the principal observations on its motion have been made, the ice must have been generally crystallized under a greater superincumbent weight than that to which it can be subsequently exposed. It is for this reason that I doubt much whether the superincumbent portion of a glacial mass can ever produce, by the compressing force (W cos α) of its weight alone, any material effect in crushing the subsequent portion.

Let us next consider what the force W sin α acting parallel to the bed of the glacier will do—i.e. supposing the previous absence of any internal constraint, what tendency will this force exert to crack and break the mass. And first let us suppose the mass to have the same amount of cohesion to the sides and bottom of the glacial valley as the constituent particles of the ice have to each other. Then in consequence of the supposed degree of plasticity of the mass, the force now considered (W sin α) will move the upper portion more than the lower, and the central more than the lateral parts, and thus the general form of the mass will be changed, and every part of it will be brought into a state of internal constraint; i.e. it will be subjected to internal pressure and tension tending to tear and crush it. By this means a plane transverse section of a glacier will be made to bulge forward into a curved surface, but it is manifest that the general tendency to produce this effect will be partly counteracted by the adhesion to the bed of the valley, and this counteracting force will be very large. But now let us conceive the adhesion to the bottom of the valley to be destroyed, while that on the flanks to the sides of the valley still remains. The bulging forward of our original plane transverse section will now be as great at the lower as at the upper surface of the glacier, and the whole of the internal pressure and tension produced in this case, and arising from the deformation of the general mass, will obviously be much greater than in the preceding case, on account of the removal of the opposing force of cohesion to the bottom of the valley. This is in fact the case in which the form W sin α would exert its maximum tendency to rupture the mass. It would act at the same kind of mechanical advantage in breaking the ice as that with which one acts when one places a stick against one's knee and pulls at both ends to break it.

If the adhesion to the sides of the valley or the obstacles it affords to the motion be considerably smaller than here supposed, but still sufficient to impede the motion of the flanks compared with the centre, the reasoning and conclusion will be still the same.

You will recollect that my early experiment on the uniform sliding of ice down an inclined plane, showed that the statical form of friction, or that exerted to impede (not the accelerated but) the uniform motion, was extremely small, so that the second of the above cases modified as in the preceding paragraph, will be the type of the actual case of a glacier which (as I suppose to be admitted) does slide over its bed. It is to the wide difference between the tendencies to rupture the mass in the two cases above discussed that I want to direct your special attention. In the case of the glacier sliding you will observe that a very large portion of the force W sin α is employed in deforming the glacier and therefore in rending and compressing it, whereas in the case of the glacier not sliding, the force employed in the same work is only the difference between W sin α and the effect of cohesion to the bed of the glacier (which must be nearly equal to W sin α and therefore only a comparatively small portion of W sin α is effective).

Hence if a glacier slide over its bed, and slide faster in the central than in the lateral portions there will be a force always tending to produce the crevasses, together with the minuter cracks and fissures which your theory of regelation, requires, indefinitely greater than if the glacier were attached to its bed by the same cohesion as that by which one part is attached to another. According to this view sliding and regelation are the essential parts of the same true glacial theory. We have nothing to do with the inconceivable property of viscosity in ice. On the contrary, according to this theory it cracks and breaks as one would suppose a brittle substance to do in an infinite variety of ways, and thus obtains its freedom of motion, while all its broken bones become healed and united again by the happy process of regelation.

By similar reasoning we see that when a glacier is forced thro' a gorge where its motion is retarded, there will be an enormous pressure acting on it à tergo, and very much greater than if the glacier did not slide pretty freely over its bed. Mere local obstacles on the sides of the valley will be as nothing to arrest the motion of the glacier under such circumstances. The ice will only be broken the more and pass on.

There is a glacial yarn for you. Examine it carefully to see if there be any flaws in it. If not I am disposed to think that it will render the theory of glacier motion pretty nearly complete as to the sufficiency of the causes assigned for it. Still experiment may do something more in pointing out the modus operandi in breaking and crushing the ice. Will it yield first to the internal tensions or to the crushing power of the internal pressures (by the bye those produced by the distortion of the mass will probably be much greater than those produced by the weight of any superincumbent on a subjacent portion) Or will the mass yield as Forbes supposes (not assuredly in the directions he supposes) by one vertical slice sliding past another? Experiment might tell us a good deal on these points. I will suggest two or three

Suppose two vessels A and B were filled with water

and placed with their open ends in contact at C. If the water

were completely frozen and the vessel A were firmly fixed

vertically, what weight suspended to B would be just sufficient to pull the vessels asunder. This would give the power of the ice to resist tension. Its extensibility might also be measured in some similar manner probably.

Possibly you may have made some experiments of this kind.

Again supposing A to be firmly fixed as before and the axes of the two vessels in the same vertical line, what force applied to B horizontally would be sufficient to rupture the ice along the section at C by making the ice in B slide horizontally past that in A. This would determine the amount of the tangential force of cohesion, as the former experiment would give the normal force of cohesion. Again if we take a short cylinder of ice what pressure would be required to crush it? What would it be also if the cylinder of ice were congeled under great pressure? as in some of your experiments.

Such experiments would probably enable us to give a very explicit answer to the question—how would the ice yield to the above forces acting upon it? Would it as I have already asked, be crushed by its own weight, or by the internal pressure superinduced by the unequal motions of the sides and centre? or would it rather yield to the superinduced internal tensions? or rather by one portion sliding past another according to Forbes's notion?

Can you make some such experiments as those suggested?[7] You might present them to the Royal Society as a fitting supplement to your former communications; and I would also write a short memoir[8] embodying my present views, with such remarks as I might deem necessary on opposing views, and as such as might be made without giving too much of a controversial character to the communication. It would at all events be no more than Forbes's paper in the Royal Transactions.[9]

My knee has made so little progress of late[10] that I have been induced to come here to place myself under the great bone quack Harrop.[11] He promises much, but how he will perform I have not yet had the time to determine.

I began this long epistle several days ago, but have not had time to finish it till today (Feb. 3).

Believe me | Yours very truly | W. Hopkins.

RI MS JT/1/TYP/2/633–38
LT Transcript Only

---

1. *'Forbes & Tyndall' in the Edinburgh*: J. Ball (anonymously), 'Forbes and Tyndall on the Alps and their Glaciers', *Edinburgh Review,* 113 (1861), pp. 221–52.
2. *The* viscous theory: see letter 1741, n. 9.
3. *Agassiz*: Louis Agassiz.
4. *'Nouvelles Etudes'*: L. Agassiz, *Nouvelles études et expériences sur les glaciers actuels, leur structure, leur progression et leur action physique sur le sol* (Paris: Victor Masson, 1847).
5. Forbes had moved from Edinburgh to become principal of the college of St Salvator and St Leonard (the United College), at St Andrews University in 1860. See also letter 1645, n. 3.
6. *a tergo*: from behind.
7. *Can you make some such experiments as those suggested?*: Tyndall recorded no further experiments at this time, and did not present a paper to the Royal Society.

8. *I would also write a short memoir*: Hopkins did present a paper: W. Hopkins, 'On the Theory of the Motion of Glaciers', *Roy. Soc. Proc.*, 12 (1862), pp. 110–20.
9. *Forbes's paper in the Royal Transactions*: see letter 1731, n. 8. See also letter 1619, n. 13.
10. *My knee has made so little progress of late*: see letters 1731 and 1746.
11. *great* bone quack *Harrop:* Henry Harrap, a professional 'rubber' of great repute in Brighton. Using massage and ointments he gained a reputation for curing joint and bone problems. He practised at 29 Grand Parade, Brighton. In the 1860s Harrap and a Mr Bennett established themselves as 'orthopaedic surgeons'.

From David Brewster 7 February 1861 1766

My Dear Mr Tyndall,

I beg leave to introduce to you my Friend the Rev[eren]d Mr. Blaikie[1] the Editor of the North British Review,[2] who is very anxious to have an article from your pen for that Journal.

I beg to thank you for the copy of your charming work on Glaciers[3] which you were so good as to send me. An article on that subject by yourself or any of your friends,[4] and founded on your own views would I am sure be acceptable.

I am | My Dear Mr Tyndall | Ever most truly yours | D Brewster

Duncliffe | Edinburgh | Feb[ruar]y. 7th 1861

RI MS JT/1/B/130
RI MS JT/1/TYP/1/161

---

1. *Rev[eren]*d *Mr. Blaikie*: William Garden Blaikie (1820–99), editor of the *North British Review* from 1860–63.
2. *North British Review*: a Scottish periodical founded in 1844 and published until 1871.
3. *your charming work on Glaciers*: *Glaciers of the Alps,* see also letter 1595, n. 4.
4. *An article on that subject by yourself or any of your friends*: no such article has been identified.

From Johann Christian Poggendorff 14 February 1861 1767

Berlin 14 Febr. 61.

Mein lieber Tyndall!

In Erwiederung ihrer Zeilen vom 10' dieses beeile ich mich Ihnen zu melden, dass ich ihre Abhandlung über die Wärme Erscheinungen der Gase mit Vergnügen entgegennehme und mich bemühen werde, sie in die Annalen aufzunehmen sobald als es nur irgendmöglich ist. Sie wissen vielleicht das Magnus sich mit einem ähnlichen Gegenstand beschäftigt hat (er hat erst neulich noch den Schluss seiner Abhandl. in der Acad. vorgelesen und die Resultate davon habe ich ins Februarheft eingerichtet); allein das soll mich nicht abhalten, auch ihre Arbeit aufzunehmen. Selten sind ja die Arbeiten zweier Experimentatoren so in allen Stücken gleich, dass sie nicht beide verdienten so in Publication gebracht zu werden.

Es hat mich übrigens sehr gefreut, durch diese Gelegenheit einmal wieder etwas von Ihnen gehört zu haben, zumal es die Umstände verhinderten, dass wir Sie im vorigen Herbst, wo ich mit Frau und Tochter eine Reise durch die Schweiz machte, am Mont blanc trafen, was namentlich meine Frau noch jetzt sehr bedauert.

Wir sind übrigens alle wohl, u. dasselbe kann ich auch von Miss Francis berichten, die, wie Sie vielleicht wissen, seit einigen Monaten bei uns wohnt. Grüssen Sie schönstens deren Bruder und seyen Sie versichert, dass Sie bei uns im besten Andenken stehen.

Ganz | der Ihre | Poggendorf

Berlin 14 Febr. 61

My dear Tyndall!

In reply to your lines of the 10th of this month,[1] I hasten to report to you, that I accept your paper on the heat expansion of gases with pleasure and will endeavour to include it in the Annalen as soon as possible.[2] You might know that Magnus was occupied with a similar subject (he read the conclusion of his paper to the Academy[3] only the other day, and I have included its results in the February issue)[4] but this will not keep me from also publishing your paper as well. Rarely is the work of two experimenters so similar in all aspects that is not warranted to publish them both.

Incidentally I was very pleased to have heard from you again due to this occasion, especially as the circumstances prevented our meeting at Mont Blanc last autumn when I travelled through Switzerland with my wife and daughter, which especially my wife still regrets even now.

Other than that, we are all well and I can report the same about Miss Francis,[5] who, as you might know, has been living with us for several months. Give the warmest greetings to her brother and rest assured that we keep you in best thoughts.

Entirely | Your | Poggendorf

RI MS JT/1/P/129

---

1. *your lines of the 10*th *of this month*: letter missing.
2. *endeavour to include it in the Annalen as soon as possible*: it appeared in the June issue: 'Ueber die Absorption und Strahlung der Wärme durch Gase und Dämpfe; und den

Physischen Zusammenhang von Strahlung, Absorption und Leitung', *Poggend. Annal.*, 113 (1861), pp. 1–53.

3. *read . . . the Academy*: Magnus's paper was read in Die Königliche Preussische Akademie der Wissenschafter zu Berlin on 7 February 1861.
4. *I have included its results in the February issue*: G. Magnus, 'Ueber die Verbreitung der Wärme in den Gasen', *Poggend. Annal.*, 112 (1861), pp. 351–52 and 497–549.
5. *Miss Francis*: sister of William Francis, the daughter of Richard Taylor by Frances Francis, governess to his daughter Sarah by his wife Hannah (née Corke), whom he had married in 1807.

From Charles Darwin 23 February [1861][1] 1768

Down, | Bromley, Kent. | Feb[ruary]. 23rd.

Dear Tyndall,

I send the letters.[2] No. 1. tells most—It is only an inference that Mr Wedgwood[3] saw congelation,[4] but it seems to me pretty clear that he did. Letters No. 2. and 3. are hardly worth your looking at. Letter 4 contains reference to protrusion of glaciers by the swelling of frozen unfiltered water. I doubt whether the letters are worth your looking at, but it has amused me to see the speculations of philosophers (my two grandfathers) nearly 80 years ago on a subject which they so little understood. It seems that Dr. D.[5] communicated the fact to Dr. Black[6] by Mr Robert D.[7] (who was my father) who lodged with Dr. Black in Edinburgh.

Dear Tyndall | Yours very sincerely | Ch. Darwin

P.S. | I send by this post, a pamphlet[8] which contains the best account, pleasantly written, of my Origin, if you chance to care to read it.

RI MS JT/1/TYP/9/2802
LT Transcript Only

---

1. The date is taken from the Darwin Correspondence Project, DCP-LETT-3067. See n. 8.
2. *I send the letters*: Charles Darwin sent Tyndall four letters from his paternal grandfather Erasmus Darwin to his maternal grandfather Josiah Wedgwood Sr. See reference in n. 1 for further detail.
3. *Mr Wedgwood*: Josiah Wedgwood (1730–95), founder of the Wedgwood company (see n. 2).
4. *congelation*: a reference to Tyndall's concept of the fracture and regelation of ice to explain glacier motion. According to this concept, when ice is broken into fragments under pressure, the fragments, becoming moist through melting, reunite when they again come into contact with a film of water between them.

5. *Dr. D.*: Erasmus Darwin (1731–1802), see n. 2.
6. *Dr. Black*: Joseph Black (1728–99), Scottish physician and chemist.
7. *Mr Robert D.*: Robert Waring Darwin (1766–1848).
8. *I send by this post, a pamphlet*: Asa Gray, *Natural selection not inconsistent with natural theology. A free examination of Darwin's treatise on the origin of species, and of its American reviewers* (London: Trübner. Boston, Mass.: Ticknor & Fields, 1861). Tyndall's name appears on the presentation list that Charles Darwin drew up for the distribution of the pamphlet.

To Jack Tidmarsh[1] 24 February 1861 1769

24th February 1861

Dear Tidmarsh

Some men appear to have a memory for names and dates, some for places, some for mere forms, and some for all. I think I am defective except in the case of forms. I can easily remember a human countenance, and before I looked at the conclusion of your letter[2] I knew by the writing from what hand it had come. You call to my mind many things which had been almost washed away by other memories. I had totally forgotten Ferguson Street,[3] and to my shame be it spoken 'the lady of the raven plume'[4] also. Even the memory of our discussions had faded almost to a blank. But your letter revives them all, and shows that behind my conscious memory lay an unconscious one.

You mention Ginty[5]—It is quite an age since I heard from him. He is now in the service of the Emperor of Brazil, and is, I believe, pushing his fortunes prosperously. His wife was in London two or three years ago, and his children are now being educated in this country. He stands I believe very high in the estimation of the emperor and his ministers; Mrs Ginty indeed informed me that his majesty sometimes does Ginty the honour of visiting him, and taking tea at his house.

Of the other men whom we used to know in former days I am in absolute ignorance, with one exception. That is Higginson.[6] He was for a long time stationed at the camp at Aldershot, where I visited him. He has often called to see me in London, and he is now stationed at Shorncliffe near Folkestone, being captain in the 16th Regt. He suffered much from illness in the East, and is a little bald—With this exception he looks young and hearty.

Hirst[7] is living quite close to me. After he had finished his articles with Carter[8] he went to Germany and remained for some years at the Universities of Marburg and Berlin. He married some years ago, but soon afterwards his wife showed signs of consumption. He took her to various parts of the Continent, in vain; and she died in Paris in the summer of 1857. I chanced to be in Paris at the time, and accompanied him to her grave. He afterwards spent much time in France and Italy, and is now first mathematical master in University College School, London.

Last summer I paid a visit for the first time to the Lake of Killarney. The weather was on the whole atrociously bad, but it could not quite disguise the extreme loveliness of the region. When well illuminated I do not think that I ever saw any thing fairer than those lakes. They seem to me to beat the English and Scotch ones in point of tenderness, and I have seen nothing in Switzerland to excel them in point of softness, though of course in grandeur they fall far behind the scenery of the Alps. On my return I paid Cork a visit. Climbed up Patrick's Hill, saw the barracks—wandered through the whole town. Some of it I had clearly retained, but some of it had been utterly washed away from my recollection. I could not fix the position of your house though I knew it was in Great George's Street.

The intellectual life of London sometimes puts a severe strain upon the mind, but on the whole I enjoy my life here extremely. I am fortunate enough to reckon the best and most distinguished men of England among my personal friends, and my work is a source of perennial delight to me. I am compelled to devote a portion of the year to bread and butter considerations, but a good portion of it is devoted purely to scientific investigation. I take an annual scamper to Switzerland and set up my health among the snowy mountains. Thus the stream of life flows on, hurrying pleasantly enough towards that ocean which all individual streamlets tend, and with which they are all ultimately to mingle.

Believe me dear Tidmarsh | Yours most sincerely | John Tyndall

I had forgotten the photograph—thanks for it. If I could get a handsome one of myself I would return the favour—but that is hardly to be hoped for.

RI MS JT/1/TYP/11/3872–73
LT Transcript Only

---

1. Jack Tidmarsh: see letter 1752, n. 1.
2. *your letter*: letter 1752.
3. *Ferguson Street*: see letter 1752, n. 9.
4. *'the lady of the raven plume'*: see letter 1752, n. 7.
5. *Ginty*: William Ginty (1820–66), a surveying colleague of Tyndall, who had emigrated to Brazil in 1854 where he worked as an agent of the Baron de Maua, contributing to such projects as the construction of the imperial gasworks and a railway which ran from Rio to Tijuca Serra.
6. *Higginson*: Thomas Charles Higginson (*c.* 1821–98), a surveying college of Tyndall in Ireland and England.

7. *Hirst*: Thomas Hirst.
8. *Carter*: Richard Carter (1818–95), the railway surveyor with whom Tyndall, Hirst, and Ginty had worked in Yorkshire.

To John Murray 9 March 1861 1770

Received from Mr. Murray the sum of twenty four pounds eleven shillings on account of the "Glaciers of the Alps." 1

John Tyndall | 9th March 1861. | £24–11–0[2]

National Library of Scotland, John Murray Archives, Ms. 41214

1. *"Glaciers of the Alps.":* Murray had published this on 6 July 1860.
2. Handwritten on the reverse: 9 March/61 | Prof. Tyndall

From Gustav Magnus 17 March 1861 1771

Berlin 17 März 61.

Mein theurer Tyndall

Es ist also in der That eingetroffen was ich befürchtete. Wir haben beide über denselben Gegenstand gearbeitet. Ich mache mir dabei einen Vorwurf daß ich in meiner Rücksicht für Sie zu weit gegangen bin. Als ich meine Versuche über die Durchstrahlung bei Anwendung von kochendem Wasser längst beendet hatte, erfuhr ich durch Clausius im Octob v. J., bei Gelegenheit des Jubilaeums der hiesigen Universität, daß Sie über die Durchstrahlung der Wärme durch die Gase arbeiteten. Ich gestehe daß ich die Ankündigung dieser Arbeit in den Proceedings übersehen hatte. Ich war zweifelhaft ob ich Ihnen schreiben und Ihnen meine Resultate mittheilen sollte. Ich überlegte mit Poggendorff, allein ich mußte, da Ihre Ankündigung bereits im Januar 1860 erfolgt war, vermuthen, daß Ihre Arbeit fast vollendet sei, und eine briefliche Mittheilung meiner Resultate erschien mir als ein Beschlaglegen auf die Priorität einem Manne gegenüber, der seit einem Jahre sich mit einer Arbeit beschäftigt hatte, die ich deßhalb auch vollendet glaubte. Ich sehe jetzt daß ich besser gethan hätte zu schreiben, denn Poggendorff sagt mir daß Sie Ihre Versuche erst im November ausgeführt haben; während der erwähnte erste Theil meiner Untersuchung vollendet war, als ich im July über die Leitung der Wärme durch die Gase las; denn ich glaube Sie kennen mich und meine Ängstlichkeit hinreichend um zu wissen, daß ich nicht von einer Leitung der Gase würde gesprochen haben, wenn ich mich nicht überzeugt gehabt hätte, daß die Verschiedenheit der Erwärmung in denselben nicht auf einer Verschiedenheit in der Durchstrahlung beruht. Deßhalb mußte diese Durchstrahlung gerade für die Wärme untersucht werden, welche für die Leitung benutzt worden war, die des kochenden Wassers. Die auffallenden Resultate welche ich wie Sie erhalten liessen mich wünschen vor der Veröffentlichung zu untersuchen wie sich die Gase bei Anwendung einer Wärmequelle von hoher Temperat verhalten und das hat mich im vergangenen Herbst beschäftigt. Es ist ein eigenthümliches Schicksal das ich habe nun schon zum dritten Male gleichzeitig mit einem Andern dasselbe Thema bearbeitet und in denselben Tagen publicirt zu haben; das Einzige was mir dabei einige Genugthuung gewährt ist daß es gerade die bedeutendsten Experimentatoren, wie Sie und Regnault sind, mit denen ich concurrire. Es ist nun ein mal geschehen und läßt sich nicht ändern. Hätten wir mehr zusammen correspondirt, es wäre nicht vorgekommen. Lassen Sie uns fortan wenigstens, wo wir ohne Zweifel beide den Gegenstand verfolgen werden, uns gegenseitig über das in Kenntniß erhalten, was ein Jeder von beiden vor hat, damit wir nicht wieder zusammen treffen. Sobald unsere beiderseitigen Arbeiten publicirt sein werden wird eine solche Verabredung leichter und bestimmter möglich sein als jetzt. Aus dem Februarheft von Poggend, werden Sie gesehen haben daß meine Versuche sich nicht auf die Dämpfe ausdehnen, dies gehört mit Ausnahme des Wasserdampfes Ihnen allein, dagegen haben Sie das Ammoniacgas nicht untersucht, wahrscheinlich weil dasselbe Ihre Röhre angegriffen haben würde. Dies Gas absorbirt aber noch viel, viel stärker als Oelbildendes Gas. Ueber die Ausstrahlung der Gase habe ich gar nicht gearbeitet, dies höchst interessante Verhalten entdeckt zu haben, gehört Ihnen ganz allein. Ich denke zunächst mich mit dem Einfluß der reflectirenden Flächen zu beschäftigen, doch habe ich vorher noch eine andere Arbeit abzuschliessen über die Leitung der Gase für Electricität, die auch sehr merkwürdige Resultate liefert.

Was Sie in den Comptes rendus von einer 15 Mal stärkeren Absorption der feuchten Luft sagen ist mir nicht verständlich, es sei denn daß Ihre Luft Nebel enthielt, oder daß vielleicht die Feuchtigkeit sich auf den Steinsalzplatten nieder geschlagen hat. Denn ich habe keinen Unterschied oder einen kaum wahrnehmbaren finden können zwischen Luft die ganz mit Feuchtigkeit bei 17°C gesättigt und trockner Luft.

Leben Sie wohl, schreiben Sie mir bald! So bald meine Abhandl die bereits in der Druckerei ist gedruckt sein wird sende ich sie Ihnen.

Von Herzen | Ihr | GMagnus.

Meine Frau und Kinder tragen mir die freundlichsten Grüße auf. Hätten Sie uns ein Mal besucht namentlich im vergangenen Jahre, die Carambulage hätte nicht statt gefunden! Kommen Sie doch bald ein Mal.

Berlin, 17 March 61

My dear Tyndall

My apprehensions have indeed been realised. We both have been working on the same subject.[1] I reproach myself for having been too ready to defer to you. At a time when my experiments on the passage of radiation using boiling water had long been finished, I heard through Clausius in October last year, on the occasion of the anniversary of our university, that you were engaged with research on the radiation of heat through gases. I confess that the announcement of this investigation in the Proceedings[2] had escaped my notice. I was doubtful if I should write to you and inform you of my results. I considered the case with Poggendorff and, as your notice had appeared already in January 1860, I supposed you had already finished the investigation and it appeared to me that a communication of my results to you, who had been engaged on the subject for a whole year, would equal a claim to priority. I perceive now that the better course would have been to write to you, because Poggendorff tells me that your experiments were made in November only, while the said first part of my investigation was already finished, when I read about the conduction of heat in gases in July. I believe you are sufficiently acquainted with me and my solicitude to know that I would not have spoken about conduction in gases, were I not convinced that the difference in the rise of temperature was not caused by a difference in the passage of radiation. It was necessary, for this reason, to examine this passage for that heat that was used for conduction, that is, that of boiling water. The striking results that I, as well as you, obtained made me desirous to examine, before publishing these results, how the gases behave with a source of heat of a high temperature, and this subject occupied me during last autumn. It is a strange fate that this is now the third time that I have been working simultaneously with another person on the same subject and also publishing the results at the same time. The only satisfaction I derive from this circumstance is that I have been competing with the most eminent experimentalists, such as Regnault[3] and yourself. It has happened and cannot be changed. If we had corresponded more frequently it would not have occurred. Without doubt, we both shall pursue the subject; in the future, let us therefore communicate to each other the object that each one of us intends to pursue, so that we do not clash again. As soon as our respective papers are published,[4] such an understanding can be more easily established than at present. You will have learned from the February issue of Poggend[5] that my experiments do not include vapours, the investigation of this part of the subject, with the exception of steam, belongs exclusively to you. On the other hand you have not examined ammonia, probably because the gas would have attacked your tube. But this gas absorbs much more than oil-building gas. On the radiation from gases no experiments have been made by me, and the interesting discoveries on this point are due to you alone. I intend to examine the influence of reflecting surfaces, but before this I have to finish another investigation on electric conduction in gases, which is also yielding most remarkable results.

Your statement in the Comptes rendus[6] that moist air has a 15 times greater absorption is not comprehensible to me, unless your air contained fog, or that perhaps a layer of moisture had settled on your rock-salt plates. For I was not able to notice a difference, or a barely perceptible one, between air that was entirely saturated with moisture at 17°C and dry air.

Farewell, write to me soon! As soon as my paper, which is already at the printers' is printed, I shall send a copy to you.

Cordially | Your | GMagnus.

My wife and children[7] charge me to send you their kindest greetings. If you had visited us, namely during the last year, the collision would not have happened! Do come soon.

RI MS JT/1/M/24

---

1. *We both have been working on the same subject*: the absorption of heat by gases.
2. *the announcement of this investigation in the Proceedings*: Tyndall (1859).
3. *Regnault*: Henri Victor Regnault (1810–78), chemist and physicist.
4. *As soon as our respective papers are published*: Tyndall (1861) and G. Magnus, 'Ueber die Verbreitung der Wärme in den Gasen', *Poggend. Annal.,* 112 (1861), pp. 351–52 and 497–549. Tyndall's paper (see letter 1767, n. 2) was also published by Poggendorff a few months after Magnus's paper. See also letters 1792 and 1793.
5. *You will have learned from the February issue of Poggend*: see letter 1767, n. 4.
6. *Your statement in the Comptes rendus*: see letter 1776, n. 3.
7. *My wife and children*: Magnus married Bertha Humblot in 1840, and they had one son and two daughters.

From Heinrich Debus — 20 March 1861 — 1772

Queenwood College | Stockbridge | March 20th 1861.

My dear John,

I received your two notes[1] in due time. I must confess I felt very much disappointed on reading the statement of your going to Cornwall.[2] Of course as you have engaged to go, you must go. How glad I should have been to have seen you here, needs no assurance. It is now nearly a year since I saw you last in this neighbourhood and when shall I see you again? I enclose the desired translation[3] and hope it will be to your satisfaction. When writing it I was and am still troubled with a severe headache, so I am afraid the English will not be very elegant.

Do not carry your theory too far or you will loose your ground.* Your friend knows that love has its duties and pains.

A happy journey, dear John, dont walk round cliffs or into deep mines or other attractive places but preserve your neck for your friends.

Good Bye | Ever yours | H. Debus.

*The ground on which your theory holds good is the inherent love and affection of your friend. par example I am in love with you.

RI MS JT/1/D/28

1. *your two notes*: letters missing.
2. *your going to Cornwall*: Tyndall went on a walking holiday to Devon and Cornwall with Francis Vaughan Hawkins and Francis Galton from 23 March to 2 April (Journal, RI MS JT/2/13c/1205–11). Tyndall's description contains one of the earliest accounts of sea-cliff climbing in Cornwall.
3. *I enclose the desired translation*: enclosure missing.

To Gustav Magnus [22] March 1861[1] 1773

My dear Magnus,—Though science is sure to be the gainer from our working independently at the same subject, it is not quite agreeable to find one's own personal claims interfering with those of a valued friend. My simple course, however, is to lay before you a brief statement of the circumstances attending the investigation of the mutual action of gases and radiant heat, feeling assured that you will draw just conclusions from the statement.

It is several years since the thought of examining the absorption of radiant heat by gases first occurred to me, but until the early part of 1859 I was so much occupied with other matters that I could not turn my attention specifically to it. In 1859 I reflected a good deal on the question, made many experiments, devised new methods, and, after some weeks of successful work, communicated in a note to the Royal Society[2] the mode of experiment and the general character of the results. This note, which was written with the express intention of setting my mind at rest as to the possible claims of other investigators, is published in the 'Proceedings' of the Society for May 26, 1859. I saw clearly that the subject would occupy me for several years. A new and vast field had in fact been laid open to experiment, and having by thought and labour fairly mastered it and obtained definite results, I felt at liberty to cultivate it calmly and quietly without fear of being interfered with by anybody. In the note referred to I state distinctly that *the investigation is in progress, a full account of it at a future day being promised to the Royal Society.*

I continued to work; and on June 10, 1859, gave a Friday evening discourse on the subject[3] before the members of the Royal Institution. The Prince Consort[4] was in the chair; and by throwing the image of the dial of my galvanometer upon a screen, I succeeded in making the experiments on the radiation and absorption of heat by gases and vapours visible to an assembly of many hundred persons. A report of the Lecture is published in the 'Proceedings of the Royal Institution' for June 10, 1859.

You tell me that you had overlooked the notice in the 'Proceedings of the Royal Society,' and that Clausius was the first to inform you in October 1860 of my being engaged in experiments on the radiation of heat through gases. I, on the other hand, did not neglect to give my results due publicity. They were communicated to De la Rive in a letter published in the 'Bibliothèque universelle' for July 1859, vol. v. page 232. My lecture of the 10th of June was also reproduced in 'Cosmos,' vol. xv. page 321, where you will find the following words:—'Ces recherches de M. Tyndall sont encore à leur début, il les continuera avec ardeur; déjà il a perfectionné sa mode d'expérimentation. Il emploie un galvanomètre d'un seul fil, avec deuz sources de chaleur aggissant sur les deux faces d'une meme pile,' &c. You will also find the subject referred to in the 'Nuovo Cimento,' vol. x. p. 196. Also in the 'Comptes-rendus,' and in other journals.[5] When I was in Paris in 1859, my experiments formed a common topic of conversation among scientific men. Foucault, Jamin,[6] Moigno,[7] Wertheim[8] and others, spoke to me about them. In fact they aroused an amount of attention far greater than I had any reason to expect.

In London it was known to everybody that I was engaged on this question; Hofmann,[9] Williamson,[10] and Frankland can all call to mind my obtaining volatile liquids from them in 1859. Faraday also recollects how often I have expressed to him my satisfaction that a new field had been opened which I might cultivate at my leisure without fear of interfering with the claims of anybody. In 1859 I had measured the absorption of oxygen, hydrogen, nitrogen, air, carbonic oxide, carbonic acid, nitrous oxide, coal gas, and olefiant gas; and I had also determined the absorption produced by the vapours of the following substances:—bisulphide and bichloride of carbon, sulphuric ether, cyanide of ethyl, chloroform, benzol, methylic alcohol, amylene, iodide of ethyl, acetate of ethyl, propionate of ethyl, formate of ethyl, iodide of amyl, chloride of amyl, ethylic ether, ethyl amylic ether, amylic alcohol, absolute alcohol, and others. *In fact in* 1859 *I had the materials for an elaborate memoir on this subject already in my hands.*

You say that I had not examined ammonia even in 1860. The fact, however is that early in 1859 I had examined it and established its great absorptive power. If you will cast your eye on my letter to De la Rive[11] published in the 'Bibliothèque universelle' for July 1859, you will see that ammonia did not escape me.

Two circumstances offered a direct hindrance to the working out of my results in 1859: the first was the desire to get the 'Glaciers of the Alps' off my hands; the second was a sudden call to undertake the duties of Professor at the School of Mines.[12] I resumed my experiments in 1860, and you will find a general statement of my results at pp. 244–45 of the 'Glaciers of the Alps,' a copy of which I had the pleasure of forwarding to Mrs. Magnus.[13] Referring to the preface to that work, you will see that it was published in June 1860; and the preface was written several months subsequently to the pages above referred to. In fact the experimental notes now before me would satisfy you that in June 1859 I had obtained, in a substantially correct form, almost all the results of which you first heard in October 1860, and which you suppose I did not obtain until the November of that year.

Believe me ever yours faithfully, | John Tyndall

J. Tyndall, *Contributions to Molecular Physics in the Domain of Radiant Heat* (London: Longman's, Green, & Co., 1872): pp. 61–63.

---

1. *[22] March 1861:* Magnus's letter to Tyndall was written on 17 March from Berlin (see letter 1771). Tyndall notes that he 'immediately sent the following reply,' which gives a date around 22 March.
2. *communicated in a note to the Royal Society:* Tyndall (1859).
3. *a Friday evening discourse on the subject:* 'The Transmission of Heat of Different Qualities through Gases of different Kinds,' *Roy. Inst. Proc.*, 3 (1859), p. 155–58.
4. *The Prince Consort:* Prince Albert, who later helped Tyndall obtain plates of rock-salt for him, important because they are transparent to radiant heat.
5. *Also in the 'Comptes-rendus,' and in other journals:* in this letter, Tyndall does not specifically mention a note in any German journal, which Magnus might have been expected to see, though he informed Clausius that he did provide one (see letter 1793, n. 12).
6. *Jamin:* Jules Jamin (1818–86), professor of physics at École Polytechnique 1852–81. See also letter 1721, n. 1.
7. *Moigno:* Abbé François-Napoléon-Marie Moigno (1804–84), mathematician and physicist. Founder of the journal *Cosmos* in 1852, and *Les Mondes* in 1862.
8. *Wertheim:* Guillaume Wertheim (1815–61), physicist.
9. *Hofmann:* August Wilhelm von Hofmann (1818–92), chemist, and the first director of the Royal College of Chemistry from 1845, until he moved to Bonn in 1864.
10. *Williamson:* Alexander William Williamson (1824–1904), professor of chemistry at University College London.
11. *my letter to De la Rive:* see letter 1617.
12. *a sudden call to undertake the duties of Professor at the School of Mines:* following the resignation of G. G. Stokes as professor of physics, Tyndall was offered the post in September 1859. His first lecture course ran from October 1859 to February 1860.
13. *Mrs. Magnus:* see letter 1771, n. 7.

From William Hopkins 25 March 1861 1774

22 Grand Parade | Brighton. | Mar. 25. 61.

My dear Tyndall,

I was exceedingly amused at the idea of your having Sir D. Brewster[1] and Blackie[2] at your feet petitioning for an article on Glaciers! Has some great physical revolution taken place in the hyperborean regions of Scotland that has melted down the moral hardness of these northern glacialists? Well, it is funny, but all right. What answer have you to the invitation. We shall hear very little more I suspect of these viscous gentlemen.

Your experiments and observations, in proving that a mass of ice changes its form and adapts itself to external conditions by actually cracking and breaking, and not by a continuous sliding of one constituent article cast another as in a viscous substance, is no unimportant step in the mechanical view of the notion. In fact, it removes the solution from dynamics to statics. We can do little with the dynamical problem of the motion of a viscous substance, but we can do much bearing on the question before us with respect to the statical pressures and tensions of a mass previously to its breaking, and therefore in determining also the directions in which it will break. Whether a glacier constituted as it is would have any considerable motion or not, if it adhered absolutely to bottom on which it rests, I know not, but I am sure of this, that the fact of its sliding along its bed must necessarily increase enormously the forces which tend to break and rupture the ice. This has never yet been pointed out, but it shows the important connexion existing between the motion due to sliding and that due to the finite and infinitesimal breaking of the mass. I think now of writing a short memoir for the Royal Society[3] on this subject without waiting for the experiments which I suggested to you in my last, but indicating them as essential for the completeness of the subject. If you can't find time to make them hereafter, I will try to make them myself.

The lecture I half promised you for the Institution[4] this spring I must decline for the present. My knee[5] though slowly and steadily improving is still but a feeble knee, and will keep me here for some time longer, though I hope ultimately to be not much the worse for the accident. I have declined one or two other lectures for the same reason. I want in the first place to write this short memoir, and then I should like much to write a general review of the whole subject for the Quarterly,[6] if Murray[7] would admit it. I don't recollect a glacial review in that journal. I don't know Murray personally. You do I presume. Can you ascertain for me whether such a review as I should propose would be accepted. I should endeavor to make it philosophical and important, and as little personal as possible.

I have not been experimenting on the transmission of heat thro' gasses. The paper of mine you allude to[8] was on the construction of a calorimeter for measuring the absolute (not merely the relative) quantity of heat which emanates from a given surface through air. I don't attempt to make a distinction between the quantity passing by radiation proper, conduction, convection &c. That is a problem of great interest and I am glad to hear that you are cogitating on it.

I hope I shall get bodily into the scientific world again next month, but I must try to be patient.

Yours truly | W. Hopkins

RI MS JT/1/TYP/2/639–40
LT Typescript Only

1. *Sir D. Brewster*: see letter 1766.
2. *Blackie*: William Garden Blaikie (1820–99), editor of the *North British Review* from 1860–63.
3. *a short memoir for the Royal Society*: see letter 1709, n. 6.
4. *The lecture I half promised you for the Institution*: Hopkins eventually gave a Friday Evening Discourse 'On the Theories of the Motions of Glaciers' on 31 January 1862: *Roy. Inst. Proc.*, 3 (1858–62), pp. 410–18.
5. *My knee*: see letter 1731.
6. *review of the whole subject for the Quarterly*: see letter 1731, n. 11.
7. *Murray*: John Murray was the publisher of the *Quarterly Review*. See also letter 1770.
8. *The paper of mine you allude to*: Tyndall's replies to this series of letters from Hopkins are missing.

To Samuel Best[1] 30 March 1861 1775

Royal Institution | Saturday | 30 March 1861

My dear Mr Best

I am very sorry that I missed you—I had just run down to the Athenaeum[2] for a basin of broth when you called. With regard to Mr Thomson's query[3] I can only say that I am sure many of my London friends would feel an interest in a good case of cleavage developed in mud by pressure.

Would you present my kind regards to Mrs Best[4] & to Miss Grace[5]

Most truly yours | John Tyndall

We had a field day at the Geological Society[6] a few days ago about cleavage.

Private Collection, Roland Jackson

1. *Samuel Best:* the Reverend Samuel Best (1803–73), a member of the RI. He was the rector of Abbots Anne, Andover, and rural dean of Andover. Educated at Eton and Cambridge, Best started a non-denominational primary school at Abbotts Ann in 1831 as well as a Friendly Society for the villagers.
2. *the Athenaeum:* The Athenaeum Club, Pall Mall.
3. *Mr Thomson's query:* not identified.
4. *Mrs Best:* Emma Best (née Duke), his wife, b. 1811 (1861 Census). They married in 1835 (Westminster Marriages Transcription).
5. *Miss Grace:* Grace Best, their daughter, b. 1841 (1861 Census). She was the eldest of three children.
6. *field day at the Geological Society:* though it was more than a month previously, rather than 'a few days', this may refer to the meeting at the Geological Society on 20 February when a paper by Roderick Murchison and Archibald Geikie was read 'On the Coincidence between Stratification and Foliation in the Crystalline Rocks of the Highlands'. Andrew Ramsay reported the discussion in his diary, noting that he had to put the audience straight when Tyndall appeared to claim Ramsay's ideas for his own. See J. C. Thackray (ed), 'To see the Fellows fight: Eye witness accounts of meetings of the Geological Society of London and its Club, 1822–1868', *BSHS Monographs,* 12 (2003), pp. 207–8. We are grateful to Caroline Lam and her colleague at the Geological Society for suggesting this.

To Auguste Arthur de la Rive 4 April 1861 1776

Royal Institution | 4th April 1861

My dear Sir

Before I sent my memoir to the Royal Society[1] the thought occurred to me of sending you a full copy of it, accompanied by a drawing of the apparatus employed. But on second consideration I imagined that the perusal of a long English memoir would be too much even for your good nature towards me. Hence the determination to wait until the memoir should be printed.

This I regret to say has not yet taken place, but the moment the first proofs reach my hand I will send you a copy. At present I send you a scanty abstract merely,[2] which will give you some general idea of the character of the enquiry.

M. Despretz[3] omitted a portion of the manuscript which I sent to him; probably because he thought it too speculative[4]—I have however throughout endeavoured to limit myself to the strict mechanical conditions of the problem; and though may modify detail, I am inclined to hope that the theory of the connexion of absorption, radiation and conduction which I have sketched out will be found substantially correct.

I thank you much for the interesting paper on glaciers,[5] which you were kind enough to send to me some time ago.

Faraday is very well—he is spending his Easter vacation at Hampton Court. I hope you are well likewise.

Believe me | most truly yours | John Tyndall

RI MS JT/1/T/1498
RI MS JT/1/TYP/1/362

1. *my memoir to the Royal Society:* Tyndall (1861).
2. *I send you a scanty abstract merely:* abstract missing. Presumably 'Note on the Transmission of Radiant Heat through Gaseous Bodies', *Roy. Soc. Proc.,* 10 (1859), pp. 37–39.
3. *M. Despretz:* César-Mansuète Despretz (1791–1863), who is listed as one of those responsible for Section V, Physique Générale, of the *Comptes Rendus.* A summary of Tyndall's paper was published as 'De l'absorption et du rayonnement de la chaleur par des gaz et des vapeurs, et sur la connexion physique du rayonnement, de l'absorption et de la conduction; par M. John Tyndall', *Comptes Rendus,* 52 (1861), pp. 364–67. The summary omits Tyndall's comments on the connection between radiation, absorption, and conduction.
4. *too speculative:* see letter 1780, n. 6.
5. *the interesting paper on glaciers:* not identified.

From Edward Sabine 17 April 1861[1] 1777

13 Ashley Place S. W. | April 16. 1861 | midnight.

Dear Tyndall,

The mean direction of the declination needle is the sum of equidistant observations of its direction in 24 hours, divided by .

The mean diurnal variation (in some definite period of time, say, a month) is the mean difference in the direction observed at every hour of the 24, from the mean direction as obtained above.

If there were no magnetic disturbances the diurnal variation at the same hour in each of the 30 days, would be a constant quantity: i.e. there would be a certain constant amount of diurnal variation (either to the east or to the west, according to the hour of the day,) in each of the 24 hours, we may call this the normal direction of the particular hour and month.
But by reason of the occurrence of magnetic disturbances, the actual direction of the needle does not always coincide with the mean or normal direction. Say for example that the normal variation at 8am is 10' of arc east of the mean direction in the day, (or of $\frac{\Sigma n \text{ observations}}{n}$); and that on a certain day the needle is seen at 8 am to point 15' instead of 10' to the east of the mean direction in the day; in such case we should say that there was a disturbance which amounted to 5' East; (an easterly deviation). Now suppose that on another

day at the same hour, ie at 8 am, the direction were only 5' east of the mean direction of the day, in such case the needle points 5' west of its normal position at that hour—and this would be called a westerly disturbance; (or a westerly deviation, if you prefer the word).
The angle comprised between the extreme east & west directions in which the needle points by reason of diurnal variation would be called by me 'the range of the diurnal variation'. The diurnal variation at any particular hour would be the difference between the monthly normal at that hour and the mean direction of the whole day; (i.e. $\frac{\Sigma n \text{ observations}}{n}$). Lamont[2] would probably call the latter 'horary variation'.

I will look out the references for the intensity in June & December & give them to you when we meet.

Sincerely yours | Edward Sabine

Wednesday morning

I have added this morning what may possibly make my meaning more clear—I returned late last night from the country.

The 'decennial variation' in any particular year is the difference between what would be the direction if there were no decennial variation, and the actual direction, so far as the latter is influenced by the causes to which the decennial variation is owing.

The annual variation in any particular month, is the difference between what would be the direction if there was no annual variation, and the actual direction, so far as the latter is influenced by the causes to which the annual variation is owing.

The 'solar diurnal variation' at any particular hour is the difference between what would be the direction if there were no solar diurnal variation, and the actual direction so far as the latter is influenced by the causes to which the solar diurnal variation is owing.

In the observations themselves all these and all other variations, from whatsoever causes they may proceed, are mixed up—To study any one of them correctly, you must eliminate the effects of the others.

In conformity with general usage, I call

Decennial variation, that variation whose period is ten years.

Annual variation, that variation whose period is a year.

Diurnal variation that variation whose period is a day.

The 'range of the decennial variation' is measured by the extremes of direction in the decennial period, due to the causes to which the decennial variation is owing.

The 'range of the annual variation' is measured by the extremes of direction which take place in the year, due to causes to which the annual variation is owing.

The 'range of the solar diurnal variation' is measured by the extremes of direction which take place in the day, due to the causes to which the solar diurnal variation is owing.

You may call, if you please to do so, the decennial variation the annual variation: the annual, the monthly variation; and the solar diurnal, the horary variation. This is a mere substitution of one term for another; in which matter, I prefer and I believe the preference is very general, the terms decennial, annual, & solar diurnal. (The solar diurnal is so called to distinguish it from another diurnal variation due to the moon.)

But the real & substantial difference between Lamont & myself, apart from a difference in our modes of designation, is, I presume, that Lamont's diurnal variation includes the effects of both solar diurnal variation & magnetic disturbances (which also have a diurnal period). In the crude observations these are mixed up together—By the process which I have adopted, each is eliminated when the other is studied. It is this process of separating the effects of different causes which in the crude observations are mixed up together, which constitutes the precautions of the methods which I have pursued, and which I believe to be the true way of analysing phenomena with a view to the discovery of their distinct causes & separate effects.

ES. | Ap[ril]. 17.

RI MS JT/1/S/22
RI MS JT/1/TYP/4/1317

1. *17 April 1861*: the letter was started on 16 April and finished on 17 April. See also letters 1760 and 1764.
2. *Lamont*: Johann von Lamont (1805–79), Scottish astronomer and physicist. Lamont spent much of his life in Bavaria, becoming assistant at the Royal Observatory at Bogenhausen near Munich in 1828. In 1835 he was appointed director of the observatory, and established it as a centre for meteorological research and measurements of the earth's magnetic field. He lectured in practical astronomy at the University of Munich from 1837, before becoming Professor of Astronomy in 1852. His best-known work was in the field of terrestrial magnetism, in which he discovered the decennial periodicity in variations of the earth's magnetic field (later the 11-year cycle), and the presence of earth currents. He became engaged in the global initiative to survey the earth's magnetic field. In 1840, he built a magnetic observatory at Bogenhausen, and participated in the collection of data for the international project between 1841–45. He also undertook surveys outside Bavaria, in places such as Denmark, France, and Spain, and devised a number of new instruments which became widely used, including the chronometer in 1850.

From Joseph Plateau 18 April 1861 1778

Gand,1 18 Avril 1861.

Monsieur,

J'ai l'honneur de vous offrir un exemplaire du mémoire que je viens de publier; il vous parviendra par la poste. Les expériences que je décris dans ce mémoire ne présentent plus aucune des difficultés de celles des séries précédentes; une fois le liquide nécessaire préparé, elles s'effectuent avec une grande facilité, et à l'instant même, et donnent de fort beaux résultats. Si vous me faites l'honneur d'en répéter quelques unes surtout celles des paragraphes 18 et 19, je suis certain que vous en verrez les résultats avec plaisir.

J'ai à cet égard une prière à vous adresser. Vous donnez quelque fois, à l'Institution Royale, des séances dans lesquelles vous répétez les expériences faites hors de l'Angleterre qui vous paraissent les plus intéressantes. Or, si celles que renferme mon mémoire actuel vous semblent dignes du même honneur, et je ne puis m'empêcher de l'espérer, vous me rendrez un bien grand service en les montrant aussi publiquement; je serais certain alors de voir en Angleterre apprécier des recherches auxquelles j'ai consacré tant de temps et de soins. Les sentiments de bienveillance que vous m'avez témoignés, et la confiance que j'ai dans la beauté et la singularité de mes résultats me donnent l'espoir que vous accueillerez ma demande, et que j'en apprendrai dans quelque temps la réalisation.

Agréez, Monsieur, l'assurance de tous mes sentiments de haute considération.

J. Plateau | Professeur à L'Université | Place du Casino 22.

P.S. J'espère publier dans quelques mois ma 6eme Série qui est déjà à peu près terminée; elle est consacrée à la théorie des systèmes laminaires du paragraphe 18 de la Série actuelle et à l'étude de leur mode de génération aussi que des conditions que doit remplir une charpente polyédrique pour donner un système parfait; vous verrez aussi, dans cette 6eme Série, des résultats bien curieux.

Ghent, 18 April 1861

Sir,

I have the honour of offering you a copy of the memoir that I have just published;[2] it will come to you by mail. The experiments that I describe in that paper do not present any of the difficulties found in the previous series. Once the necessary liquid is prepared, it is easy to do them, and immediately, and they give very good results. If you would do me the honour of repeating some of them, especially those from paragraphs 18 and 19, I am sure that you will be pleased by the results.

I have a request to ask of you. Sometimes you give sessions at the Royal Institution, in which you repeat experiments done outside England that you find the most interesting. However, if the ones described in my current memoir seem to you worthy of the same honour, I cannot help but hope that you will do me the great favour of showing them in public.[3] Then I will certainly see England appreciating the research into which I put so much time and effort. The kindness that you have shown me and the confidence that I have in the beauty and uniqueness of my results, give me hope that you will welcome my request and that I will learn about its implementation sometime soon.

Please accept, Sir, the assurance of my highest regards.

J. Plateau | Professor at the University | 22 Place du Casino

P.S. I hope to publish in a few months the 6th Series,[4] which is already almost finished. It is dedicated to the theory of laminar systems of paragraph 18 in the current Series, and to the study of their mode of generation as well as the conditions under which a polyhedral structure can create a perfect system. You will see in the 6th Series as well some very interesting results.

RI MS JT/1/P/89

1. *Gand:* Ghent.
2. *the memoir that I have just published:* probably Plateau, J., '5th Series.' *Mém. Acad. Roy. Sci. Belg.,* 33 (1861).
3. *I cannot help but hope that you will do me the great favour of showing them in public:* Tyndall (and possibly Frankland) may have later demonstrated them to the RI, though no reference has been found; see letter 1875. See also letters 1811, 1862, 1864, and 1879.
4. *the* 6th *Series:* Plateau, J., '6th Series.' *Mém. Acad. Roy. Sci. Belg.,* 33 (1861).

To Heinrich Debus [19 April 1861][1] 1779

Friday m[ornin]g

My dear Heins,

It gives me great pleasure to inform you that yesterday[2] the Council of the Royal Society did itself the honour of choosing you among the fifteen to be recommended for election.[3] You may take it for granted that your election is secured.

It gratified me much to observe the very strong support which you received. You were very near the head of a list of I believe forty five candidates.

Now old fellow light your cigar, and take a pleasant tranquil puff, and enjoy the thought of being what you ought to have been long ago. The colleague of your friend

John

Tom[4] was also chosen. Greville Williams,[5] Roscoe,[6] and Simpson[7] I am sorry to say did not come in—but I think they are sure to be elected next year.

RI MS JT/1/T/342
RI MS JT/5/11/1376

1. *[19 April 1861]*: see n. 2.
2. *yesterday*: the Council meeting, at which Tyndall was present, was held on Thursday 18 April 1861.
3. *for election*: as a Fellow of the Royal Society. Debus was elected FRS at the Council Meeting held on 6 June 1861 (see also letter 1696).
4. *Tom*: Thomas Hirst was also elected in 1861.
5. *Greville Williams*: Charles Greville Hanson Williams (1829–1910), chemist, was elected in 1862.
6. *Roscoe*: Henry Enfield Roscoe (1833–1915), chemist, was elected in 1863. See letter 1748, n. 1.
7. *Simpson*: Maxwell Simpson (1815–1902). He was an Irish chemist, who had studied in Bunsen's laboratory in Marburg. He was elected in 1862. His wife, Mary (née Martin) was the sister of Thomas Hirst's wife, Anna, who died in 1857.

From George Gabriel Stokes 7 May 1861 1780

Lensfield Cottage Cambridge | 7th May 1861

My dear Dr Tyndall,

Your paper[1] was ordered to be printed last Thursday. I directed Mr White[2] to get estimates for the plate so that we may be getting on with the plate. I have got the MS which I took partly to help me in making some remarks from the referees[3] and partly to save expense in case you should wish to make any more than a few trifling changes which could be made on the proof sheets.

I was one of the referees, and I have the following remarks to make

I.

It does not seem to me that there is any real discrepancy between Franz[4] and yourself at least nothing more than is fairly attributable to the difference in the sources of heat and similar differences in the mode of conducting the experiments. Franz says <u>nothing about drying the air</u> and other gases except in the case of nitrous acid gas which as ordinarily produced is liable to be charged with water and in which after all the drying was so imperfect that Franz speaks of the interior of the tube being moistened by it. He doubtless took for granted that the small quantity of moisture present would go for nothing. You on the other hand have discovered that while dry air is very transparent watery vapour is highly absorbent. Hence Franz's results for air should be compared, not with yours for dry air, but with yours for the air of the laboratory, and the amount of absorptions which you both make out, though not equal, are quite comparable with each other.

On the other hand I cannot imagine that the cause you suggest can have materially affected Franz's experiments. He appears to have taken a series of observations in which a screen was alternately interposed and removed; and as this occupied some time any such merely temporary cause of disturbance as that you mention could hardly fail to be detected by the irregularity which it would produce in the results.

II.

p.15 [Explanation of the apparent (but temporary) decrease of absorption when the air was exhausted] "due to the aqueous vapour precipitated" .

I don't suppose the result depends upon the precipitation of vapour, (though probably some was precipitated) but upon the high emissive power of water as compared with air. The absorbing and emitting power of water is likely to be much the same in the elastic and in the liquid state, so that the same effect I imagine would have been produced (making allowance for the smaller quantity of water present) had the quantity present been too small for any to be precipitated. Indeed precipitation, as far as it goes, seems likely to produce an antagonistic effect, on account of the latent heat given out.

III.

P. 16 end: "so that the heat emitted by the copper plate might cross a vacuum before entering the experimental tube, thus retaining the quality which belonged to it at the moment of emission". Add except in so far as it might be interfered with by the absorption of heat by the rock salt" otherwise the statement is too absolute.[5]

IV.

The only serious point I have to remark about[6] is the connexion or supposed connexion between good conduction and bad radiation, bad conduction and good radiation.

Now in the first place is it a physical law that good conductors are bad radiators and bad conductors good radiators? I confess I doubt it. It is true the metals conduct well and (at least when polished) radiate badly. But the metals are linked together by so many physical and chemical properties that the different metals must not be looked on as so many independent witnesses. And what do you say to charcoal (graphite) Being black it is almost certainly a good radiator, better I should suppose than white porcelain. But as it conducts electricity pretty well it probably also conducts heat well.

In focusing the electric light for a short time only with a graphite pole I have been struck with the high temperature of the graphite at the end remote from the seat of incandescence. A vitreous substance would not have become sensibly hot in the time so far from the seat of the high temperature.

And now for the reasoning. This seems to me to assume that a given molecule loses the main part of its motion in the course of one or two vibrations. For you suppose that in the case of a good radiator a molecule does not lose much motion by communication to the adjacent molecules because it has already lost it by communication to the ether, which acts as a buffer. This I cannot understand except on the condition I mentioned. But if the fraction of the vis viva of a molecule which is lost in a single vibration is very small, so that a number of successive vibrations are nearly equal, it seems to me that the loss by communication to the ether and the loss by communication to other molecules must take place sensibly independently of each other. For what happens in the course of a single complete vibration must decide whether the communication in either way shall be comparatively rapid or not.
Now the phenomena of interference, in which one undulation is found to interfere with another which precedes or succeeds it by a great number of periods. (Fizeau and Foucault have traced interference up to a retardation of somewhere about a couple thousand undulations.[7] This seems to show that in the cause exciting these undulations there must be great regularity, that the motion of a vibrating molecule must be sensibly the same for a great number of consecutive vibrations so that only a very small fraction of the motion can be lost in one vibration. But if so it seems to me that the loss by communication to the adjacent molecules (conduction) and the loss by communication to the ether (internal radiation) must go on sensibly independently of each other; so that if the amplitude of excursion were diminished in the ratio of 1 to $1-c$ by the first alone and 1 to $1-r$ by the 2nd alone, in one vibration, c and r being very small fractions it would be diminished in the ratio of 1 to $\overline{1-c}\;\overline{1-r}$ by both together or in n vibrations in the ratio of 1 to $(1-c)^n\,(1-r)^n$ and I can see no reason why a diminution of c should imply an increase of r or the converse.

V.

p. 82 'A lead bullet suspended from a string and made to oscillate will do so more quickly than &c' more or less quickly according to circumstances.[8] Thus if two bullets were joined so that the line joining the centres was in the direction of oscillation they would oscillate more quickly than a single sphere. But the time of oscillation of a lead bullet oscillating as a pendulum would be increased by the resistance of the water (supposed for simplicity to be perfectly smooth) only in the proportion of $\sqrt{s}$ to $\sqrt{s+½}$ (s being the specific gravity of lead, about 11½) or 1 to $1+\frac{1}{4s}$ nearly or 1 to $1+\frac{1}{46}$. The effect on the time would be less and less in proportion to the lightness of the fluid—compared with the solid. Now taking account of the small inertia which we must attribute to the ether and the persistence of the luminous vibrations of ponderable molecules I don't think the time of vibrations can be sensibly affected by the greater resistance.

Supposing the vibration to be sufficiently broadly based I should prefer regarding the slow vibrations and the greater resistance to distinct consequences of a common cause (the greater size and unwieldiness) to supposing the latter to be the cause of the former.

As to Sir W. H's paper[9] next Thursday (tomorrow) is Ascension Day; 16th Croonian Lecture[10] 23rd Whitsun week no meeting 30th virtually promised to Dr Smith.[11] I am afraid therefore I cannot have time for the proof while he is in Town.

The other referee has also made remarks about which I have written to him.

Yours very truly | G.G. Stokes
Dr Tyndall F.R.S.

RI MS JT/1/S/221
RI MS JT/1/TYP/4/1365–69

1. *Your paper:* the manuscript (RS PT/63/1) which became Tyndall (1861).
2. *Mr White:* Walter White.
3. *the referees:* the referee in addition to Stokes was William Thomson. He wrote a detailed report in scrawled pencil (RS RR/4/273) and a shorter follow-up note to Stokes on 19 April 1861 (RS RR/4/272). His report begins: 'I have read Prof. Tyndall's paper "on the absorption of radiant heat" (Bakerian) with much interest and have no hesitation in advising its publication in the Transactions. The results as to the great absorption of radiant heat by olefiant gas and the vapours experimented on are (so far as I have the means of judging) perfectly novel, and constitute a most important contribution to science. The illustration of the theory of exchanges which the experiments on radiation by heated gases and vapours are striking and instructive'. Thomson picked up particularly on the discrepancies between Tyndall and Franz (see n. 4) and on the assumptions behind Tyndall's 'calculated results', both of which Stokes highlighted.
4. *Franz:* (Johann Carl) Rudolph Franz (1811–1902) had published a paper making claims about the absorption of heat by various gases, at variance with those of Tyndall (R. Franz, 'Ueber die Diathermanität einiger Gasarten und gefärbten Flüssigkeiten', *Poggend. Annal.,* 94 (1855), pp. 337–56). Tyndall explained why he believed that Franz's experimental arrangement was at fault (Tyndall (1861), pp. 27–28). Thomson also believed that Franz's results were invalid, but they raised questions for him about Tyndall's experimental arrangement, though not such as to invalidate his conclusions (RS RR/4/273).
5. Tyndall deleted 'thus retaining the quality which belonged to it at the moment of emission', but did not add the suggestion of Stokes to the final manuscript.

6. *The only serious point I have to remark about:* Tyndall held for some time to the idea that good radiators of heat were poor conductors and *vice versa.* See letter 1781, n. 8.
7. *Fizeau and Foucault have traced interference up to a retardation of somewhere about a couple thousand undulations:* Hippolyte Fizeau (1819–96) and Jean Bernard Léon Foucault (1819–68) showed that the speed of light was slower in water, when compared to speed in air, providing evidence for the wave theory of light. The undulations of light were considered to move slower due to the presence of water. See H. Fizeau, *C.R. hebd. Acd. Sci.,* 29 (1849), pp. 90–94 and 132, and L. Foucault, 'Sur les vitesses de la lumière dans l'air et dans l'eau', *Ann. de Chim. et de Phys.,* 3rd *series* 41 (1851), pp. 129–64.
8. Tyndall deleted this paragraph from the final manuscript, and made many other detailed changes following the comments of Stokes and Thomson.
9. *Sir W. H's paper:* this is perhaps Sir William Rowan Hamilton (1805–65), but there is no paper by Sir W. H. in *Roy. Soc. Proc.* or *Phil. Trans.* in 1861–63.
10. *Croonian Lecture:* the Croonian Lecture was given on 16 May by Charles Edward Brown-Sequard (1817–94), 'The Croonian Lecture: On the Relations between Muscular Irritability, Cadaveric Rigidity, and Putrefaction', *Roy. Soc. Proc.,* 11 (1860), pp. 204–14.
11. *Dr Smith:* Edward Smith (c. 1818–74). His paper was read on 30 May: 'On the Elimination of Urea and Urinary Water, in Their Relation to the Period of the Day, Season, Exertion, Food, and other Influences Acting on the Cycle of the Year', *Roy. Soc. Proc.,* 11 (1860), pp. 214–18.

To George Gabriel Stokes [14 May 1861][1] 1781

Tuesday,

My dear Mr Stokes

I have read your remarks on my paper.[2]

With regard to the precipitation of aqueous vapour I have introduced the word "probably"[3] which may qualify the statement sufficiently—I have myself no doubt that this was the cause, as at the time I made a series of contemporaneous experiments in a second table, and saw the precipitation occur. I have myself applied the explanation you suggest to other vapours.

The phrase "thus retaining the quality they possessed at the moment of emission" I have entirely struck out.[4] Melloni's results,[5] however, would justify it.

With regard to Franz[6] I fear the case must stand as I have stated it. His experiments are undoubtedly wrong. Just consider the case of carbonic acid[7] which he actually makes less absorbent than oxygen. Even granting to him the watery vapour of an English November day he could not properly explain the result.

Your remarks on the connexion of absorption and conduction refer to the last two pages of the paper. If you are decidedly of opinion that these should be withdrawn I have no objection to omit them.[8] I[9]

<Hereafter LT Transcript Only>

am however quite willing to risk the publication of the paper as it stands.

Is your experiment with the graphite a fair one? is it the heat of conduction that you felt at the remote end of the carbon pole? At the point of contact of metal and carbon there is usually a considerable resistance, to which in some measure at least the heating of this end must be due. I will not be positive but I think I could find you a "vitreous substance" that would conduct heat far better than graphite.[10]

I know for a certainty that a cube of the carbon used by Bunsen in his battery, and thro' which electricity passes with the utmost freedom, is almost impervious to heat. This however is a very porous kind of carbon.

However these are points which a few experiments will place beyond all doubt. I may say the objections you have urged and a good many more occurred to my own mind before I wrote the memoir.

Shall I transfer it to Taylor & Francis?[11]

very truly yours | John Tyndall

Stokes Correspondence, Cambridge University Library/MS.Add.7656/T597/1(2)–2(2)
RI MS JT/1/T/1432
RI MS JT/1/TYP/4/1467

---

1. *[14 May 1861]*: this letter replies to letter 1780. 14 May was a Tuesday.
2. *your remarks on my paper*: see letter 1780.
3. *the word "probably"*: Tyndall added the word to the manuscript (RS PT/63/1, p. 15) and it appears in the published paper. See Tyndall (1861), p. 5.
4. *The phrase "thus retaining the quality they possessed at the moment of emission" I have entirely struck out*: see letter 1780, n. 5.
5. *Melloni's results*: Macedonio Melloni had published his major work in 1850, *La thermochrôse ou la coloration calorifique* (Naples: Baron, 1850).
6. *With regard to Franz*: see letter 1780, n. 4.
7. *carbonic acid*: carbon dioxide.
8. *I have no objection to omit them*: Tyndall omitted the paragraph in his manuscript making the strong case that good radiators are bad conductors and vice-versa, but retained a long section 'On the Physical Connexion of Radiation, Absorption, and Conduction'. See letter 1780, n. 6.
9. At this point, the fragment from the Stokes Correspondence, Cambridge University Library, ends. The rest of the letter is held at the Royal Institution.
10. *conduct heat far better than graphite*: see letter 1780. Stokes took issue with Tyndall's comments on the relationship between conduction and radiation.
11. *Taylor & Francis?*: the publishers of *Phil. Trans.*

From George Gabriel Stokes 16 May 1861 1782

Lensfield Cottage Cambridge | 16 May 1861

My dear Dr Tyndall,

I think you must have misunderstood me[1] about precipitation of aqueous vapour, which you say you saw. I never doubted the fact of precipitation. What I said was that I looked on precipitation on the one hand and the apparent obstruction of radiant heat on the other as common consequences of the cold produced by expansion, and that it would not be correct to speak of precipitation as the cause of the observed effect unless it was through that precipitation that the cold took effect on the thermo-multiplier, whereas it appeared to be rather in spite of precipitation that the effect was produced. In fact, in order that your explanation, making precipitation the cause, should be valid, it would be necessary that water should offer decidedly more obstruction, weight for weight, to the passage of heat in the state of liquid than in the state of vapour; in short that a column or layer 1/100 inch thick should offer decidedly more obstruction than the same expanded into vapour, though the length of tube occupied by the vapour at say 60° F and a pressure of ½ inch mercury would be about 1000 inches. Experiment has not shown this to be true or false; and the analogy of coloured bodies would lead us I think to expect that there would probably be no great difference in the absorption according as the water happened to be in the liquid or aeriform state.

On the other hand your experiments indicate a perfect explanation of the observed result by attributing it to the cold produced by expansion taking effect by the dysdiathermancy (if I may coin such a word) and consequent good radiation of the substance HO (water or vapour as it may happen to be).

As to the connexion between radiation and conduction,[2] I feel that it is not for me to dictate to, hardly even to advise, one who has done such capital work. I can only say that while the generality of the alleged physical connexion seems to me to need further examination, the reasoning by which it is attempted to account for the connexion seems to me altogether inconclusive; perhaps chiefly because it points to a state of things so different from what optical considerations would lead us to regard as probable.

As to the heating of the coke pole,[3] I paid no particular attention to the thing at the time, and cannot say that the cause is not what you suggest. But there is no use conjecturing about what a few experiments would settle.

As to the tables "Calculated Results", the law assumed in the calculation is so evidently a physically impossible one[4] that any reader I think would take it for what it was worth and no more, nor would he I imagine suppose that the author took it for a law of nature. If the incident rays were homogeneous the intensity would decrease in a geometric ratio, or would be expressed by Ar^x,

r being a fraction less than 1, and x the length of path, or (in case of variable density) equivalent path i.e. length with a standard density answering to the actual length with the variable density. If the rays were partly non-absorbable, and the absorbable rays were all absorbed at the same rate, the intensity would be expressed by $A + Br^x$. But all these are only approximations to the truth; for we have an infinite number of kinds of rays differing in their rate of absorption so that we should have for the intensity
$[A + Br^x + Cr'^x + Dr''^x +]-$
We cannot of course determine an infinite number of constants from experiment. Hence wherever we stop there must be something arbitrary in stopping there; and the formula $A + Br^x$ (equivalent to that proposed by Thomson)[5] is not rigorously correct any more than your formula $A - Bx$, though the former is nearer the truth than the latter, and would be strictly correct under a conceivable mode of distribution of the rates of absorption, which the latter would not.

I think you had best send the paper to press at once, pleasing yourself as to the last chapter.[6]

Yours very truly | G.G. Stokes
Dr Tyndall F.R.S.

RI MS JT/1/S/222
RI MS JT/1/TYP/4/1370–71

---

1. *I think you must have misunderstood me:* see letter 1781.
2. *As to the connexion between radiation and conduction:* see letter 1780, n. 6 and letter 1781, n. 8.
3. *As to the heating of the coke pole:* this is not mentioned in letter 1781, which implies that Tyndall may have sent a second response, now missing.
4. *the law assumed in the calculation is so evidently a physically impossible one:* in addition to giving the observed absorption, Tyndall calculated the absorption 'on the supposition that it is proportional to density'. See Tyndall (1861), p. 12.
5. *equivalent to that proposed by Thomson:* Thomson explained his mathematical approach, and his objection to Tyndall's assumptions, in his letter to Stokes of 19 April 1861 (RS RR/4/272).
6. *pleasing yourself as to the last chapter:* see letters 1780, n. 6, and 1781, n. 8.

To George Gabriel Stokes 19 May [1861] 1783

Sunday, 19th May

My dear Mr Stokes.

Being engaged so much in lecturing and other matters just now, my mind is thrown a little out of gear, and indeed my time so filled that I could not, when I received them, devote the attention they deserved to your remarks on my paper:[1]—I have looked over them again this morning and have felt much interested by them.

As far as I can revive my impressions[2] I figured the propagation of heat thro' a solid in this way:—Imagine a prism, say of rock crystal, with a source of heat applied to one of its terminal surfaces, a definite amount of motion being thus imparted to the terminal superficial stratum of molecules. The question is how is this motion disposed of?

I suppose one portion of it (A) to be in the first instance communicated to the ether which divides the stratum from its next neighbour; a portion of this portion being transferred to the molecules of the second stratum, thus assisting conduction.

I suppose a second portion (B) to be transferred directly, which is pure conduction.

The second portion of (A) I suppose to be transmitted to the ether external to the prism, and thus lost as regards conduction.

The larger the ratio of A to B the greater I suppose is the loss just alluded to, and the shorter the distance to which the motion can be transmitted from particle to particle thro' the mass. In fact the motion is not there to be transmitted, having been wasted in space. Bad conduction is the diversion of the motion from the molecules to the ether.

I suppose the ratio of A to B to depend mainly upon the form of the molecule. If it be smooth, round and heavy, (A) will be less and (B) greater than if it be rough or pitted, and specifically light.

Supposing, the terminal stratum of our prism once set in motion, that the source of heat is suddenly withdrawn, the stratum being permitted to exhaust its motion; a portion of the latter would be communicated to the ether behind the stratum, and there take the form of radiant heat.

It appears to me that the copiousness of this radiation would depend on precisely the same quality as that which renders the component A large, or which in other words diminishes conduction.

Thomson refers to you[3] in connexion with Bunsen & Kirchoff.[4] I want to know more about this matter, for I have a dim remembrance that once as I sat at your side in the Philosophical Club[5] you mentioned Foucault's experiment[6] to me in connexion with the subject of Fraunhofer's lines.[7] Do you remember this conversation? Can you refer me to any thing published on the subject? Of course I shall be only too willing to act on Thomson's suggestion.

very truly yours | John Tyndall

RI MS JT/1/T/1421
RI MS JT/1/TYP/4/1468–69

---

1. *your remarks on my paper*: see letters 1780 and 1782.
2. *As far as I can revive my impressions*: this refers to the difference of opinion between Tyndall and Stokes on the relationship between radiation and conduction of heat. See letter 1780, n. 6 and letter 1781, n. 8.
3. *Thomson refers to you*: William Thomson.
4. *Bunsen & Kirchoff*: Tyndall was preparing his Friday Evening Discourse on 7 June 'On Solar Chemistry', *Roy. Inst. Proc.,* 3 (1858–62), pp. 387–96, in which he gave full credit to Kirchhoff for discovering the relationship between the dark lines in the solar spectrum and the bright lines in the spectra of heated metals.
5. *Philosophical Club*: an elite dining club of the Royal Society, established after the reforms to the Royal Society in the 1840s, in addition to the existing Royal Society Club. Tyndall had been elected in 1855. See T. E. Allibone, *The Royal Society and its Dining Clubs* (Oxford: Pergamon, 1976).
6. *Foucault's experiment*: Stokes had written a letter to the *Philosophical Magazine* in which he reported an observation that Foucault had made to him: 'On the Simultaneous Emission and Absorption of Rays of the same definite Refrangibility; being a translation of a portion of a paper by M. Léon Foucault, and of a paper by Professor Kirchhoff', *Phil. Mag.*, 19 (February 1860), pp. 193–97. Foucault's paper was published in *L'Institut* for 7 February 1849, having been brought forward at a meeting of the Philomatic Society on 20 January. Kirchhoff's communication 'On Fraunhofer's Lines', dated 20 October 1859, was made to the Academy of Sciences in Berlin on 27 October and printed in the *Monatsbericht,* p. 662.
7. *Fraunhofer's lines*: see letter 1595, n. 10.

From George Gabriel Stokes 22 May 1861 1784

Lensfield Cottage Cambridge | 22nd May 1861

Dear Dr Tyndall,

I have only time at present just to say that there is no doubt reflection on the surfaces of the globules of water would produce the effect you describe.[1] To you who had seen the thing the explanation may have appeared obvious; but when I read of precipitation I pictured to my mind's eye moistened ends and a few sparse globules. I think the word dim or dimness introduced in correcting the proof[2] would lead the reader instantly and infallibly to draw in his mind a correct picture of the phenomenon, and then there can be no question that the assigned cause is at least a vera causa whether or not it be the actual cause which produced the difference in the results.

Yours very truly | G. G. Stokes

RI MS JT/1/S/223
RI MS JT/1/TYP/4/1372

1. *the effect you describe:* see letter 1782, but this also implies that there may be a missing letter from Tyndall.
2. *dim or dimness introduced in correcting the proof:* on the assumption that this relates to the precipitation of water, Tyndall did not introduce the word 'dim or dimness' as suggested.

From James Joseph Sylvester[1] 26 May 1861 1785

26th May | 1861

My dear Sir,

Mr Derffel[2] formerly assistant professor of physics and mathematics under Petzval[3] at Vienna now earning his bread in England as a teacher of music in which he is eminent as a composer and performer is anxious to obtain cards of admission to the lectures at the Royal Institution on the Friday evenings.

As a scientific foreigner if this privilege could be accorded to him I should be much obliged by your obtaining it for him. He is an excellent mathematician to my certain knowledge. William Francis whom you well know was the person who originally introduced him to me. If you are able to comply with this intercession, you would perhaps be kind enough to furnish him with the means of admission by a communication made to him—direct at his lodgings 23 Queen St May Fair. Englishmen are well treated by their scientific friends abroad. I am sure you will agree with me that we ought to endeavour to return in kind the liberality we meet with especially in the case of a man like Derffel thrown out of his proper sphere by an adverse current of circumstances.

Believe me | Yours very truly | J. J Sylvester

RI MS JT/1/S/291
RI MS JT/1/TYP/4/1508

1. *James Joseph Sylvester:* (1814–97), professor of mathematics at the Royal Military Academy at Woolwich from 1855–70.
2. *Mr Derffel:* Joseph Derffel (1823–84), pianist, composer and physicist. Formerly a private

lecturer of higher mathematics and physics at the Polytechnic School of Vienna, before becoming assistant professor of physics and mathematics at the University of Vienna. Derffel resigned from this position to follow his interest in music. He left Vienna for England, where he became a well-known pianist and composer (*Niederrheinische Musik-Zeitung,* 13 (March 1865), pp. 69–70).

3. *Petzval:* Joseph Petzval (1807–91), Professor of Mathematics at the University of Vienna from 1836–77.

From Heinrich Debus 2 June 1861 1786

Queenwood College. | near Stockbridge. | Hampshire. | June 2$^{\underline{th}}$ 1861.

My dear John,

Although I have nothing in particular to say yet I feel inclined to write, and so I sit down in order to send you a few lines.

The other day whilst walking up and down the corridor, before going to lecture, I stepped into N° 10[1] and there I found on the table the Roy. Inst. lectures delivered by you and others some years ago.[2] Naturally, I turned to your lecture, and when I read the old familiar style I took the book away with me and read the whole of your work and Faraday's.[3] I feel transported to days gone by and pictures and images, thoughts and arguments, which had almost been obliterated from my mind began to revive and regain their original freshness. I was pleased with your conclusion and thought it eloquent. When I hear such questions mentioned, as: Is language an original gift of the Creator, or is it an artificial production of the human mind, &c I cannot help thinking that men who can propose such questions are not of the highest calibre. And if I come to consider the value, pains and time devoted to philosophical stories I must think that people do not know exactly the requirements of the age, the country or the mind. Language is a <u>Form</u>, a mode, how the mind expresses the results of its labors so as to be intelligible to other men. I am certain that it has as such its laws, like Logic and other abstractions from the working of our nature, and that it is a connecting link between mind, matter, and mind. But the realities of the world, the substance created, the things that be, their motions, transitions, relations, in fact the substance of all, is one Kingdom, it is the land where mathematics and the sciences hold their sway, and where twattle has no mask to delude the world. Nature is the mind of the Creator revealed to mortal eye, read it, be its true exponent, and you shall know more of Him then all the schools can teach you.

I saw a few days ago your name in the paper as one of the guests of my

Lord Mayor at the Mansion House.[4] Then it has come to this at last, you have turned a companion of Aldermen and turtle-soup[5] and are reckoned one of the great of the land. To be serious, I liked to see that my Lord Mayor recognizes such a body as the Roy Soc and the Roy Acad. It is a great sign and will no doubt bring good fruit. After this week, which commenced to day, our examination commences, and another half year has gone down in the gorge of time. I have not yet decided where to go during the holidays. If it is possible for you to arrange, then we must have another day together either in this part of the country or when I come to London. George has quite recovered[6] and appears as jolly as usual. Do you feel inclined to come to the concert?[7]

Now I have written you a long letter just as thoughts appeared in my mind and you must receive its shortcomings in an indulgent manner. With my kind remembrance to Tom

yours always | H. Debus.

RI MS JT/1/D/29

---

1. *No 10*: not identified, but perhaps one of the classrooms. Plans of Queenwood College have not been found.
2. *the Roy. Inst. lectures delivered by you and others some years ago*: this series of seven weekly lectures, given between 29 April and 10 June 1854, were a consequence of Faraday's intense concern about the popularity of spiritualism and table-turning. He criticised the system of education that could produce such credulity, and commissioned with Bence Jones a series of seven Lectures to highlight the wider issues. The series was opened by William Whewell, speaking on 'The Influence of the History of Science upon Intellectual Education', *Lectures on Education, delivered at the Royal Institution of Great Britain* (London: J W Parker & Son,1854).
3. *your work and Faraday's*: Tyndall gave his lecture 'On the Study of Physics as a Branch of Education' on 20 May 1854 (see ref n. 2, pp. 171–211). Faraday's lecture 'Observations on Mental Education' was given two weeks earlier, on 6 May (see ref. n. 2, pp. 39–88).
4. *one of the guests of my Lord Mayor at the Mansion House*: Tyndall dined at the Mansion House on 25 May 1861 (Journal, RI MS JT/2/13c/1217).
5. *turtle-soup*: a reference to the reputation at big dinners of guests dining on turtle-soup.
6. *George has quite recovered*: possibly George William, son of George Edmondson, headmaster of Queenwood College.
7. *Do you feel inclined to come to the concert?*: the date of the concert, which ended the term at Queenwood and which Tyndall often attended, is unknown. Though Tyndall went to the Isle of Wight from 16–19 June, he does not mention going via Queenwood in his journal.

From George Gabriel Stokes 7 June 1861 1787

Lensfield Cottage Cambridge | 7 June 1861

My dear Dr Tyndall,

I returned home last night after an absence of about 10 days. I would have written again in answer to your further questions and remarks[1] only that I thought I should have met you at the R.S. and should have been able to speak about it.

You asked me one explicit question which I have not yet answered.

I do not recollect my conversation with you at the Phil Club. I first heard of Foucault's experiment[2] from Foucault himself whom I met at dinner at Dr Arnots[3] when he came to London to receive the Copley Medal.[4] I was greatly struck with it. He told me that no one in Paris seemed to think anything of it except Babinet.[5] Believing the line to be due to soda[6] I asked him about the poles and he said that they were of carbon and that there had been various salts on the poles. In a conversation shortly afterwards with Thomson I informed him of Foucault's exper^{t}. When he asked me how I accounted for the appearance of a bright and dark line in the same place I gave the explanation you will find in the note in the Phil Mag Vol. 19 p. 196[7] and said I believed it was due to sodium. He asked me if there were any other examples and I mentioned an observation of Sir David Brewster (Rep. Brit. Ass.)[8] who found bright lines in the spectrum of burning potassium (or rather deflagrating nitre) agreeing exactly with certain dark lines in the red in the solar spectrum. Thomson if I recollect right at once jumped to the conclusion that we should find all or most of the solar lines coincident with bright lines in artificial flames.[9] I was not prepared to leap so far but thought that many of the non-atmospheric lines might be due to gaseous substances in the solar atmosphere having the selective absorbing power of NO_4, I, Br &c I never published anything on the subject nor did I look on my remark as more than a (to me) obvious inference from a very remarkable discovery due to Foucault.

I agree with you that the coincidence in the n$^{\underline{os}}$ for radiation and conduction is too remarkable to be accidental. I am not however satisfied with the explanation[10] but we shall I hope have time to talk about this.

Yours very truly | G.G. Stokes.

RI MS JT/1/S/224
RI MS JT/1/TYP/4/1373–73a

1. *your further questions and remarks:* letter 1783.
2. *Foucault's experiment:* see n. 7 and letter 1783, n. 6.

3. *Dr Arnots:* Neil Arnott (1788–1874), Scottish physician, natural philosopher and inventor, and author of a successful textbook *Elements of Physics,* published from 1827 onwards.
4. *Copley Medal:* the premier medal of the Royal Society, awarded to Foucault in 1855.
5. *Babinet:* Jacques Babinet (1794–1872), physicist.
6. *soda:* sodium.
7. *Phil Mag Vol. 19 p. 196:* see letter 1783, n. 6.
8. *an observation of Sir David Brewster (Rep. Brit. Ass.):* D. Brewster, 'On Luminous Lines in Certain Flames corresponding to the Defective Lines in the Sun's Light', *1842 Brit. Assoc. Rep. Transactions of the Sections,* 12 (1843), p. 15.
9. *we should find all or most of the solar lines coincident with bright lines in artificial flames:* see letter 1783, n. 4 and n. 6.
10. *I am not however satisfied with the explanation:* see letters 1780, n. 6, and 1781, n. 8.

To Juliet Pollock[1] 17 June 1861 1788

17th June 1861.

The sea holds jubilee this sunny morn
And I with heart content upon its verge
Join in the laughter of the breaking waves.
And glad, right glad the sympathetic land,
Shaking her hazel tresses in her mirth
While all her copses tremble into song.
The thickset trees which crowd the Undercliff
The scented woodbine on the neighbouring knoll—
The foxglove shaking all its purple bells—
And roses blushing mid the tender green—
All blend into a bouquet for the sight:
But not for sight alone, for beauty sends
Its finer essence down into the heart.
Lady! my friend—thou surely wilt not frown,
If lingering here I miss that other joy
Of meeting thee and thine to-morrow night!

J.T. | Monday 17th June 1861.

RI MS JT/1/T/1538
RI MS JT/1/TYP/6/1978

---

1. Tyndall noted this poem he sent to Juliet Pollock in his journal (RI MS JT/2/13c/1219). He had gone to the Isle of Wight on 16 June. He returned on 19 June and slept for the first time in the house at 14 Waverley Place that he shared with Thomas Hirst.

To Miss Guthrie[1] 20 June 1861 1789

I know the ice-machine of Harrison;[2] wherein the ice is produced by the vaporization of ether. It is truly an excellent apparatus, but I was not aware that he had reduced it to domestic dimensions. Excuse this official sheet: I am in the midst of the Council of the Royal Society[3]—listening even at this moment to the secretary reading the minutes of the last meeting.

Maggs Auction Catalogue 2006, 1390, no. 179

1. *Miss Guthrie:* probably Anne Leonora Guthrie (1807–93). She was the half-sister of Mary Dawes and lived at 4 Berkeley Street until moving in 1858 to Pool House, Hereford. See N. Denison, *Richard Dawes: Education Pioneer and Dean of Hereford* (The Somborne & District Society, 2017). This is a fragment of a letter, taken from an auction catalogue.
2. *the ice-machine of Harrison:* James Harrison (1816–93), the Scottish-born and Australia-domiciled inventor who later tried unsuccessfully to launch a refrigerated meat trade from Australia. Harrison had made his ice machine available to Tyndall for his experiments relating to the structure and motion of glaciers. See 'On Some Physical properties of Ice', *Phil. Trans.*, 148 (1858), pp. 223–24.
3. *I am in the midst of the Council of the Royal Society:* Tyndall was a member of the Council from 1860–2, his second of four tenures.

From Edward Sabine 21 June 1861 1790

June 21. 1861.

Dear Tyndall.

I find an account of the comparative retardation of the Pendulum in Hydrogen Gas and Atmospheric Air in the Phil Trans. for 1829 p 227. (Exp[t] Feb[y] 9 & 10)[1]—I made similar experiments in other gasses, but cannot at the moment refer to the account of them. (Olefiant Gas[2] was one of those experimented upon.)

On the subject of the facts of the solar diurnal magnetic variation you will find the facts tolerably lucidly stated in the Introduction to the 2[d] vol of the St Helena Obs[ns] p.p. cxvii to cxxi.[3]

You asked me some time ago about the annual variation of the Total magnetic force of the Earth.

You will find in page CXXIV of the 3[d] vol. of the Toronto Obs[ns],[4] the results showing that at that station in the Northern hemisphere the Total Force is decidedly greater in the months of November to February than in May to August—And in the 1st vol, Hobarton p. lxxxvi[5] the conclusion that at that station in the Southern hemisphere the Total Force is also greater from October to February than from May to August. The latter conclusion was drawn from the Obs[ns] which I possessed when that vol. was printed: I have subsequently received more from Hobarton, and have them now preparing for publication.

I purpose to enter then more fully into the question as regards Hobarton, tho' I have no reason to suppose that the conclusion will be other than the one referred to in vol. 1—but it will probably be in my power to state somewhat more precisely for Hobarton (as I have done for Toronto) the approximate precise value of greater intensity of the Earths Magnetism in December than in June.

faithfully yours | Edward Sabine.

RI MS JT/1/S/23
RI MS JT/1/TYP/4/1320

---

1. *in the Phil Trans. for 1829 p 227. (Expt Feby 9 & 10)*: E. Sabine, 'On the Reduction to a Vacuum of the Vibrations of an Invariable Pendulum', *Phil. Trans.,* 119 (1829), pp. 207–35.
2. *Olefiant Gas*: ethene, a gas that was of great interest to Tyndall in his radiant heat experiments, as it was a strong absorber of heat.
3. *Introduction to the* 2d *vol of the St Helena Obs*ns: *Observations made at the Magnetical and Meteorological Observatory at St. Helena with discussions of the observations at St. Helena, The Cape of Good Hope, The Falkland Islands, Carlton Fort in North America, and Pekin. Vol. II. 1844 to 1849* (London: Longman, Brown, Green, and Longmans, 1860), pp. cxvii–cxxi.
4. 3d *vol. of the Toronto Obs*ns: *Observations made at the magnetical and meteorological observatory at Toronto in Canada. Vol. III. 1846, 1847, 1848* (London: Longman, Brown, Green, and Longmans, 1857), p. cxxiv.
5. *1*st *vol, Hobarton*: *Observations made at the magnetical and meteorological observatory at Hobarton, in Van Diemen Island and the Antarctic Naval Expedition. Vol. I. Commencing with 1841. With abstracts of the observations from 1841 to 1848 inclusive* (London: Longman, Brown, Green, and Longmans, 1850), p. lxxxvi. Sabine appears to have given an incorrect reference. Variations in the Total Force are described on pp. lxvii–lxxii.

From William Snow Harris 26 June 1861 1791

6 Windsor Villas | Plymouth | 26 June 1861

Dear Tyndall,

Fearing you might be induced to leave London, and having more regard for you, than for myself relative to your visit here. I sent on a telegram to prevent your coming away to morrow. The Westerly channel drift has set in upon us since I last wrote to you, with very unsettled weather—there is much sea outside the headlands and a strong S. W. wind: occasionally fine—I saw a new gale springing up this morning. Barometer rapidly falling—so I sent the telegram. Now, 3 p.m., we have it home, heavy rain and wind.

I need not assure you how terribly disappointed I am at not seeing you and so we all are. A very celebrated person Madame Thellon,[1] one of the greatest Artistes in Europe, is to dine with us tomorrow and I looked to a pleasant evening for you previous to our starting along the coast Friday—well if all this is <u>not</u> on the cards—we must get a new deal.

Now I think you really must endeavor so to manage matters as to make me compensation for all this terrible disappointment and see what you can do to carry out your charitable intention to spend a few days with me. The weather will again settle in fine in a day or two I dare say—About the 6th July you would see the launch of a fine ship if you could come about that time. Perhaps you could arrange for a few days before you go to the Continent[2] I am certain you would be greatly charmed with this place and its vicinity.

Always sincerely and truly yours | W. S. Harris

Professor Tyndall | F.R.S. &c.

RI MS JT/1/TYP/2/475
LT Transcript Only

---

1. *Madame Thellon*: Madame Thellon was an English female vocalist best known for her singing at the Théâtre de la Renaissance in Paris. See *The Theatrical Observer; and Daily Bills of the Play,* no. 5377 (15 March 1839).
2. *Perhaps you could arrange for a few days before you go to the Continent*: Tyndall left for Switzerland on 31 July 1861, and does not seem to have made a trip to the West Country prior to that to visit Snow Harris.

From Johann Christian Poggendorff 28 June 1861 1792

Berlin 28 Juni 61.

Theuerster Freund!

In Folge Ihres letzten Briefes habe ich Ihnen durch Asher gleich ein Paar Abdrücke von der Übersetzung Ihrer schönen Abhandlung gesandt, einerseits um Ihnen zu zeigen, daß ich die Abhandlung wirklich in die Annalen aufgenommen habe, und andrerseits um Ihnen Gelegenheit zu geben etwas einige Übersetzungsfehler zu verbessern und mir anzuzeigen. Ich will hoffen, daß Sie diese Abdrücke richtig erhielten, und zugleich, daß Sie nicht viele und keine groben Fehler finden mögen.

Diejenigen, welche Sie mir anzeigten, könnte ich natürlich nicht wieder verbessern, da die Abhandlung beim Empfange Ihres Briefes bereits gedruckt war. Ich werde diese Fehler jedoch am Schluße des Bandes anzeigen.

Ich habe übrigens Ihre schöne Arbeit mit Vergnügen gelesen, und wenn ein solches Zusammentreffen wie das mit Freund Magnus allerdings für die betreffenden Personen auch gerade nicht angenehm seyn kann, so muß es doch für die Wissenschaft für einen Gewinn angesehen werden, wenn wichtige Resultate sogleich mehrseitig bestätigt werden.

Daß meine Frau Ihr interessantes Buch richtig, wenn gleich etwas verspätet erhielt, werden Sie aus deren eignen Zeilen ersehen. Auch ich habe mir erlaubt, etwas hineinzulesen und mich an seinem Inhalt erfreut. Es ist uns beiden eine liebe Erinnerung sowohl an die Schweiz, als vornehmlich an Sie selbst. Möchten Sie uns doch bald einmal wieder mit einem Besuch erfreuen. Es ist bei uns so schlimm nicht, wie die Times ihren Lesern glauben machen will!

Wenn mir der Himmel bis nächstes Jahr Leben u. Gesundheit erhält, so komme ich vielleicht zur nächsten Industrie-Ausstellung nach London und es würde mich dann natürlich ungemein erfreuen, wenn ich Sie u andere Freunde dann sehen könnte.

In Hoffnung darauf, | ganz der Ihre | Poggendorff

Berlin 28 June 61.

Dearest friend!

Following your last letter,[1] I have sent some copies of the translation of your splendid article directly through Asher,[2] on the one hand to show you that I have really included your paper in the Annalen,[3] on the other hand to give you the opportunity to correct and inform me about any translation mistakes. I hope that you received these copies well and also that you may not find many mistakes, let alone substantial ones.

Naturally, I can no longer correct those that you might indicate to me, as the paper will already have been published by the time your letter arrives. However, I will indicate those errors at the end of the volume.

By the way, I have read your fine work with pleasure, and if a clash such as that with friend Magnus cannot be exactly pleasant for the people involved,[4] one must see it as a gain for science when important results are confirmed by different sides.

That my wife received your interesting book,[5] albeit with some delay, you will see from her own lines. I also allowed myself to read a little of it and took pleasure in its contents. To both of us, it is a lovely memory of Switzerland and especially of you. May you soon delight us again with a visit. It is not as bad here as the Times wants its readers to believe![6]

If heaven keeps me alive and healthy until next year, I might attend the next industrial exhibition in London[7] and naturally I would be tremendously delighted if I could see you and other friends then.

In anticipation, | entirely your | Poggendorff

RI MS JT/1/P/130

1. *your last letter:* letter missing.
2. *Asher:* Adolphus Asher.
3. *I have really included your paper in the Annalen:* see letter 1767, n. 2. See also letters 1771 and 1793.
4. *and if a clash such as that with friend Magnus cannot be exactly pleasant for the people involved:* this dispute over priority for the discovery of absorption of radiant heat by gases, and the role of water vapour, ran for many years but was marked by cordiality on both sides.
5. *your interesting book: Glaciers of the Alps.*
6. *It is not as bad here as the Times wants its readers to believe!: The Times* published two letters from its readers in early May, in which it was suggested at length that English travellers in Prussia were 'in the position of a prisoner of war in a hostile land'. Writing from Berlin, it appears that Poggendorf is referring to these horror stories: Viator. 'English Travellers in Prussia.' *The Times,* (9 May 1861), p. 6, and E.S. 'The English in Prussia.' *The Times,* (10 May 1861), p. 9.
7. *I might attend the next industrial exhibition in London:* the 1862 International Exhibition was held in London from 1 May to 1 November 1862. Poggendorff and his daughter came over and were hosted by Tyndall at 14 Waverley Place (see 29 September 1862, Johann Poggendorff to Tyndall, RI MS JT/1/P/133).

To Rudolf Clausius 29 June 1861 1793

My dear Clausius,

It is a very long time since a letter has passed between you and me,[1] and I believe I am to blame for this long silence. I have many times thought of writing to you; but I have been so hard worked and my health at intervals has been so poor, that when I could write I had not the time, and when I had the time I was too unwell. Now however the weight of my work has been accomplished and I lose no time in endeavouring to make atonement for my neglect of duty.

I ought long ago to have sent Prof. Escher his book[2]—indeed I am quite ashamed when I think of the great length of time which I have kept it from him. But the fact is that the discussions of the glacier question interested a great many scientific men in England, and made them anxious to become acquainted with Rendu's memoir. For their sakes I delayed sending it. It was also my intention to make use of it in some remarks which I thought of writing with reference to Prof. Forbes's last pamphlet:[3] but I dislike personal discussion so much that I thought it better not to publish any formal answer to Forbes, but to wait until my book should reach a second edition,[4] and then make my statement. Tomorrow, or the day following I hope to send away Prof. Escher's volume, and I hope you will have the kindness to return my warmest thanks to him for his great kindness in permitting me to keep it so long.

Along with it I will send three copies of my book on the Alps,[5] one for you, one for Mousson[6] and one for Escher; I hope they will be accepted as a testimony of my sincere regard, and that I shall be excused the neglect of not having sent them many months ago.

I had some correspondence with Wiedemann and Vieweg[7] regarding a German translation of the book: but like every thing else I have neglected answering the last letter which I received upon the subject, and thus it remains undecided.

During the past half year I have been able to do but little original work; my lectures having occupied all my time. But I have just sent in the proof of my paper on the Absorption and Radiation of heat by gases and vapours.[8] A translation of it will soon appear in Poggendorff,[9] and I hope it will interest you when you read it. What splendid things Kirchhof and Bunsen have been doing of late![10]

Magnus it appears has been working at the same subject as I have. I had a letter from him some time ago[11] which appeared to have been written under feelings of disappointment. If I understood him right he thought that his results were obtained before mine. The fact is that all my results were obtained in 1859, a year before Magnus touched the subject: I took possession of the field by the note I published and circulated in France, Italy, Switzerland and Germany;[12] and having done so waited until my apparatus had reached a high degree of perfection before I published in full. So that as regards both my published and my unpublished work I was first in the field, and indeed had no idea that any body else had entered it. The results of Magnus are of great value, for a subject so important ought not to be permitted to rest upon the testimony of any single individual.

I have given up the glaciers for the present, but have not given up Switzerland. Early in August I start;[13] and hope after spending some time at the Aeggischorn[14] to go on to the Monte Rosa district, and perhaps try to climb the Matterhorn.[15] In company with a friend I attempted it last year,[16] and had we time I think we should have surmounted it. This year it will probably be ascended before I arrive, as several of the best climbers in England start in July to make the attempt.

I gave a lecture some days ago on Kirchhof's researches[17] and managed to make the reversal of the sodium line manifest to 500 people. I first steeped my coal points in chloride of sodium: they gave the yellow band on an otherwise continuous spectrum. In front of the electric lamp I placed a hot flame, and in the flame a bit of sodium, it was volatilysed, and cut out the yellow line, leaving an intense dark band in its place. It is quite a glorious experiment: according as the sodium flame was absent or present the yellow line or the black line appeared.

Goodbye | Kindest greetings to my Zürich friends | & believe me | very sincerely yours | John Tyndall | 29th June 1861

RI MS JT/1/T/180

---

1. *It is a very long time since a letter has passed between you and me:* the last extant letter from Tyndall to Clausius is 1664, dated 7 December 1859.
2. *I ought long ago to have sent Prof. Escher his book:* Arnold Escher von der Linth. The book referred to may be a copy of Rendu's *Théorie des Glaciers de la Savoie* (see letter 1708, n. 4), which Clausius sent to Tyndall after Mousson (see n. 6) had taken it from Escher von der Linth's library in his absence. See also letter 1619, n. 13.
3. *Prof. Forbes's last pamphlet:* see letter 1731, n. 4.
4. *until my book should reach a second edition:* in the event a second edition was not published until 1896, after Tyndall's death.
5. *three copies of my book on the Alps: Glaciers of the Alps,* see also letter 1595, n. 4.
6. *Mousson:* Albert Mousson (1805–90), professor of physics at the University of Zürich.
7. *I had some correspondence with Wiedemann and Vieweg:* letters missing.
8. *I have just sent in the proof of my paper on the Absorption and Radiation of heat by gases and vapours:* Tyndall (1861).
9. *A translation of it will soon appear in Poggendorff:* see letter 1767, n. 2. See also letters 1771 and 1792.
10. *What splendid things Kirchhof and Bunsen have been doing of late!:* Tyndall is referring to their work on spectra. See letter 1747, n. 3.
11. *I had a letter from him some time ago:* letter 1771.
12. *the note I published and circulated in France, Italy, Switzerland and Germany:* France: 'Sur la transmission de la chaleur de qualité différente à travers les diverses espèces de gaz, par M. John Tyndall', *Cosmos,* 15 (1859): p. 321–25; Italy: 'Della trasmissione del calore di diverse qualita attraverso al diversi gas, di J. Tyndall', *Il Nuovo Cimento,* 10 (1859): p. 196; Switzerland: 'Sur la diathemansie des gaz, Lettre de M. John Tyndall a M. le prof. Auguste de la Rive', *Bibliothèque Universelle,* 5, (1859): p. 232 (letter 1617). The German note was not published until 1861, hence explaining why Magnus may not have been aware of it: 'Note on the transmission of radiant heat through gaseous bodies', *Die Fortschritte der*

Physik im Jahre 1859, 15 (1861), pp. 368–69. A summary also appeared in *Journal de Pharmacie et de Chimie,* 37 (1860), pp. 204–8 and in *Phil. Mag.*, 19 (1860), pp. 60–61.
13. *Early in August I start:* Tyndall left on 31 July 1861.
14. *after spending some time at the Aeggischorn:* The Eggishorn stands by the Aletsch Glacier.
15. *perhaps try to climb the Matterhorn:* Tyndall was unsuccessful in 1861. His guide, Johann Bennen, declared the attempt impossible, and they did not even set foot on the mountain.
16. *In company with a friend I attempted it last year:* see letter 1726, n. 13.
17. *I gave a lecture some time ago on Kirchhof's researches:* see letter 1783, n. 4.

To Heinrich Debus [June 1861][1] 1794

Royal Institution, Saturday

My dear Heins

Excuse me for not having written to you before but I have been dreadfully busy and indeed I overlooked your question[2] about the books of the Royal Society.

As Fellow[3] you are entitled to the loan of the books of the library, and also entitled to give a note to a friend so that he also may borrow the books.

Write therefore to Walter White Esq[r] Assistant Secretary and he I doubt not will send the books to you at once.

Do <u>not</u> buy a Magneto-electric machine—you can illustrate the subject without it. You can make small coils and use the coils of your magnet. With a good sized galvanometer you can thus render the phenomena of induced currents visible to a large audience.

If you have <u>lots</u> of money buy a Magneto electric machine, but if you have not <u>lots</u> buy something else.

Yours ever | John Tyndall

RI MS JT/5/11/1365
LT Transcript Only

---

1. *[June 1861]:* it is assumed this letter follows the election of Debus as Fellow of the Royal Society (see n. 3).
2. *overlooked your question: letter missing*
3. *As Fellow:* Debus was elected FRS on 6 June 1861. See letter 1779 in which Tyndall notes that the Royal Society had formally recommended Debus's election at a Council meeting on 18 April 1861.

To the Lord President and Vice-President of the Committee of Council on Education[1] 1 July 1861 1795

MY LORDS,

I HAVE this day re-perused the Report of the Illumination of the Picture Galleries of the South Kensington Museum,[2] bearing date the 20th of July 1859, and beg to state that I see no reason to deviate from the opinion embodied in that report.

With reference to the question raised by the letter of Mr. Braidwood,[3] to which my attention has been directed, I have to state that, with the observed temperatures of those portions of the gallery which are most exposed to the action of the gas before me, it is my opinion that no danger can arise from such elevations of temperatures as the gas applied as at present is competent to produce.

Such elevations of temperature are indeed very much less than those effected by the solar heat, in a room roofed by glass, upon an ordinary summer's day. In presence of these facts, I do not hesitate to express the decided opinion that no danger is to be apprehended from the mode of illumination now in use at South Kensington.

I have, &c. | (Signed) JOHN TYNDALL.

Thursday, 11th July 1861.

Report of the Commission on the Heating, Lighting, and Ventilation of the South Kensington Museum together with minutes of evidence and appendix [4206], *House of Commons Papers,* (1868–69), p. 90.

---

1. *Lord President and Vice-President of the Committee of Council on Education:* The Committee of the Privy Council on Education was created on 10 April 1839 'to superintend the application of any sums voted by Parliament for the purpose of promoting public education'. The Lord President was Granville George Leveson-Gower, 2nd Earl Granville.
2. *Report of the Illumination of the Picture Galleries of the South Kensington Museum:* [106] Lighting picture galleries by gas. Report of the commission appointed to consider the subject of lighting picture galleries by gas, *House of Commons Papers,* 15 (1859 Session 2). Tyndall had been a member of the Commission chaired by Michael Faraday, with August Hofmann, Richard Redgrave and Capt. Francis Fowke, RE. It concluded that 'there is nothing innate in coal gas which renders its application to the illumination of picture galleries objectionable'.
3. *the question raised by the letter of Mr. Braidwood:* James Braidwood, Superintendent of the London Fire Engine Establishment, had written to the principal librarian of the British Museum on 11 April suggesting that it was too risky to light the galleries of the British Museum by gas: British Museum (Lighting by Gas): Copy of any reports made to Trustees

of the British Museum with reference to lighting that institution by gas, and any resolution of the Trustees in relation thereto, *House of Commons Papers,* 34 (1861). The Trustees passed a resolution on 13 April saying they were of the opinion that the Museum should not be open at any hour that required gaslight.

From John Frederick William Herschel 21 July 1861 1796

Collingwood July 21. 1861.

My dear Sir

Accept my best thanks for your valuable paper the Bakerian Lecture on the diathermancy &c of Gases[1]—which is an immense step in the Physics of Aeriform Bodies. —Also for your two lectures on Scientific teaching and on Solar Chemistry.[2]

On this latter allow me to regret that while naming Talbot & Miller,[3] as having early described the bright bands peculiar to certain flames you have omitted to notice the results which I obtained as early as 1822 and described in a paper printed in the Transactions of the RS Edin[4] for that year and the further indications of a similar nature in §'s 520–524 of my treatise on Light in the Encycl. Metropol published in 1827.[5] In that Treatise §520 mention is made of Brewster's & Talbot's results and those of 'others' (pursuant to a principle adopted throughout that article of never citing my own works—but in some such oblique form of expression). Of the indications given in the subsequent Sections, whatever is not to be found in the then published papers of Brewster & Talbot is my own—and the expression used in §524 'the colours thus communicated by different bases to flame afford in many cases a ready & convenient way of detecting extremely minute quantities of them &c' taken in conjunction with the distinct announcement of <u>Bright lines</u> <u>appropriate to certain of them</u> might have led an enquiring person at an earlier period than the present to much that is now so strikingly brought out. I was then so circumstanced as to be unable to prosecute the enquiry but I always considered it a very promising one—and you will excuse me if I say that I do think it a little hard, seeing that <u>every</u> other name connected with the subject—Brewster, Talbot, Miller, Masson, Willingen, Plücker, Wheatstone, Foucault, Stokes, Thomson,[6] is mentioned in this admirable lecture (which must have created quite a sensation and will be accepted as a resumé of the history of the subject) mine should never once have been alluded to. —

On the general subject of Kirchoff's theory of the <u>absorption</u> of the peculiar coloured light of a body by the flame of that body there occurs to me a remark which may or may not be worth your consideration. —You (or Kirchoff) assimilate this absorptive action to that of glass on heat emanating from bodies heated below incandescence. —You say that it only does not transmit the rays because it disperses them—i.e. so much is scattered quaquaversum that little is shot forward directly. —Granted. —But apply this to the Sun—quaquaversum means in all directions outwards into space. (Inwards & backwards also—but this will be re-absorbed & radiated out again—so that ultimately all will find its way out in some direction or other). —Now the totality of these quaquaversum radiations amounts to a radiation quaquaversum of the totality—ie to an emanation some how or other (however circuitously) from the Sun's general spherical surface equally in all directions into Space of the whole light just as if it had never been absorbed.

There is a little paper w[hic][h] I sent to the Lond & Edin Phil Mag printed in their 3[rd] series Vol 3 N° 18 Dec. 1833[7] on the absorption of light (by coloured media) which perhaps may be found to contain something in point on this matter.

Excuse this rambling letter & believe me yours very t[rul]y | JFW Herschel

PS.[1] Vegetable colours are bleached by those rays which they reflect most abundantly not those which they absorb in preference

P.S.[2] Is it not rather startling to assume that molecules which are particularly fitted to vibrate in one rigidly definite time & which are in the air of such vibration should yet par excellence stifle & destroy vibrations in that identical definite time when propagated into them ab extra? It is clear there is a link wanting in the chain of the argument.

RI MS JT/1/H/94
RI MS JT/1/TYP/2/505–6
Royal Society Letters (RMD 5 August 2011), 23_335 (this is Tyndall's own transcription)

---

1. *your valuable paper the Bakerian Lecture on the diathermancy &c of Gases:* Tyndall (1861).
2. *your two lectures on Scientific teaching and on Solar Chemistry:* the first is probably Tyndall's 'Elementary Magnetism', a lecture to primary school teachers given at South Kensington on 30 April 1861. It was published in the 4th and subsequent editions of *Fragments of Science.* The second is his discourse on 7 June 1861 'On the Physical Basis of Solar Chemistry' (see letter 1783, n.4).
3. *Talbot & Miller:* see W. H. F. Talbot, 'Some Experiments on Coloured Flames', *Edin. J. Sci.,* 5 (1826), pp. 77–81. William Allen Miller (1817–70) made many discoveries in optical experiments.
4. *paper printed in the Transactions of the RS Edin:* J. F. W. Herschel, 'On the Absorption of Light by Coloured media, and on the Colours of the Prismatic Spectrum exhibited by certain Flames; with an Account of a ready Mode of determining the absolute dispersive Power of any Medium, by direct experiment. In a letter to Dr Brewster, Read November 18. 1822', *Transaction of the Royal Society of Edinburgh,* 9 (1823), pp. 445–60.

5. *my treatise on Light in the Encycl. Metropol published in 1827:* J. F. W. Herschel, 'Light', *Encyclopedia Metropolitana,* 4, (1845), pp. 341–586. Written and circulated in 1827.
6. *Brewster, Talbot, Miller, Masson, Willingen, Plücker, Wheatstone, Foucault, Stokes, Thomson:* David Brewster, William Henry Fox Talbot (1800–77), William Allen Miller, Antoine Philibert Masson (1806–50), Wolkert Maarten Simon van der Willingen (1822–78), Julius Plücker (1801–68), Charles Wheatstone (1802–75), Léon Foucault, George Gabriel Stokes and William Thomson. For two of Brewster's key papers, see 'Observations on the lines of the Solar Spectrum, and on those produced by the Earth's Atmosphere, and by the Action of Nitrous Acid Gas', *Transaction of the Royal Society of Edinburgh,* 12 (1834), pp. 519–30 and 'On a new analysis of Solar Light', *Edin. J. Sci.,* 5 (1831), pp. 197–206.
7. *a little paper w*h *I sent to the Lond & Edin Phil Mag printed in their* 3rd *series Vol 3 No 18 Dec. 1833:* J. F. W. Herschel, 'On the Absorption of Light by Coloured Media, Viewed in Connexion with the Undulatory Theory', *Lon. & Ed. Phil. Mag.,* 3 (1833), pp. 401–12.

From Richard Dawes 22 [July] 1861[1] 1797

Deanery, Hereford. | Sept. 22d 1861.

My dear Tyndall

Many thanks for your so kindly sending me now and then abstracts of your lectures, of which I have received two just now, and which I always peruse with great profit and with great pleasure. In fact you always interest me.

With,[2] of our Parish school, is much pleased by your sending him a copy of the one on Friday 7th of June at the Royal Institution.[3] He is a most valuable little man and doing us a great deal of good here.

And now when are we to have the pleasure of seeing you and Frankland[4] down here, it would give both Mrs Dawes and myself the greatest pleasure to get you down for a week when you are tired of work and a healthy excursion into the country would do you both good. We could run down to Merthyr Tydfil and see the mining district there, and to the neighbourhood of Brecon and what are called our black mountains, where you would both imagine yourselves in some Alpine district. At the end of this week I am going away for some ten days, but after that I believe we shall be here all the summer and autumn. At one time we had thought of going to the seaside for a short time but this wet weather makes us inclined to give it up. I can only say we do hope that you will contrive a visit here when it perfectly suits Frankland and yourself. Are you going to the Alps this summer? Write and tell me when it would suit you best, September or October or earlier, but recollect whenever you can come you will have a most hearty welcome. Both of you are old friends whom I highly esteem, and I don't want you to forget me.

Tell Frankland you have heard from me, but I most likely shall write to him today if not interrupted by a Chapter meeting. I am wondering how the alterations about Addiscombe will affect him—I hope for good, and that some permanent appointment for him may arise out of it.[5]

Mrs Dawes sends her kindest regards, and | Believe me, my dear Tyndall, | Always and sincerely yours | R. Dawes.

We have not seen London since July 1859.[6]

RI MS JT/1/TYP/1/322b
LT Typescript Only

1. *22 [July] 1861:* letter 1798, dated 24 July, seems to be the reply. The typescript gives the month as September, but the context of the reply, Tyndall's June Discourse and Dawes's enquiry about his movements 'this summer' imply that 22 July is likely.
2. *With*: George Henry With. See also letters 1753, 1754 and 1758.
3. *the one on Friday 7*th *of June at the Royal Institution*: the title of Tyndall's lecture delivered on that day was 'On the physical basis of Solar Chemistry'. See letter 1783, n. 4 and [Anon], 'Royal Institution', *Morning Post,* (11 June 1861), p. 3.
4. *Frankland*: Edward Frankland.
5. *alterations about Addiscombe . . . some permanent appointment for him may arise out of it*: Addiscombe House, near Croydon in Surrey, was the Royal Indian Military College of the East India Company that prepared cadets for the Engineers, Artillery, or Infantry. Frankland was appointed as lecturer in chemistry and physics in 1859. But following the transfer of the government of India from the Company to the Crown, after the Indian Mutiny, it was announced on 6 June 1861 that the College would close. It is not clear what permanent appointment Dawes had in mind, although Frankland subsequently increased his involvement with the RI, and became professor of chemistry in 1863. See C. A. Russell, *Edward Frankland: Chemistry, Controversy and Conspiracy in Victorian England* (Cambridge: Cambridge University Press, 1996), pp. 208–17.
6. This line is handwritten at the end of the typescript.

To Richard Dawes 24 July 1861 1798

Royal Institution | 24th July 1861

My dear Mr Dean,[1]

Thank you many times for your goodnatured letter.[2] At the present moment I am in the midst of my preparations for Switzerland, from which I hope to return with the strength of a giant. It is tremendously hard, believe me, to get away from London. Bread & butter considerations first cast a hundred entanglements round a man, his health drives him away periodically, and the remnant of his time he grasps like a miser, knowing that into it the real performance of his life is to be crammed. Be assured of it, it would give me great delight to join you at Hereford, but a week for me is like a five pound note from a man whose whole worldly possessions are limited to thirty pounds. Now just fancy yourself with thirty pounds in the world. Would you not clutch it with miserly tenacity? I wish you were nearer London—I wish I could run down to you on Saturday and run up again on Monday—But those scandalous Whigs in placing you at Hereford[3] diminished the pleasure of many a decent man. Still I will not say nay to your kind invitation—If I can find any means of satisfying my conscience that it is allowable I will run down to you for a week during the autumn.

Goodbye and present my very kindest remembrances to Mrs Dawes.

Believe me, dear Mr Dean | Yours ever truly | John Tyndall.

Private Collection, via Frank James

RI MS JT/1/TYP/1/322

1. *Mr Dean*: Richard Dawes was Dean of Hereford from 1850–67.
2. *Thank you many times for your goodnatured letter*: probably letter 1797.
3. *But those scandalous Whigs in placing you at Hereford*: The Whig government of 1846–52 was led by the Prime Minister Lord John Russell.

To John Frederick William Herschel 24 July 1861 1799

Royal Institution | 24th July 1861

My dear Sir John

Let me mingle with my thanks for your letter[1] an expression of sincere regret that my ignorance of the existence of your paper in the Edinburgh Transactions[2] should have caused me to overlook your rightful claims as regards the spectra of coloured flames.

I trust you will allow the fact of my having been compelled to prepare this lecture in the interval of other pressing duties to weigh with you in estimating the magnitude of my sin of omission.

A slight reparation is still in my power. The lecture is to be republished in the Philosophical Magazine for next month;[3] and I shall this day see the printers with a view of having due reference made to your researches.

Were it not that I am in the midst of my preparations for Switzerland,[4] and therefore scientifically out of gear it would give me pleasure to say something on the subjects to which you have drawn my attention. I think however I ought to also absolve Kirchhoff; for the responsibility of dealing with the subject in the manner indicated in the report rests entirely on my own shoulders.

Believe me | Most sincerely yours | John Tyndall

RS HS 17.388

1. *thanks for your letter:* letter 1796.
2. *your paper in the Edinburgh Transactions:* see letter 1796, n. 4.
3. *The lecture is to be republished in the Philosophical Magazine for next month:* 'Proceedings of the Royal Institution:–Prof. Tyndall on the Physical Basis of Solar Chemistry', *Phil. Mag.*, 22 (1861), pp. 147–56. Tyndall added a note on p. 156. He also penned an open letter to Herschel in October, published in the November issue (letter 1814).
4. *in the midst of my preparations for Switzerland:* Tyndall left for Switzerland a week later.

From Charles Darwin 24 July [1861][1] 1800

2. Hesketh Crescent, | Torquay | July 24.

My dear Tyndall,

I write a single line to say that it was not of the least consequence your retaining the letter[2] for as long as you liked

I shall be most happy at any time to send you the letters again, if you like to publish any passages.

We are all here idling for health-sake and enjoying the beautiful scenery and sea air.

By the way the other day at dinner we were all sticking bits of ice together by their points, marvelling at the phenomenon[3] and talking of you.

Yours most sincerely | Ch. Darwin.

RI MS JT/1/TYP/9/2803
LT Transcript Only

1. *24 July [1861]*: the date is taken from the Darwin Correspondence Project, DCP-LETT-3217, given by reference to the Darwins' stay in Torquay during the summer of 1861.
2. *retaining the letter*: see letter 1768, n. 2.
3. *sticking bits of ice together by their points, marvelling at the phenomenon*: see letter 1768, n. 4.

To William White Cooper[1] 25 July 1861 1801

25th July 1861

My dear Mr. Cooper,

It was quite a pleasure to me to receive your note & to be informed by it that Mrs. Cooper[2] and yourself had not forgotten me. Unfortunately, I have an examination on Saturday one on Monday & a final one on Tuesday.[3] Like a heathen I must employ Sunday in looking over the papers of my Saturday's exam; as I must get off to Switzerland on Wednesday morning.[4] Indeed, yours is the third invitation which has converged on Sunday next.[5]

Were it not so near the moment of my departure, and had I not my hands so very full it would give me real pleasure to join you at your country residence.[6]

Believe me | yours most truly | John Tyndall

RI Uncatalogued, box 864.

---

1. *William White Cooper:* William White Cooper (1816–86), appointed surgeon-oculist in ordinary to Queen Victoria in 1859.
2. *Mrs Cooper:* Mary Elizabeth (*b.* 1823/4), daughter of Samuel Poyser of Derby, married William on 9 December 1845 (*ODNB*). The Coopers often invited Tyndall to dine, and had introduced him to Richard Owen (Journal, 10 February 1855, RI JT/2/13c/727). According to Sir D'Arcy Power, William White Cooper was the last medical man to fight a duel, though the details of the duel are unknown (see http://livesonline.rcseng.ac.uk/biogs/E001044b.htm, accessed 29 May 2017).
3. *an examination on Saturday one on Monday & a final one on Tuesday:* i.e. 27, 29 and 30 July. It is not clear what these were. Tyndall had examined at Chelsea Hospital on 22 June (Journal, RI MS JT/2/13c/1219) presumably for the Commissioners of Council of Military Education but he was also examining for University of London at the time.
4. *I must get off to Switzerland on Wednesday morning:* Tyndall left on Wednesday 31 July.
5. *the third invitation which has converged on Sunday next:* it is not known what the other two were.
6. *your country residence:* the 1871 Census shows the White Coopers at Fern Acres Cottage, Fulmer, Eton, Buckinghamshire.

To Rudolf Clausius 26 July 1861 1802

26th July 1861

My dear Clausius,

Accept my best thanks for your kind invitation,[1] which it would give me great pleasure to accept were it in my power to do so. But I long ago pledged myself to a party of my friends; some of whom are now on their way to the Æggeschhorn, and some of whom are to accompany me, and my engagement is to go direct to the mountains.

In fact I am anxious to expel towns and science and every thing that could remind me of London from my brain. I am heartily tired and want rest.

Do you not intend to visit the neighbourhood of Zermatt? I shall be wandering about the Monte Rosa district for many days.

I had a pleasant letter from Magnus a few days ago.[2] He and M[rs] Magnus and their daughter[3] intend to visit Switzerland this year.

My plans of returning are not yet complete: I shall probably finish with Chamonix;[4] but I may return by Interlaken, and if so should be greatly delighted to see you there.

Will you return my best thanks to M[rs] Clausius[5] for her kindness to me, and with best regards to her & you.

believe me dear Clausius | Yours ever truly | John Tyndall

RI MS JT/1/T/181

1. *your kind invitation*: letter missing.
2. *I had a pleasant letter from Magnus a few days ago*: letter missing. The most recent extant letter is 1771 from 17 March 1861.
3. *Mrs Magnus and their daughter*: see letter 1771, n. 7.
4. *I shall probably finish with Chamonix*: it appears that Tyndall did not go to Chamonix. He left Zermatt on 31 August (Journal, RI MS JT/2/13c/1254) and according to letter 1809 he was at the Manchester meeting of the British Association, which took place from 4–11 September.
5. *Mrs Clausius*: Clausius married Adelheid Rimpham, who died in childbirth in 1875, leaving him to raise their six children.

From Johann Joseph Benen (Bennen) [July 1861][1] 1803

HOCHGESCHÄTZTER HERR TYNDALL,—Indem ich mit Herrn Tugget (50) fünfzig Tage auf Reisen war und Heute erst nach Laax gekommen bin, habe ich Ihren werthen Brief von 22 Juni auch nur erst Heute erhalten; so dass ich denselben Ihnen auch nur Heute gleich beantworten konnte. Wo ich Ihnen mit Vergnügen melde, dass ich immerhin bereit sein werde Sie zu begleiten wann und wohin Sie nur wünschen.

Herr Tyndall! Ich mache Ihnen meine Komplimente für das gute Zutrauen zu mir und hoffe noch an der Zeit gekommen zu sein um wieder Gelegenheit zu haben Sie bestens und baldigst zu bedienen.

Mit Hochschätzung und Empfehlung, | Ihr Diener, | Benen.

Dear Mr Tyndall

As I was travelling with Mr Tugget[2] for (50) fifty days and just returned to Laax[3] today, I also only received your esteemed letter of 22 June[4] today; so that I was also only able to answer it directly today. I respond with pleasure that I will surely be ready to accompany you when and wherever you wish.

Mr Tyndall! I give my kind regards for your good confidence in me and hope the time is come that I shall soon have the opportunity to serve you to the best of my abilities.

With my respect and compliments, | your servant, | Benen.

J. Tyndall, *Mountaineering in 1861,* p.7.

1. *[July 1861]:* dated in the context of the letter he received. See n. 4.
2. *Mr Tugget* : Francis Fox Tuckett (1834–1913). Fox Tuckett had made the first ascent of the Aletschorn on 18 June 1859 with Bennen, Peter Bohren and Victor Tairraz. He was one of the most prolific mountaineers of the period.
3. *Laax:* Bennen's home village in the Rhône valley, near Fiesch.
4. *your esteemed letter of 22 June:* letter missing.

To Michael Faraday 21 August 1861 1804

Monte Rosa Hotel, | Zermatt | 21st. Aug. 1861

My dear Mr. Faraday.

A quiet hour has succeeded the storm of noises raised by departing tourists, and this hour I purpose to make pleasant to myself by writing to you. I reached Thun[1] on the 2nd of Aug. and met there at the Railway station my faithful and favourite guide Bennen;[2] proceeded thence to Meyringen,[3] and instead of going on by the usual route to the Grimsel, quitted that route by entering the Urbachthal at Hof, crossing the Gauli glacier, scaling the ridge which separates this glacier from that of the Lauteraar, and proceeding by the latter and the Unteraar glacier to the Grimsel. From the Grimsel I passed to the Aeggischhorn, and thence by the glacier of the Aletsch to the Bel Alp; a fine promontory which commands most glorious views. We accomplished an act of mercy on the glacier by dragging with great effort a cow out of a crevasse, in which the poor brute had been firmly jammed. From the Bel Alp I proceeded to Randa, in the valley of St. Nicholas, and here my new and really earnest work commenced.

There are two grand mountains, which above all others have of late years excited the attention of our best climbers; which have been tried on various occasions by able and competent men, but which have hitherto successfully resisted all attempts to reach their crests; these are the Matterhorn or Mont Cervin, and the Weisshorn. I have always regarded the latter as the noblest mountain in the Alps and from its position and magnitude it forms a grand and striking object from every point of importance in Switzerland. Randa rests at the base of this mountain; the summit of which is however far withdrawn behind the pine-clad pedestal on which it stands. Having to pass so near it the thought of trying the mountain naturally occurred to me. A preliminary inspection enabled us to fix on a resting place where the night previous to the attempt might be passed. This was underneath the ledge of an overhanging rock, at an elevation of about nine thousand feet above the sea. I had two pairs of coverlets sewn together so as to form two sacks, and into these at the proper hour I managed to creep in succession. My guides dug up the clay a little so as to soften my bed, and here at 8. P.M. on the evening of the 18th. of August I lay down.

The weather was most glorious and the scene around me surpassingly grand. At the opposite side of the valley was the range of the Mischabel, with its two grand peaks the Grabenhorn and the Täschhorn. Next came the Alphubel, next the Alleleinhorn[4] and Rympfischhorn,[5] both encased in an enamel of glittering snow; then came the Cima di Jazzi, then the mass of Monte Rosa clasped by its glaciers and brightly illuminated from bottom to top. The craggy face of the Lyskamm turned towards us was half in gloom and half in glory: the two twins Castor and Pollux were most singularly illuminated—across their waists was drawn the deep shadow of a portion of the Breithorn, while their summits and bases glowed like burnished gold. The mottled crags of the Breithorn itself were irregularly splashed with the sunlight. As the sun descended more and more this scene, all of which was in view from the cleft in which I lay, assumed the most sublime appearance. The highest summits occupied the circumference of a semicircle;—they burned with glory, and here and there long stretches of light over the lustrous snowfields linked the illuminated peaks together. These seemed hung in heaven like a chain of enormous opals, fit to form the necklace of an archangel.

But as the sun sinks the moon, within one day of being full, is travelling up the opposite face of heaven, and finally appears exactly behind the peak of the Alleleinhorn. She rises to the summit, clears it, and bears away splendidly thro' the crimson tinted sky. The definition of the peak rendered the motion perfectly visible, and made the orb appear like a vast balloon ascending majestically in the air. I lay with my face towards the moon and gazed long upon the glorious satellite until finally my face and eyes became so chilled that I was fain to protect them by a light handkerchief. The inflammation which might be expected to succeed such a chilling process when carried too far, is doubtless sufficient to account for the "blindness" which the innocent moonbeams have usually laid to their charge. As the cold of the night increased I squeezed myself more and more underneath my ledge thus lessening the area of clear sky against which my body could radiate its unrequited heat. Nothing could be more solemn than the night. Up from the valley came the muffled thunder of the Vispbach, caused by the rush of its waters and the incessant grinding of its stones. Over the lofty Dom flashed in succession the stars of Orion, until finally the entire constellation hung aloft. Higher up was the effulgent moon whose light gleamed in succession from the snow fields and pyramids; those only whose position suited the direction of the rays sending back the silvery gleams. The twins were the last to occupy this position, and long they held it, shining with spectral beauty until the moon set behind the hills.

We rose betimes; breakfasted however too early and had to wait an hour for the advancing day. A faint illumination overspread the east, and with this promise of the coming sun we packed our wine and provisions together and at a quarter before 4 A.M. commenced our expedition. We rounded a shoulder of the mountain, a stretch of glacier lay before us, but previous to entering upon it I disburthened myself of my shooting jacket, leaving it on the mountain side and proceeding forward in my shirt sleeves. The mountain in all its grandeur is now fairly before us, and our aim is to gain the lowest end of one of its arêtes. We tramp over the first snow, cut our way through the mounds and chasms of the glacier, reach a Bergschrund[6] which cuts a snow couloir up which we have to ascend in two; cross it, cut steps up the slope, gain the rocks and are soon on the terminal crags of the arête. Thence we cross the saddle which tops the couloir, reach the rocks at the opposite side which we find most fantastically riven and in many cases extremely loose and dangerous. Forward however we cheerily clamber, tortuously winding round the crags, or scaling them amain. The arete is notched and hewn into rude turrets by frost and weather, while the fragments of this colossal sculpture are strewn confusedly around. We work forward for a couple of hours, and then see two moving objects on the glacier far beneath us. We at first thought them chamois, but soon pronounced them men. The telescope removed all doubt; one of them carried a knapsack and the other an axe. Our expedition had caused great excitement among the athletes of Randa, and here were two of them following in our track, and prepared perhaps to dispute with us the ascent of their local Titan. This possibility, however, did not cause us much uneasiness.

Coming to a place at which a deep gap was hewn in the arête, flanked right and left by two vertical towers of rock, we produce our rope, and manage to let each other down into the gap. We round the opposite turret and soon after quit the arête, working our way across the ridges and couloirs below it. The latter are filled with clear hard ice, the cutting of which is exceedingly labourious. We again gain the arête, over which the snow folds itself at intervals, and along the snow we work our way. On the right side of the arête the mountain is tremendously precipitous, and in many places the steepness on both sides is appalling. We reach the end of one of the subdivisions of the ridge, from which to the next highest rock, stretches a line of snow which is as narrow [as] a wall, and bevelled off to a sharp edge at the top. How to cross this catenary I knew not, for I had no notion that it could bear a man's weight. The practical knowledge of my guide however enabled him to estimate its sustaining power. He walked along it turning out his toes, so as to place his feet across the arête, and I and my second guide followed. On each side the precipices were tremendous, and I smiled grimly as I stood on the middle of the wall and thought what a slip might accomplish. 'Had the snow been less perfect I would not have attempted it for the world' was Bennen's remark after we had crossed. The theory of the thing was quite clear. The snow was a little moist, of exceedingly fine grain, and hence its attachments when pressed down were innumerable. Thus it attained its marvellous rigidity. This accomplished we clamber steeply upwards, cautiously avoiding the looser rocks. With all our care however we sometimes dislodge vast masses which bound like demons on the slope to our left, loosen others by their fall and set the mountain in a roar as they whiz and thunder to the snow fields four thousand feet below us. Our climbing is a most extraordinary process:, there is scarcely a position possible to my body into which it was not folded up at one time or another during the day. Sometimes it was a fair pull upwards, sometimes an oblique twist round the corner of a rock tower; sometimes it was the grip of the finger ends in a fissure and the lateral shifting of the whole body in a line parallel to the crack. Many times I found myself with my feet highest and my head lowest. The action of the human arm is truly wonderful, and never before this day was I so deeply impressed with the marvellous combination of strength and flexibility in this member of our frame. The glimpses of the summit which we obtained after three hours climbing inspired us all, and our hopes condensed almost to certainty. But when, at the end of six hours, the summit still towered clear and distant in the heaven above us, our feelings became considerably qualified. The constant and copious expenditure of mechanical force began to tell upon us all. Wenger[7] complained of his lungs, an expression of deep weariness shaded Bennen's face, while I was half bewildered and stupified by the incessant knocking about. I however kept close watch on myself lest the callous indifference which the exertion had produced should degenerate into carelessness; and I felt all along that in case of sudden emergency I had a fair residue of power to call upon. I wished also to assure my self of Bennen's self control, and once on observing the sickly despair which filled his eyes as he looked towards the summit, I told him plainly that he must not persist on my account; that the moment he ceased to feel confidence in himself he ought to turn back. 'I am quite sure of myself' was his reply 'Of course I feel this work, let me try the effect of a little food.' He did so and was greatly refreshed. Looking more confidently towards the mountain he exclaimed *'Herr! mir müssen ihn haben'*[8]—and on we went. It was amusing and instructive to observe how by a kind of instinct we paid out our muscular energy so as to make its sensible loss a minimum; pausing after each stretch of the limbs, and gathering up strength for the next strain. I thought of Englishmen in battle, of the quality which made them famous; it was this very habit of never yielding—of fighting on when every hope seemed cut away, and I was resolved that in the present instance this quality should suffer no debasement. An hour and three quarters additional are consumed in this slow sapping of difficulty. A pure white snow cone rose still above us but clearly within reach. Ten times I was assured by both men that it was the highest point; but I had scaled so many pets and eminences, which promised to be the last that I feared to deliver myself up to the belief in this one. We gain the top of a prism of granite; from this a sharp snow arête, runs to a little point. This Bennen soon gains, moves a little way down on the other side so as to afford me room. In an instant I press the very highest snowflake of the mountain and the prestige of the Weisshorn is for ever gone!

Every alpine traveller knows how prominently the Weisshorn lifts its noble peak, and how strikingly visible it is from every point of importance in Switzerland. Conversely every point of importance must be visible from the Weisshorn. We formed the centre of an Alpine circle of unparalleled grandeur. Switzerland, Savoy, Italy, and the Tyrol all spread their mountain treasures before us. I opened my note-book to write a few words concerning the view, but I was absolutely unable to do so. There was something incongruous, if not profane, in allowing the descriptive faculty to meddle with that which belonged to the soul alone, so I resigned myself up to the silent contemplation of the scene, completely overpowered and subdued by its unspeakable magnificence. As I write to you my friend this glorious picture reconstructs itself in my imagination, and long years must pass ere its splendid proportions and gorgeous colourings can fade from the canvas of my memory.[9]

Ever yours | John Tyndall

I am off to Breuil to-morrow, and afterwards may visit Chammouni.[10] Kindest regards to Mrs Faraday[11] and Miss Barnard.[12]

RI MS JT/1/TYP/12/4104
Faraday Correspondence, 6:4047

---

1. *Thun:* a town in the canton of Bern, Switzerland.
2. *Bennen:* Johann Joseph Bennen (1824–64), an Alpine guide from Laax.
3. *Meyringen:* Meiringen, municipality in the canton of Bern, Switzerland.
4. *Alleleinhorn:* now Allalinhorn.
5. *Rympfischhorn:* now Rimpfishhorn.
6. *Bergschrund:* deep crevasse, often very dangerous for climbers.
7. *Wenger:* Ulrich Wenger, an Alpine guide from Grindelwald.
8. "*Herr! mir müssen ihn haben*": Sir! We must have him (in Helvetic High German).
9. *my memory:* See *Mountaineering in 1861: A Vacation Tour* (London: Longman, Green, Longman and Roberts, 1862) for Tyndall's published account of the ascent of the Weisshorn (specifically pp. 41–66). This letter was published as 'Ascent of the Weisshorn,' *Illustrated London News,* 7 September 1861, pp. 240–41 (see letter 1809).

10. *Chammouni:* now Chamonix, see also letter 1633, n. 2.
11. *Mrs Faraday:* Sarah Faraday, née Barnard (1800–79), wife of Michael Faraday.
12. *Miss Barnard:* Jane Barnard (1832–1911), Faraday's niece.

To Thomas Archer Hirst 21 August 1861 1805

21st August 1861 | Zermatt

My dear Tom

I forward you the only letter found at the Poste Restante. Yours pleased me extremely.[1] I was pleased to find you so well lodged, and so high up above the air of these confounded valleys; and also much pleased with the vivid and vigorous description of the thunderstorm. No doubt the desire of the physician is to hold you, but a cure, if you could be sure of it would be cheaply purchased at 20 days imprisonment. I should say that an Englishman would be a Godsend there, for I do not imagine the Bad is in the least possessed by the children of Albion.[2]

It is near the dinner hour & I have not time to say much; tomorrow morning I cross the Theodule[3] with Ramsay[4] & his family. My thoughts as you know were at the Weisshorn; knowing I should pass near it on my way to Zermatt: it was the second great mountain in Switzerland—next to the Matterhorn in prestige. Well on Monday last—at about 2 P.M. your friend John led by the brave Bennen after a day of tremendous toil reached the top, and the trodden crown of the mountain now bears the handle of Bennen's axe; with a red pocket hand kerchief flailing from it like the Union Jack of England.[5]

I shall be at Breul[6] for some days & know not where afterward.

Yours affectionately | John

RI MS JT/1/T/654

1. *Yours pleased me extremely:* letter missing.
2. *the children of Albion:* Albion is an ancient name for Great Britain.
3. *Theodule:* The Théodule Pass connecting Zermatt (Switzerland) and Breuil-Cervinia (in Italy).
4. *Ramsay:* Andrew Crombie Ramsay (1814–91), Professor of Geology, Royal School of Mines and local Director, Geological Survey.
5. *after a day of tremendous toil reached the top:* see letter 1804 for a fuller account of this climb.
6. *Breul:* Breuil.

From John Francis Julius von Haast[1] 27 August 1861 1806

Christchurch Province of Canterbury |
New Zealand 27 Aug[ust] 1861

Sir,

I have the honour of presenting to you by the mail a little volume,[2] the title of which: Report of a topographical and geological exploration of the western districts of the Nelson Province New Zealand undertaken for the Provincial Government by Julius Haast Esq. will sufficiently explain its contents. If you should find time, to look through it, I hope you will judge it with indulgence, for as a fearless traveller yourself, you will surely perceive the great difficulties, under which I laboured. The contents of this little work will show you, that I have named many mountains and rivers after the great savants of Great Britain. I will not apologize for the absence of your name amongst them, for I had before starting on my hazardous journey made arrangements with the Canterbury Government to enter into their service as Geologist, after having finishing my work in Nelson and I purposely reserved your name for this part of the country.

Being under engagement to survey the Southern Alps, a magnificent mountain chain of an altitude of 10 to 13500 feet and covered with perpetual snows, from which in every direction glaciers descend, I thought it my duty to reserve your name for a mountain giant only, intending there to name it after one to whose labour Science owes so much in the investigations of these beautiful phenomena.

I chose the huge pyramidal peak in this astral chain from which I found several large glaciers descend, feeding the two main source branches of the Rangitata and the Southern source branches of the Rakaia and I am certain you would enjoy it very much should you gaze upon M^t^. Tyndall[3] in all its stern and savage beauty, even the base of which has never been tread on by a human foot before. At the close of my 1 year by which time I shall have finished a great deal of my work in the Alps, I shall begin to construct my maps and I will take the first opportunity of sending you a copy, so that you may see where Mt Tyndall lies.[4] The investigation of the Southern Alps is a most pleasing task, presenting numberless points of the highest geological interests; in the case you wish my information concerning their numerous glaciers, I shall be only too happy, as far as it is my power to do so, to comply to your wishes.

Believe me Sir | Yours very faithfully | Julius Haast.

Professor | John Tyndall FRS | etc etc. | Royal Institution of Great Britain | London.

BL Add MS 53715 0001 to 0003

1. *John Francis Julius von Haast:* (1822–87), German geologist then in the employ of the provincial government of Canterbury, New Zealand.
2. *A little volume:* Julius Haast, *Report of a topographical and geological exploration of the western districts of the Nelson province, New Zealand: undertaken for the provincial government by Julius Haast, Esq.* (Nelson: C. and J. Elliott, 1861).
3. *Mt. Tyndall:* Mount Tyndall, 2,496m (8,189 ft), Otago, New Zealand.
4. *so you may see where Mt Tyndall lies:* a full account of the expedition is found in, Julius von Haast, *Geology of the Provinces of Canterbury and Westland, New Zealand* (Christchurch, 1879).

From Richard Dawes 9 September [1861][1] 1807

Deanery, Hereford, | Sept. 9th

My dear Tyndall,

I was glad to find from a note I have just had from Frankland that we may look forward to the pleasure of seeing you here on Friday next, but I am sorry you think of so short a stay. I just write a hasty line to say it will be quite convenient to receive you on Thursday the 12th, by which time the Manchester Meeting[2] will have ended, should it be convenient to you to come, and you might like to hear the oratorio of the Messiah[3] in the Cathedral on Friday. On whichever day you can come, be in time for dinner,

Believe me, my dear Tyndall, | Always sincerely yours | R. Dawes.

I take the chance of this finding you directed to the meeting of the British Association.

RI MS JT/1/TYP/1/322a
LT Typescript Only

1. *[1861]*: LT's note. This accords with internal evidence (see n. 2).
2. *Manchester Meeting*: thirty-first meeting of the BAAS, held at Manchester, 4–11 September 1861.
3. *the oratorio of the Messiah*: George Frideric Handel's oratorio *Messiah,* composed 1741 and performed in Hereford Cathedral on Friday 13 September as part of the Hereford Music Festival. See [Anon], 'Hereford Music Festival,' *Morning Post,* 16 September 1861, p. 6.

From Thomas Archer Hirst 10 September 1861 1808

St Moritz | Sepr 10th 1861

My dear John

After leaving you in the Visp valley I remained the night in Zermatt and next morning crossed the Theodule to Breuil. I then descended into Italy as far as Chivasso near Turin, turned eastwards & through Novara to Arona on Lago Maggiore. I traversed the Lake and pushed on to Lugano where I was fortunate enough to meet Ramsay and hand over to him his letter which contains an acceptable sum of money. I remained a day at Lugano but finding it too hot and relaxing I pushed on across the lake of Como and up the valley of the Valteline[1] which I followed as far as Tirano and then turned to the left over the magnificent Bernina Pass into the Engadine valley. I arrived at St Moritz on Sunday Evening the 1st and have been here ever since. I intend to remain as long as possible and then to come straight home so as to arrive in Town before the 25th. During the 10 days I have been here my legs have improved greatly I have some hopes of returning with them quite healed. I know not whether to ascribe this good effect to the Chalybeat waters[2] of which I drink daily, to the fresh air at an elevation of 3400 ft to the wholesome diet or to the daily excursions which I make to altitudes varying from 6 to 10 thousand feet. Perhaps all these circumstances have had their share in producing the effect in question. The treatment I have here been advised to follow differs altogether from that of Leucherbad[3] and is far better suited to my sluggish temperament. I have nothing to do here with warm baths & drinks everything is cold. Every morning at 6 I take a cold shower bath and drink the water whose temperature is about 40°F. Every night I bind up my legs in linen cloths steeped in cold water; around them I then wrap flannel and go to sleep as soon as they begin to tingle with the warmth of my own body which is then prevented from escaping. Next morning circulation is again stimulated by holding them for a while under the shower of cold water. In this manner I have hardened myself and no longer shiver as I did when I first reached St Moritz. I have not met Steiner,[4] but have made many other pleasant German, Swiss & English acquaintances. This is the first time I have used pen and ink since I came for it is the first wet day. I will leave all further news until we meet and only here express the hope that you have returned strong & that we shall spend together a winter of good health and solid work.

Yours ever | T.A. Hirst.

I will write to Mrs Tunstall[5] a day or two before my arrival.

RI MS JT/1/H/246

1. *Valteline:* Valtellina, a valley in the Lombardy region, Northern Italy.
2. *Chalybeat waters:* mineral spring waters containing iron.
3. *Leucherbad:* Leukerbad, a municipality in the Valais canton, Switzerland. Famous for its thermal baths.
4. *Steiner:* possibly Jakob Steiner (1796–1863), Swiss mathematician.
5. *Mrs Tunstall:* Hirst and Tyndall's housekeeper at 14 Waverley Place. See *Hirst Journal* RI MS JT/2/32c *c.*8 June 1861.

From Michael Faraday 11 September 1861 1809

Royal Institution | 11th. Sept. 1861.

My Dear Tyndall

Lest I should not catch you to day I write to thank you for your admirable letter[1] (for I have not been able to thank you before), and say how thankful we were for it. I sent it off at once to the Illustrated London News, where it appeared last Saturday.[2] I hope you do not disapprove. Shortly after it was gone, I wished it had been sent to the Times; and I don't know what made me choose the other, except some association, mingling with a wandering memory, cause it. So you have been at Manchester,[3] where I durst not go. I conclude you are as well as you can be and that is just as we wish it. We are stopping at Hampton Court now but I hope we shall meet much somewhere soon.

Ever yours | M. Faraday

RI MS JT/1/TYP/12/4152
Faraday Correspondence, 6: 4056

1. *your admirable letter:* see letter 1804.
2. *where it appeared last Saturday:* John Tyndall, 'Ascent of the Weisshorn,' *Illustrated London News,* 7 September 1861, pp. 240–41.
3. *have been at Manchester*: the thirty-first meeting of the BAAS, was held at Manchester, on 4–11 September 1861.

To Juliet Pollock 4 October [1861][1] 1810

Friday 4th Oct.

My dear Mrs Pollock

Many thanks for allowing me to send you the manuscript.[2]

It is simply a sketch of my tour this year. I have written it to an imaginary friend just to allow myself a little freedom of expression—You can fancy it addressed to yourself if you like—though really the X is what it represents to mathematicians—an unknown quantity.

I will send you the rest as soon as it is finished though if what I now send pass the remainder is only of secondary importance.

Autumn is here! I left my trees green, I find them sere; their leaves are now shaken down and whirled about by the wind.

Goodbye | ever yours | John Tyndall

RI MS JT/1/T/1173

---

1. *1861*: Tyndall refers below to the manuscript of his 'tour'. This is almost certainly a reference to *Mountaineering in 1861,* published 16 April 1862 (see also letter 1804, n. 9).
2. *Send you the manuscript*: for confirmation that Tyndall received comments on the manuscript of *Mountaineering in 1861* from Juliet, see letters 1866 and 1870.

From Joseph Plateau 7 October 1861 1811

Gand | 7 Obre 1861

Mon Cher Monsieur Tyndall

Je me suis empressé de satisfaire à votre demande en vous envoyant un second exemplaire de ma 5ème Série; vous l'avez sans doute reçu. J'ai maintenant l'honneur de vous adresser trois exemplaires de ma 6ème Série, qui vient de paraître; l'un de ces exemplaires vous est naturellement destiné, et j'espère que vous auriez l'extrême obligeance d'offrir, de ma part, le second à la Société Royale, et le 3ème à l'Institution Royale.

Dans cette 6ème Série, je continue et j'étudie, par la théorie <*1 word missing*> expérience, le sujet commencé dans la 5ème; vous y trouverez toute la suite des expériences, si variées et si curieuses, à effectuer avec les charpentes en fil de fer.

D'après le désir que vous avez bien voulu m'a exprimer j'ai commandé à notre fabricant toute la collection des charpentes propret à répéter les expériences de ces deux Séries; j'y ai fait joindre cinq supports en fil de fer pour y <*1 word missing*> les charpentes après les avoir retirées du <*1 word missing*> de, supports qui sont représentés dans la figure 6 de la 5ème Série. J'oxiderai toutes les charpentes avant qu'elles vous soient envoyeés, de sorte qu'elles seront prêtes à servir. La collection comprendra, en autre, trois charpentes qui ne sont pas décrites dans la 6ème Série. L'une d'elles est destinée à montrer que trois *lames planes* aboutissant les unes aux autres suivant une *arête* liquide unique forment nécessairement un système d'équilibre stable, vous comprendrez cette charpente à la première vue, les autres donnent lui à de jolis systèmes dont je n'ai point parlé.

La construction de toutes ces charpentes est déjà très avancée, et je pense qu'elles pourront vous être envoyées dans une quinzaine demoi, de sorte que vous les aurez à temps pour la rentrée de la Société Royale, si vous persistez dans *<1–2 words missing> [1 word illeg]* intentions à mon égard; mais pour que la caisse vous parviendra d'une manière sûre et sans retard, veuillez avoir la bonté de m'écrire votre adresse exacte; et comme nous sommes peu habitués au genre d'écriture des personnes de votre nation, je vous prie de vouloir bien tracer cette adresse en caractères très lisibles, afin qu'il ne puisse y avoir encore erreurs.

Quant au vase à parois planes en verre, accompagné de ses accessoires, vous le recevrez de même dès qu'il sera terminé, et je presserai la chose.

Agréez, mon Cher Monsieur Tyndall, l'assurance de tous mes sentiments d'affection.

J. Plateau | Professeur à l'Université | Place du Casino 22.

P.S. J'ai une autre prière à vous adresser. Quand vous aurez donné suite à votre bonne promesse, peux-je espérer encore de votre obligeance que vous voudrez bien m'informer du résultat, soit par un mot de lettre, soit en m'envoyant sous bande un article de Journal qui rendrait compte de la Séance, soit de toute autre manière?

Ghent | October 7th 1861

My dear Mr. Tyndall,

I hasten to fulfil your request by sending you a second copy of my 5th Series;[1] which you have undoubtedly received. Now I have the honour of sending you three copies of my 6th Series,[2] which had just been published. One of those copies is naturally meant for you, and I hope that you would be so kind as to offer, on my behalf, the second one to the Royal Society, and the 3rd one to the Royal Institution.

In this 6th Series, I continue to study, through the theory *<1 word missing>* experiments, the subject begun in the 5th. You will find there all further experiments, so varied and so interesting, which would be done with the wire framing.

According to the wishes that you have kindly expressed to me, I ordered the entire collection of clean frames from our manufacturer in order to repeat the experiments of those two series. I joined five support wires to *<1 word missing>* there the frames after having removed them *<1 word missing>* of, supports that are represented in figure 6 of the 5th Series. I will oxidise all of the frames before they are sent to you so that they will be ready for use. The collection will include, in addition, three frames that are not described in the 6th Series. One of them is meant to show that three flat blades, leading one after another following a unique liquid edge, necessarily create a stable equilibrium system. You will understand that frame at a first glance. The others provide good systems, which I have not spoken about yet.

The construction all these frames is already very advanced, and I think I could send you them in a couple of weeks, so that you will have them in time for the start of the Royal Society. If you will stay in *<1–2 words missing> [1 word illeg]* intentions towards me; but in order for the box to arrive safe and without delay, would you be so kind as to write me your exact address.[3] And because we are not very used to the style of writing by people of your country, I ask you to kindly trace the address in clearly legible characters, so that we may avoid any errors.

As for the vessel with flat sides of glass, together with its accessories, you will receive it also as soon as it will be finished, and I will expedite the matter.

Please accept, my dear Mr. Tyndall, the assurance of my highest and affectionate regards.

J. Plateau | Professor at the University | 22 *Place du Casino*

P.S. I have another request for you. When you would have fulfilled your kind promise, could I hope to again rely on your kindness to keep me informed of the result; either by letter, or by sending me wrapper-packed an article of the Journal that would report on the Session, or in any other way?

RI MS JT/1/P/90

1. *A second copy of my* 5th *series:* J. Plateau, 'Recherches expérimentales et théoriques sur les figures d'équilibre d'une masse liquide sans pesanteur', *Mém. Acad. Roy. Sci. Belg.* 33 (1861), separate pagination. For the letter accompanying the first copy, see letter 1778.
2. *three copies of my* 6th *Series:* J. Plateau, 'Recherches expérimentales et théoriques sur les figures d'équilibre d'une masse liquide sans pesanteur', *Mém. Acad. Roy. Sci. Belg.* 33 (1861), separate pagination. See letter 1778, where Plateau tells Tyndall about the contents of this new series.
3. *write me your exact address:* letter missing.

From Anne Wynne[1] 8 October 1861 1812

Corfu| 8th. October 1861

My dear Dr. Tyndall,

It was at Coire on the 13th September that Colonel and I read with great delight in Galignani's[2] of your ascent of the Weisshorn. I do like you to perform those feats and you have joy in them, joy in all the grandeur and joy in standing where no one else has stood. And since our return here we have received the book[3] which you so kindly sent us above a year ago and for which we greatly thank you. We thank you for your continued kindness in thinking of us and we thank you for the very book itself. I could not lay it down until I had come to the end of the narrative part. I love to read of those sort of things and you write of them well.

I do wish we could see you somehow, but there is nothing to bring you here, and when we go away we go only to the Tyrol, and not to those break-neck places of it where you love to haunt, and where it would be the delight of Edward and Frank[4] to go too.

They have good hands and active feet, tho' I cannot tell whether they could ever do what you do. Frank is at school at Buxton in Derbyshire and our only way of seeing him is by going to Germany and getting him to meet us there. This year we all from here, Col. Wynne,[5] Lucy[6] and I, went to the Tyrol; and Frank came there to us, and Eddy too got a months' leave, and we had the happiest time all together, wanting only Harry.[7]

We were very happy, and we were in such a nice place at Rentle near *[word illeg];* at an inn where they treated us so fairly so kindly and in so friendly a way, that the place being one of such beauty far and near, we could not bear to leave it, and stayed there till within ten days of the term of Frank's and Eddy's holidays.

Then we made a little tour by Landeck to the Stelvio and over it; and by Bozen over the [Brenner Pass] into Innsbruck Munich and Stuttgart, where we broke up: and Col. Wynne and I returned here by the way of the Splügen, Milan and Venice.

You know we don't see things with the knowledge of them which you have, so of course in our ignorance we lose much, but we see with great pleasure. Eddy, Frank and Lucy too were here and there and everywhere; and I too, pursued the beautiful scenes into the mountains as far as my head would allow me, but that is not far. Lucy has done what not every young lady of eighteen would do. She has stayed behind us in Germany for the sake of learning much which she cannot get at here. She is not at a school, but in a German family of ladies, at Cannstatt near Stuttgart, and so far the plan answers admirably. The family likes her and she likes it, and she is learning much and is eager on it and very happy.

And I am glad to have her away from this garrison town which is not I think a nice place for young girls, tho' there are many simple nice young English girls here, Lucy's friends, of whom nobody would know that they had lived in a garrison town. Still we are glad to have Lucy away from the place and eager on learning and living the simple German life in that nice intellectual kindly religious family, they are Lutherans and Lucy for the time goes where they go.

This place of Corfu is a vexing place—one may have a happy home in a place which does not itself please one, and our home is a happy one and my husband strong and well.

But this place is a vexing place, a grieving place. O so ill governed, not governed at all, and England in utterly a false position here.[8] This 'Protection' by us of these islands is a ridiculous thing and mischievous lowering to England oh so painfully. All do not take it to heart as I do but all the English here say the same thing—there is nothing else to say—'a ridiculous government ridiculously administered'. Either let us have these islands and govern them, or let us give them up and have nothing to say to them.

As it is they are no good to us, we enrich them, ask no taxes from them, let them do just as they like, and so they despise England, an idle chattering race they are whose only occupation is intrigue, a two tongued people. They would be a better people, a more hospitable people, if like other people they were taxed and then made to work. A happier people in that case perhaps they would not be, for they are an Oriental race in their love of ease and idleness. They have not stirring pursuits.

I won't write more about this vexing position of England here. I had not meant to do it at all. I want to say that I hope you are well, and that you take proper care of yourself. I wish you had a sister living with you to take care of you when you are in London. Is de Chaillet's book true?[9] It is very interesting and makes one think, makes one reflect about more than the gorillas; viz.: the long continued suffering condition of Africa, and the remedies for it now.

I wonder I have not written you a merrier letter for I have only this public grief to grieve me—and I have just now been reading the merriest letter which Colonel Wynne has been writing to Frank, curious that I should have employed the same time in exhaling angry feelings against the Ionians—It is only because of England—that is the reason I grieve so.

Dear Mr Tyndall Col. Wynne wants me not to send this letter as he says he wants to write too. So he may, another time.

I am your truly | affectionate friend | A. Wynne | sent 15th Oct.

RI MS JT/1/TYP/5/1857–59

---

1. *Anne Wynne:* Anne Wynne (d. 1864), married to Colonel George Wynne (see n. 5).
2. *Gallignani's: Galignani's Messenger,* an English-language newspaper published in Paris.
3. *the book:* most likely *Glaciers of the Alps.*
4. *Frank:* Frank George Wynne, third son of Anne and George Wynne.
5. *Col. Wynne:* Colonel George Wynne (1804–90), then stationed in Corfu at Fort Abraham.
6. *Lucy:* Lucy Wynne, only daughter of Anne and George Wynne.

7. *Harry:* Henry Le Poer Wynne (d. 1874), eldest son of George and Anne Wynne.
8. *England in utterly a false position here:* The Ionian Islands had been a protectorate of the United Kingdom since 1815. Sovereignty was transferred to Greece in 1864.
9. *de Chaillet's book:* Paul du Chaillu, *Explorations and Adventures in Equatorial Africa* (London: John Murray, 1861).

To Heinrich Debus 14 October 1861 1813

Royal Institution | 14th, Oct, 1861.

My dear Heinrich

I was very glad to receive your letter,[1] which is some compensation for your absence from the Isle of Wight. However you were wise not to come, for in the first place we had to get from Southampton to Cowes in a storm, and secondly after we have arrived there we were compelled by the rain to keep within doors. You must indeed be hardly worked at present, and the pay of £20 for the additional lectures appears to me incomprehensibly small. But you are a man of a physical head and the pleasure which the subject gives is in part compensation. Nevertheless £20 for two whole days a week is less than any thing of the kind I ever heard of in England before. If Queenwood cannot give more, be content; if it can you ought to have it.

I continue to be exceedingly well in health, though I work in rather an unwholesome atmosphere. In fact I feel rather to be getting young than getting old. My hairs are beginning to turn gray, but my body is usually full of vigour. Indeed it is many years since I have been so strong. My work is going on daily—It is tremendously difficult and sometimes drives me to the brink of despair. In the laboratory where I work there are three classes of actions on Radiant Heat. The first due to the pure oxygen and nitrogen alone is expressed by the number 1; the second due to aqueous vapour is expressedly the number 40 and the third due to the remaining foreign constituents of the air is represented by the number 27. This will give you some idea of the delicacy of my apparatus, and of the immense difficulty of bringing it into such a condition. Our atmosphere is a practical vacuum to rays of heat: and all that has been hitherto ascribed to the air is mainly due to the aqueous vapours. It is at least 40 times more potent than the whole air itself. Don't mention these results. I am daily fighting with difficulty and usually do not quit the laboratory from morning till night. My progress is exceedingly slow, but by constancy I hope to get something done. I was under the impression that Mr E.[2] had got a good man for lecturing; but I think you may congratulate yourself in getting rid of that ill conditioned individual to whom you allude. Tom[3] and I had a glorious walk yesterday. We got through about 18 miles of the most beautiful country imaginable. If you desire to know Mr Harcourt's[4] address I can get it at once by writing to my friend Cook,[5] the editor of the Saturday Review. If you want the address let me know and I will obtain it for you.

My mornings are occupied in writing out an account of my Swiss tour for publication:[6] so that my days are as full as your own.

goodbye | always Yours sincerely | John Tyndall

RI MS JT/5/11/1333

1. *your letter:* letter missing.
2. *Mr. E:* this is likely to be George Edmondson, headmaster of Queenwood College.
3. *Tom:* Thomas Hirst.
4. *Mr. Harcourt:* possibly Augustus George Vernon Harcourt (1834–1919) (*ODNB*).
5. *my friend Cook:* John Douglas Cook (*c.* 1808–68), newspaper editor (*ODNB*). See also letter 1743, n. 1.
6. *An account of my Swiss tour for publication:* Tyndall was preparing the manuscript of *Mountaineering in 1861* published on 16 April 1862.

To Juliet Pollock [19 October 1861][1] 1814

Saturday Morning 7.A.M.

My dear Mrs Pollock,

Welcome home! When your kind husband[2] called to see me yesterday I had some hope of being able to speak my welcome last night instead of writing it now. But I was thus employed:—

The moon was bright and I returned swiftly from St. John's Wood[3] to the Institution. My men were gone; but I could not afford to lose the chance of observing the moon's radiation. I have a platform on the house top over seventy feet above the floor of the laboratory. I ascended to this platform fifteen times from the laboratory—now seventy multiplied by fifteen makes a height of one thousand and fifty feet! This I climbed last night between 8.15 and 10.30 o'clock.

The result opposed all my notions and ought to make me humble. I have often ridiculed the idea of the moon's beams being cold; but last night the result was exactly equivalent to this. My face was colder when I looked at the moon than when I looked upwards at the same inclination towards any other part of the heavens. Of course in such a case I did not trust my sensations—indeed they could tell me nothing, but I have hard figures for the result which is quite certain.

Now ask your husband to account for this; and do not mention it to any body else. I am almost certain that you will puzzle him.

Yours always | John Tyndall.

RI MS JT/1/TYP/6/2218

1. *[19 October 1861]:* Tyndall made his measurements of lunar radiation the evening before (see letter 1815).
2. *your kind husband:* William Frederick Pollock.
3. *St. John's Wood:* a district of northwest London in the city of Westminster (where his house was).

To John Frederick William Herschel 21 October 1861 1815

Dear Sir John,

I am anxious to address this note to you upon a subject which you have in great part made your own, because I fear that neither in my book upon the Alps, nor in my recently published papers, have I made due reference to your estimable researches on Solar Radiation.[1] I have been for some time experimenting on the permeability of our atmosphere to radiant heat, and have arrived at the conclusion that true air, that is to say, the mixture of oxygen and nitrogen which forms the body of our atmosphere, is, as regards the transmission of radiant heat, a practical vacuum. The results from which the opacity of air has been inferred are all to be ascribed to the extraneous matters diffused in the atmosphere, and mainly to the aqueous vapour. The negative results recently obtained[2] by that eminent experimenter, Professor Magnus of Berlin, have induced me to reinvestigate this point; and the experiments which I have made not only establish the action of aqueous vapour, but prove this action to be comparatively enormous. Here is a typical case:–On the 10th of this month I found the absorptive action of the common air of our laboratory to be made up of three components, the first of which, due to the pure air, was represented in magnitude by the number 1; the second, due to the transparent aqueous vapour, was represented by the number 40; while the third, due to the effluvia of the locality and the carbonic acid of the air, was represented by the number 27. The total action of its foreign constituents on the day in question was certainly sixty-seven times that of the atmosphere itself; while the aqueous vapour alone exerted an action at least forty times that of the air.

I have also to communicate to you some results of lunar radiation which connect themselves with your speculations. On Friday the 18th of this month, I made a series of observations on the moon[3] from the roof of the Royal Institution. From six concurrent experiments, I was compelled to infer that my thermoelectric pile lost more heat when presented to the moon than when turned to any other portion of the heavens of the same altitude. The effect was equivalent to a radiation of *cold* from our satellite. I was quite unprepared for this result, which, however, you will at once perceive, may be an immediate consequence of the moon's *heat.* On the evening in question a faint halo which surrounded the moon, and which was only visible when sought for, showed that a small quantity of precipitated vapour was afloat in the atmosphere. Such precipitated particles, in virtue of their multitudinous reflexions, constitute a powerful screen to intercept the terrestrial rays, and any agency that removes them and establishes the optical continuity of the atmosphere must assist the transmission of terrestrial heat*. I think it may be affirmed that no sensible quantity of the obscure heat of the moon, which, when she is full, probably constitutes a large proportion of the total heat emitted in the direction of the earth, reaches us. This heat is entirely absorbed in our atmosphere; and on the evening in question it was in part applied to evaporate the precipitated particles, hence to augment the transparency of the air round the moon, and thus to open a door in that direction for the escape of heat from the face of my pile. The instrument, I may remark, was furnished with a conical reflector, the angular area of which was very many times that of the moon itself.

I remain, | Yours very faithfully, | John Tyndall

October 21, 1861

* I was going to add "into space;" but the expression might lead to misapprehension. My experiments indicate that the absorption of water is a *molecular* phenomenon. If we suppose the aqueous vapour of the atmosphere to be condensed to a liquid shell enveloping the earth, the experiments of Melloni[4] would lead us to conclude that such a shell would completely intercept the obscure terrestrial rays. And if the vapour be equally energetic, our atmosphere would prevent the *direct* transmission of the obscure heat of the earth into space. On this point, however, I wish to make some further observations.

J. Tyndall, 'Remarks on radiation and absorption', *Phil. Mag.*, 22 (1861), pp. 377–78.

---

1. *researches on Solar Radiation:* Herschel had investigated solar radiation from the 1820s, famously designing an instrument, named by Herschel an actinometer, to measure the sun's direct heat.
2. *The negative results recently obtained:* for a published account, see G. Magnus, 'On the Passage of Radiant Heat Through Moist Air', *Phil. Mag.*, 23 (1862), pp. 249–52.
3. *Observations of the moon:* Tyndall also mentions these observations to Heinrich Debus. See letter 1820.
4. *Experiments of Melloni:* Macedonio Melloni (1798–1854), Italian physicist. Melloni commenced important experiments on radiant heat early in his career, with the first results appearing in M. Melloni, 'Memoir sur la transmission libre de la chaleur rayonnante par cifferants corps solides et liquides', *Annales de Chimie et de Physique*, 53 (1833), pp. 5–73.

In a later published account Tyndall makes reference to: M. Melloni, 'On the nocturnal cooling of bodies', *Scientific Memoirs* 5 (1852), pp. 453–76; 530–52 (first published in *Annales de Chimie et de Physique,* 22 (1848), pp. 129–60, 467–96). See J. Tyndall, 'On the relation of aqueous heat to aqueous vapour', *Phil. Trans.,* (1863), p. 12.

To Editor of *The Times* 21 October 1861 1816

Sir,

The following observation may interest a large class of the readers of *The Times;* I therefore send it to you to be disposed of as you think fit:

Many years ago Mr William Hopkins of Cambridge pointed to the state of the rocks over which glaciers had passed as conclusive evidence that these vast masses of ice move bodily along their beds. Those rocks are known to have their angles rasped off, and to be fluted and scarred by the ice which has passed over them. Such appearances, indeed, constitute the entire evidence of the former existence of glaciers in this and other countries, discussed in the writings of Venetz,[1] Charpentier, Agassiz, Buckland,[2] Darwin, Ramsay,[3] and other eminent men.

I have now to offer a proof of the sliding of the ice exactly complementary to the above. Suppose a glacier to be a plastic mass, which did not slide, and suppose such a glacier to be turned upside down, so as to expose its under surface; that surface would bear the impression of its bed, exactly as melted wax bears the impression of a seal. The protuberant rocks would make fallows of their own shape in the ice, and the depressions of the bed would be matched by protuberances of their own shape on the under surface of the glacier. But, suppose the mass to slide over its bed, these exact impressions would no longer exist; the protuberances of the bed would then form longitudinal furrows, while the depressions of the bed would produce longitudinal ridges. From the former state of things we might infer that the bottom of the glacier is stationary, while from the latter we should certainly infer that the whole mass slides over its bed.

In descending from the summit of the Weisshorn on the 19th of August last I found, near the flanks of one of its glaciers, a portion of the ice completely roofing a hollow, over which it had been urged without being squeezed into it. A considerable area of the under surface of the glacier was thus exposed, and the ice of that surface was more finely fluted than ever I have observed rocks to be. Had the tool of a cabinetmaker passed over it, nothing more regular and beautiful could have been executed. Furrows and ridges ran side by side in the direction of the motion, and the deeper and larger ones were chased by finer lines, produced by the smaller and sharper asperities of the bed. The ice was perfectly unweathered, and the white dust of the rocks over which it had slidden, and which it had abraded in its passage, still clung to it. The fact of sliding has been hitherto inferred from the action of the glacier upon the rocks; the above observation leads to the same inference from the action of the rocks upon the glacier. As stated at the outset, it is the complementary proof that the glacier moves bodily over its bed.

Your obedient servant, | John Tyndall.

Royal Institution, Oct. 21

RI MS JT/6/5/1

J. Tyndall, 'An Alpine observation', *The Times,* (22 October 1861), p. 10.

---

1. *Venetz:* Ignace Venetz (1788–1859), Swiss engineer whose 1833 work *Mémoire sur les Variations de la temperature dans les Alpes de la Suisse* suggested that Europe used to be covered by glaciers.
2. *Buckland:* William Buckland (1784–1856), geologist who served as Reader in Mineralogy at Corpus Christi College, Oxford, and would go on to become Dean of Westminster. Convinced by Agassiz's glaciation theory, he served as the proponent of that theory in England (*ODNB*).
3. *Ramsay:* Andrew Crombie Ramsay (1814–1891), Professor of Geology, Royal School of Mines and later Director General of the Geological Survey (*ODNB*).

To Heinrich Debus 24 October 1861 1817

Thursday, 24 October | 1861.

My dear Heinrich.

I am in want of a few volatile liquids: can you give me any. Boiling point up to 120°C. would answer.

Have you any body in your laboratory that would draw dry air for several days, by means of an aspirator over a quantity of fragments of onions placed in a glass tube, and who would determine what they had lost in weight by this process?

Have you any body who would perform this calculation for me:[1] given the aqueous vapour of our atmosphere 1 per cent. Supposing the said atmosphere to fill a tube 3 feet long, and suppose the aqueous vapour to be condensed to a liquid plate of the same diameter as the tube, what would be the thickness of the plate of liquid?—Some of your physical pupils will perhaps work this out for me. I want it much.

Can you devise any means of getting a surface coated with soot without involving hydro carbons in the pores of the carbon?

Do you know of any experiments[2] that have been made on musk to determine the loss of weight in a given time. They say it will continue to scent places for 20 years without losing any weight.

Answer me these questions if you can, and I will send you by and by an account of some results[3] which will surprise you.

Yours ever | John Tyndall

RI MS JT/1/T/262
RI MS JT/1/TYP/7/2369

---

1. *Have you any body who would perform this calculation for me:* for Tyndall's own results after carrying this out himself, see letter 1820.
2. *Do you know of any experiments:* response from Debus missing or not sent.
3. *an account of some results:* see letter 1820.

To George Gabriel Stokes 24 October 1861 1818

Dear Mr. Stokes

Mr. Matteucci's[1] paper on "the secondary electro-motor power of nerves &c", appears to me worthy of publication in the Philosophical Transactions.[2]

John Tyndall |24th Oct. 1861[3]

RS RR/4/170

---

1. *Mr. Matteucci:* Carlo Matteucci (1811–68), Italian physicist and neurophysiogist interested in the electrical properties of animal tissues, such as the electrical current of the heart.
2. *Publication in the Philosophical Transactions:* published as C. Matteucci, 'Electro-physiological researches. Eleventh Series. On the Secondary Electromotor Power of Nerves, and its Application to the Explanation of certain Electro-Physiological Phenomena', *Phil. Trans.* 151 (1861) pp. 363–72
3. *24th Oct. 1861:* note on reverse reads: '1861 Matteucci by Tyndall', possibly by Stokes.

To George Gabriel Stokes 24 October 1861 1819

24th Oct. 1861

My dear Mr. Stokes.

I have read the paper of Sir Snow Harris[1] which was referred to me by the Council.

I do not think the publication of the first 30 papers of this memoir would materially advance our knowledge of electricity. They might I think be condensed with advantage to one tenth of their present volume.

In this condensed state they might form an introduction to a series of experiments which should consist of an expansion of the concluding portion of Sir Snow Harris's enquiry. The experiments to which he refers at the end of his papers would, if expanded, furnish material for a memoir well worthy of publication in the Philosophical Transactions.[2]

Sir Snow Harris evinces an intimate acquaintance with the labours of the old electricians, but he sometimes permits those of the modern workers to escape his notice. The 7th volume of the Philosophical Magazine for example contains a masterly paper[3] by the lamented Kohlrausch[4] on the very subject which Sir Snow Harris has taken up. In this paper our author makes no reference whatever, and he appears to be quite unaware of its existence.

I would recommend that the paper be placed in the hands of its author with the recommendation to pursue the experiments touched upon toward the conclusion of the memoir.

Yours very truly | John Tyndall

RS RR/5/95

1. *Sir Snow Harris:* Sir William Snow Harris (1791–1867) natural philosopher with scientific interests in electricity and magnetism (*ODNB*).
2. *well worthy of publication in the Philosophical Transactions:* the paper was not published in the *Phil. Trans.*
3. *a masterly paper:* R. Kohlrausch, 'Theory of the electric residue in the Leyden jar', *Phil. Mag.,* 7 (1854), pp. 315–20; 412–26; 476–89.
4. *Kohlrausch:* Rudolf Kohlrausch (1809–58), German physicist and associate professor at the University of Marburg. The Kohlrausch function, named after the discharge of a capacitor, stems from his article on the Leyden jar.

To Heinrich Debus [Late October 1861][1] 1820

Most heartily do I thank thee oh! Heinrich for thy instructive letters to me.[2] Please thank Flight[3] for me for that useful calculation.

No doubt a good deal of aqueous vapour would be mixed up with the scent of the onion.[4] I experimented thus. I passed air first over the onion, and then over fragments of potato cut in a similar manner. I deducted the offset of the potato from the total effect of the onion, and in doing so I assumed that I deducted the aqueous vapour.

Could you not apply the same method? At all events you can do this—As far as aqueous vapour is concerned the large Spanish onion and the small common onion will act alike, but as far as the odours are concerned the small common one is very much stronger than the Spanish are. Could you not send air over both, and see whether they lost alike or lost differently: if differently the balance wd be due to odour of the common onion, if sensibly alike it would shew that the quantity of matter producing this odour must be very small. My object is to get some idea of the quantity of matter which produces the effects that I have observed.

Would not this be a good experiment! Musk you say has been thought to give its scent to a process of decay due to oxidation. Suppose I make pure hydrogen the vehicle for conveying the scent into the tube if the scent be due to oxidation I ought to get no action if it be independent of oxidation I ought to obtain the same effect with hydrogen as with air—This experiment it appears to me would decide whether the scent was due to oxidation or not.

Thank you very much for your offer of the liquids[5]—I will seek what Frankland Williamson[6] & Hoffmann[7] can give me, and if they have not all I require I will write to you.

Our atmosphere hardly stops radiant heat at all. It is sensibly a vacuum, and those observations from which the opacity of the air has been inferred are entirely due to the extraneous matters in the atmosphere & mainly to the aqueous vapour.

The aqueous vapour on the 10th of this month exerted an action at least 40 times that of the whole air.[8]

During last full moon I experimented on her radiation and found my thermo-electric pile when turned on the moon lost more heat than when turned on any other part of the heaven of the same altitude. It was practically a radiation of cold. I have written a short letter to Herschel on these subjects which will appear in the next Phil. Mag.[9]

goodbye Heins | Yours ever | John Tyndall

Alternierende Ströme[10] are, I should say, what you obtain by induction in making and breaking contact—they are in opposite directions.

RI MS JT/1/TYP/7/2369–70

1. *[Late October 1861]*: the letter refers to Tyndall's moon experiments which were carried out on 18 October and follows a letter sent to Debus on 24 October (see letter 1817). A later letter followed in early November (see letter 1823).
2. *thy instructive letters to me:* letters missing.
3. *Flight:* Walter Flight (1841–1885), mineralogist taught by Tyndall and Debus at Queenwood (*ODNB*).

4. *scent of the onion:* see letter 1817.
5. *Your offer of liquids:* Tyndall had asked Debus for some volatile liquids in a previous letter (see letter 1787).
6. *Williamson:* Alexander William Williamson (1824–1904), British chemist (*ODNB*).
7. *Hoffman:* August Wilhelm von Hofmann (1818–92), German organic chemist.
8. *an action at least 40 times that of the whole air:* on 10 October 1861 Tyndall noted in his laboratory notebook: 'it is certain that the foreign contaminants in the air of the laboratory exert at least 67 times the absorptive power of the air itself, and the aqueous vapour alone exerts about 40 times the absorption of the air'. Experimental Diary, 10 October 1861, RI MS JT/3/46.
9. *in the next Phil. Mag:* John Tyndall, 'Remarks on radiation and absorption,' *Phil. Mag.,* 22 (1861), pp. 377–78.
10. *Alternierende Ströme*: alternating current.

To Michael Faraday 2 November 1861 1821

Saturday 2nd. Nov. 1861

My dear Mr. Faraday

I have been desirous of writing to you for some time past, but have been held back by the fear that you might consider me obtrusive; the same fear almost holds me back now. I wished to say a word or two regarding my position here, and to state to you my feelings on a subject which has of late been often in my thoughts. I do not know whether Dr. Bence Jones committed any breach of trust in reading to me the letter in which you resign the Juvenile lectures,[1] but if he did I am sure that you will forgive him, and that he will forgive me for alluding to it here. Your resignation probably throws this work into my hands for the present; I shall do it as my duty, but apart from the question of duty, I wish to say that, if the lectures numbered fifty instead of six, and if the reward were Zero instead of fifty pounds, I would cheerfully undertake them if it relieved you of the slightest anxiety regarding them. And with reference to the concluding portion of your letter, where you place yourself in the Managers hands as regards your further connexion with the Institution, let me say this much, even at the risk of being deemed impertinent. The fact of your being here was my chief motive in coming to this place, and it has continued to be my chief privilege ever since. The Royal Institution has given me many pleasures and many advantages—so many indeed that there is no situation in this country to which I could look with the least envy; but the main gratification and greatest pleasure of my life hitherto has been my association with you. All this may seem very much out of place, but I wish to make a clean breast of it, by saying that if my position here caused you to do anything calculated to weaken your connexion with the Institution, I should not think myself justified in remaining. The day that you resign I shall resign also.

Believe me my dear Mr Faraday | Ever truly yours | John Tyndall

RI MS JT/1/TYP/12/4112–13
Faraday Correspondence, 6: 4089

---

1. *Juvenile lectures*: otherwise known as the 'Christmas Lectures', which Faraday himself had set up in 1825, and which he had given annually in the 1850s.

From Joseph Tairraz[1] 3 November 1861 1822

Chamonix | le 3 9bre 1861

Monsieur le professeur Tyndall,

Je viens vous informer par ces quelques lignes que je n'ai pu accomplir la commission que vous m'aviez confiée par la lettre d'Auguste Balmat et de laquelle je m'étais chargé avec tant de plaisir—

Monsieur le Docteur Pitschner avec lequel j'étais engagé au moment qu'Auguste Balmat me fit voir votre lettre ne voulut à aucun prix me céder 2 jours pour m'acquitter de votre commission. Il m'avait promis cependant que quand il ferait son ascension au Mont Blanc de m'accorder le temps nécessaire pour faire mes recherches. Mais huit jours après époque de son ascension il ne put lui-même à cause de la neige et du froid aller que jusqu'au grand plateau. Je dus donc y renoncer (ne pouvant l'abandonner en pareil lieu—ne pouvant à cause de la quantité de neige et du mauvais temps réaliser seul avec un porteur cette ascension—J'attendis toujours que quelque voyageur vienne me frayer un peu la route. L'occasion ne se présenta plus. Je dus donc abandonner avec grand regret une chose que j'aurai faite pour vous et pour moi avec tant de plaisir.

Votre très dévoué et obéissant serviteur | Tairraz Joseph, photographe

Chamonix | November 3rd 1861

Professor Tyndall,

With these few lines here, I would like to inform you that I could not accomplish the commission that you entrusted me through Auguste Balmat's letter, and which I took on with much pleasure—

Doctor Pitschner,[2] with whom I was engaged when Auguste Balmat showed me your letter,[3] did not want under any circumstances to give me 2 days to fulfil your commission. However, he promised me that when he would make his ascent of Mont Blanc, he would give me the time needed to do my research. But eight days after the time of his ascent, he could not go further than the Grand Plateau, himself, because of the snow and the cold. Thus I had to give it up (I could not leave it in such a place—because of the amount of snow and bad weather, not being able to climb the ascent alone with a porter—I waited still for some traveller to come and clear the route for me a bit. The occasion did not present itself. Thus, it is with much regret that I had to abandon something that I would have done for you and for me with so much pleasure.

Your most devoted and obedient servant | Tairraz Joseph, photographer

RI MS JT/1/T/2

1. *Joseph Tairraz:* (1827–1902), Swiss mountain guide and photographer.
2. *Doctor Pitschner:* Wilhelm Pitschner (1822–71), German geologist and Alpinist. Pitschner led scientific missions to Mount Blanc in 1859 and 1861, sponsored by the Prussian government.
3. *your letter:* letter missing.

To Heinrich Debus [4 November 1861][1] 1823

Monday

My dear Heins

It is not at all necessary for you to come to London on the anniversary[2]—of course we sh[d] be glad to see you, but it would cost you time and money, so therefore I could not advise you to come.

I feel perfectly satisfied that any mode of experiment which seems satisfactory to you as regards the determination of the loss by scent in the case of the onion[3] would appear equally so to me—

Perhaps as I have made the experiment with the potato it would be as well to do it in the way I suggested.[4]

Yours in great haste | John Tyndall

RI MS JT/1/T/341
RI MS JT/1/TYP/7/2399

1. *[4 November 1861]:* this letter follows letter 1820. The most likely date is Monday 4 November.
2. *on the anniversary:* the Anniversary Meeting of the Royal Society on 30 November each year. Debus had been elected FRS in June.

3. *determination of the loss by scent in the case of the onion:* see letter 1817 and 1820 for Tyndall's commentary on his experiments with the scent of an onion.
4. *the way I suggested:* see letter 1820.

From Michael Faraday 4 November 1861 1824

ROYAL INSTITUTION OF GREAT BRITAIN | 4 Nov. 1861.

My dear Tyndall,

Your letter of Saturday night[1] has raised thoughts and feelings which are often with me, and are always grateful; and though I felt as one who knew your mind before, still, the act by which it was manifested (not revealed) was not the less pleasant and acceptable. I never doubted from the first that as we came together so we have continued, save that the bond of union, and its cause, have grown in strength from then till now. There is a loyalty between us which I am sure of on your part, and, (though a man may doubt himself) I do not doubt on my own.

Ever, my dear Tyndall, | Your affectionate friend, | M. Faraday

RI MS JT/1/TYP/12/4153
Faraday Correspondence, 6: 4090

1. *Your letter of Saturday night:* letter 1821.

From Edward Frankland 7 November 1861 1825

Chemical Theatre, St. Bartholomew's Hospital, E.C. |
November 7, 1861

My Dear Tyndall,

On throwing the spectrum of lithium upon the screen yesterday, I was surprised to see a magnificent blue band. At first I thought the chloride of lithium must be adulterated with strontium; but on testing it with Steinheil's apparatus[1] it yielded normal results without any trace of a blue band. I am just now reading the report of your Discourse in the 'Chemical News,'[2] and I find that you have noticed the same thing. Whence does this blue line arise? Does it really belong to the lithium or are the coke-points or ignited air guilty of its production? I find three blue bands with chloride of sodium, but they have not the definiteness and brilliancy of the lithium band. When lithium wire burns in air, it emits a splendid crimson light; plunge it into oxygen, and the light changes to bluish white. This seems to indicate that a high temperature is necessary to bring out the blue ray.

Ever yours sincerely, | E. Frankland

P.S. I have just made some further experiments on the lithium spectrum and they conclusively prove that the appearance of the blue line entirely depends upon temperature. The spectrum of chloride of lithium ignited in a Bunsen's-burner flame does not disclose the faintest trace of the blue line: replace the Bunsen's burner by a jet of hydrogen (the temperature of which is higher than that of the Bunsen's burner), and the blue line appears, faint, it is true, but sharp and quite unmistakable; if oxygen be now slowly turned into the jet, the brilliancy of the blue line increases until the temperature of the flame rises high enough to fuse the platinum and thus put an end to the experiment. —E. F.

(On the occasion referred to by Dr. Frankland, it was a general impression among the chemists present at the lecture that I had used the word lithium for strontium throughout the evening. This induced me to ask Dr. Miller[3] to test my chloride of lithium which he found quite pure. I afterwards showed the blue band, the splendour of which is unrivalled, to my class at the School of Mines. The coal-points without the lithium show nothing of the kind; with the lithium the band always appears. Either therefore the substance itself is so altered by the exalted temperature that new periods of oscillation are possible to it, or the medium in which it vibrates is so changed in elasticity as to permit of the same thing. The observation appears to be one of considerable significance. I would also draw attention to the experiment by which the absorption of the yellow band by the sodium flame was effected on the same occasion, as one of the most striking class experiments in the whole range of optics. It is very easily performed, a band 18 inches long and ¾ of an inch wide being quite attainable within ordinary lecture-room limits. A salt flame 10 feet thick produced no such effect. Dr. Miller, I am informed, repeated this experiment with success before an evening meeting of the British Association at Manchester (see Phil. Mag. vol. xxii. p. 154).[4]—J. T.)

E. Frankland, 'On the blue band of the lithium spectrum', *Phil. Mag.*, 22 (1861), pp. 472–73.

---

1. *Steinheil's apparatus:* Karl August von Steinheil (1801–70), German physicist and astronomer who was professor of mathematics and physics at the University of Munich, 1832–70. He developed an apparatus that was used for spectral analysis.
2. *Report of your Discourse:* see 'Proceedings of Societies', *Chemical News* 4 (1861), pp. 61–64.
3. *Dr. Miller:* William Allen Miller (1817–70), Professor of Chemistry at King's College (*ODNB*).

4. Reference to J. Tyndall, 'On the physical basis of solar chemistry', *Phil. Mag.*, 22 (1861), p. 154.

From John Frederick William Herschel 10 November 1861 1826

Collingwood | Nov 10/61.

Dear Sir

Many thanks for your communication of "Remarks on Radiation & Absorption".[1] The fact of air being non absorptive and vapour highly so is indeed a very capital one and must afford a key to much that has hitherto been inexplicable in meteorology. The dissipation of cloud & clearing of sky by the Lunar rays is, as you very justly observe a palpable consequence of her heat being entirely absorbed aloft (as I always imagined it was) As to the actual production of cold on your radiation instrument by directing its axis on the moon you will excuse me if I don't quite see the force of your reasoning. The moon's heat dissipates the vapour not only in that angular space of the sky as seen from the focus which is occupied by her disk (a circle ½° diameter) but also and equally over an immense area all around.

If the result be one you <u>can certainly</u> rely on and unequivocally proved by many experiments—the only explanation I can think of is that we receive radiant heat in considerable quantity from the ultra-lunar spaces. This the moon intercepts over the visible area of a circle ½° diam. while all the heat it sends is stopped aloft so that from that half degree <u>none</u> reaches the Earth's surface.

But if I recollect rightly Ritchie[2] <u>did</u> find some small heating power in the moon's rays.* P. Smyth[3] on Tenerife[4] indicates also some very small effect—but then he was at 10,000 feet altitude.

I remain | Dear Sir | Yours very truly | J. F. W. Herschel

P.S. If vapor absorbs heat and air does not then the vapour should be heated by the Sun's rays & not the air (except by conduction from the vapor) Therefore the S.G.[5] of the vapor would be diminished relatively—the ascensional power increased (see Fourier?[6] on the ascent of clouds[7]) and may it not be that an atmosphere of attenuated vapor may extend beyond the aerial one?

(*I see on reference to his paper[8] that he only projected an expt. whether he ever made it, I cannot remember.)

RI MS JT/1/H/95
RI MS JT/1/TYP/2/508
Royal Society Letters (RMD 5 August 2011), 23_347 (this is Tyndall's own transcription)

1. *communication of "Remarks on Radiation & Absorption.":* see letter 1815.
2. *Ritchie:* William Ritchie (c. 1790–1837), Professor of Natural Philosophy at the RI from 1831 until his death (*ODNB*).
3. *P. Smyth:* Charles Piazzi Smyth (1819–1900), Astronomer Royal for Scotland and Professor of Natural Philosophy, University of Edinburgh (*ODNB*).
4. *on Tenerife:* C. Piazzi Smyth, *Tenerife, An Astronomer's Experiment: or, Specialties of a Resident Above the Clouds* (London: Lovell Reeve, 1858), pp. 208ff.
5. *S.G.:* Specific Gravity, the ratio of the density of the vapour to that of air.
6. *Fourier:* Jean-Baptiste Joseph Fourier (1768–1830), French mathematician and physicist.
7. *ascent of clouds:* most likely referring not to Fourier but to Augustin-Jean Fresnel. See A. Fresnel, 'Sur l'Ascension des nuages dans l'atmosphère', *Annal. Chem. et Phys.* (1822), pp. 260–63.
8. *reference to his paper:* William Ritchie, 'On a new photometer with its applications to determine the relative intensities of artificial light &c', *Phil. Trans* (1825), pp. 141–47.

To Nevil Story-Maskelyne 15 November 1861 1827

15th Nov. 1861.

My dear Maskelyne,

Is there any possibility of procuring a largish diamond[1] for a simple experiment on radiant heat. The ray used would no more injure it than a ray of light; and the point is of considerable theoretic importance.

I wrote in despair to the Trustees[2] yesterday asking for some salt. Never was mortal baffled & perplexed as I have been. My work is tremendously difficult and if I am not helped it will conquer me.

Yours ever | John Tyndall

Natural History Museum, DF 15/2

1. *procuring a largish diamond:* for further correspondence with Story-Maskelyne regarding requests made to the British Museum, see letter 1844.
2. *I wrote in despair to the Trustees:* letter to the Board of Trustees of the British Museum missing.

To George Gabriel Stokes 15 November 1861 1828

15th Nov. 1861.

My dear Mr Stokes

Would you kindly favour me with your opinion[1] on the following point.

A flame plays against one of the vertical faces of an exhausted cubical vessel; being competent to keep that face at a temperature of 200°.

The opposite face of the cube is kept at the constant temperature of 0°.

Half an atmosphere of air is now admitted into the cube; and convection sets in between the hot parts & the cold ones.

The amount of chilling of the hot face is capable of accurate measurement: let it be measured & noted.

An additional half atmosphere is now admitted in to the cube, making the total amount there an atmosphere. The chilling is neither augmented nor diminished by the additional half atmosphere.

Now inasmuch as we have in the case of the whole atmosphere double the number of particles loading themselves with heat & carrying it off, and inasmuch as twice the number only do the same work as once the number am I not justified in concluding that the <u>average</u> velocity of the carrying particles in the case of the half atmosphere is double their velocity in case of the whole atmosphere.

Yours very sincerely | John Tyndall

RI MS JT/1/T/1396
RI MS JT/1/TYP/4/1476

1. *favour me with your opinion*: for Stokes' reply, see letter 1829.

From George Gabriel Stokes 19 November 1861 1829

Lensfield Cottage Cambridge | 19 Nov$^{\underline{r}}$ 1861

My dear Dr. Tyndall,

I was employing my leisure time to finish an investigation relating to a paper of Balfour Stewart's,[1] about which I have written to him, or I would have sooner replied to you.

According to what appears to me to be most probable I think that you <u>are</u> justified in supposing that the equality of cooling points to a doubly quick circulation in the air of half density. However I have not yet sufficiently closely considered the thing. I will write to you again[2] or speak about it when we meet.

Yours very truly | G.G. Stokes

RI MS JT/1/S/225

1. *Balfour Stewart's:* Balfour Stewart (1828–87), physicist and meteorologist, then Director of Kew Observatory (*ODNB*).
2. *I will write to you again:* no additional correspondence on this subject has been identified.

To John Frederick William Herschel 21 November 1861 1830

21st Nov. 1861

Dear Sir John

Here follows an account of the observations on Lunar Radiation referred to in my letter in the last number of the Philosophical Magazine.[1]

On the roof of the Royal Institution I had a platform erected, so high as to enable me to sweep the heavens with my thermo-electric pile, without impediment from the chimney-pots of London. The pile was supported on a vertical stem with which the axis of the instrument could be made to enclose any required angle. Wires passed from the pile to an excellent galvanometer placed in the laboratory at a vertical distance of 72 feet below the platform on the roof.

On directing the pile towards the heavens so that the reflection of the instrument embraced no terrestrial object the chilling of the face turned skyward was so great,[2] and the consequent galvanometric deflection so large that it was quite hopeless to operate upon the needle in this position. The zenith is also intensely cold at noon. To move it a single degree would have required many hundred times the quantity of heat or cold necessary to urge it through one of the lower degrees of the galvanometer. To render it manageable I therefore operated in the following manner.

The galvanometer was a differential one, that is to say, two wires ran side by side round the astatic needle at the instrument. The end of one of these wires were connected with the pile on the roof; the ends of the second wire I connected with a second pile, which was turned toward a vessel kept at a constant temperature by boiling water. The direction of the current caused by the heat below was opposed to that caused by the cold above, and the needle was thus compelled to take up its position among the lower degrees of the scale.

I then ascended to the roof, fixed my pile at its proper angle to its stem, and directed it <u>off</u> the moon. I descended and observed the galvanometer—the needle oscillated between 10° and 20°; its mean position was therefore 15°.

I reascended and turned the pile <u>on</u> the moon in descending I found the needle oscillating between 30° and 40° the mean position being 35°.

The ascending and descending were repeated six times in succession and the following results were obtained:

Deflections

	off the moon	on the moon
1	15°	35°
2	27°	40°
3	33°	40°

The numbers all show cold, the deflection being much as would be produced by a refrigeration of the face of the pile turned towards the sky, and we find that the deflection where the pile was on the moon is in all cases less than when it was off. In other words less heat was lost by the instrument when directed off the moon, than when directed towards it. This, as I have stated, is equivalent to a radiation of cold from our satellite.

The note in which I referred to these observations was probably written without due reflection on the circumstances. In fact I had been reading your excellent essay on Meteorology,[3] and some of its passages brought my sins of omission so forcibly to my mind that I wrote the letter off hand. I pictured the matter thus:—supposing the earth to be surrounded by an atmospheric veil of precipitated vapour, the depth to which the moon's heat could clear away that veil would be greatest along the line drawn from the earth's centre to the moon, and would diminish from this normal all round, thinning off to nothing where the incidence became sufficiently oblique. The escape of terrestrial heat as far as the atmosphere is concerned would be less impeded along the normal than in any other direction. Theoretically speaking, when I turned my pile off the moon I turned it towards a portion of the atmosphere, where the moon's heat was less effective in rendering the sky clear, but whether the difference could be measured is another question. The observations made during the last solar eclipse tend, I believe, to push the height of our atmosphere enormously beyond the limits hitherto assigned to it; were the limits within which precipitated vapour can form similarly expanded the explanation which I have given might apply; but not to an atmosphere limited as we have hitherto supposed ours to be.

I attach but little importance to the explanation as in all probability the disturbing elements at hand were able to mask the influence of the moon. On referring to a map I find that on turning my pile off the moon I turned it more over the mass of London, and the diminution of the cold may have been due to the temperature of the air suspended over this vast focus of artificial heat. I will not speak with confidence as to the influence of this or any other cause till I have time to investigate to the question thoroughly: at present I am working hard at an enquiry of another kind.

Let me complete this letter by asking you to suspend your judgment on the points of difference between my friend Prof. Magnus & myself:[4] in a very few weeks[5] I hope to have something interesting to communicate to you on this and kindred subjects.

Believe me dear Sir John | Yours very faithfully | John Tyndall

I should be extremely obliged to you if you would favour me with line saying whether you see any objection to the publication of this letter (I have a copy) in the next number of the Philosophical Magazine.[6]

Is it not at all events clear that a hot body may so modify a medium on which it acts that a second body shall have its temperatures lowered in the presence of the hot one!

RS HS 17.389 (this is Tyndall's own transcription)
RI MS JT/1/TYP/2/509–11

1. *last number of the Philosophical Magazine*: see letter 1815. John Tyndall, 'Remarks on radiation and absorption,' *Phil. Mag.*, 22 (1861), pp. 377–78.
2. *the chilling of the face turned skyward was so great*: in the published version of this letter (see n. 6 below) Tyndall here adds a note.
3. *your excellent essay on Meteorology*: this is thought to be John Herschel, *Meteorology: From the Encyclopaedia Britannica* (Edinburgh: Adam and Charles Black, 1861).
4. *points of difference between my friend Prof. Magnus & myself*: for a detailed account of this disagreement, see Irena Maria McCabe, 'Second Best as a Researcher, Second to None as a Populariser? The Atmospheric Science of John Tyndall' (PhD Thesis: University College London, 2012), pp. 146–65.
5. *in a very few weeks*: Tyndall did not communicate again with Herschel on this subject within the promised timeframe.
6. *the publication of this letter*: published as John Tyndall, 'Observations on lunar radiation,' *Phil. Mag.*, 22 (1861), pp. 470–72. The published version shows a number of significant changes of wording and has an additional note as indicated above. It was not presented as a letter to Herschel.

From John Frederick William Herschel 22 November 1861 1831

Collingwood | Nov 22/61.

My dear Sir

You do not say[1] when your exp[erimen]ts were made—but I conclude naturally that you took the opportunity of a full moon. (Interesit multum) Now I cannot believe that the surface of the full moon can possibly radiate cold—for I do not see how it can possibly be of a temperature much lower than boiling water seeing it has been long exposed (the central part of the disc) to unmitigated sunshine. On the other hand I can easily believe that the new moon may be very intensely cold though not so cold as the celestial spaces beyond, yet more cold than the upper region of our atmosphere.

But now let us consider the geometrical relations of the matter.

Suppose ASB the Earth's surface
C its centre asb the surface

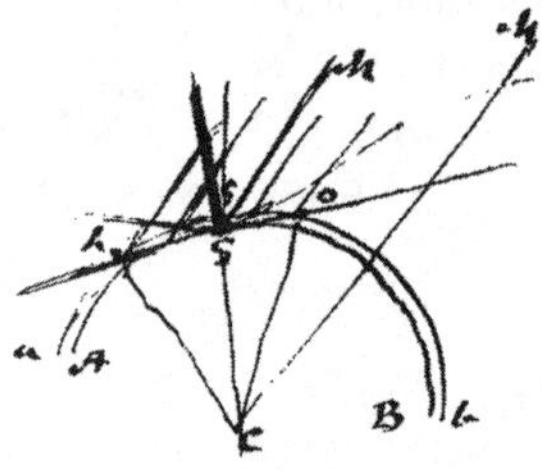

of the atmosphere. CM the
direction of the moon from C

nearly identical (with the small difference only of parallax) with SM from S At 80 miles the density of the air is only 1–100,000,000th of its density at the surface. Suppose then we take 80 miles for the height of the atmosphere—then the extreme span of the visible horizon from hS, to So will correspond to a terrestrial arc of 10° = hCS = SCo on either side of S—but as there is no use in trying to observe down to the very horizon, I suppose we may limit the effective span or the terrestrial arc over which the portion of upper atmosphere observed extends, to 5° on each side of S—which would make 10° of difference of incidence of the moon's rays on the extreme edges m and n of that portion of the atmospheric shell exposed.

Suppose the moon at full and at 50° of altitude as seen

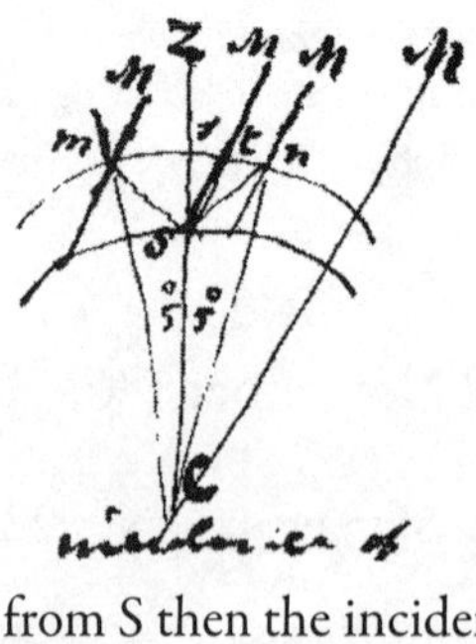

from S then the incidence of its rays on m will be 5° and at n, 35°

and at t the point exposed when the instrument is directed to the moon about 39° 3'.

As the whole heat cannot be supposed to be absorbed at the very outside, the differences of these angles are all assuredly materially in excess of the really effective differences—and when reduced accordingly I should doubt very much whether any difference of radiation could be expected. And there ought to be a gradation in passing from n through the zenith to m,—a gradual diminution of heat—so that the direction of the moon itself StM should be colder than the direction Sn, but warmer than the zenith, and that direction

(SZ) again warmer than Sm which I do not think agrees with your result as you seem to indicate the moon as the coldest place in the sky.

After all however, the station is so very unfavourable a one and the distribution of the London smoke and hot air and vapour so very uncertain that I should not be disposed to rely on any conclusions drawn from other than a long series of experiments—if then. —Only the pure and equable air of a station far from cities—or in some such atmosphere as that of Egypt or Malta would quite satisfy me on a point of such delicacy.

Any how I think it may be safely stated that if the instrument indicates a maximum of cold in the direction of the moon herself—the cause must lie elsewhere than in her influence in warming the upper regions of the air.

Believe me my dear Sir | Yours v[er]y t[rul]y | JFW Herschel.

PS. I don't know that any body has ever investigated the law of barometric pressure in an atmosphere of pure vapour without any air. Very peculiar considerations would apply in such a problem.

RI MS JT/1/H/96
RI MS JT/1/TYP/2/512–13
RS HS 23_348 (this is Tyndall's own transcription)

1. *You do not say when:* referring to Tyndall's opening letter on his experiments on lunar radiation. See letter 1830.

To John Frederick William Herschel 23 November 1861 1832

Royal Institution of Great Britain | 23rd Nov. 1861

Dear Sir John

Let me say one word more[1] & then I will not pester you until this matter is completely investigated.

When I said that the effect I observed was equivalent to a radiation of cold by the moon I entertained no idea that there was any such radiation—but the effect was such as if the moon radiated cold.

From the astronomer Royal's lecture at Manchester[2] it might be inferred that the atmosphere extended enormously beyond the limits which you have assumed—but then you would urge the tenacity of the air at those distances—I will here only say that at present I am able to measure with certainty the radiation of a particular description of vapour possessing the $\frac{1th}{1012,500,000}$ of an atmosphere of tension. As soon as my work is complete I will give you a fuller account of my experiments.

Yours most truly | John Tyndall

RS HS 17.390

1. *Let me say one word more*: Tyndall is here replying to Herschel's response to his experiments on lunar radiation. See letters 1830 and 1831.
2. *the astronomer Royal's lecture at Manchester*: George Biddell Airy (1801–92), the Astronomer Royal, was President of Section A (Mathematics and Physics) during the meeting of the BAAS at Manchester, September 1861.

From John Frederick William Herschel 24 November 1861 1833

Collingwood | Nov. 24/61.

My dear Sir,

What you state[1] about the tenacity of a vapour whose radiation you have succeeded in measuring is truly astonishing. One thing only occurs, (without calling in question either the delicacy or the precision of your means of exploration) & that is by what means it is possible to assure yourself that such is the degree of tenacity of the matter subjected to it—i.e. within 3 or 4 places of figures?

Au reste—as the air at 80 miles is only $\frac{1}{100{,}000{,}000}$ of the density at the surface, it would attain your degree of tenacity at only 10 additional miles of altitude and every 10 miles more will add another 0 to the denominator of the fraction. Still that some matter, vapour, or whatever else it may be at 100 or even at a much more considerable elevation, no one who has attended to the phenomena of the Aurora Bor. can doubt—but as to its having density enough to scatter the Sun's light and form the Corona about the moon in an eclipse, I should be disposed to demur.

If you are satisfied that, in turning your explorator off the moon, on all sides, round its center, so as exactly to exclude the moon from its field of View, and no more, indications of greater inward radiation or increased warmth are unequivocally manifested, I can no how see how it can be accounted for by the cause you assign—or in any other way than from radiation from extra-lunar space or starry matter capable of penetrating the upper strata of our air, while the radiation from the moon is stopped by the upper air & that of the space behind the moon, in that direction, by the body of the moon—.

Be that as it may however it is evident that you are opening up a highly interesting chapter in the book of Thermotics—and will I doubt not prosecute it with effect.

Believe me | Dear Sir | Yours very truly | JFW Herschel

RI MS JT/1/H/97

1. *What you state:* this is the final reply in a series of letters about Tyndall's experiments on lunar radiation. See letters 1830, 1831 and 1832.

From François Marcet[1] 30 November 1861 1834

Dear Professor Tyndall,

I took the liberty of sending you a few days ago a paper of mine[2] published in the Bib-univ on the nocturnal radiation of the earth—The latter part alone has any claims to novelty, consisting of some experiments I have made lately on the effects of nocturnal radiation on the lake of Geneva—I have made an extract of it in English & send it enclosed, in case you should think it worth inserting in the Philosophical Magazine.

I have to thank you for your paper[3] on the absorption and radiation of heat by gases & vapours, which I received last week, & your more recent lecture[4] on the Physical basis of solar Chemistry, which interested me much. The contents of the first paper I need hardly tell you I was already acquainted with since you were kind enough last year to show me some of your results at the Royal Institution.

We much regretted seeing nothing of you at Geneva on your return from the mountains. Your last expedition was most interesting, but as far as I could judge, full of danger. I have not heard that you attempted the Matterhorn, as you had an idea of doing last year, at least, if I understood you rightly, so I suppose you have thought better of it or put it off to another year.[5]

Believe me, dear Dr Tyndall | Yours ever truly | F Marcet

Geneva Novr. 30th. 1861

RI MS JT/1/M/49

1. *François Marcet:* (1803–83), Anglo-Swiss physicist.
2. *a paper of mine:* François Marcet, 'Remarques sur un mémoire de Charles MARTINS relative à l'accroissement nocturne de la temperature avec la hateur danes les couches sur les effets du rayonnement nocturne au-dessus des grandes surfaces d'eau', *Bibliothèque Universelle: Archives des Sciences Physiques et Naturelles* 12 (1861): 267–87. English extract published in *Phil. Mag.*, 23 (1862), pp. 9–12.
3. *your paper:* Tyndall (1861).
4. *your more recent lecture:* John Tyndall's lecture 'On the Physical Basis of Solar Chemistry,' delivered at the RI on Friday, 7 June 1861.
5. *put it off to another year:* Tyndall had contemplated climbing the Matterhorn in August but decided against it. The first successful ascent was in 1865. Tyndall made it to the summit in 1868. See Roland Jackson, *The Ascent of John Tyndall*, pp. 224–40.

To Heinrich Debus [early December 1861][1] 1835

Wednesday

My dear Heinrich

Anderson[2] sent you all the points and scraps of points that we possess: They are a poor stock, but the shortest of them, they being hard, will burn for some time.

Now listen to me. I am working like a slave, but must have a day or two to strengthen my brain. I am putting what would be the finest experimental enquiry ever made,[3] if I had time to complete it—into order before I commence preparing for my Xmas lectures.[4] But it will never do to attack these while feeble.

Now you heathen, suppose I go down to Queenwood on Saturday[5] would you walk with me over the world on Sunday—If they give me lodgings at Queenwood I would make an effort to remain for the concert which will be on Tuesday.

I will take work with me which I might do on the mornings of Monday and Tuesday. I will walk, run, jump or stand on my head during the afternoons and thus put my body in order.

Mrs Edmondson[6] knows what a flabby good for nothing fellow I am, and how necessary it is for such a culprit to be careful in small things—ask her if I bring down some oatmeal with me whether I could not have a little porridge in my bedroom at 8 A.M.

I am living on this at present and thus far it has given me the strength of a Hercules! Answer me quick! | John

With the porridge I can digest chalk.[7]

RI MS JT/5/11/1363
LT Transcript Only

---

1. *[early December 1861]:* see n. 3–6.
2. *Anderson:* Charles Anderson (1790–1866), lecture and laboratory assistant at the Royal Institution 1832–66.
3. *the finest experimental enquiry ever made:* refers to Tyndall's extensive experiments on radiant heat made around this time.
4. *My Xmas lectures:* Tyndall first delivered Christmas lectures in 1861 and did not do so again until 1863. See letter 1821 for discussion of Faraday's decision to no longer deliver the 'juvenile' or Christmas lectures and Tyndall's offer to take them on.
5. *suppose I go down to Queenwood on Saturday:* Hirst records that Tyndall was in Queenwood on Tuesday 17 December. *Hirst Journal*, 22 December 1861 (RI MS JT/2/32c/1598)

6. *Mrs Edmondson:* Anne Edmondson, wife of George Edmondson, headmaster of Queenwood College. Anne was very active in the management of the college, to Tyndall's dismay, and died on 19 November 1863 (*Hampshire Advertiser,* 21 November 1863, p. 2).
7. *I can digest chalk:* for his indigestion.

From Arthur Pepys Whately [4 December 1861][1] 1836

DEAR TYNDALL,

I have great pleasure in informing you that you were last night unanimously elected to the new Vice Presidentship of the Alpine Club. The Committee have long been desirous of including you among their number, and have, on more than one occasion, been prevented from requesting you to allow yourself to be put in nomination, only by feeling that your engagements were such as must prevent you from attending the meetings, while the numbers of the Committee and officers were not sufficiently large to allow of any absent members. This difficulty has at last been met by the creation of a new Vice Presidentship for the express purpose of connecting our more distinguished members more closely with the governing body of the Club than has hitherto been possible. And you will I hope understand that though nothing could give us greater pleasure than to see you at the Committee meetings (for which you will receive due notice) yet that you are not to be considered as being under any obligation whatever to take any more active part in the Club affairs than your own convenience may suggest.[2]

I am, | Very truly yours, | A. P. WHATELY

Life and Work, pp. 388–89

---

1. [*4 December 1861*]: the minutes of the Annual Meeting of the Alpine Club record Tyndall's election as Vice President on 3 December 1861. Minutes, Annual Meeting, Alpine Club Library Archives.
2. *Any more active part:* this did not provide Tyndall with sufficient reassurance: see letter 1849.

To [Editor of the *The Athenaeum*][1] 7 December 1861 1837

If the editor of the Athenaeum is not already tired of the subject he will perhaps be so kind as to find the accompanying note a corner in his next impression.

John Tyndall

[enclosure]

ASCENT OF THE WEISSHORN.

December 7, 1861.

Will you allow me to say one word in reference to a sentence in this day's *Athenaeum,*[2] as it may help a brave man in his vocation. After accomplishing the last dead lift which placed my two companions and myself[3] on the highest rock of the Weisshorn, we paused a moment to take a breath. The summit was within a few paces of us—a little pyramid of spotless snow, with a sharp ridge leading to the top of it. The man who had been in the rear of our little party during the day here stepped forward, and set his foot upon the ridge, with the apparent intention of going to the top. His progress was arrested by a rather emphatic exclamation of "Back!—not you." Johann Joseph Bennen—the best and bravest climber I have ever met—had been my leader throughout. Him I ordered to the front, and he it was who first planted a foot upon the summit of the Weisshorn.[4]

JOHN TYNDALL.

Matterhorn Museum, Zermatt (covering letter)
The Athenaeum, 14 December 1861, p. 808 (enclosure)

---

1. *To [Editor of the The Athenaeum]:* no salutation is given on the covering letter. The editor from 1853–69 was William Hepworth Dixon (1821–79), historian and explorer.
2. *a sentence in this day's* Athenaeum: the sentence reads: 'the first man on the top was . . . Dr. Tyndall'. [Anon.], 'Review: *The Alps, or Sketches of Life and Nature in the Mountains*', *Athenaeum* 7 December 1861, p. 759.
3. *my two companions and myself:* Tyndall was accompanied by Bennen and Ulrich Wenger.
4. *first planted a foot upon the summit of the Weisshorn:* They reached the summit on 19 August 1861. It was the first ascent of the Weisshorn.

From Heinrich Gustav Magnus 10 December 1861 1838

Berlin d. 10 Decemb 61

Mein theurer Tyndall!

Meine Briefe sind zwar nicht so geschrieben daß sie sich zur Veröffentlichung eignen, indeß habe ich nichts dagegen daß Sie davon publiciren was Ihnen gut dünkt, ich bin überzeugt Sie werden nichts drucken lassen was sich nicht für die Öffentlichkeit eignet. Ich bitte Sie aber sehen Sie sich mit Ihrer Publication vor. Sie sagen zwar in Ihrem Briefe an Sir John Herschel nicht gerade zu daß meine Versuche unrichtig sind, aber da Sie anführen daß Sie Ihre Resultate bestätigt gefunden haben, so liegt darin ausgesprochen daß die meinigen falsch sind. Sie begreifen daß ich mich dabei nicht beruhigt habe. Da auch ich meine früheren Resultate bestätigt fand, so habe ich ferner nachgeforscht welches der Grund Ihrer so überaus auffallenden Werthe sein könnte und ich habe gefunden daß dieser in der von Ihnen angewandten Methode liegt. Steinsalzplatten sind zu solchen Versuchen nicht anwendbar, sie ziehen zu leicht Feuchtigkeit an. Legt man eine Steinsalzplatte unter eine Glocke, unter welcher sich gleichzeitig ein Gefäß mit Wasser befindet, so wird die Steinsalzplatte so naß daß die Salzlösung abtropft. Legt man die Platte dann in die Luft, so wird sie wieder trocken. Ich brauche Ihnen nicht zu sagen wie atherman die Lösung ist. Ich habe Versuche ausgeführt mit einer Röhre die mit solchen Platten verschlossen war. Sobald feuchte Luft hineinkommt nimmt die Menge der durchgehenden Wärme sehr bedeutend ab. Eine kleine Notiz über diese Versuche liegt fertig redigirt um sie der Academie mitzutheilen doch möchte ich nicht daß durch dieselbe unser Verhältniß gestört würde. Ich hänge nicht daran sie zu publiciren und will sie deßhalb gern, wenn Sie es wollen, unterdrücken. Nur daran liegt mir dass die Wahrheit zu Tage komme und daß das Publicum erfahre welche Angaben richtig sind. Leid sollte es mir außerdem sein wenn ein Dritter sich in die Angelegenheit mischte, was, wenn lange gezögert wird leicht geschehen könnte. Ob die hygroscopische Beschaffenheit des Steinsalz allein die auffallenden Resultate herbeigeführt hat, die Sie beobachtet haben, will ich nicht behaupten, es wäre möglich daß noch andere Umstände mitwirkten. Aber so viel weiß ich daß diese Beschaffenheit einen großen Einfluß übt.

Es thut mir leid daß diese Correspondenz in eine Zeit fällt, in der Sie so beschäftigt sind, daß Sie kaum zu einer Wiederholung der erwähnten Versuche mit dem Steinsalz kommen werden, aber dennoch muß ich Sie bitten mir zu schreiben. Ich wiederhole daß ich gern, wenn Sie es wünschen, die ganze Notiz unterdrücken will, aber dann müßten Sie so gut sein, mir sofort zu schreiben.

Stets Ihr | GMagnus.

Berlin, 10 Decemb 61

My dear Tyndall!

Although my letters are not written in such a way that they are suitable for publication, I do not oppose that you publish those parts that seem good to you, I am certain that you will not send to publication anything that is not suitable for the public. But I ask you to be careful with your publication. In your letter to Sir John Herschel[1] you do not say exactly that my experiments are incorrect, but as you mention that you found your results confirmed, this implies that mine are wrong. You will understand that I could not put my mind at rest. As I also had found my previous results confirmed, I further investigated the reasons for your very conspicuous values and found them in the method that you employed. Mineral salt plates cannot be used in these experiments; they absorb moisture too easily. If one places a mineral salt plate under a bell below which one simultaneously places a container with water, the mineral salt plate will get so wet that the saline solution will drip. If one then exposes the plate to air, it will dry again. I do not have to tell you how atherman[ous] the solution is. I carried out experiments with a tube that was sealed with such plates. As soon as moist air flows in, the amount of heat passing through decreases very noticeably. A short communication on these experiments is ready to be sent out to the Academy,[2] but I do not want that our relationship will be affected by it. I am not attached to publishing it[3] and am therefore willing to suppress it, if you want. My only intention is to bring the facts to light and to let the public know which information is correct. I would also regret if a third party intruded into this matter, which could easily happen if one hesitates too long. I do not want to claim that the hygroscopic properties of mineral salt were solely responsible for the particular results that you observed; it is possible that other circumstances played a role. But as far as I know, this property exerts great influence.

I am sorry that this correspondence comes at a time when you are so busy that you will not likely get to repeat the above mentioned experiments with the mineral salt, but I nevertheless have to ask you to write to me. I repeat that I will gladly hold back the entire notice if you wish, but in that case you would have to be so kind as to write to me immediately.[4]

Always your | GMagnus.

RI MS JT/1/M/26

1. *In your letter to Sir John Herschel:* see letter 1815.
2. *the Academy:* Die Königlich-Preußische Akademie der Wissenschaften
3. *I am not attached to publishing it:* for the published account see H. Magnus, 'Ueber den durchgang der strahlenden wärme durch feuchte luft und über die hygroskopischen eigenschaften des steinsalzes', *Annalen der Physik und Chemie* 190 (1861/2), pp. 635–39. Tyndall's response was contained in: J. Tyndall, 'Remarks on recent researches on radiant heat,' *Phil. Mag.* 23 (1862), pp. 252–66. This follows a translation of Magnus's above-mentioned paper.
4. *write to me immediately:* no reply has been found.

From William Hopkins 13 December 1861 1839

Park House | Cambridge | Dec. 13/61.

Dear Tyndall,

I am engaged to write an article for the quarterly on Glaciers: I hope to do it for the Spring number.[1] At present I am writing a memoir on the subject for the Royal Society.[2] I had nearly finished it when at Manchester, but there

I lost it, and to my immense annoyance have been obliged to write it all again. However it is nearly finished. Stokes is away just now but will soon be back and I will see him about Graham's paper[3] which I have not yet seen.

You mentioned Forbes's intention of giving a course of lectures on glaciers at St. Andrews. I should like to see them if printed[4] before I write the article for the Quarterly.

In the memoir I am writing, I have deduced my general results as to internal pressures, tensions, &c from a more general mathematical analysis of the problem, and am putting the whole together, to show how completely my mathematical results and the sliding of glaciers harmonize with your own views. I am not making the paper more controversial than I can help, but I am obliged to have a few swipes at Forbes to justify my own views. There never was such a set of gross mechanical blunders foisted on the public as by Forbes. Sir David Brewster told me at Manchester that he (Sir D.) was a convert to my, (or rather to our), views.

Yours truly | W. Hopkins. | If any thing glacial comes in your way, let me know.

RI MS JT/1/TYP/2/641
LT Typescript Only

1. *I hope to do it for the Spring number*: a review by Hopkins of several books on glaciers eventually appeared as: [Anon.] [William Hopkins], 'The glacial theory', *Quarterly Review,* 114 (1863), pp. 76–125.
2. *a memoir on the subject for the Royal Society*: William Hopkins, 'On the theory of the motion of glaciers,' *Phil. Trans.,* 152 (1862), pp. 677–745.
3. *Graham's paper*: Thomas Graham, 'Liquid diffusion applied to analysis', *Phil. Trans.,* 151 (1861), pp. 183–224.
4. *I should like to see them printed*: a published version of these lectures has not been found.

From William Hopkins 18 December 1861 1840

Cambridge | Dec. 18/61

Dear Tyndall,

In writing my memoir on Glacial motion,[1] I want more precise information on one point than I can gain from existing books. I want the means of judging how far the veined structure in the lower part of a glacier is a structure transmitted from a higher part (where it was first generated) but modified as to the forms of the lines of structure by the subsequent motion of the glacier—or, whether it is more especially due to the direct and immediate action of the pressure to which the ice becomes subjected in the localities which any assigned portion of it occupies during its motion. The glacier proceeding from the Géant[2] with its dirt bands must afford I think one of the best tests to decide this question. I have no doubt as to your theory of the dirt bands.[3] According to it they must be formed in positions very nearly coincident with the original veins or lines of structure at the foot of the ice-fall. Now since your dirt curves must necessarily be transmitted curves, deformed into long loops by the motion of the glacier, the same must probably hold with respect to the lines of structure if they remain in coincidence with the dirt bands. Is such the case or not? Forbes's theory of them[4] would make them necessarily so, but would not make them, I imagine, (I never can thoroughly understand any of his theories) so entirely transmitted forms as yours. And there it is manifest that his observations are loose and inaccurate.

The question of the transmission or instantaneous formation of the veined structure has an important bearing on any theory of the longitudinal structure under the central moraine arising from the confluence of two glaciers. If that structure be generated by the more immediate action of the physical causes to which it is referable at the junction, then Forbes's theory cannot possibly be true: but may have a little toe, as it were, to rest upon if the structure may be considered as transmitted from the confluent glacier. The glacier of the Aar seems the most important case. Can you tell me how far below the Abschwung[5] the longitudinal structure is fully developed? Also how far in the case of the junction of the Léchaud and the Géant glaciers? Certain cases of sudden beginnings and sudden terminations, of portions of the glacier where the structure is forcibly developed, afford an argument against the fact of transmission, at least to any considerable distance.

Yours truly | W. Hopkins.

RI MS JT/1/TYP/2/642–43
LT Typescript Only

---

1. *my memoir on Glacial motion*: letter 1839, n. 2.
2. *the Géant*: a glacier on Mont Blanc.
3. *your theory of dirt bands*: this was communicated in J. Tyndall, 'On the veined structures of air-bubbles and dirt-bands, and remarks upon glacier theories', *Phil. Trans.*, 149 (1859), pp. 279–307 and *Glaciers of the Alps.*
4. *Forbes's theory of them*: first articulated in J. D. Forbes, *Travels Through the Alps of Savoy* (Edinburgh: Black, 1843).
5. *the Abschwung*: a buttress of the Schreckhorn at the confluence of the Lauteraar and Finsteraar glaciers. Louis Agassiz had a hut erected there.

From Heinrich Gustav Magnus 19 December 1861 1841

Berlin 19 Dec 61.

Die Thermosäule ist fertig! Viel früher als ich erwartet habe! Kleiner hat sich ausgezeichnet! Er hatte indeß nur einen conischen Reflector gemacht. Ich habe, in Folge Ihrer Bestellung, noch einen zweiten bestellt der heut oder in den nächsten Tagen fertig sein soll. Mit diesem kostet die Säule 28 [thaler]. Der 2 Reflector ist ungeheuer theuer, er vermehrt den Preis um 3 Thaler, denn mit <u>einem</u> Reflector kostet sie nur 25 [thaler]. Ich sende Ihnen die Säule durch Asher. Ich halte dies für den sichersten Weg. Sie haben über die Art der Sendung nichts bestimmt. Ebenso bitte ich die 28 [thaler] durch Asher für Kleiner zugehn zu lassen.

Ich werde heut die erwähnte Notiz in der Acad. vortragen, und Ihnen sobald sie gedruckt ist ein Exemplar senden. Ebenso darf ich Sie wohl um ein Exemplar Ihrer Abhandlung bitten, sobald sie gedruckt sein wird.

Ich komme auf die Sache nicht noch ein Mal zurück, aber begreifen kann ich nicht wie der Wasserdampf, der bei 15°C nur 13,5 Millim. Spannung hat und also nur 56 von dem Volumen der Luft einnimmt, eine 40 Mal größere Absorption üben soll als diese. Doch Sie haben sich dieses gewiß vollkommen überlegt.

Wie schrecklich ist der Tod des Prinzen Albert! Dr Becker, sein früherer Secretair und Bibliothekar ist von hier nach London gereist. Sollten Sie ihn sehn so bitte grüßen Sie ihn und sagen Sie ihm, seine Sachen stünden alle unverändert. Ich habe in diesen Tagen nicht in das Laboratorium kommen können weil ich einen schlimmen Fuß habe. Ich schreibe deßhalb dies vom Sopha und bitte daher die schlechte Schrift zu entschuldigen.

Mit vielen Grüßen für Faraday und Allen die sich meiner erinnern

Ihr | GMagnus.

Berlin 19 Dec 61.

The thermopile is ready! Much earlier than I had expected! Kleiner[1] has excelled himself! He had, however, only made one conic reflector. Following your request, I ordered a second one that should be ready today or in the next days. With this, the thermopile will cost 28 [thaler]. The 2nd reflector is incredibly expensive, it increases the price by 3 Thaler,[2] as with <u>one</u> reflector it costs only 25 [thaler]. I will send you the thermopile through Asher.[3] I consider this is the safest way. You did not specify anything about the type of dispatch. I also ask you to send through Asher the 28 [thaler] for Kleiner.

I will present the mentioned report[4] today at the Acad. and will send you a copy as soon as it is printed.[5] Similarly may I ask you for a copy of your paper[6] as soon as it is printed.

I will not return to the matter but I cannot comprehend how steam, which has only 13,5 millim. tension at 15°C and thus takes up only 6 of the volume of air is supposed to exert an absorption 40 times larger than the latter. But I am sure that you have considered this thoroughly.

How awful is the death of Prince Albert![7] Dr Becker, his former secretary and librarian travelled from here to London. If you happen to see him, please give him my regards and tell him that his things remain unchanged. I am not able to go to the laboratory these days because I have a sore foot. I am therefore writing this from the couch and ask you to forgive the bad handwriting.

With many greetings to Faraday and everyone who thinks of me

Yours | GMagnus

RI MS JT/1/M/27

1. *Kleiner:* instrument maker and/or dealer in Berlin.
2. *Thaler:* Prussian currency.
3. *Asher:* Adolphus Asher. Asher and Company had shops in Berlin and London and was a major supplier to the British Library.
4. *the mentioned report:* see letter 1838.
5. *Send you a copy:* see letter 1838, n. 3. A translation appeared in *Phil. Trans.*, 23 (1862), pp. 249–52.
6. *Your paper*: J. Tyndall, 'Remarks on recent researches on radiant heat', *Phil. Mag.*, 23 (1862), pp. 252–66.
7. *the death of Prince Albert:* Albert, Prince Consort and husband of Queen Victoria died on 14 December 1861 from typhoid fever.

To George Gabriel Stokes 21 December 1861 1842

21st Dec. 1861

My dear Mr. Stokes,

I have a paper ready to be placed in your hands:[1] I lecture on the subject of it on Friday evening January 17th at the Royal Institution.[2]

Last year it was not deemed necessary to bring my paper forward at the Royal Society before it formed the subject of my Friday evening lecture. But then it was the Bakerian Lecture[3] & that may make a difference.

Would you kindly inform me what the arrangement will be[4] this year so that I may hold myself prepared for it?

I had no opportunity of talking to you about the convection question but

I see no escape from the conclusion that, with the same difference of temperature, the velocity of the particles of air at half an atmosphere of tension, is to the velocity at a whole atmosphere of tension is as 64: 34.

This retardation in the case of the denser gas I can refer to no other cause than internal friction.

I do not like to trust myself even in simple matters when I am hard pressed as I am at present—and I sh[d] therefore be much obliged to you for your opinion on the following point:—Supposing two gases of the specific gravity S and S' at the common tension of 1 atmosphere to be allowed to stream thro' a small aperture into a vacuum what will be the relative volumes discharged in a given time supposing the elastic force to have no resistance save the inertia of the gas itself to overcome—that is to say supposing friction internal & external to be entirely absent?

Yours very truly | John Tyndall

RI MS JT/T/1/397

---

1. *a paper to ready to be placed in your hands*: published as Tyndall (1862).
2. *at the Royal Institution*: Tyndall lectured on Friday 17 January at the RI on the subject of the transmission of heat through gases. Abstracted as J. Tyndall, 'On the absorption and radiation of heat by gaseous matter', *Roy. Inst. Proc.* 3 (1858–62), pp. 404–7.
3. *Bakerian lecture*: an annual lecture established through the bequest of Henry Baker FRS (1698–1774) and delivered by one of the Fellows of the Royal Society. Tyndall's lecture was published as Tyndall (1861). This was his second Bakerian lecture. Over his lifetime, Tyndall was invited to give four Bakerian lectures, more than anyone since.
4. *What the arrangement will be*: for Stokes' response, see letter 1843.

From George Gabriel Stokes 24 December 1861 1843

Lensfield Cottage Cambridge | 24 Dec. 1861

My dear Dr. Tyndall,

Although there is a general direction that papers be read in the order of their presentation full power is given to the President[1] to "order otherwise". If it were convenient to you to bring forward the subject at the Royal Institution on Jan. 17th that would be sufficient reason, I feel sure, for taking your paper out of its turn.

The 9th is engaged, or rather the papers for it have been announced, though I hardly expect they will occupy the evening, but I don't like to take your paper without its having been announced. If the 16th will suit I have no doubt it can be taken then. The sooner you send in your paper the better that it may be marked as received, and so the order of reading may deviate the less from that according to which it would be taken in its proper turn. Please not to forget that we shall want an abstract.

I don't know that it is necessary for reception at the R.S. that the paper should not have been read elsewhere (or rather the subject of it given) before it is read at the R.S. but it is better that it should come first before the R.S., or at any rate that it should have been received by the R.S. some time before the subject was given elsewhere. In that case it may well be deemed not the author's fault that it was not also read.

I have a meeting to attend in a few minutes, and as I don't like to lose the post I will send off this as it is and write again[2] about the question of convection in gases.

Yours very truly | G. G. Stokes

RI MS JT/1/S/226

1. *given to the President:* referring to the President of the Royal Society.
2. *write again:* see letter 1846.

To Nevil Story-Maskelyne 26 December 1861 1844

26th Dec 1861

My dear Maskelyne

I have received the reply of the Trustees[1] to my application,[2] and have been directed to apply to you. What I require for my researches are two plates of salt three inches in diameter and an inch thick. I would engage if required to do so, and if they are not used out in obtaining results ten million times their value, to return the said plates to the British Museum, to be applied to the future advancement of science as the Trustees shall deem desirable.

Yours very truly | John Tyndall

N.S. Maskelyne Esqr

Natural History Museum DF15/2

1. *the Trustees:* The Trustees of the British Museum.
2. *my application:* letter missing.

From William Hopkins 27 December 1861 1845

Cambridge | Dec. 27/61

Dear Tyndall,

It will be my object in the glacier paper I am writing[1] to fit and frame the theories of glacial motion together, as far as they harmonize into a consistent whole. I define solidity carefully and call ice solid (not confounding solidity and rigidity as Forbes has always tried to do unjustifiably); and I define viscosity and say that ice cannot be viscous, maintaining that your experiments of the crushing and regelation of ice make it no longer justifiable to call ice viscous. I then introduce my own investigations of the internal tensions and pressures, but worked out by a more general mathematical analysis than heretofore. I don't attempt to popularize, but to scientific-ize, the subject. I have applied my results to explain the formation of fissures in various cases; also to explain the pressure theory of the veined structure. I have also endeavoured to explain the different ways in which the internal actions may overcome the cohesion, and to point out the mechanical experiments still wanting to render this part of the subject more complete. I also show what I believe to be the impossibility of Forbes's theory of the veined structure being true. I give a brief account of the sliding of glaciers, in doing which I must necessarily expose Forbes's ignorance of the nature of the forces called into play by that motion. I am also anxious to make it understood how much effectiveness is given by this sliding movement to the internal pressures and tensions in dislocating the mass. Forbes has attempted to do this by appealing to hydrostatic pressure of the water contained in the pores of the ice. I must necessarily betray his ignorance of this problem, and also his entire misconception of the vis-viva.[2]

The paper contains more things that have been done before those things that are new, but they have never been put together systematically before. My purpose is to give the subject the most scientific and demonstrative form it will at present admit of. I don't wish to be controversial, nor do I intend as you will see from this sketch to be historical. I wish to reserve the popular and the historical views for my review in the Quarterly,[3] and I should be most thankful to you for any jottings down you may send to me. I shall have my memoir by me 10 days or a fortnight longer, and should be very glad of any further distinct facts about the veined structure. How far is the longitudinal structure continued down the centre of the Aar glacier? The question of transmission is a difficult one. Do the dirt sands on the Mer de Glace coincide with the structure both in the lower and higher parts of the glacier? It appears to me that accurate observations and measurements are still wanting on this subject. Have you heard any thing more about Forbes's lectures at St. Andrews?

Yours truly, | W. Hopkins.

RI MS JT/1/TYP/2/644–45
LT Typescript Only

---

1. *the glacier paper I am writing*: published as W. Hopkins, 'On the theory of the motion of glaciers', *Phil. Trans.*, 152 (1862), pp. 677–745
2. *his entire misconception of the vis-viva*: the operative force of glaciers (its mass multiplied by the square of its velocity).
3. *my review in the Quarterly*: published as W. Hopkins, 'Glacial theories', *Quarterly Review*, 114 (1863), pp. 76–125.

From George Gabriel Stokes 27 December 1861 1846

Lensfield Cottage Cambridge | 27th Decr 1861

My dear Prof Tyndall,

On the 24th I began a letter to you in continuation of my last,[1] and then Xmas intervened, and yesterday I was attending to another matter. The consideration of all the circumstances which have to be attended to in your war with air, in successive experiments, of different pressures is a complicated matter and demands further consideration. I find I must look out your letter and consider the thing more maturely.

But you have proposed one simple question which I may as well answer at once. If a gas discharges itself from a vessel at one given pressure (or the atmosphere) into another at another given pressure through a hole in a thin plate (in order to get rid as far as may be of the effects of internal friction) then the velocity of discharge varies inversely as the square root of the density under a standard pressure, the effects of internal friction being assumed to be insignificant compared with those of inertia. In comparing gases of different nature we should find this law of efflux on the supposition that the internal friction may be neglected.

Yours very truly | G. G. Stokes

P.S. For fear I should have misunderstood the object of your question which may be to compare the rates of discharge of the same gas under given circumstances I may mention that the usual formula for the vel[ocit]y, of discharge (v) is

$$v = \sqrt{\frac{2\pi}{D} \log_e \frac{p_1}{p_2}}$$

where

π is a standard pressure

D the density under pressure π

the pressure in the vessel from which the gas comes

——-into which it goes

and the density of the gas moving with the vel[ocit]y v is deemed to be that statically belonging to the pressure .

In this formula the effect of internal friction and of the cold produced by expansion are not taken into account nor those of the 'contraction of the vein' unless in so far as may be done by taking instead of the actual area of the orifice an area reduced in a ratio to be determined by experiment.

The formula gives results differing widely from exper[imen]t when is small. See a memoir by Barré de St. Venant & Wantzel Journal Polytechnique Cah.[2] XXVII p. 85[3] referred to in my report on Hydrodynamics. Rep. Br. Ass^{n} for 1846 p. 13.[4]

Yours very truly | G. G. Stokes

RI MS JT/1/S/227

1. *in continuation of my last:* letter 1843.
2. *Cah:* cahier.
3. *memoir:* B. Wantzel, 'Mémoire et expérience sur l'écoulement de l'air', *Journal de l'École Polytechnique,* 27 (1839), pp. 85–122.
4. *for 1846:* G. G. Stokes, 'Report on recent researches in hydrodynamics', *Brit. Assoc. Rep.,* (1846), pp. 1–20.

To George Gabriel Stokes 29 December 1861 1847

29th Dec. 1861

My dear Stokes,

You have, I am glad to see, received the first and largest instalment of my paper[1]—I sent it off at once on receiving your last letter,[2] but was unable at the time to write & tell you that 40 or 50 pages still remain in my hands. I shall be ready for the 16th if you desire it, but it will be rather hard work to hold forth at the R S on the 16 and at the R I on the 17th. The Xmas lectures moreover leave me but little time for preparation.

The 'convection' part of the paper I have not yet sent to you as I wished to look over it once more. In addition to it I have short sections on the deportment of ozone & of aqueous vapour & a critical analysis of the last papers of Magnus.[3] I will send them all to you very soon.

In your note of the 27th[4] you touch the point which chiefly interests me, when you speak of the aperture in a thin plate being resorted to to get rid as far as possible of internal friction.

Mr Graham[5] you know has made two series of experiments;[6]—in one he drives his gas through a small aperture in a thin plate, in the other through long capillary tubes. The former process of transfer he calls Effusion, the latter Transpiration—the velocities of discharge being totally different in both. Now it appeared to me that the physical difference between transpiration & effusion was that the former dealt with external resistance mainly and the latter with internal resistance. Suppose the capillary tube to be of such a size as just to admit two atoms of the gas to pass through it abreast: these atoms being urged by the same force have no tendency to slide past each other, and the consequence would be that the resistance offered to their passage would be wholly derived from their rubbing against the walls of the canal. Two men equally swift, equally strong & equally desirous of getting forward, would, if jammed abreast in a narrow passage, pass through it from beginning to end, without any tangential sliding of man past man.

In the presence of a small aperture however the particles immediately in front of the orifice are driven swiftly past their fellows right & left, and internal friction ought to come prominently into play. Effusion is like the crush at a theatre door where man elbows himself past man, whereas when the crowd has got past the portal, and into the passage the elbowing ceases, effusion being changed for transpiration.

Graham finds that the velocity of transpiration of any gas is proportional to the pressure, and inversely proportional to the length of the capillary tube. In effusion on the contrary he finds that augmenting the density has but little influence, a double pressure producing only a slightly augmented discharge. How can this be explained otherwise than by assuming that for the same gas the internal resistance augments approximately as the pressure. It is a case of E/R where E represents the elastic force & R the resistance. As in the voltaic current you may augment E, but if R be augmented in the same proportion the discharge remains the same. Miller[7] describes the effect of pressure in these words. 'Change in the density of the gas has but little influence on the rate of effusion, the volume effused in a given time being nearly uniform whatever the amount of condensation or rarefaction.'[8] And again with reference to transpiration he says—'The rate of transpiration for the same gas increases directly as the pressure.... This is a very remarkable result & stamps the process of transpiration with a character quite unlike that of effusion.'[9] (Chem. Phys. Pl. I pp 89–91).

What strikes me principally is this, that in my experiments on convection I find that augmented density plays the same part as in Graham's experiments on effusion—but in my case the ineffectiveness of the greater density is I think due to internal friction, which augments approximately as the density. The question is do not the results of Graham admit of the same explanation?

Yours very sincerely | John Tyndall.

RI MS JT/T/1/398

---

1. *my paper:* full version published as Tyndall (1862) and noted as received 9 January and read 30 January.
2. *your last letter:* letter 1846.
3. *last papers of Magnus:* as recommended by Tyndall, these had appeared in translation in two issues of the *Phil. Mag.* in July and August, 1861 (G. Magnus, 'On the propagation of heat in gases', *Phil. Mag.,* 22 (1861), pp. 1–11; 85–106). The German original had been published in April of that year (G. Magnus, 'Ueber die verbreitung der warme in den gasen', *Annalen der Physik und Chemie,* 112 (1861), pp. 497–548).
4. *in your note of the 27*th*:* letter 1846.
5. *Mr Graham:* Thomas Graham (1805–69).
6. *has made two series of experiments:* as described in, T. Graham, 'On the motion of gases', *Phil. Trans.,* 136 (1846), pp. 573–632.
7. *Miller:* William Allen Miller (1817–70).
8. *". . . condensation or rarefaction":* W. Miller, *Elements of Chemistry: Theoretical and Practical. Part I Chemical Physics* 2nd ed. (London: John Parker and Sons, 1860), p. 89. The final clause, 'whatever the amount of condensation or rarefaction', does not appear in first edition.
9. Miller, *Chemical Physics* (see n. 8), p. 90.

1862

From George Gabriel Stokes 1 January 1862 1848

Lensfield Cottage Cambridge | 1 Jany 1862

Dear Prof[esso]r Tyndall,

I delayed answering your last letter[1] because I wanted first to look into Graham's papers.[2]

His results as to effusion convince me that in that case the velocity of discharge is regulated almost entirely by inertia to be overcome, and not by internal friction. If the gas were perfectly smooth to itself, I don't believe the vel[ocit]y of effusion would be much increased. Graham finds that the experiments follow these laws.

(1) The vel[ocit]y of effusion into a vacuum (estimated with reference to the density the gas has in the reservoir) of a given gas is independent of the pressure or very nearly so.

(2) With different gases the velocity varies inversely as the square root of the density under a given pressure.

Now these are precisely the laws which would have to be obeyed by perfectly frictionless gases, in which the velocity of efflux would be regulated solely by inertia, whereas we have reason to believe that the 2nd law (whatever might become of the first) would <u>not</u> be obeyed if the vel[ocit]y of effusion were regulated by internal friction mainly; for we have reason to believe that in different gases the coefficient of friction depends on the nature of the gas, and not simply on its specific gravity.

I have attended much to internal friction, and have succeeded in solving the problem (and an uncommonly tough one it was) of determining the motion of a sphere or cylindrical rod oscillating as a pendulum in a liquid or gas, taking into account the internal friction in the fluid. I send you a copy of my paper;[3] I have no spare copy of my first paper on internal friction,[4] in which I have obtained in a particular way the general equations; but this does not signify, for I take for granted you don't want to go into the mathematics, and I send it that you may cast your eye over some of the tables giving the comparison of theory and experiment, especially those at pp. 59 and 62 which are the most important. The quantity *[η]* in a frictionless fluid would be 2 for a long cylinder, and 1½ for a sphere, irrespective of the diameters. All the rest is unaccounted for on the Common theory (that in which a fluid in its motion is supposed perfectly smooth) and the theory which I have applied shows that they (the differences) ought to be accounted for if the value of one disposable constant (*[r']*, the index of friction) be properly assumed; and when the great difference in size and shape of the pendulums is taken into account, and it is remembered that the formula applied contains no second disposable constant whereby a seeming agreement between theory and observation might be brought about, the agreement will I think be deemed highly satisfactory.

Now one supposition made in deducing the formula is that there is no finite slipping of the fluid over the surface of the solid—in other words that the in receding from the surface of the solid the velocity of the fluid changes in a continuous manner from that of the solid to that of the fluid at some distance from the solid. This supposition, which seemed extremely probable a priori, is in accordance with an experiment of Sir James South's[5] mentioned in the introduction to my paper, and is confirmed by the remarkable agreement between the result of calculation and observation exhibited in my tables. Now if we may extend the same supposition to transpiration tubes, i.e. that there is no abrupt sliding of the fluid over the side of the tube, the velocity of transpiration will be determined by internal friction. The fluid will in fact move in cylindrical shells (the tube being supposed of circular bore) with a velocity increasing from *o* at the surface to a maximum in the middle. A slice of any shell

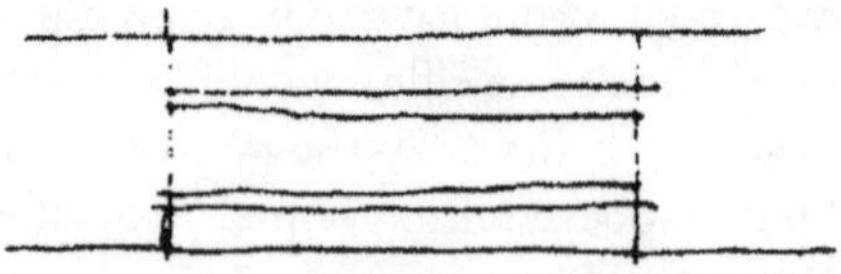

will be urged onwards by the difference of pressure at its two ends, and by the friction of its neighbor inside, and retarded by the friction of its neighbor outside. The velocity in comparing different points of a cross section of the tube will vary as $a^2 - r^2$ (Camb. Phil. Trans. VIII, 305) where r is the distance from the axis, and a the radius of the bore. The successive positions of a cross line of particles will be like

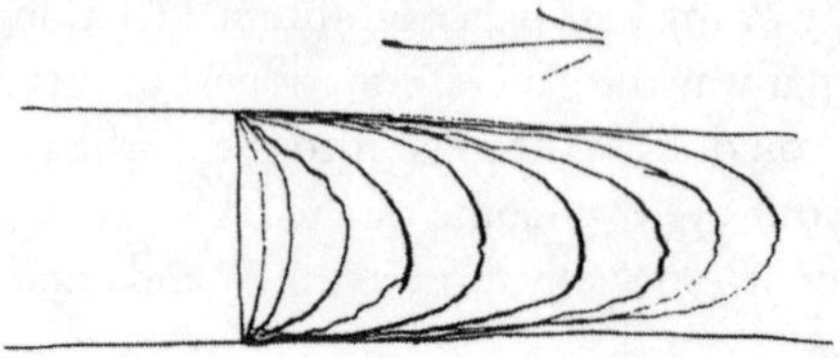

and it is evident that on this supposition of the absence of <u>abrupt</u> sliding over the side of the tube the rate of transpiration <u>must</u> depend on <u>internal</u> friction only, since we may imagine a solid cylinder of fluid comprising all but the whole to move within a pipe composed of the same fluid.

If there be any finite slipping of the fluid, its effect would doubtless be far more sensible in transpiration experiments than in pendulum experiments. Its existence would be best exhibited by the law of the discharge as regards the diameter of the bore. The laws as to difference of pressure and length of tube would be equally obeyed in both cases, i.e. whether the friction be mainly internal or external. The discharge per second ought to vary on the former supposition as the 4th power on the latter as the cube of the diameter. M. Poiseuille[6] (Savants Etrangers IX 433)[7] found the law of the 4th power to be obeyed in the transpiration of water and alcohol through capillary tubes. Helmholtz and Piotrowski[8] (Sitzungsberichte of Vienna Acad. XL, 607)[9] found that in order to reconcile theory and observation with reference to the motion of a hollow sphere filled with liquid & also to account more exactly for Poiseuille it was necessary to admit a small finite slipping. An experiment of Coulomb's[10] which I have mentioned (p. 74 of paper sent) in discussing his results would seem to indicate that in that case the slipping was insensible. A direct exper[imen]t of Helmholtz and Piotrowski's showed however that a small but sensible difference was produced according as water was in contact with glass or silver.

Now a consideration, on the one hand of certain capillary phenomena relating to solids not wetted by the liquid, and on the other of the ignition of a mixture of H and O by sponge platinum, and allied phenomena, make it highly probable that of the two forms of fluid, the liquid and the gaseous, the former is that in which, if in either, slipping is more likely to occur. But we have seen reason to think that even in the transpiration of liquids the effect of slipping is slight, so that in the transpiration of gases it is probably insensible. If so the rate of transpiration is regulated by <u>internal</u> friction, which is also confirmed by what Graham found that the <u>nature</u> of the tube (e.g. whether it was glass or copper) was a matter of indifference.
As far as may be gathered from a single experiment of Gen[era]l Sabine's (see p. 2) we may say that in a given gas at a given temperature the internal friction varies as the density, or $\mu \alpha \rho_1$, or μ^1 the "index of friction" is constant.

In your convection exper[imen]t the hot face is found to lose a given amount of heat in a given time whether the air be at whole or half pressure. In the latter case the same amount of heat entering the air would produce a double expansion in a given volume or (what I take it comes nearer the actual circumstances) the same expansion in a double volume. In either case the total ascensional force due to gravity would be the same as in air of whole pressure, and on the supposition of a double vel[ocit]y, we should have, to make equilibrium with this, the same total force as for full pressure, the force being halved for a given velocity in consequence of the reduced density, and doubled for a given density in consequence of the augmented velocity *

Yours very truly | G.G. Stokes

The memoir by book post.

P.S. I was forgetting to write about your paper.[11] I don't think the paper need be read on the 16th but to prevent any objection being raised I think the paper ought to be in our hands complete by the 16th. The title could then be read for form's sake if thought desirable.

* My notion is that the heat enters the air by way of radiation and absorption. The transparency to heat w[hic]h you found for air does not disprove this because your heat was necessarily sifted by rock salt, and in examining rock salt Melloni's[12] heat was sifted by air. Any heat with respect to which rock salt and air are both black would not show itself in either Melloni's exper[imen]ts or yours.

RI MS JT/1/TYP/4/1377–80

1. *Your last letter*: letter 1847.
2. *Graham's papers*: T. Graham, 'On the motion of gases: part I', *Phil. Trans.,* 136 (1846), pp. 573–632. T. Graham, 'On the motion of gases: part II,' *Phil. Trans.,* 139 (1849), pp. 349–92.
3. *a copy of my paper*: G. G. Stokes, 'On the effect of the internal friction of fluids on the motion of pendulums', *Transactions of the Cambridge Philosophical Society,* 9 (1851), pp. 8–106.
4. *my first paper on internal friction*: G. G. Stokes, 'On the theories of the internal friction of fluids in motion, and of the equilibrium and motion of elastic solids', *Transactions of the Cambridge Philosophical Society,* 8 (1845), pp. 287–319.
5. *Sir James South*: Sir James South (1785–1867), astronomer (*ODNB*)
6. *Poiseuille*: Jean Léonard Marie Poiseuille (1797–69), French physicist.
7. *(Savants . . . 433)*: *Mémoirs des Savants Estrangers* (9) (1846).
8. *Piotrowski*: Gustav von Piotrowski (1833–84), Polish-Austrian physicist.
9. *(Sitzungsberichte . . . 607)*: 'Uber Reibung von tropfbarer Flüssigkeiten', *Sitzungsberichte der Kaiserlich Akademie der Wissenschaften,* 40 (1860), pp. 607–58.
10. *Coulomb's*: Charles-Augustin de Coulomb (1736–1806), French natural philosopher.
11. *write about your paper*: Tyndall (1862). See letter 1847, n. 1.
12. *Melloni's*: Macedonio Melloni, (1798–1854), Italian physicist.

To Arthur Pepys Whately [early January 1862][1] 1849

MY DEAR WHATELY,

When I accepted the vice presidentship[2] of the Alpine Club I fully hoped to be able to attend your meetings and to fulfil to some extent at least the duties of my position.

For a long time previous however I had entertained the idea that I ought to withdraw from the Club, as I felt that I was a mere cumberer of the ground.

This idea has become stronger than ever within the last few days. I am and shall be oppressed with work and hence it is utterly impossible for me to attend your meetings or to do anything which could practicably promote the interests of your body.

I therefore with regret ask you to accept my resignation, and I beg of you and those of my friends whose kindness I could discern in the act of my election to pardon me for temporarily accepting a position which I am in no way qualified to fill.

My brain at present is rushed almost to its bearing limits: Do not ask me therefore to correspond further on this matter. In accepting my resignation without any expostulation[3] you are doing me a kindness and behaving as my friend.

Believe me, dear Whately,[4] | Yours most truly, | JOHN TYNDALL

Life and Work, p. 389.

---

1. *[early January 1862]*: see letters 1836 and 1850 for date range rationale.
2. *When I accepted the vice presidentship*: letter missing. See letter 1836 for communication informing Tyndall of his appointment to the position.
3. *without any expostulation*: on the other possible motives behind Tyndall's resignation, see *The Ascent of John Tyndall,* pp. 224–25.
4. *Whately*: see letter 1850 for Whately's reply.

From Arthur Pepys Whately [Early January 1862][1] 1850

DEAR TYNDALL,

It is hardly within my authority to accept resignations of officers of the Alpine Club,[2] but I will lay your letter[3] before the Committee at their next meeting, which will not, however, take place for the next two or three weeks. I need hardly say that I shall do this with very great regret, on more grounds than one, and it would afford me great pleasure to hear that a reconsideration of your determination had relieved me from this most unpleasant duty.

I am, | Sincerely yours, | A.P. WHATELY

Life and Work, pp. 389–90

1. [*Early January 1862*]: The Committee of the Alpine Club met on 27 January 1862 (27 January 1862, Committee Minutes, Alpine Club Library Archives). Whately notes in this letter that he writes two or three weeks before that meeting.
2. *Alpine Club*: see letter 1836 for appointment of Tyndall as Club Vice-President.
3. *your letter*: letter 1849.

To George Gabriel Stokes 3 January 1862 1851

Royal Institution | 3rd. Jan. 1862.

Dear Mr. Stokes

I have received your letter[1] & read it with profit—Accept my best thanks for it. Long may you be able as the years run round to wield at the commencement of each period a pen as effectively as you do now.

Yours truly | John Tyndall

One word with regard to the influence of the rocksalt plates. Take a large surface—uniformly heated. Present your thermo-electric pile to it furnished with a conical reflector blackened within—The production of the cone intersects the surface in a circle, & this circle is the available radiating space. All parts outside of it send rays against the internal surface of the cone & such rays are there absorbed. Retreat from the surface—say to ten times the first distance—the available radiating surface increases as the square of the distance but the deflection of the galvanometer remains perfectly constant. This is a proof that the heat diminishes in intensity exactly as the square of the distance increases. When the air is very dry you may push this to any reasonable laboratory distance. Now the law could not thus clearly come out if the air had exerted a sensible absorption.

RI MS JT/T/1399

1. *your letter*: letter 1848.

From George Gabriel Stokes 4 January 1862 1852

Lensfield Cottage Cambridge | 4 Jany 1862

Dear Prof. Tyndall,

I have been further considering[1] Graham's results, and have arrived at a rather striking conclusion.

Theory and experiment cannot be compared accurately as to absolute results, on account of the irregularities as to form and area of a section of the bore of a capillary tube. I have however taken one of Graham's experiments (Phil. Trans. 1846 p.603 Table XVII)[2] where the diam[ete]r, of the tube, in a part measured, is found to be .0222 in. For want of better data I have assumed the bore uniform and circular throughout, though as the tube was made up of different parts joined together by fusion, and as it was several times bent, and as Graham remarks the bore was probably narrowed at the bends, the resistance is doubtless much underrated by the supposition of a uniform and circular bore of .0222 in. diameter.

Assuming the resistance to be due to internal friction, and the index of friction

$$\mu^1 \left(= \frac{\mu}{\rho} \right)$$

to have the numerical value ν deduced from Baily's[3] pendulum experiments viz.

$\nu\mu^1$=.116 inch,

I have deduced the times of transpiration from theory on two suppositions as to the relation of μ to the density.

That $\mu \alpha \rho$ or μ^1 is constant

That μ is independent of ρ

It will be seen (see table next page) that the law of discharge is given almost exactly on the second supposition; and the absolute value is not so very different from the observed when it is considered that the resistance was doubtless much under-rated in assuming the base uniform, and that the numerical constant on which the whole depends was got from such a totally different class of experiments, namely exper[imen]ts on the effect of air on the time of vibration of a pendulum.

Range of wire gauge	Time by Exp[t]	Time μ^1 const.	by theory μ const.	<u>Time by theory</u> Time by Exp[t] μ^1 const.	μ const.
28 to 24	213	72	127	.338	.596
24 to 20	227	85	135	.375	.592
20 to 12	549	239	325	.435	.592
12 to 8	388.5	189	228	.488	.587

Seeing that the effects of contractions of the bore, or even changing it without contraction to an elliptical form, and those of effusion so far as they may have been operative in a sensible manner, would both be to augment the observed times, it is hard to say how much, if any, of the observed deficiency (about 2/5ths of the whole) would remain to be explained (by a small finite slipping or otherwise) if the two named causes of retardation were got rid of.

The same conclusion, that μ is independent of ρ, I deduce independently from Graham's two laws that the velocity of transpiration through a very long capillary of uniform bore into a vacuum, estimated at the density at the high pressure side, varies directly as the pressure and inversely as the length of the tube.

It is remarkable that Maxwell arrived at the same conclusion[4]—that μ is independent of ρ—in the speculations in which he assimilates gases to collections of enormous numbers of perfectly elastic balls moving about and impinging in all possible ways. It seemed to him however almost too paradoxical to admit.

As to the absorption of heat by air my notion was that possibly air changed <u>very rapidly</u> with a change of quality (refrangibility) of the radiant heat from highly opaque to highly diathermanous, so that it was only a <u>proportionately small</u> fraction of the whole initial radiation with respect to which air was <u>moderately</u> diathermanous, and which alone could be rendered sensible in such an experiment as that you mention. Likely or unlikely, such a view is I think conceivable.

Your conclusion about the transmission of heat (radiant heat) by a smoked plate of rock salt is so strange that one wishes for rather fuller details, to obviate a possible objection that the heat passed instantly through the salt, warmed the soot, and that radiated in all directions. If the plate were placed either very near the source or very near the pile such an effect might be produced to a sensible degree, but hardly if the plate were placed midway, a tolerable space intervening between the source & the pile. If it be true transmission, the galvanometer ought to tell the same tale wherever the plate were placed, supposing it large enough (or the source sufficiently limited) to intercept all the rays which fall on the pile. Or the amount of heat which is either scattered or absorbed and re-emitted by the soot might be measured separately by interposing a screen which should allow the pile to see the plate but not the source, and correcting for the angle of obliquity (multiply by the secant for the emitted part)

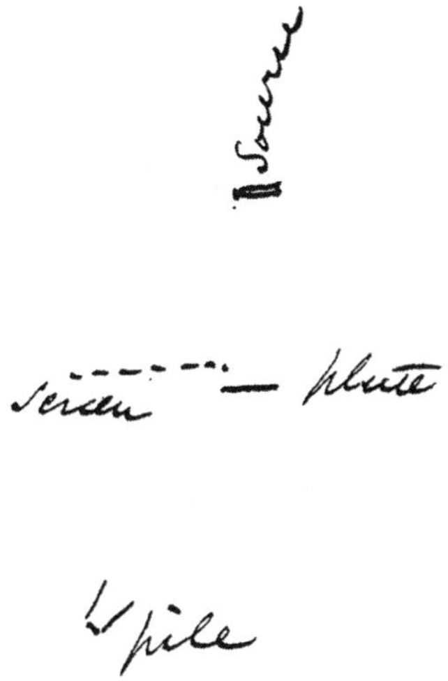

Yours very truly | G.G. Stokes

turn over

P.S. If the observed heat be scattered that would still imply diathermancy of soot but the effect could not be observed in the same way. To prove that it was not absorption and re-emission it would I think be satisfactory to show that a piece of silver foil blackened by soot on both sides and interposed as a screen in place of the blackened salt produced no, or only a trifling deflection of the galvanometer.

RI MS JT/1/S/229
RI MS JT/1/TYP/4/1381–83

---

1. *I have been further considering*: see letter 1848.
2. *(Phil. Trans. 1846 p.603 Table XVII)*: T. Graham, 'On the motion of gases,' *Phil. Trans.*, 136 (1846), pp. 573–631.
3. *Baily's*: Francis Baily (1774–1844), British astronomer and stockbroker (*ODNB*)
4. *Maxwell arrived at the same conclusion*: J. Maxwell, 'Illustrations of the dynamical theory of gases, part I', *Phil. Mag.*, 19 (1860), pp. 19–32.

From George Gabriel Stokes 6 January 1862 1853

Lensfield Cottage | Cambridge | 6 Jan. 1862

Dear Prof[r] Tyndall,

Some of Graham's results,[1] especially the great difference in the rate of transpiration he obtained with a long capillary of tapering bore discharging into a vacuum ascending as the narrow or wide end of the bore was foremost, (P.T. for 1846 p. 598)[2] lead me to think that the resistance in his tubes was not sensibly proportional to the first power of the velocity (mean velocity throughout the section at that point) but that a part depending on the square of the velocity was also very sensible, though the former was predominant. Now if the observed results were deemed to be produced by the first power of the vel[ocit][y] by an unknown function μ of the density, the really existing term depending on the square of the vel[ocit][y] being most important at the end of the tube near the receiver, where the density is less and the vel[ocit][y] greater, would cause us to exaggerate the value of μ at a low density.

Til more experiments shall have been made, directed to this particular end, I don't think we can say anything as to the variation of μ with ρ.

Yours very truly | G.G. Stokes

RI MS JT/1/TYP/4/1384

1. *Some of Graham's results*: see letters 1847, 1848, 1851, and 1852 for previous discussion of Graham's experiments.
2. *(P.T. for 1846 p. 598)*: T. Graham, 'On the motion of gases: part I', *Phil. Trans.*, 136 (1846), pp. 573–632.

To John Francis Julius von Haast 8 January 1862 1854

Royal Institution | Albemarle Street London | 8th January 1862

My dear Sir

I reviewed your kind & flattering letter[1] and your interesting little book[2] a few days ago. For both of them pray accept my best thanks. When I commenced my observations upon glaciers I had very little idea of the tidings of my work reaching so far as New Zealand, and I can unfeignedly assure you that I consider it a very high reward to find that a scientific explorer in that distant region thinks my work worthy of recognition.

It has been now my custom for some years to spend annually a few weeks in the Alps. This year I could only afford myself a brief holiday but I accomplished the ascent of one splendid mountain which had hitherto baffled all attempts made to scale it—I mean the Weisshorn. It is close upon 15000 feet high & is certainly the most magnificent cone in Switzerland. Of all the really difficult mountains the Matterhorn now alone remains, and the man who reaches its summit, will, I think, perform a more difficult piece of work than any hitherto accomplished.

I have written out an account of the whole expedition of this year—it will be published by Longman in the course of the spring[3] and as soon as it appears I shall do myself the pleasure of sending you a copy.

Were it not so very far I should be tempted to go to New Zealand and make an excursion among these never trodden peaks. My friend D[r] Hooker has also imbued me with a strong desire to go to India, but I fear I shall not be able to accomplish the journey. At all events if I cannot go to New Zealand it will be a pleasure to me to reflect on your kindness, and to remember that the most agreeable recompense to which a mountaineer could aspire[4] has through that kindness been secured to me.

Wishing you health and strength to pursue your explorations

Believe me dear Sir | Very faithfully yours | John Tyndall

Alexander Turnbull Library, National Library of New Zealand MS 0037–143

---

1. *your kind & flattering letter*: letter 1806.
2. *interesting little book*: J. Haast, *Report of a topographical and geological exploration of the western districts of the Nelson province, New Zealand: undertaken for the provincial government by Julius Haast, Esq.* (Nelson: C. and J. Elliott, 1861).
3. *Published by Longman in the course of the spring*: *Mountaineering in 1861* was published by Longmans on 16 April 1862.
4. *to which a mountaineer could aspire*: referring, presumably, to Haast's naming a New Zealand mountain after Tyndall. See letter 1806.

To Miss Moore[1] 9 January 1862 1855

Royal Institution | 9th Jan. 1862.

Thanks my dear Miss Moore for your kind note—the attention of the little fellows[2] quite cheered me up & I got through the ordeal quite as well as I expected.

The Yankees[3] though mad are not devoid of that shrewdness which sometimes associates itself with madness—What a set of vulgarians they are—No doubt there are choice & refined spirits among them, but their position amid such awful scum cannot be enviable.

Kindest regards to your mother[4] and sister, and long may you be able to write to me & cheer me as you have hitherto done.

Yours ever | John Tyndall

RI MS JT/1/T/1074

1. *Miss Moore*: Harriet Moore (1801–84) or Julia Moore (1804–1904).
2. *the attention of the little fellows*: Tyndall delivered a course of six lectures for a juvenile audience at the RI, with the last one being given on Tuesday 7 January 1862.
3. *The Yankees*: those on the Union side of the American civil war. The historical event presumed here might be the *Trent* affair that had threatened to precipitate a war between the United Kingdom and America, then in the throes of the civil war.
4. *your mother*: Harriet Moore, wife of surgeon and vaccine promoter James Carrick-Moore. She had two daughters then living: Harriet Jane Moore, a celebrated watercolour artist and Julia Moore.

To Thomas Archer Hirst [10 January 1862][1] 1856

Friday m[orni]ng

My dear Tom.

I have delivered your message to M^rs^. Tunstal.[2] Debus is here with me at present—My health is quite good—all I need is to live according to the law of Reason to keep it good. The Juveniles are past;[3] and thank the gods without a failure. I at first thought I should have no audience. The American affair[4] & the Prince Consorts death[5] turned peoples minds away from science—But to my surprise the audience amounted to nearly 500—The felicitations have been general—Everybody says that every body was gratified and thus this piece of work has rolled pleasantly off my hands.

Debus will be here when you return on Sunday.[6] Until then believe me

Yours affectionately | John

RI MS JT/1/T/972

1. *[10 January 1862]*: see n. 3 and 6 for rationale.
2. *Mrs Tunstal*: Hirst and Tyndall's housekeeper at 14 Waverley Place.
3. *The Juveniles are past*: Tyndall gave his last of six Christmas Lectures for a juvenile audience on Tuesday 7 January 1862.
4. *The American affair*: this may refer to the *Trent* affair that had threatened to precipitate a war between the United Kingdom and America, then in the throes of the civil war. See letter 1855, n. 3.

5. *The Prince Consorts death*: Prince Albert died on 14 December 1861.
6. *when you return on Sunday*: Hirst did return on Sunday 12 January and found Debus with Tyndall. See *Hirst Journal*, 19 January 1862, RI MS JT/2/32c/1600.

From Mary Susanna Stokes[1] 11 January 1862 1857

Dear Mr. Tyndall,

I have to thank you for your most agreeable as well as valuable 'reminder of the Montanvert.' 1857 gave me my only peep at snowy mountains,[2] and I have often longed since to go again, & look with hope to some long vacation when we may take all our children to be happy with us, & all the better to enjoy it from having your little book.[3]

Wishing you a happy new year, &, to use a North of Ireland word, many happy moutainy days.

Believe me yours truly | Mary S. Stokes

Lensfield Cottage | Cambridge Jan 11.

RI MS JT/1/S/285

---

1. *Mary Susanna Stokes*: (1824–99), wife of George Gabriel Stokes (1819–1903) and daughter of the astronomer Rev Thomas Romney Robinson, was raised in Carrickmacross, County Monaghan, Ireland.
2. *1857 gave me my only peep at snowy mountains*: Tyndall met Mary Stokes and her new husband on 21 August 1857 on the glacier as he descended to Chamonix. See Journal, RI MS JT/2/13c/1029.
3. *your little book*: *Glaciers of the Alps* (see letter 1595, n. 4).

From Nevil Story-Maskelyne 20 January [1862][1] 1858

Jan 20

My dear Tyndall

I made the requisite application for authorisation to have the salt specimen removed for cutting; but I am informed that the 'small piece' contemplated by the Trustees in their acquiescence in your request[2] is not the same thing as two pieces 'considerable in size for this material' which was what I had reported you to require. The result is that the matter is to be again discussed at the Board of the Trustees & my proposal for the cutting the man & retaining the property in the portion, to be considered.

I am sorry for the delay to you.

Hereafter when some future Tyndall wants a piece more than twice as big as you do I or my successor may have to regret our grand and as I believe unique specimen having been divided into two pieces.

It seems like cutting up for a pocket glass a disc of perfect flint glass big enough for an exceptionally large Telescope.

You will not share this feeling.[3] The true question however is how is our spur[4] best to embrace this aim of science.

Natural History Museum, DF15/2

1. *[1862]*: this is a reply to letter 1844.
2. *your request*: letter 1844.
3. *Your will not share this feeling*: for Tyndall's reply, see letter 1859.
4. *Our spur*: or 'our spar'.

To Nevil Story-Maskelyne 22 January 1862 1859

22nd. Jan. 1862.

Dear Maskelyne

Your 'lens' illustration is altogether defective,[1] but if even shown to be so, another, I doubt not would be found; and unhappily for me I have not the time necessary to argue the case out.

I send by this post a letter to the Trustees;[2] it is rather with a view of excusing my own boldness than with the hopes of removing the obstacles which oppose the granting of my request.[3] Do not I beg of you give yourself any further trouble on my account in connexion with this subject.

Yours truly | John Tyndall

Natural History Museum DF15/2

1. *Your 'lens' illustration is altogether defective*: refers to Maskelyne's analogy found in letter 1858.
2. *the Trustees*: Trustees of the British Museum. The letter is missing.
3. *my request*: see letter 1844.

From Hector Tyndale 23 January 1863 1860

Sandy Hook, Maryland, | (Opposite Harper's Ferry, Virginia)
January 23[d] 1862

My dear John

It really gave me great pleasure to receive your short note of <u>no date</u>.[1] November last probably, as I received it almost a month since, having come through the channel of home. It did me good to hear of Hirst and yourself again, for although an American in England (if he has any character at all) like an Englishman in America, shows his worst points & excites more antagonisms than he deserves, perhaps, and making the application to myself, I am yet more than <u>only</u> glad to hear from you both, as from those friends much sundered many ways, but friends at bottom. Faults which roil nationalism become rather pleasantly salient when not too near! Since the writing of yours our aggregated vanities have been nearly at war. But, I presume our bit of leek-eating has averted it.[2] But <u>our</u> national stomach has been damaged by the meal, and the stomach is called a proud one. And just now every American is in earnest. We have been looking foolishly for sympathy and, properly, got only Manchester morality[3] instead. It is true that what England has done to us makes no show of right for us to do to England—because each nation has it's height chalked against the doorpost of History and no one should be tried by any higher, or lower, standard than it's own mark has made. We were wrong, <u>by our own claims</u>, in taking the American traitors from an <u>English</u> ship[4] even tho' England set us the example, and tho' it is somewhat hard to take a slap in the face from the dirty hands of such thieving vagabonds as those rebel commissioners are. The men have been given up however, and chyle or chyme[5] <u>must</u> be made from the swallowing. Compensations, I trust, do not fail at this point of the World.

Hirst and yourself must have a good time of it now—taking your own brandy and 'mutton-chops' off your own mahogany! I should not mind dropping in upon you of a night and 'trying it on' with you. Be good enough to drink my health and consider me present in the spirit—not in the <u>brandy</u> unless it is good. And then, the luxury of smoking your own window curtains! with no rising ghost of a neat housewife to groan out—<u>pah</u>! after the exit has been made. Why, your 'own vine and fig tree'[6] is a fool to it!

You speak of John Brown.[7] Now there are queer things abroad. I am now stationed here, being Senior Major of the 28th Regm't Penn[a] Vols:, in command of this Post with a battalion of the Regm't: and this Post is immediately opposite (across the Potomac river) from Harper's Ferry,[8] a strategical point in the rebel Slave State of Virginia! It has struck me sometimes, the poetical justice of the thing—to say nothing of political justice of my commanding & dominating there! A part of my Regiment thrashed a body of them at Harper's Ferry in Octr. last. You remember that Harper's Ferry was the scene of Brown's attempt to carry off Slaves, or return them to freedom. This war is desolating all the Southern Country and here upon the border land. Three pretty towns opposite all vacant and deserted—the hand of the Ruler is heavy indeed. Daily the rebel Pickets (we write it so here) and their Scouts prowl around the hills, very beautiful hills, of Virginia, but they keep away from their dens: Somebody said that the Hand would be laid heavily upon Virginia for the hanging, & the manner of it, of John Brown—and Virginia has borne the brunt of this war so far, especially these adjacent Counties. But this can't interest you. About a month after our, Mrs Tyndale's[9] and my, arrival at home I received the appointment I now hold and have been here, or near here, for about six months. You must always remember that our 600.000 men, had to be organized and readied as an Army.

Mrs Tyndale sent her remembrances to Mr Hirst and yourself as also did Mrs Mitchell[10] to you. I hear from home every few days. With kindest regards to Hirst and wishes that he will write me too and if so address as usual as my position will change.

I am very truly affectionately | Your Cousin & friend | Hector Tyndale

RI MS JT/1/T/51

---

1. *note of no date:* letter missing. Perhaps also a swipe at Tyndall, who frequently didn't date his letters.
2. *our bit of leek-eating has averted it*: refers to American back-tracking which brought to an end the *Trent* affair.
3. *Manchester morality*: the civil war threatened to cut off the supply of cotton from the American South. Manchester morality refers to efforts to secure the interests of Britain's cotton industry.
4. *American traitors from an English ship*: Tyndale here refers to the two confederate diplomats taken by force from RMS *Trent* by the Union.
5. *chyle or chyme*: different products of digestion.
6. *'own vine and fig tree'*: a Biblical allusion to Micah 4:4 and here referring to Tyndall's settled and enviable domestic circumstances.
7. *John Brown*: John Brown (1800–59), American abolitionist.
8. *Harper's Ferry*: a town in Jefferson County, West Virginia. The town changed hands eight times during the Civil War.
9. *Mrs Tyndale's*: Julia Tyndale, née Nowlen (1823–97), wife of Hector.
10. *Mrs Mitchell*: not identified.

From Johann Rudolf Albert Mousson 7 February 1862 1861

Verehrtester Herr,

Empfangen Sie <allem> voraus meinen herzlichsten Dank für das schöne Geschenk, das Sie mir letztes Jahr mit Ihrem Buche „The glaciers of the Alps" gemacht haben. Ich ersuchte Freund Clausius Ihnen meine Freude und meinen Dank zu hinterbringen; hoffentlich hat er es nicht unterlassen. Leider fühle ich mir die Kräfte nicht mehr, Ihren kühnen Gängen durch das Gebirge in der Wirklichkeit zu folgen; nicht um so geringeren Genuß gewährt es mir aber Sie in Gedanken zu begleiten und Ihren reichen Bemerkungen und Belehrungen zu folgen. Für mich ist es entschieden, daß Sie den Schlüßel zu der räthselhaften Bandstruktur wirklich gefunden und damit gleichsam den Schlußstein zu unserer Kenntniß über die Gletscher geliefert haben. Außerdem aber enthält Ihr Werk eine Fülle interessanter Beobachtungen und Beme[r]kungen, die einen ganz neuen Reiz auf die Reisen im Alpengebirge werfen; denn Niemand versteht wie Sie die strengen Grundsätze der Wissenschaft in eine so klare, einfache, allgemein faßliche Form zu bringen.

Doch zum Gegenstand dieser Zeilen, der nichts weniger ist, als eine sehr angelegentliche Bitte im Namen unserer naturhistorisch-physicalischen Gesellschaft, deren Ehrenmitglied Sie sind. Gleichzeitig mit dem gegenwärtigen Briefe lege ich ein kleines Schriftchen an Sie auf die Post, welches Sie mit den Verhältnißen und der Persönlichkeit eines jüngern Mannes, Dr Alexander Schlaefli, bekannt machen wird, der, nachdem er durch einen 6 jährigen Aufenthalt in der europeischen [sic] und asiatischen Türkei sich dazu vorbereitet hat, nun im Begriff steht von Bagdad aus eine bedeutendere Expedition nach Zanzebar und der Ostküste Africa's zu unternehmen. Er will sich dort bleibend als Arzt unter den Arabern, deren Sprache und Sitten er vollkommen kennt, niederlaßen, um daneben für Naturgeschichte und Meteorologie, so viel die Umstände gestatten, zu wirken. Dr Schlaefli, jetzt 30 Jahr alt, hat in Zürich studirt, ist tüchtiger Arzt und Naturforscher, voll warmen Eifers für seine wissenschaftlichen Zwecke, dabei klug und verschloßen, zäh, abgehärtet wie kaum einer. Er wird von Zürich aus für seine wissenschaftlichen Expeditionen ausgerüstet und unterstützt und sammelt für die schweizerisch-polytechnischen Sammlungen. Die schweizerische Bundesbehörde hat offizielle Schritte gethan, um ihm als Reisenden des schweizerisch-polytechnischen Museums englische ministerielle Empfehlungen an die Consuln und Residenten der von ihm besuchten Punkte zu verschaffen, was uns denn wirklich verheißen ist. Unsere Gesellschaft, welche die Tüchtigkeit und Zuverläßigkeit des jungen Mannes, der ihr corr. Mitglied ist, kennt, hat die Leitung des Unternehmens durch eine Comission aus Prof.

Heer, Escher v. d. L. und mir, übernommen und sucht ihrem Reisenden nun von allen Seiten Empfehlungen und moralische Unterstützung zu verschaffen. Escher hat es übernommen an Prof. Murchisson zu schreiben und ich nehme die Freiheit mich an Sie zu wenden, da Sie wie ich glaube Secretair der geographical and geological Society sind, um, wenn es irgend möglich wäre eine offizielle Empfehlung von Seiten einer jener Gesellschaften zu erwirken. Bei den Consuln und Residenten, den da und dort zerstreuten wissenschaftlichen Reisenden, den englischen Etablissements, den Angestellten des Seedienstes gilt ein solches Wort der Empfehlung mehr als alle Privatbriefe. Ja wie fern eine solche Begünstigung für einen Mann der nicht direkt in englischen Diensten steht, erhältlich ist, werden Sie natürlich entscheiden müssen; von unserer Seite würden wir natürlich den höchsten Werth darauf legen und Ihne[n] zu *[wärmstem]* Danke verpflichtet sein. Sind ja doch die Früchte solcher Unternehmungen, ich meine die wissenschaftlichen, nicht die materiellen, ein Gemeingut aller Völker und verdienen daher auch von weitern Kreisen aus Anerkennung und Unterstützung. Wir hoffen, die bewährte Liberalität der englischen Gesellschaft auf dem Felde der Wissenschaft werde ihnen gestatten über die Grenzen der *[schweizer]* Nationalität herauszutreten, um der kleinen befreundeten, in ihren Mitteln leider beengten Schweiz unter die Arme zu greifen. —

Wenn Sie eine solche moralische Unterstützung für möglich halten, so haben Sie die Güte, sich mit Prof. Murchisson [sic] über die Art, wie *[1 word illeg.]* es geschehen könnte, ins Reine zu setzen. Unsere Gesellschaft wird Ihnen recht sehr für Alles, was Sie thun können, verbunden sein. —

Genehmigen Sie die Verleihung wahrer Hochachtung und Ergebenheit.

Alb. Mousson | Prof.

Zürich | den 7. Februar | 1862.

P.S. die Punkte, welche Dr Schläfli auf seiner Reise berührt, sind: Bagdad, Bassorah, Abu-Schehr, Mascat, Aden, Mombassa und Zanzebar, dann, da er mit Küstenschiffen fährt, noch andere weniger bedeutende Punkte.

Most esteemed Sir,

Before <everything else>, accept my warmest thanks for the lovely gift that you gave me last year with your book 'The Glaciers of the Alps'. I asked our friend Clausius to mention my delight and my thanks to you; hopefully he did not neglect to do so. Unfortunately, I no longer feel the energy to follow your bold journeys through the mountains in reality; it does not afford me any lesser pleasure, though, to accompany you in my thoughts and to follow your rich remarks and advice. For me, it is settled that you have really found the key to the mysterious veined structure and with this have provided the copestone, as it were, to our knowledge of glaciers. In addition to this, though, your work contains a wealth of interesting observations and remarks that cast a whole new allure on travels in the mountains of the Alps; for no-one understands how to render the rigid principles of science into such a clear, simple and generally comprehensible form like you do.

But to return now to the purpose of these lines, which is no less than a very pressing request on behalf of our Natural History and Physical Society, whose honorary member you are.[1] At the same time as the present letter, I am posting a small little work to you, which will make you acquainted with the circumstances and character of a younger man, Dr Alexander Schlaefli,[2] who, after preparing himself for it by means of a 6-year visit to the European and Asian parts of Turkey, is now about to undertake a more important expedition, starting out from Baghdad, to Zanzibar and the east coast of Africa. He intends to settle there, living as a doctor among the Arabs, with whose language and customs he is completely familiar, in order to do work in natural history and meteorology at the same time, as much as circumstances permit. Dr Schlaefli, now 30 years old, has studied in Zürich, and is a competent doctor and naturalist, full of warm enthusiasm for his scientific aims, yet smart and quiet, tough, and hardy like barely anyone. He is being equipped and supported from Zürich for his scientific expeditions, and is collecting for the Swiss polytechnic collections.[3] The Swiss federal authority[4] has taken official steps to provide him, as a traveller of the Swiss-Polytechnic Museum,[5] with English ministerial recommendations to the Consuls and Residents of the points visited by him, which has actually been promised to us. Our Society, which is familiar with the competence and reliability of the young man, who is its Corr. Member, has taken over the management of the enterprise by means of a committee consisting of Prof. Heer,[6] Escher v. d. L.[7] and me, and is now seeking to provide its traveller with recommendations and moral support from all sides. Escher has taken on the job of writing to Prof. Murchison,[8] and I am taking the liberty of turning to you, since you, as I believe, are Secretary of the Geographical and Geological Society,[9] in order, if it were in any way possible, to obtain an official recommendation on the part of one of those societies. With the Consuls and Residents, the scientific travellers scattered here and there, the English establishments, and the employees of the naval service, such a word of recommendation is worth more than all private letters. Of course, to what extent such a favour is obtainable for a man who is not directly in English service is something you will naturally have to decide; from our side, we would naturally set the greatest score on it and owe you the *[warmest]* debt of gratitude. The fruits of such undertakings, I mean the scientific, not the material ones, are of course the common property of all peoples, and therefore deserve acknowledgement and support from wider quarters too. We hope that the proven liberality of the English Society in the field of science may allow them to step out over the borders of *[Swiss]* nationality in order to help out small, friendly Switzerland, which is unfortunately straitened in its means. —

If you think such moral support is possible, then be so kind as to clear things up with Prof. Murchison about how *[1 word illeg.]* it could happen. Our Society will be very much obliged to you for everything that you can do. —

Permit me to give you my true respect and devotion,

Alb. Mousson | Prof.

Zürich | 7 February | 1862.

P.S. the points at which Dr. Schläfli will be stopping on his journey are: Baghdad, Basra, Bushehr, Muscat, Aden, Mombasa and Zanzibar, and then, since he is travelling by coastal ships, some other less important points.

RI MS JT/1/M/146

1. *our Natural History and Physical Society, whose honorary member you are*: Société de Physique et d'Histoire Naturelle de Genève. Tyndall had been elected in early 1859. See letters 1607, 1610, 1611 and 1614.
2. *Dr Alexander Schlaefli*: Alexander Schläfli (1832–63). Naturalist and doctor, educated in Zürich, Paris and Jena. He spoke nine languages and travelled widely, including around the Black Sea, Mesopotamia and India. He died of dysentery in Baghdad in October 1863 (*Historisches Lexicon der Schweiz,* http://www.hls-dhs-dss.ch/textes/d/D32096.php accessed 13 September 2018).
3. *Swiss polytechnic collections*: not identified, but presumably collections of the Federal Polytechnical School in Zürich, which had opened in 1855.
4. *Swiss federal authority*: not identified.
5. *Swiss-Polytechnic Museum*: see n. 3.
6. *Prof. Heer*: Oswald Heer (1809–83), geologist and naturalist, who became Professor of Botany at the University of Zürich in 1851.
7. *Escher v. d. L.*: Arnold Escher von der Linth.
8. *Prof. Murchison*: Roderick Murchison.
9. *you, as I believe, are Secretary of the Geographical and Geological Society*: in this, Mousson was mistaken. We do not have Tyndall's reply.

From Joseph Plateau 11 February 1862 1862

Gand, 11 Février 1862

Mon Cher Monsieur Tyndall,

M. Van Hese est prêt à vous expédier la caisse contenant le vase à parois planes avec les petits appareils accessoires. Ce vase et ces appareils ne sont pas exactement conformes aux figures que j'ai données dans mes 1ère, 2ème, et 4ème Séries; je dois donc vous donner à cet égard des explications.

Commençons par le vase: La plaque de verre qui forme le couvercle, au lieu de présenter deux ouvertures n'en présente qu'une au milieu et, en outre, cette plaque n'est pas munie d'un rebord métallique, de sorte qu'on peut la faire glisser sur le bord supérieur du vase; quand il s'agit d'introduire dans celui-ci soit le bec de la petite seringue, soit l'une des spatules, soit une autre pièce quelconque destinée à modifier une figure liquide par exemple un fil de fer droit pour réunir des sphères d'huile qu'une seule, e&., on fait tourner et glisser un peu la plaque couvercle, de manière à laisser libre un espace triangulaire vers l'un des angles du vase, ou bien on fait glisser simplement cette même plaque de façon à laisser ouverte une rainure longitudinale le long du bord de l'une des faces du vase, et c'est par l'une ou par l'autre de ces ouvertures que l'on introduit la pièce en question. M. Duprez qui a fait pour moi une grande partie des expériences, a trouvé cette disposition beaucoup plus commode que celle qui est représentée dans ma 1ère Série.

Passons aux petits appareils. Les expériences, telles que je les ai décrites, se font sans trop de difficultés dans le cabinet, parce qu'alors on est libre de prendre toutes les précautions contre les causes perturbatrices; mais lorsqu'on veut répéter ces mêmes expériences dans une leçon publique, les difficultés deviennent bien plus grandes: si les densités de l'huile et du mélange alcoolique ont été parfaitement réglées avant l'entrée des élèves, leur égalité s'altère en peu de temps, lorsque la salle est remplie, à cause du changement de la température, et il faut alors modifier le mélange alcoolique et qui exige beaucoup de temps et d'embarras. Heureusement, il y a un moyen simple d'éviter ces inconvénients, moyen qui m'a été suggéré par M. Duboscq. Si le mélange alcoolique ayant sensiblement la densité convenable, on y verse une petite quantité de mélange à 160 de Beaumé, mélange qui s'étend en couche au fond du vase, que l'on mêle un peu, à l'aide de la spatule de bois, cette couche avec la partie inferieure du liquide qui remplit le reste du vase, que l'on verse ensuite, sur le liquide une petite couche d'alcool pur, et qu'enfin on mêle de même cette dernière imparfaitement avec la partie supérieure du liquide, on aura un mélange dont la densité ira en décroissant depuis le fond du vase jusqu'à la surface libre du liquide et ce mélange demeurera en cet état pendant très longtemps. Si donc, avant d'établir cette hétérogénéité, on avait formé une figure d'huile, et si les choses sont disposées de manière que l'on puisse enfoncer plus ou moins dans le mélange le système solide auquel adhere cette figure d'huile, on pourra toujours trouver une couche alcoolique dans laquelle il y aura égalité entre les deux densités; si donc la température vient à s'élever de sorte que l'huile tende à descendre, il suffira d'abaisser un peu le système pour rétablir la régularité de la figure; si au contraire la température s'abaisse, ce qui donne à l'huile une tendance à monter, il suffira de soulever suffisamment le système. Pour vous mettre à même d'employer ce procédé dans vos leçons j'ai fait construire tous les systèmes solides de manière qu'ils puissent se suspendre à la tige qui traverse le bouchon métallique du couvercle du vase. Avec un peu d'habitude, je suis convaincu que vous réussirez parfaitement les expériences au moyen de ce procédé.

Voici maintenant l'énumération des petits appareils en question; Je n'ai fait faire que ceux qui sont relatifs aux expériences principales.

10 Un disque vertical percé à son centre, pour une expérience du §17 de la 2ème Série.

20 Trois systèmes solides, pour la réalisation du cylindre liquide stable, du cylindre liquide instable, et pour montrer l'impossibilité d'avoir à la forme cylindrique quand le rapport entre la distance des bases et leur diamètre surpasse trop la limite de la stabilité (2ème Série, §46). Dans ces trois systèmes, le diamètre des disques en regard est de 3 centimètres, dans le premier, le rapport est 3, dans le second 3, 4, et dans le troisième 4. La valeur exacte de la limite de la stabilité du cylindre est *[π],* c'est-à-dire le rapport de la circonférence au diamètre ou 3, 14; Le premier système est donc en deux de la limite, et vous permettra de réaliser sans peine un cylindre d'huile parfaitement stable et permanent. Avec le second, pour lequel le rapport est un peu superieur à la limite vous realiserez encore un cylindre en employant la petite manœuvre indiquée dans une note du paragraphe cité plus haut. Enfin, avec le troisième, pour lequel le rapport surpasse très notablement la limite, il vous sera impossible d'approcher même de la pièce cylindrique.

30 Un appareil du même genre mais à disques plus petits, pour la réduction de l'onduloïde renflé en son milieu et sans cylindre intérieur (4ème Série, §§ 5 et 10).

40 Un cylindre solide pour la formation d'un semblable onduloïde terminé rigoureusement aux cercles de gorge de deux étranglements consécutifs, et parfaitement stable (4ème Série, § 4 et 5).

50 Un système composé de deux anneaux plats, pour la réalisation de l'onduloïde étranglé en son milieu (4ème Série, § 13).

60 Deux anneaux attachés chacun à un fil de fer droit, servant à étendre la masse d'huile attachée à l'un des disques ou des anneaux précédents jusqu'à l'autre disque ou l'autre anneau.

70 Une spatule en fer pour rendre homogène, par une agitation douce, l'huile qui constitue une figure liquide; une spatule en bois, pour opérer le mélange du liquide alcoolique; une autre est recouverte de coton pour conduire les masses d'huile sans qu'elles y adhèrent; une petite seringue en verre pour l'absorption de l'huile (2ème Série §9); Je pense que le bec de cette seringue est trop long, ce qui l'empêcherait de fonctionner facilement; mais s'il en est ainsi vous pourrez facilement faire raccourcir ce bec.

80 Fil de fer droit portant en son milieu un petit disque pour la rotation d'une masse d'huile et sa conversion en un anneau (1ère Série, §§ 8 à 12).

90 Deux tiges de différentes longueurs pouvant traverser le bouchon métallique et à la partie inférieure de chacune desquelles peut s'adapter chacun des appareils des N°s 1, 2, 3, 4, 5, et 8 précédents. L'une de ces tiges est munie d'une manivelle et l'autre d'un simple bouton.

Vous avez reçu, en outre, avec les charpentes polyédriques deux systèmes d'anneaux, dont le grand suivant à la formation du caténoïded'huile plein, (4ème Série §§ 14 à 21), du caténoïde d'huile laminaire, des lames planes qui en résultent, des anneaux liquides qui se forment quand on perce celles-ci, et de la conversion de ces anneaux en petites masses déparées (5ème Série *[*§§*]* 2 à 4 et 4ème Série *[*§*]* 35). Vous avez reçu en même temps un système de disques au moyen desquels on obtient une portion renflée du nodoïde (4ème Série §§ 25 à 32). Enfin votre charpente cubique peut se suspendre dans le vase en verre, de manière que vous pouvez réaliser un cube d'huile plein, et sa conversion en système laminaire d'huile (2ème Série § 30 à 33).

Ainsi que je vous l'ai annoncé dans ma dernière lettre, vous trouverez, dans la caisse, un morceau de savon de Marseille, un morceau de savon de Rimmel, quelques pipes, et enfin un petit flacon de liquide glycérique. Quand on emploie l'une de ces pipes pour la 1ère fois, il faut la laisser tremper l'orifice pendant quelques minutes dans le liquide glycérique, pour que la terre poreuse dont elle est formée s'en imprègne, et, chaque fois qu'on emploie de nouveau la même pipe; il faut en essuyer la partie nouvelle de liquide glycérique à moins qu'on ne vienne immédiatement de s'en servir.

J'arrive au liquide. Celui que je vous envoie est en bien petite quantité, et n'est point préservé au savon de Marseille; en voici les motifs: je n'avais pas de bon liquide au savon de Marseille; j'ai donc voulu vous en préparer; j'ai fait d'abord de vains efforts pour simplifier le procédé, indiqué, pour l'hiver, dans mon § 13; comme je ne réussissais pas, j'ai tâché donc préparer en suivant de point en point les prescriptions du procédé dont il s'agit, mais, je ne puis m'expliquer pourquoi j'ai encore complètement *échoué.* Comme les gelées étaient alors survenues, il aurait été inutile de recommencer. Cependant, afin que vous puissiez constater les grandes persistances, je me suis décidé alors à vous envoyer la petite quantité que j'avais d'un liquide préparé d'une manière un peu différente, et qui vous donnera des bulles persistant quatre ou cinq heures, si vous les faites avec les précautions que je vous ai indiquées dans ma lettre précédente.

Dans ce liquide, le savon est remplacé par de l'oléate de soude, ce qui est bien préférable, comme je l'ai trouvé; ce même liquide est un peu ancien maintenant, mais peu de jours après sa préparation, il m'avait donné des bulles persistant au delà de onze heures, et je ne doute pas que, mieux préparé encore, il ne puisse donner des bulles durant vingt-quatre heures. Je vous dis ceci en confidence, car je n'ai pas encore publié ces résultats.

Je m'étonne de n'avoir reçu aucune nouvelle de vous depuis la lettre que j'ai eu l'honneur de vous adresser le 11 Décembre dernier, lettre dans laquelle je répondais à celle où vous m'annonciez vos insuccès. Au commencement du Décembre la température était fort douce, et si vous avez fait de nouveaux essais avec des savons anglais, en suivant les conseils que je vous donnais vous avez dû réussir, puisque j'ai réussi ici avec le savon de Rimmel. Maintenant, si vous n'avez rien fait, il ne vous reste que deux partis à prendre, à cause du froid extérieur: ou bien faites les expériences devant la Société Royale avec une simple solution de savon, ce qui vous permettra de réaliser parfaitement les figures laminaires, seulement avec une très petite durée mais avec la faculté de les refaire aussi souvent que vous voudrez, ou bien attendez l'été, pour préparer du liquide glycérique avec le savon de Marseille que vous trouverez dans la caisse.

Si vous prenez ce dernier parti, je vous engage à employer, au lieu de savon, de l'oléate de soude que les chimistes avec lesquels vous êtes en relation à Londres pourront sans doute vous préparer suffisamment pur. Dans ce cas, dissolvez une partie d'oléate solide dans 50 parties d'eau distillée, et mêlez soigneusement deux volumes de glycérine de Price avec 3 volumes de cette solution, puis attendez 4 ou 5 jours; si votre oléate est assez pur, il ne se formera aucun dépôt et le liquide vous donnera après ce temps des persistances extraordinaires. Vous pourriez essayer cette préparation déjà présent mais j'ignore s'il elle réussit en hiver, c'est une tentative à effectuer.

J'espère que je recevrai bientôt une lettre de vous, et je vous prie, en attendant, d'agréer l'assurance de tous mes sentiments d'affection.

J. Plateau | Professeur à l'Université | Place du Casino 22.

P.S. Mr Van Hese vous écrira immenainment pour vous annoncer l'envoi de la caisse, et vous envoyer alors quantité généreuse

Ghent, 11th February 1862

My dear Mr Tyndall,

Mr Van Hese is ready to ship the box containing the vessel with flat sides, with the small accessory apparatuses. This vessel and those devices do not entirely comply with the figures that I showed in my 1st, 2nd, and 4th Series.[1] Thus, I will give you some explanations about it.

Starting with the vessel: the glass plate which forms the lid, instead of having two openings, has only one in the middle, and additionally, that plate does not have a metal rim, so that it can slide on the upper edge of the vessel, when we place into it either the spout of the small syringe, or one of the spatulas, or whichever other piece that is meant to change a liquid figure, for example a straight wire to gather the spheres of oil as one, etc; we slightly rotate and slide the top plate in order to leave a triangular space at one of the corners of the vessel, or even we simply slide that same plate in a way that will leave a longitudinal groove open along the edge of one of the vessel's sides, and it is through either one of those openings that we can insert the aforementioned piece. Mr Duprez,[2] who did a large part of those experiments for me, found this setup to be much more convenient than the one that was presented in my 1st Series.

Let us move onto the subject of the small devices. The experiments, as I described them, can be done without too much trouble in an office because there you are free to take all the precautions against disruptive causes. But when we wish to repeat those same experiments in a public lecture, the difficulties increase: If oil densities and the alcohol mixture were perfectly regulated before the students enter, their equality is altered in a short time, when the room is filled, because of the change in temperature, and then the alcohol mixture has to be modified and which requires much time and trouble. Fortunately, there is a simple method to avoid these drawbacks, by a means that was suggested to me by Mr Duboscq.[3] If the alcohol mixture has roughly the proper density, then wefill a small quantity of a mixture at 160 of Beaumé,[4] a mixture that settles at the bottom of the vessel, and which we mix a little using a wooden spatula, that layer with the lower part of the liquid that fills the rest of the vessel. Afterwards we pour on the liquid a small layer of pure alcohol, and finally we mix it as well, imperfectly, with the top part of the liquid. This will give us a mixture whose density will be decreasing from the bottom to the top free surface of the liquid, and this mixture will remain in this state for a very long time. Thus, if before establishing heterogeneity we formed an oil figure, and if those things are arranged in such a way as to allow us to know more or less the solid system to which adheres this oil figure within the mixture, we could always find an alcohol layer in which there would be equality between the two densities. So if the temperature increases to cause the oil to descend, it will be sufficient to lower the system a little in order to restore the regularity of the figure. If on the contrary the temperature lowers to cause the oil to rise, it would be sufficient to elevate the system enough. To make this procedure available for use in your lessons I constructed all the solid systems in a way that they could be suspended on the rod that passes through the metallic plug on the lid of the vessel. With a little practice, I am convinced that you will succeed perfectly in the experiments using this method.

Now here is the list of the small apparatuses in question. I included only those relating to the main experiments.

10 A vertical disc pierced in the centre, for an experiment of §17 of the 2nd Series.[5]

20 Three solid systems, to achieve the stable liquid cylinder, the unstable liquid cylinder, and to show the impossibility of having a cylindrical form when the ratio between the distance of bases and their diameter exceeds the limit of stability (2nd Series, §46). In those three systems the diameter of the appropriate discs is 3 centimetres. In the first, the ratio is 3; in the second 3, 4; and in the third 4.

The precise value of the limits of the cylinder's stability is π, that is to say that the ratio of the circumference to diameter or 3, 14. The first problem is thus within two of the limit, and it will allow you to easily achieve a perfectly stable and permanent cylinder of oil. With the second one, for which the ratio is a bit higher than the limit, you accomplish again a cylinder using the manoeuvre shown in the note cited in the above paragraph. Finally with the third one, for which the ratio significantly surpasses the limit, it would be impossible for you to approach even the cylindrical part.

30 A similar device but with smaller discs, for reducing the thickened onduloid in the middle and without an inner cylinder (4th Series, §§5 and 10).

40 A solid cylinder to form a similar onduloid strictly ending at the circles of the gorge of two consecutive narrow passages, and perfectly stable (4th Series, §4 and 5).

50 A system made up of two flat rings, for achieving the onduloid that is constricted in the middle (4th Series, §13).

60 Two rings, each attached to a straight wire, used to extend the oil mass that is attached to one of the discs or one of the previous rings, up until the other disc or the other ring.

70 An iron spatula to homogenize, by soft stirring, the oil that constitutes a liquid figure. A wooden spatula to be used for the mixture of liquid alcohol. Another covered with cotton to direct the oil masses without them adhering to it. A small glass syringe for the absorption of oil (2nd Series §9). I think that the syringe's nozzle is too long, which prevents it from working smoothly, but if it is you could easily shorten its nozzle.

80 A straight wire having in its centre a small disc for the rotation of oil mass and its conversion into a ring (1st Series, [§§] 8 to 12).

90 Two rods of different lengths that can pass through the metal cap, and the inner part of each of them can fit into each of the devices of the previous N°s 1, 2, 3, 4, 5, and 8. One of those rods has a crank and the other has a simple button.

In addition, you have received the polyhedral structures with two systems of rings, whose main purpose at the formation of the full oil catenoid, (4th Series §§14 to 21), of the laminar oil catenoid, of the flat gap that results, the liquid rings that are formed when those are pierced, and of the conversion of those rings into small and soiled masses (5th Series [§§] 2 to 4, and 4th Series § 35). At the same time, you have received a system of discs with the help of which we can obtain a bulging portion of the nodoid (4th Series §§25 to 32). Finally, your cubic frame can be suspended in a glass vessel, such that you can achieve a full oil cube, and its conversion into a system of laminar oil (2nd Series §30 to 33).

As I have told you in my last letter, you will find in that box a piece of Marseille soap, a piece of Rimmel soap, some pipes, and finally a small flask of glycerin liquid. When we use one of those pipes for the 1st time, it must be left to soak the opening for a few minutes in the glycerin liquid, so that the porous earth from which it is formed, is permeated by it and each time that we use again the same pipe, the new part in the glycerin liquid should be wiped clean unless we come to use it immediately.

Now I come to the liquid. The one that I am sending you is of a rather small quantity, and is not preserved with the Marseille soap. Here are the reasons why: I did not have a good enough liquid from the Marseille soap. Thus, I wanted you to prepare it; first I made vain efforts to simplify the process, as indicated for winter, in my §13. Since I failed, I tried to prepare it by following step by step the requirements of the process in question, but, I could not explain why I have still completely failed. Since the frosts have occurred by then, it would have been useless to do it again. However, to ensure that you could see the great persistence, I decided to send you the small amount that I had of the liquid that was slightly differently prepared, and which would give you persistent bubbles for four or five hours, if you would follow the precautions that I mentioned in my previous letter.

In this liquid the soap is replaced by sodium oleate, which is much better, as I found. This liquid is slightly old now, but a few days after its preparation it gave me persistent bubbles for more than eleven hours. I have no doubt that, with an even better preparation, it could give me bubbles for the duration of twenty-four hours. I tell you this on trust because I have not published these results yet.

I am surprised to have not received any news from you since the letter that I had the honour of writing to you on December 11th.[6] It was a letter in which I answered to the one where you told me about your failures. In the beginning of December the temperature was rather warm, and if you have made new attempts with the English soaps, by following the advice that I gave you, you should have succeeded, since I succeeded here with the Rimmel soap. Now, if you have not done anything, you are left with only two courses of action because of the cold outside: to do the experiments in front of the Royal Society with a simple soap solution, which would allow you to perfectly perform the laminar figures, only very shortly but with the ability to redo them as much as you wish; or to wait for the summer in order to prepare the glycerine liquid with the Marseille soap that you will find in the box.

If you should take the latter course, I urge you to use sodium oleate instead of soap, which the chemists with whom you are familiar in London could do for you in a sufficiently pure form. In that case, dissolve one part of the solid oleate in 50 parts of distilled water, and carefully mix two volumes of glycerine of Price with 3 volumes of that solution; then wait for 4 or 5 days. If your oleate is pure enough, it will not form any deposits and after this the liquid will give you extraordinary persistence. You could try this already existing preparation, but I do not know of it to succeed in winter, it is an attempt to make.

I hope that soon I will receive a letter from you, and I ask you in the meantime, to accept the assurance of my warmest wishes.

J. Plateau | Professor at the University | 22 Place du Casino

P.S. Mr Van Hese will write to you to tell you about the shipment of the box, and to send you a generous amount.

RI MS JT/1/P/91

---

1. *my 1st, 2nd, and 4th Series*: refers the following publications: J. Plateau, 'Mémoire sur les phénoménes que présente une masse liquide libre et soustraite a l'action de la pesanteur', *Mém. Acad. Roy. Sci. Belg.,* 16 (1843), separate pagination; and J. Plateau, 'Recherches expérimentales et théoriques sur les figures d'équilibre d'une masse liquide sans pesanteur', *Mém. Acad. Roy. Sci. Belg.,* 23; 30 (1849; 1858), separate pagination.
2. *Duprez*: François Duprez (1807–84), Belgian physicist and at this time assistant to the by then blind Joseph Plateau.
3. *Duboscq*: Louis Jules Duboscq (1817–86), French experimentalist, photographer and instrument maker.
4. *Beaumé*: the Baumé scale, developed by Antoine Baumé in 1768, is a measure of liquid density.
5. *§17 of the 2nd Series*: referring to section 17 of Plateau, 'Recherches expérimentales et théoriques', (1849), pp. 23–26. An English translation of this paper was published in the *Annual Report of the Board of Regents of the Smithsonian Instituion,* (1863), pp. 207–85. Subsequent volumes of the same journal (1864–66) carried English translations of series 3–6.
6. *writing to you on December 11th* : letter missing.

To Juliet Pollock 18 February 1862 1863

18th Feb 1862

My dear Mrs. Pollock

It was very kind of you to come & see me, and your presence would have been a pleasure to me—Possibly I had adjourned to the Turkish Bath,[1] as Tuesday is one of my Turkish days. I wish like you I could get out of town for a day or two. 'My head is ne'er out of school'[2] and I would gladly escape from my lessons for a little while.

I am writing out my present course of lectures[3]—nothing like the written lectures has yet appeared and I think the publication of them will do good.

The first proofs of my little Alpine book[4] have come to hand: with a little polishing it will read smoothly enough.

I wish you a pleasant time at the Wells: give Moss[5] ten kisses for me. As soon as you come back I shall be ready for any amount of W——Y![6]

Yours always | John Tyndall

RI MS JT/1/T/1135
RI MS JT/1/TYP/6/1980

---

1. *Adjourned to the Turkish Bath*: likely the baths at 74 Jermyn Street.
2. *'My head is ne'er out of school'*: from Thomas Hood's poem 'A retrospective review'. T. Hood, *The Complete Poetical Works* (Boston: Phillips, Simpson and Co., 1859), pp. 487–90.
3. *my present course of lectures*: Tyndall delivered a course of twelve lectures on heat at the RI, starting 23 January and running to 10 April 1862. These were published as J. Tyndall, *Heat,* 2nd ed. (London: Longman, Green, Longman, Roberts & Green 1863).
4. *The first proofs of my little Alpine book*: published on 16 April as *Mountaineering in 1861.*
5. *Moss*: Juliet's youngest son, Maurice Emilius (1857–1932).
6. *W——Y!*: presumably Whiskey (Irish spelling), a drink Tyndall clearly enjoyed (see letter 1717).

From Joseph Plateau 19 February 1862 1864

Gand, 19 février 1862

Mon Cher Monsieur Tyndall,

Mr Van Hese est arrivé trop tard au dek avec la caisse, le vaisseau était parti, et il en résultera, dans l'envoi un retard de dix à quinze jours, en attendant, je viens vous mettre à même de réussir infailliblement les expériences devant la Société Royale.

Vous avez reçu, j'espère, ma lettre du 11 de ce mois, lettre dans laquelle je vous donne les explications relatives aux appareils accessoires qui accompagnent le vase à parois planes. Dans cette même lettre je vous disais que n'ayant pas réussi, à cause de l'hiver, à vous préparer un bon échantillon de liquide glycérique au savon de Marseille. Vous trouveriez dans la caisse un petit flacon enfermant une liquide glycérique beaucoup meilleur où le savon était remplacé par de l'oléate de soude. J'ajoutais que je vous engageais à envoyer cette dernière préparation qui réussirait peut-être par le *[1 word illeg]* températures. Or je viens de faire moi-même cet essai, en me plaçant, à dessin, dans les conditions de température les plus défavorables, et j'ai complètement réussi. J'ai fait ensuite un second essai dans des conditions analogues et j'ai de même complètement réussi. Il n'y a donc plus à cet égard le moindre doute, le liquide à l'oléate de soude peut se préparer sans aucun égard aux températures, et c'est là le véritable liquide glycérique: non seulement il réussit toujours, mais il donne des lames d'une énorme persistance, qui peut aller jusqu'à onze ou douze heures. Celui dont je vous parlais dans ma lettre du onze est déjà trop ancien et ne vous donnera que quatre ou cinq heures de persistance, mais vous trouverez dans la caisse un second petit flacon que j'y ai fait mettre depuis, qui est intitulé Liquide Glycérique N° 2, et qui vous donnera les grandes persistances en question.

L'essai fait par M. Frankland est donc, vous le voyez, bien loin d'être suffisant; je vous engage avec les plus vives insistances à préparer le liquide à l'oléate de soude pour les expériences devant la Société Royale; ces expériences se feront alors avec une extrême facilité et un succès complet. Voici maintenant les renseignements nécessaires pour vous mettre à même de préparer le liquide en question. D'abord j'obtiens de la manière suivante l'oléate de soude suffisamment pur:

On se procure de l'acide oléique impur qui s'obtient dans les fabriques de bougies stéariques, il est brun et d'une odeur fétide; vous le trouverez probablement à la fabrique de M. Price chez les pharmaciens de Londres. On dissout deux parties, en poids, de soude caustique dans quinze parties d'eau distillée, on y ajoute trois parties de l'acide oléique en question, on mêle, on chauffe jusqu'à cent dégrées centigrades, on entretient cette température pendant une dizaine de minutes en mêlant de temps en temps, puis on laisse refroidir l'oléate de soude formé vient surnager en couche solide; on l'enlève, on le place dans un linge enveloppé de plusieurs papiers absorbants et on le compresse fortement dans une presse. En cet état, il est encore trop impur; on le redissout, à chaud, dans douze fois son poids d'eau distillée, on laisse plus au moins refroidir et on filtre; puis on y introduit du sel marin en petits cristaux et on agite; l'oléate de soude se sépare de nouveau; et il faut ajouter assez de sel marin pour séparer alors tout l'oléate. Celui-ci vient encore surnager en couches solides, qu'on enlève et que l'on compresse comme précédemment; enfin on conserve l'oléate de soude aussi obtenu dans un flacon parfaitement bouché, pour qu'il n'absorbe pas l'humidité de l'air.

Il est de toute nécessité que le sel marin employé soit bien pur, c'est-à-dire débarrassé des sels étrangers (probablement de magnésie) qu'il contient toujours; sans cela quand on redissout l'oléate dans l'eau pour l'employer il se forme toujours un précipité blanc qui passe à travers les filtres.

Quelque soit du reste l'acide oléique dont vous vous servirez, la préparation est la même.

Maintenant pour former le liquide glycérique, dissolvez une partie en poids de cet oléate de soude sec dans cinquante parties d'eau distillée, faites cette solution à chaud, et, quand elle sera refroidie, mêlez deux volumes de glycérine de Price avec trois volumes de cette même solution d'oléate, agitez fortement pour bien mêler, puis laissez reposer pendant quatre ou cinq jours; vous aurez alors un excellent liquide glycérique. Du reste, en recourant à l'obligeance de M. Frankland, il vous procurera peut-être, soit par le procédé au-dessus soit par un autre, de l'oléate de soude d'une pureté suffisante. Encore une fois je ne puis assez vous engager à préparer le liquide dont il s'agit; vos peines seront bien compensées par l'excellence des résultats.

Agréez, mon Cher Monsieur Tyndall, l'assurance de tous mes sentiments d'affection | J. Plateau

P.S. Le sel marin se purifie par des cristallisations successives. Si vous faites la préparation ci-dessous, suivez de point en point mes indications. Je vous autorise parler du liquide à l'oléate à la Société Royale.

Ghent, 19th February 1862

My dear Mr. Tyndall,

Mr. Van Hese arrived too late at the deck with the box.[1] The ship already left and it will result in a delay of ten to fifteen days for the shipment. In the meanwhile, I have just put you in a position of flawlessly succeeding with the experiments in front of the Royal Society.[2]

I hope you have received my letter from the 11th of this month,[3] a letter in which I explained to you the accessory apparatuses that are coming with the planar-walled vessel. In that same letter, I told you that in the case of failure because of winter, you should prepare a good sample of liquid glycerin with Marseille soap. You will find in the box a small flask with a much better glycerin liquid, where the soap was replaced by sodium oleate. I added that I urged you to send in my last letter, this last preparation which could perhaps succeed by the *[1 word illeg]* temperatures. Now I have just done this test myself, by placing myself in the most unfavourable temperature conditions, and I fully succeeded. Thus, there is no doubt about it, a liquid with sodium oleate could be prepared regardless of the temperature, and that is the true glycerin liquid: not only that it always succeeds, but it also gives layers with enormous persistence, which could continue up until eleven or twelve hours. The one that I mentioned in my letter of the eleventh,[4] is already too old and will give you only four or five hours of persistence, but you will find in the box a second small flask that I put since then, labeled Glycerin Liquid N° 2, which will give you much better persistence.

The attempt made by Mr. Frankland[5] is, as you see, sufficient enough, and I am most definitely urging you to prepare the liquid with sodium oleate for the experiments in front of the Royal Society. Those experiments will then be completed very easily and very successfully. Here is the necessary information to allow you to prepare this liquid. First, I achieve a sufficiently pure sodium oleate in the following manner:

We procure it from impure oleic acid that is obtained from factories of stearic candles, it is brown and foul smelling. You can probably find it at Mr. Price's[6] factory from the pharmacists of London. We dissolve two parts, by weight, of caustic soda in fifteen parts of distilled water, and then we add three parts oleic acid. We mix it and we heat it to one hundred degrees Celsius; we continue keeping this temperature for about ten minutes, while mixing occasionally. Then, we allow the formed sodium oleate to float up in a solid layer; we remove it and place it in linen wrapped in several paper towels, and we compress it powerfully in a press. In this state, it is still too impure. We reduce it, while hot, in distilled water twelve times its weight; we leave it to cool more or less and to get filtered. Then, we add some sea salt in small crystals and we stir. The sodium oleate separates from the piece, and enough sea salt should be added to separate all of the oleate. That will also float up in solid layers that we will remove and compress as previously. Finally, we keep the sodium oleate obtained as such, in a perfectly sealed flask, so that it would not absorb the moisture in the air.

It is absolutely necessary that only pure sea salt will be used, in other words, free of foreign salts (probably magnesium) that it always contains. Without this when the oleate is re-dissolved in the water for use it always forms a white precipitate that passes through the filters.

As for the rest of the oleic acid that you would use, the preparation is the same.

Now, to produce the glycerin liquid, dissolve one part, by weight, of that dry sodium oleate in fifty parts of distilled water, heat that solution, and when it will cool down again mix two volumes of glycerin from Price with three volumes of the same solution of oleate. Stir vigorously to mix it well, and then let it sit for four or five days; you will have an excellent glycerin liquid. As for the rest, by making use of Mr. Frankland's kindness, he will perhaps procure for you, either by the procedure abovementioned or in another way, sodium oleate of sufficient purity. Again, I cannot urge you enough to prepare the liquid in question; your troubles will be well compensated by the excellent results.

Please accept, my dear Mr. Tyndall, the assurance of my warmest regards | J. Plateau

P.S. Sea salt is purified by successive crystallizations. If you do the preparation below, follow my indications step by step. I authorize you to tell the Royal Society about the liquid oleate.

RI MS JT/1/P/92

1. *box*: of instruments etc: see letter 1862.
2. *succeeding with the experiments in front of the Royal Society*: no record found.
3. *my letter from the 11th of this month*: letter 1862.
4. *my letter of the eleventh*: letter 1862.
5. *The attempt made by Mr Frankland*: no record found.
6. *Mr Price's factory*: Price's Patent Candle Company, a large candle manufacturer that also produced soaps, lubricating oils and glycerine. Price's was a household name.

To Miss Moore[1] [5 March 1862][2] 1865

My dear Miss Moore

Suppose I do! I think it will be very pleasant. I am a 'party' for I was once addressed by a Yankee as 'that party'!

Yours ever | John Tyndall

Wednesday,

RI MS JT/1/T/1089
RI MS JT/1/TYP/3/868/1

1. *Miss Moore*: Harriet Moore (1801–84) or Julia Moore (1804–1904).
2. *[5 March 1862]*: in pencil on manuscript, probably by LT.

To Juliet Pollock [11 March 1862][1] 1866

Imagine[2] Moss[3] stretched upon the floor with his head and shoulder propped by one little arm, the other in the air warding off a descending blow, over him his irate father wielding a birch rod, while with his eyes streaming with tears the boy exclaims 'O Pa! don't strike so hard'—I am Moss in this position.

He would in the first place declare his readiness to do almost every thing you have suggested[4]—He will put in 'direct', modify that strong statement;[5] he will abolish Socrates. Nevertheless if you would deign to listen to a prostrate little boy he would in connexion with the Greek say that he still believes his allusions to be correct. Previous to Socrates & Christ men had certain notions regarding their relationship to the Deity, which notions they were unwilling to see disturbed. Both Socrates & Christ did disturb these notions, & for this they suffered. That Socrates was slain by the personal malice of the Sophists may be true, but then the charge against him shows that they were afraid to aver their real motives. They had the populace to excite with the fear that the relationships towards the Gods in which they had been accustomed to believe were in danger. Whether the dreaded change was to be brought about by the introduction of new deities, or by an ethereal rather than a physical philosophy, is a matter of entire indifference to the point in question.

The tendril & the pole shall be taken care of.[6] The mention of extreme unction was simply made to avoid quoting the text. The word 'protestant' placed before 'clergyman' will set this matter right. Moss's chastener misestimates his boy's character if he supposes that Moss had the least idea of being 'Smart' in his allusion to 'winking images'. Smartness, in truth, is a thing for which poor Moss entertains a rather hearty contempt, and he will abolish with right good will every expression that could possibly lead people to think that he wished to appear 'smart'.

Nor is it within the capabilities of the boy's constitution to manufacture a 'sneer'. He confesses frankly that when he heard the Bishop of Oxford[7] define the quantity of rain which was indicative of the displeasure of the Almighty, his mind and heart recoiled from what he believed, and continues to believe, mere Paganism. The position of the bishop at the time was not in the boy's eyes a dignified one, but a transcendently presumptuous one. Nevertheless it is written 'thou shalt not speak evil of the Ruler of the people',[8] & I will cleanse my writing[9] from every thing that could be regarded as a sneer.

And now Moss will endeavour to set upon his legs and stand firmly there. The question that he has touched is a purely scientific question. Who discovered the connexion between the various agencies of Nature, and who but a scientific man has a right to affirm that the law by which a cloud forms and a shower descends is as strong a law, nay part of the same law as that by which a river rolls down hill? Science is incompetent to prove miracles impossible; but she is competent to state that the demand periodically made for fine weather & for rain cannot be complied with without a miracle. This is within the clear & distinct province of a scientific man. If left unaided by him when will Theologians reach the conception? never till doomsday—They will go on as heretofore 'asking amiss',[10] and continue to find, as heretofore, their prayers a nullity. You may deem their faith harmless if not serviceable—I deem it harmful and bad; and I see it cropping out under a hundred noisome forms in society. The region in which prayer is operative is the human heart: retire to that citadel and you are safe from the attacks of science, but come out from it into that system of things which it is the vocation of science to explore, and confusion will infallibly attend your operations. You have no business there & were your army of 'defenders' a million times what it is you would eventually be powerless in this position.

J. T. | 11/3/62

RI MS JT/1/T/1539

---

1. *[11 March 1862]*: date from LT Transcript. The letter was written before *Mountaineering in 1861* was published (on 16 April 1862).
2. *Imagine*: salutation and possibly other content missing from original and transcript.
3. *Imagine Moss*: Juliet's youngest son, Maurice Emilius (1857–1932)
4. *you have suggested*: Tyndall is referring here to appeals from Juliet to moderate his discussion on miracles and the efficacy of prayer in the manuscript for *Mountaineering in 1861.* The letter in which Juliet makes these appeals is missing. However, when Juliet read the published version, 'without the fetter of a critic's task to bind me to . . . close verbal

attention' she described the chapter on miracles as 'perfect now, and . . . more likely to do good in its quiet, sober tone than a more impetuous protest'. See letter 1887.

5. *He will put in 'direct', modify that strong statement*: this seems to refer to the statement, 'the idea of direct personal volition mixing itself in the economy of nature is retreating more and more' (*Mountaineering in 1861,* p. 34).
6. *The tendril & the pole shall be taken care of.*: it is not clear what the problem was in the manuscript but Tyndall writes in the published version that, 'We may err in linking the imperishable with the transitory, and confound the living plant with the decaying pole to which it clings.' (*Mountaineering in 1861,* p. 34).
7. *The Bishop of Oxford*: Rev. Samuel Wilberforce (1805–73), Anglican Bishop of Oxford. Wilberforce had instructed clergy in the Archdeanery to pray for fine weather to assist the harvest the previous summer.
8. *'Thous shalt not speak evil of the Ruler of the people'*: Biblical quotation from Acts 23: 5.
9. *I will cleanse my writing*: no direct reference to the Bishop of Oxford appears in *Mountaineering in 1861.*
10. *'asking amiss'*: biblical quotation from James 4: 3.

To George Gabriel Stokes 15 March 1862 1867

15th March 1862

My dear Prof. Stokes.

I have read the paper of Mr. Tomlinson[1] on the motions of camphor[2] referred to me by the Council.

Many of Mr. Tomlinson's experiments are very ingenious, and the account of them is clearly and often graphically written. If published separately, the paper would probably form a brochure[3] which would obtain a great many interested readers.

I think however the labours expended upon the subject, and the minuteness of detail into which Mr. Tomlinson enters, are not sufficiently rewarded by his results, if judged from a scientific point of view. And I suppose this is the point of view to be adhered to in considering the fitness of a paper for the Philosophical Transactions.

I do not entertain a doubt that by varying the conditions of experiment, a fresh reinvestigation taking up the subject would obtain many new & beautiful effects in addition to those described by Mr. Tomlinson. In fact, I should consider it extremely probable that Mr. Tomlinson himself would, if he employed his experimental ingenuity, which is very considerable, for a sufficient time be able to bring before the Royal Society, many papers on this same subject; & all containing accounts of experiments most pleasing to the eye, as many of those really are which he now describes.

Since reading the accounts of Weber's experiments on kindred subjects,[4] which though similar to and indeed in some cases identical with those described by Mr. Tomlinson, are not once referred to in his paper. I have myself made numberless experiments in the spreading of films over the surface of water. I have obtained images of what Mr. Tomlinson calls 'cohesion figures' ten feet in diameter, and in some cases exhibiting most beautiful molecular effects & the most curious motions. They may be obtained almost without aid, and I once entertained the idea of describing some of them in the Philosophical Magazine, but it did not, I confess, occur to me to send them to the Royal Society for publication in the Philosophical Transactions.

If I cannot recommend the publication of Mr. Tomlinson's paper as it now stands it is not on account of any defects in his experiments; but I think the subject not of sufficient importance, and that the addition which Mr. Tomlinson has made to our knowledge is not sufficiently considerable to render the publication of the paper in extenso desirable. As far as experimental philosophers are concerned I think in its present state it would excite weariness rather than interest. I would not recommend its rejection by the council; but I would strongly recommend its curtailment to a fraction of its present volume. Thus altered I think it might appear in the Transactions of the Royal Society.

Believe me | Yours very truly | John Tyndall

RS RR/4/266

---

1. *Mr. Tomlinson*: Charles Tomlinson (1808–97), lecturer in experimental science at King's College School (*ODNB*).
2. *camphor*: a terpenoid or naturally occurring waxy and flammable organic chemical.
3. *would probably form a brochure*: while not published in the *Phil. Trans.*, Tomlinson's paper was published as one of three essays in 1863. C. Tomlinson, *Experimental Essays* (London: Virtue Brothers & Co., 1863).
4. *Weber's experiments on kindred subjects*: Wilhelm Eduard Weber (1804–91), German physicist who, along with his brother Ernst Heinrich, had worked on fluid dynamics amongst other subjects.

To George Gabriel Stokes 15 March 1862 1868

15th March 1862

My dear Mr. Stokes,

I have read Mr. Walker's[1] paper, and though I cannot say that I have mastered its more minute details, which would be difficult without seeing the observations actually performed, I have been sufficiently informed by my perusal of it to enable me to recommend its publication in the Philosophical Transactions.[2] I think the assiduity Mr. Walker has displayed in collecting his results worthy of the especial encouragement of the Royal Society;

The question as it stands is just in that position which excites a desire to know more about it. The operator must have the means of varying the conditions of his experiments, and must not be confined to telegraph lines. Preliminary experiments would doubtless indicate the best mode of creating a permanent apparatus for the systematic observation of the currents. Probably such experiments would involve but little outlay. This I think certain that a working apparatus many thousand times more delicate than that employed by Mr. Walker might be devised, and with such an apparatus the deflections which he has observed might be obtained with an infinitesimal fraction of the amount of wire which he uses.

At all events, I should hardly think of erecting a permanent or expensive apparatus without ascertaining beforehand the real necessities of the case. I should not be surprised if a single paddock, or even a good sized yard were found sufficient for the collection of the desired currents; and if it is proved to be the case the advantage of being able to change the conditions of experiments promptly would be immense.

Yours very truly | John Tyndall

RS RR/4/277

1. *Mr Walker's*: Charles V. Walker (1812–82), electrical engineer (*ODNB*).
2. *its publication in the Philosophical Transactions*: appeared as C. Walker, 'On magnetic calms and earth-currents', *Phil. Trans.*, 152 (1862), pp. 203–20.

To Heinrich Debus 24 March [1862][1] 1869

24th March

My dear Heinrich[2]

I shall be delighted to see you on the 3rd. —The testimonial of Wurtz[3] is very nice, but it is not to be compared with that of Frankland—one is shadow in comparison with the solid substance of the other.

I have not yet decided regarding my Easter trip: I think I shall run down to Portland to see my friend Eddy Wynne[4] & have a rush about the County there. We will talk of the subject when we meet.

You certainly are very strongly supported as regards Aberdeen.[5] If the chair were to go by testimonial I should say that it would surely be yours. I suppose the decision will soon be made known.

I am working very hard—Lecturing three courses a week,[6] and engaged on two books[7] at the same time. One of the latter is now complete & will soon be published.[8]

Goodbye till we meet | Yours ever | John Tyndall

RI MS JT/1/T/309
RI MS JT/5/11/1375

1. *[1862]*: Debus applied for the Chair of Chemistry in Aberdeen in 1862. See n. 5.
2. *My dear Heinrich*: Debus crossed out before 'Heinrich' in handwritten original.
3. *Wurtz*: Charles Adolphe Wurtz (1817–84), the Alsatian French organic chemist.
4. *Eddy Wynne*: Edward Toler Wynne (1837–89) son of George Wynne, military officer and head of the Irish Ordnance Survey division that Tyndall worked for in the late 1830s.
5. *as regards Aberdeen*: the Chair of Chemistry had become vacant with the death of the previous incumbent, Andrew Fyfe, on 31 December 1861. The University Court appointed James Brazier on 8 April 1862, noting that he had already discharged the duties of the Chair for a number of years. [Anon.], 'Professor of Chemistry,' *Aberdeen Journal*, 9 April 1862, p. 5.
6. *Lecturing three courses a week*: Tyndall was giving a course of 12 evening lectures on heat at the Royal Institution. The course began on Thursday 23 January. He was also delivering a series of 10 evening lectures on light at the Government School of Mines. The third course has not been identified.
7. *engaged on two books*: *Mountaineering in 1861* (see letter 1804, n. 9) and J. Tyndall, *Heat* (London: Longman, Green, Longman, Roberts and Green, 1863).
8. *will soon be published*: *Mountaineering in 1861* was published on 16 April 1862, see also letter 1804, n. 9.

To William Frederick Pollock 24 March 1862 1870

24th March 1862

My dear Pollock,

The final revise of my little book[1] is now in my hands and I cannot allow it to quit them without saying to you that I believe your criticisms of the miracle chapter are of the very greatest importance as regards the popularity of the little thing. I am sure I should have altered many things myself, and indeed substantially made all the changes which you have suggested before letting it out of my hands. When I am hard pressed I write with a slight infusion of temper; but I am <u>not</u> perverse, and a little patient consideration afterwards, if I can only spare the time it requires, clears all the clouds away.

Whatever I have to say on the subject in question shall be said upon its merits and with the most earnest desire to offend nobody. I think as long as I adhere to this spirit and confine myself to what clearly connects itself with Natural Philosophy, the public of England will not bring me to the Stake. At least I am sure if they do your wife[2] will protest against the act.

Yours ever | John Tyndall

RI MS JT/1/TYP/6/1984
LT Typescript Only

1. *my little book*: *Mountaineering in 1861* published 16 April 1862.
2. *your wife will protest against the act*: for Tyndall's response to Juliet Pollock's reflections on his chapter on miracles and prayer, see letter 1866.

From Richard Thomas Lewis 25 March 1862[1] 1871

To Professor Tyndall
1 Lowndes Terrace, Knightsbridge, S.W.,
March 25, 1862.

Sir,

The enlarged appearance of the sun and moon at rising or setting has long been a subject of controversy; and although astronomers and others have proved by micrometrical measurements that the apparent diameter of those bodies is *actually* largest when they are on the meridian (since they are nearest to us then), the popular mind refuses to believe that the illusion is only mental, and hitherto no illustration has been found sufficiently satisfactory to convince those who are sceptical in the matter. The best illustration brought forward as yet, to my knowledge, was that which appeared some time since in 'Recreative Science,'[2] in an article upon "The Rising and Setting of the Sun Physiologically considered". The writer had, one dark night, been gazing for some time upon the glowing furnace of a lime-burner, after which, on going out into the dark, he saw the luminous image of the fire in the air before him as a large circle; but happening to look up, this image apparently shrank up to about half its former diameter. This, however, does not appear to satisfy many persons; and since your most interesting lecture of the 16th ult.[3] was the means of my discovering a far more forcible and complete illustration of this singular effect, I venture to trespass upon your time by detailing it to you. On my return home after that lecture I proceeded to make sundry experiments upon binocular vision, amongst which were the following:—

Centring the eyes upon a distant point A, I introduced a stereograph of the moon at such a distance from the eyes that the axis of each passed centrally through the picture taken for it; the result was a perfect and beautiful stereographic projection of the moon at A, apparently solid and spherical (as fig. 1). Then, without in any way altering the distance between the eye and the slide, I centred the

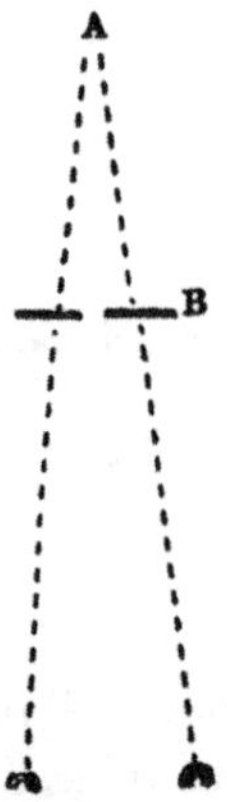

eyes upon a point nearer to me than B, in such a way that the

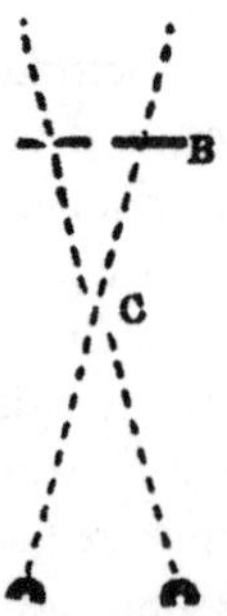

axis of the right eye passed through the centre of the left picture, the left also passing through the right: the result was a perfect view of the moon, but in this case *inside out,* as you termed it (the moon appearing quite concave), the image being seen *in the air at* C, and appearing in a most perceptible degree *smaller* than when in the former case I saw it apparently at A, nearly double the distance. Now as the distance between the real object and the eye was in both cases identically the same, the image upon the retina must have been of the same size in each experiment, the apparent enlargement of the object in one case and contraction in the other being purely mental. This singular effect may have been noticed by yourself before this; but in case it should not have come under your notice, I thought you would perhaps excuse the liberty I take in addressing this communication to you. Of course it would be quite out of place and equally unnecessary for me to explain the philosophy of the matter here, or the difficulties connected with first attempts to break the eye of the habit of focusing itself according to the amount of convergence; but for the delight with which I, as a young student of astronomy, made the above observation, as also for my future independence of all stereoscopes, I feel greatly and entirely indebted to the course of lectures upon light. In performing the second of the two experiments named, so perfect a perspective inversion is seen in the case of landscapes, that one feels for the moment endowed with the fabled optical powers of the lynx, men being seen most distinctly through brick walls, houses, or even hills.

Hoping, Sir, that you will excuse the liberty I have taken I thus addressing so long a communication to you,

Believe me to be, Sir| Yours very respectfully| Richard T. Lewis.

'On the Changes in the Apparent Size of the Moon', *Phil. Mag.,* 23 (1862), pp. 380–82.

---

1. 25 *March 1861*: sent on this date, published in May issue of the *Phil. Mag.*
2. *in Recreation Science*: T. Zeno, 'On the rising and setting of the sun physiologically considered', *Recreative Science,* 3 (1862), pp. 110–20. This first appeared in July 1861 and was later incorporated into a combined volume of the journal published in 1862.
3. *your most interesting lecture of the 16th ult.*: Lewis was attending Tyndall's ten lectures on light delivered at the Royal School of Mines. It was the lecture on 16 March that prompted Lewis to write to Tyndall. See also letter 1872.

To Richard Thomas Lewis[1] 27 March 1862 1872

27th March 1862

Dear Sir,

Your experiment is an excellent one;—so good indeed that I am disposed to have your letter[2] sent to the Philosophical Magazine for publication.[3]

You remember the hollow & the solid frustum of a pyramid which I referred to in my lectures.[4] The upper section of the frustum changes in apparent area exactly as your norm changes.

Your solution has been already revealed by Sir David Brewster—but your treatment is evidently quite original & does you very great credit

I am dear Sir | Yours very faithfully | John Tyndall

Royal Society MM/9/19

1. *Richard Thomas Lewis*: (1839–1916), amateur microscopist, entomologist and temperance advocate. See Emily J. Lewis, 'Obituary notices: Richard Thomas Lewis FRMS', *Journal of the Quekett Microscopical Club*, 13 (1916–18), pp. 201–3.
2. *your letter*: see letter 1871.
3. *sent to the Philosophical Magazine*: appeared as R. Lewis, 'On the changes in the apparent size of the moon', *Phil. Mag.*, (1862) pp. 380–82.
4. *in my lectures*: Lewis was attending Tyndall's ten lectures on light delivered at the Royal School of Mines, starting 22 February 1862. It was the lecture on 16 March that prompted Lewis to write to Tyndall. See letter 1871.

From Wilhelm Martin Logeman[1] 27 March 1862 1873

Monsieur

J'ai reçu avant hier votre honorée du 24, mais je n'ai pas voulu engager M. Juncker à vous envoyer sa note sur les aimants avant de m'avoir convaincu par l'expérience que surtout le grand aimant repondit pleinement à mon attente. Il a éprouvé un revers dans la fabrication: un des lamelles s'est cassé lorsqu'on la dressait après la trempe. Hier cependant il a pu en faire l'essai en présence de M^r van Breda et de moi. Il y a suspendu sans grande précaution, un poids de à peu près 280, deux cents quatre vingt livres anglais (avoir du poids naturellement). Il s'occupe maintenant de le polir et il vous l'enverra avec les deux autres, qui portent 50 et 24 livres, dans la semaine prochaine. En attendant, il vous envoie sa note, *[1 word illegible]* pour acquit, comme vous le désirez.

Je vous donnerai avis du jour exact de l'expédition. La caisse qui contiendra les trois autres caisses dont chacune contient un aimant vous parviendra de Rotterdam et sera déclarée contenir des instruments de physique. Peut être qu'il vous sera possible d'empêcher que les agents de la douane n'ouvrent la caisse autrement qu'en notre présence. Je sais par expérience que de cette manière seulement vous pourrez être bien sûrs de recevoir les aimants en bon état.

Agréez monsieur, les salutations empressées de votre dévoué serviteur
W. M. Logeman.
Haarlem 27 mars 1862

Sir,

On the day before yesterday I received your letter from the 24th, but I did not wish to engage Mr Juncker[2] to send you his note on the magnets before I was myself convinced through the experiment that especially the large magnet fully met my expectations. He experienced a setback in the manufacturing: one of the lamellae broke when we put it up after the tempering. Yesterday, however, he was able to test it in the presence of Mr van Breda[3] and me. He suspended it without much precaution, a weight that is approximately 280, two hundred and eighty English pounds (naturally avoirdupois[4]). Now he is busy polishing it, and he will send it to you along with the other two, which are 50 and 24 pounds, next week. In the meanwhile, he sends you his note, *[1 word illegible]* for granted, as you wished.

I will notify you of the exact date of the shipping. The box that contains the three other boxes, of which each contains a magnet, is sent to you from Rotterdam and will be declared as containing instruments for physics. Perhaps you will be able to have the Custom agents opening the box only in our presence. I know from experience that only in this way can you be sure of receiving the magnets in good condition.

Please accept, Sir, eager greetings from your devoted servant

W. M. Logeman.

Haarlem, 27th March 1862

RI MS JT/1/L/31

---

1. *Wilhelmus Martinus Logeman*: (1821–94), son of an instrument maker. Logeman started a factory for the manufacture of steel magnets in 1849. He later became a physics teacher after the factory failed in 1863. Logeman worked for Jacob van Breda and was closely connected to a number of periodicals such as *Album der Natuur* and *Wetenschappelijke bladen.* See P. Molhuysen en P. Blok (eds), *Nieuw Nederlandsch Biografisch Woordenboek. Part 9.* (Leiden: A.W. Sijthoff, 1933), pp. 616–18.
2. *Mr. Juncker*: not clear in original and not identified.
3. *Mr van Breda*: Jacob Gisbertus Samuel van Breda (1788–1867). Van Breda began his scientific career as Professor in botany, chemistry and pharmacy at the Athenaeum in Franeker in 1816. He later became Professor in botany, zoology and comparative anatomy at Ghent. Forced to flee during the Belgian Revolution, he gained a temporary post at Leiden in 1831, and later a chair in zoology and geology in the same city in 1835. From 1839 onwards he left Leiden to become secretary of the *Hollandse Maatschappij van Wetenschappen* (Holland Society of Science) at Haarlem, and director of the Teylers museum. He became the chairman of the commission set up to create a geological map of the Netherlands in 1852. See P. Molhuysen en P. Blok (eds), *Nieuw Nederlandsch Biografisch Woordenboek. Part 1.* (Leiden: A.W. Sijthoff, 1911), pp. 457–58.
4. *avoirdupois*: measurement system using pounds and ounces.

From William Snow Harris[1] 27 March 1862 1874

6 Windsor Villas | Plymouth | 27 March 1862.

My dear Tyndall,

You will be, I am sure, sorry to hear that I have had a long and suffering illness, of, now, seven months duration, owing to an attack of Iritis, and inflammation of the left eye, and altho' now progressing favourably, my medical friends will not permit me the free use of my eyesight—hence I am unable to address you in my own handwriting.

My son, who I hope one day to present to you, is a candidate for admission into the Athenaeum Club, and will probably come on for ballot about the end of May[2] or the beginning of June: may I venture to hope that you will honor him with my your signature upon the Ballot Certificate, in common with many of my other influential friends?

Allow me to state for your information that he is at present Resident Engineer in Superintendence of the new forts in progress at Portsmouth:—he is the inventor of a new method of atmospheric railway of which he made a beautiful working model: he is the proposer of a new harbor of Refuge and Coaling Depot in Plymouth Sound, for H.M. Ships—which has received the approbation of the best Engineers in the country and relative to which he printed a very interesting paper which was most favourably received by the Admiralty.

He has received, as you may well suppose, a good classical mathematical and physical education: he is much devoted to physical science and I can vouch for, this being a perfect gentleman, having the highest sense of honour.

As the object of these signatures is to assure the Club of the respectability of the candidate, I trust you will see no objection to comply with my request on my son being made known to you.

The Ballot Certificate will lie on the table a week before the day of election of which I will give you due notice.

This late attack of mine has been a sad affair for me; it has thrown back my scientific pursuits dreadfully and prevented me from getting a new work on electricity into the press.[3]

I look forward with hope and expectation to the pleasure of a visit from you sometime for the summer in order that you may see my cabinet of electrical instruments: I think I have some of the most beautiful instruments for electrical research which have as yet been constructed: I mean of course instruments for illustrating the laws of statical electricity. I am quite satisfied that the scientific work so far as the received opinions go, are very ill-formed upon this very interesting subject.

I should be very loath to leave this world without some provision for the security of those Instruments in their proper place, and have an idea that the geological museum would be a fitting place for their reception: upon this question, and the general laws of electrical force I should like to have a conference with you—could you manage to come and see us some time for this summer?

Believe me to be | My dear Tyndall | Your very sincere friend | W. Snow Harris.

P.S. I hope to be in town about the middle of May.

RI MS JT/1/TYP/2/476–77

---

1. *William Snow Harris*: Sir William Snow Harris (1781–1867), natural philosopher known for his work in electricity and magnetism. He was elected FRS in 1831 and lectured at the RI in the 1830s (*ODNB*).
2. *come on for ballot about the end of May*: Thomas Harris was elected a member on 26 May 1862, proposed by his father and seconded by William Fairbairn (FRS). Tyndall signed his Candidate's Certificate for Ballot. We are grateful to Jennie de Protani, archivist at the Athenaeum Club, for this information.
3. *a new work on electricity into the press*: the work appeared posthumously, with a memoir of the author, in 1867. W. Harris, *A Treatise on Frictional Electricity* (London: Virtue and Co., 1867).

From Joseph Plateau 28 March 1862 1875

Gand, 28 Mars 1862

Mon Cher Monsieur Tyndall,

Je viens vous remercier bien vivement pour toutes de peines que vous vous êtes données et pour la répétition de mes expériences devant la Société Royale, veuillez transmettre aussi mes remercîments à M. Frankland.

Je regrette seulement que votre lettre soit si courte et si peu explicite, et j'attends de vous une prochaine lettre plus détaillée.

En premier lieu, M. Frankland, dont la spécialité est la chimie, a-t-il pu énoncer quelques vues théoriques, a-t-il fait connaître les lois qui régissent mes systèmes laminaires, lois qu'en votre qualité de système vous auriez certainement développées et sans lesquelles ces expériences ne constituent que des phénomènes curieux, une sorte de jeu d'enfant?

En second lieu, je ne comprends pas bien l'emploi de la lumière électrique pour éclairer ces mêmes systèmes; cette lumière, en effet, n'*émane* que d'un foyer de peu d'étendue et des lois, chaque lame doit? simplement en montrer l'image brillante, comme le ferait une petite glace, sur une très petite portion de sa surface; comment donc les systèmes ont-ils pu, en cette manière montrer les couleurs qu'ils étalent?

En troisième lieu, vous ne me dites pas de quel liquide vous vous êtes servi; était-ce de la simple eau de savon, ou bien du liquide glycérique? Dans ce dernier cas, le liquide était-il fait au savon de Marseille, à un autre savon, ou à l'oléate de soude, et quelle était la persistance des lames?

Enfin vous ne me dites pas si vous avez reçu la dernière caisse de M. Van Hese, caisse contenant le vase à parois planes en verre, avec ses accessoires, du savon de Marseille, du savon de Rimmel et deux échantillons de liquide glycérique à l'oléate de soude. Vous ne me parlez pas non plus de ma dernière lettre, dans laquelle je vous indiquais la préparation de ces derniers liquides, et je vous donnais les explications relatives au vase et aux petits appareils qui l'accompagnent.

J'espère que vous viendrez bien me répondre dans peu de temps, et me donner les renseignements sur ces différents points.

Lorsque j'ai publié ma 5ème Série, l'arrivée de la saison chaude m'avait empêché de compléter l'étude du liquide glycérique et de sa préparation en hiver; maintenant, j'ai profité des derniers froids pour achever cette étude, et je suis arrivé à connaître parfaitement le liquide en question, toutes les influences qui le modifient et sa bonne préparation par les basses températures. Comme vous me faites espérer que vous voudrez bien répéter les expériences à l'Institution Royale, et comme vous avez maintenant du savon de Marseille, je vais vous indiquez les résultats auxquels je suis arrivé, et vous pourrez aussi vous passer d'oléate de soude et preparer, à coup sur, un excellent liquide au savon de Marseille quand même le froid reviendrait, et surtout s'il revient.

Ainsi que je vous l'ai déjà dit, le liquide glycérique le plus parfait se formerait avec de la glycérine de Price, de l'eau distillée et de l'oléate de soude pur, dans des proportions exactes; mais ce liquide ne serait pas un mélange, ce serait une combinaison. Maintenant, quand on forme le liquide en employant du savon de Marseille, il se forme encore une combinaison de glycérine, d'eau et d'oléate de soude, le stéarate et le margarate sont séparés; mais une portion des deux derniers sels demeure dissoute et à l'état de mélange dans le liquide; une autre portion apparait sous la forme d'un précipité blanc. Cette action chimique s'effectue lentement, et, à des températures basses, elle exige trois ou quatre jours. Quand on prépare le liquide en été, d'après le premier procédé de mon paragraphe 13, il n'y a, à cause de l'élévation de la température, qu'une petite portion du stéarate et du margarate qui apparaissent sous forme de précipité, et quand on a laissé monter celui-ci et qu'on a recueilli le liquide limpide, ce liquide contient encore une grande quantité de stéarate et de margarate dissous, qui demeurent dans le liquide à l'état de mélange, et qui, comme substances inactives, rendent ce liquide moins bon; aussi n'obtient-on-alors qu'une persistance maxima de 4 heures. Mais quand on prépare le liquide aux basses températures par le procédé que je vous indique plus loin, la plus grande partie du stéarate et du margarate sont précipités, et le liquide limpide recueilli ne contient presque plus que de l'oléate de soude avec la glycérine et l'eau; aussi donne-t-il alors des persistances qui peuvent atteindre 8 heures. Telle est la théorie du liquide glycérine au savon de Marseille, théorie que mes nombreuses expériences me font regarder comme bien véritable; Voici maintenant le procédé pour la préparation en hiver. Il faut prendre du savon bien humide; celui dont je me sers a été conservé dans une cave; alors la proportion d'une partie de ce savon sur 40 parties d'eau distillée est la plus convenable. Si le savon est plus au moins desséché, la proportion doit sans doute être un peu moindre, mais je ne puis rien préciser à cet égard. Après avoir fait, à chaud, la solution de savon, l'avoir laissée refroidir jusqu'à ce qu'elle ne soit plus que légèrement chaude, puis l'avoir filtrée, on mêle un volume de glycérine de Price avec 2 volumes de cette solution, on agite fortement et longtemps, puis on expose le flacon à une température de 60 à 100 centigrades, et on l'y laisse pendant 4 jours. Le liquide est alors devenu fortement troublé, on le filtre, toujours à ces basses températures, puis on y ajoute un 9ème de son volume de glycerine de Pricede sorte qu'alors la proportion de glycérine, en volume, est à celle de la solution de savon comme 2 à 3. On laisse encore, pendant quatre jours le liquide exposé aux mêmes basses températures; si ces températures ne sont pas descendues au-dessous de ce qu'elles étaient lors de la filtration précédente le liquide ne s'est plus troublé; S'il y a eu un abaissement plus grand, le liquide présente un nouveau trouble et on le filtre encore. La préparation est alors terminée, et le liquide est excellent, beaucoup meilleur que celui que l'on préparait en été, par mon premier procédé.

Comme vous voulez bien me promettre de faire encore à mes expériences, l'honneur de les répéter devant l'Institution Royale, mais comme la température extérieure ne s'abaissera plus beaucoup sans doute, vous pourrez, si vous n'avez pas un très bon liquide, et si vous voulez en préparer par le procédé ci-dessus, placer le mélange dans une cave, la plus fraîche dont vous pourriez disposer. Je suis persuadé que vous réussiriez aussi, quoique un peu moins bien qu'en hiver. Si vous le faites, suivez exactement les proportions que je vous indique. J'ai eu soin d'envelopper d'étain le savon que je vous ai envoyé; j'espère qu'il se sera conservé assez frais.

J'ai parlé à M. Van Hese; il y a plus de 15 jours qu'il vous a envoyé son compte; il ne comprend pas ce qui a pu être cause qu'on ne vous l'ait pas remis, et il a fait immédiatement des démarches pour que ce retard soit réparé.

J'attends votre réponse avec impatience, et vous prie d'agréer l'assurance de tous mes sentiments d'affection.

J. Plateau | professeur à l'Université | Place du Casino 22

Ghent, 28th March 1862

My dear Mr. Tyndall,

I wish to thank you graciously for all the trouble that you have taken, and for repeating my experiments in front of the Royal Society. Please also convey my thanks to Mr. Frankland.

I only regret that your letter was so short and so inexplicit, so I will be waiting for a more detailed letter from you next.[1]

Firstly, Mr. Frankland, who specializes in Chemistry, did he have the opportunity to state certain theoretical views? Did he make known the laws governing my laminar systems, laws that in your quality of a system, you would certainly have developed, and without which the experiments that we consider interesting phenomena would be a kind of child's play?

Secondly, I do not understand the use of electric light to illuminate these systems; this light, in fact, emanates only from a hearth of small extent, and laws, each blade must simply show the brilliant image, like a small mirror would do it, on a very small portion of its surface. So how could the systems, in this manner, show the colours that they display?

Thirdly, you have not told me which liquid you have used. Was it the simple soap water, or was it the glycerine liquid? In the latter case, was the liquid made from Marseille soap, from another soap, or from sodium oleate, and what was the persistence of the layers?

Finally, you have not told me if you have received the latest box from Mr. Van Hese,[2] a box containing the glass vessel with planar walls, with its accessories, Marseille soap, Rimmel soap, and two samples of the glycerine liquid with sodium oleate. You also did not speak to me about my last letter, in which I indicated the preparation of the latter liquids, and where I gave you some explanations regarding the vessel and the small apparatuses that come with it.

I hope that you will be kind enough to answer me soon,[3] and that you will inform me on these points.

When I published my 5th Series,[4] the arrival of the hot season prevented me from completing the study of the glycerine liquid and of its preparation in winter; now, I took advantage of the latest cold weather to accomplish this study, and I came to know perfectly the liquid in question, all of the influences that change it, and its good method of preparation for lower temperatures. As you give me hope that you would be kind enough to present these experiments to the Royal Institution,[5] and as you now have the Marseille soap, if you specify the results that I received, and if you could also skip the sodium oleate and prepare, suddenly, an excellent liquid from Marseille soap, even when the cold would come back, and especially if it returns.

As I have already mentioned, the most perfect glycerine liquid is formed with glycerine from Price,[6] distilled water, and pure sodium oleate, in exact proportions.

But this liquid will not be a mixture, it will be a combination. Now, when we form the liquid we use Marseille soap, and it is again created from a combination of glycerine, water, and sodium oleate; the stearate and the margarate are separated; but one portion of the two last salts remains dissolved and in a state of a mixture in the liquid. Another portion appears in the form of a white precipitate. This chemical action takes place very slowly, and in low temperatures it requires three or four days. When we prepare this liquid in the summer, according to the first procedure of my paragraph 13, there is only a small portion of stearate and margarate that appear in the form of a precipitate, because of the increase in temperature. And when we leave it to rise and collect the clear liquid, this liquid still contains a large amount of dissolved stearate and margarate, which remain in the liquid in a mixed state, and as inactive substances they make this liquid worse. Also, we should not forget of a maximum persistence of 4 hours. But when we prepare the liquid at lower temperatures by the procedure that I will explain further down, the largest part of the stearate and the margarate is precipitated, and the clear liquid that is collected does not contain almost anything aside from sodium oleate with glycerine and water. It will also give persistence that could reach 8 hours. Such is the theory of the glycerine liquid from Marseille soap, a theory that my numerous experiments showed me as well-confirmed. Now here is the procedure for the preparation in winter. We must take a very humid soap; the one which I use was kept in a cellar. Thus, the proportion of one part of this soap over 40 parts of distilled water, is most suitable. If the soap is more or less dry, the proportion should probably be slightly less, but I cannot be exact in this regard. After having made the hot soap solution, leave it to cool down until it is only slightly warm. Then to filter it, we mix one volume of glycerine from Price with 2 volumes of this solution. We stir vigorously for a long time, then we expose the vial to a temperature of 60 to 100 degrees Celsius, and we leave it there for 4 days. The liquid then becomes very cloudy; we filter it, always at lower temperatures, then we add a 9th of its volume of glycerine from Price, in a way that the proportion of glycerine, in volumes, is to the soap solution as 2 to 3. We leave the liquid expose to those same low temperatures again for four days. If those temperatures have not fallen below what they were since the previous filtration, the liquid is not cloudy anymore. If there was a greater reduction, the liquid would be cloudy again and we should need to filter it again. Thus the preparation is complete and the liquid is excellent, much better than the one that we prepared in the summer according to my first procedure.

Since you are kind enough to promise me to do my experiments again, the honour of repeating them in front of the Royal Institution, but since the temperature outside will probably not drop much anymore, you could, if you do not have a very good liquid, and if you would want to prepare it according to the aforementioned procedure, place the mixture into a cellar as soon as you can. I am sure that you will also succeed, though not as good as in the winter. If you are doing this, follow the proportions that I indicated closely. I took care to wrap the soap that I sent you in tin. I hope that it will stay cool enough.

I spoke with Mr. Van Hese. More than 15 days have passed since he sent you his account.[7] He does not understand what could have been the cause of it not being given to you, and he immediately took steps to amend the delay.

I very much look forward to your reply,[8] and I ask you to accept the assurance of my warmest wishes.

J. Plateau | Professor at the University | 22 Place du Casino

RI MS JT/1/P/93

---

1. *more detailed letter from you next*: letters missing or not written.
2. *the latest box from Mr Van Hese*: see letter 1862.
3. *kind enough to answer me soon*: letter missing or not written.
4. 5th *Series*: J. Plateau, 'Recherches expérimentales et théoriques sur les figures d'équilibre d'une masse liquide sans pesanteur', *Mém. Acad. Roy. Sci. Belg.*, 33 (1861), separate pagination.
5. *present these experiments to the Royal Institution*: no record found.
6. *Price*: Price's Patent Candle Company.
7. *since he sent you his account*: letter missing.
8. *your reply*: letter missing or not written.

From Sir Philip de Malpas Grey-Egerton[1] 31 March 1862 1876

28B Albemarle Street
March 31 1862

Dear Tyndall

I see in my County Newspaper[2] received this morning that a magnificent block of 'pure white rock salt' has been obtained from the mines of Messrs. Fletcher and Rigby of Marston near Northwich.[3] It is stated to be 4½ and is destined for exhibition at the International.[4] I advise you to keep your eye upon it, and if it should be pure enough for your purposes I will do what I can to procure you a cut of it.

Believe me | yours very truly | [Philip Grey–]Egerton

RI MS JT/1/E/100

---

1. *Sir Philip de Malpas Grey-Egerton*: (1806–81), 10th baronet, politician and palaeontologist (*ODNB*).
2. *my County Newspaper*: possibly referring to [Anon.], 'Contributions from Cheshire to the

great exhibition', *Cheshire Observer and General Advertiser: for Cheshire and North Wales*, 29 March 1862, p. 5.

3. *Northwich*: town and civil parish in Cheshire, England.
4. *exhibition at the International*: The International Exhibition or Great London Exposition ran from 1 May to 1 November 1862 and was based adjacent to the gardens of the Royal Horticultural Society.

To Rudolf Clausius[1] 5 April 1862 1877

I wish I could possess myself of all the writings of Dr. Mayer on Heat. Where do you think I could obtain them?[2]

RI MS JT/5/11
LT Transcript Only

---

1. *To Rudolf Clausius*: letter fragment.
2. *Where do you think I could obtain them?*: see letter 1892 for Clausius's reply.

From John Frederick William Herschel 12 April 1862 1878

Collingwood | Hawkhurst | Kent | April 12/62

Dear Sir

I have to thank you for two papers on radiant heat[1] both extremely interesting—you have made a grand step in meteorology in showing that the dry air is perfectly transcalescent[2] & that the invisible moisture is what stops the sun's heat. In your 2[n]d paper you express a wish to procure a good crystal of Rock Salt—I shall send you per railway a very fine specimen of which voici l'histoire.[3]

After the publication of my paper in the Phil Trans 1840[4] I was very desirous to disengage myself from the influence of glass prisms & lenses & ascertain if possible whether in reality my insulated heat-spots α β γ δ ε in the specturm were of solar or terrestrial origin. Rock Salt was the obvious resource and after many & fruitless endeavours to obtain sufficiently large & pure specimens, as the late Dr Somerville[5] was so good as to send me (as I understood from a friend in Cheshire) the very fine block which I now forward. It is however much cracked but, I have no doubt pieces large enough for lenses & prisms (especially if cemented? together) might be got from it.

But I was not prepared for the working it—evidently a very delicate & difficult process, (I proposed to dissolve off the corners &c and as it were lick it into shape) & though I have never quite lost sight of the matter I have not yet been able to do anything with it. Meanwhile I put it by. On looking at it a year or two after I was dismayed to find it had lost much by deliquescence. Accordingly I potted it up in salt in an earthen dish with iron rim—placed it on an upper shelf in a room with an Arnott Stove where it has remained ever since.

If you should find it of any use I would ask you if possible to repeat my expt. as described, & settle that point which has always struck me as a very important one.

Will you permit me a remark on your 1st paper. You speak of air being dynamically heated by letting external air rush into a vaccuum (on the Joulian &c theory of vis viva about which I am very far from being satisfied). Now—when a receiver is exhausted & air let in till the indraft ceases & then the stop-cock is closed for some minutes & then again opened, more air rushes in and so two or three times. Now certainly it may be said that in entering though latent heat at first disappears during the expansion, yet this is restored on the subsequent filling in & consequent compression, and that the heat developed by the shock is then placed in evidence & keeps the air expanded till an equilibrium (momentary) is attained.

But this ignores the vessel. Query, is not the true explanation that, the entering air being excessively chilled by expansion takes up heat from the vessel. As the elasticity approaches that of the external air this heat is given out & sustains part of the pressure & it is not till after closing the stop cock & letting the cold vessel again take up the lost heat that the deficiency of air becomes sensible and a partial vacuum results? Such has always at least been my view of this phenomenon.

I remain | dear Sir | yours very truly | JFW Herschel

RI MS JT/1/H/98

1. *Two papers on radiant heat*: J. Tyndall, 'On the absorption and radiation of heat by gaseous matter. Second memoir', *Phil. Trans.*, 152 (1862a), pp. 59–98. J. Tyndall, 'Remarks on recent researches in radiant heat', *Phil. Mag.*, 23 (1862b), pp. 252–66.
2. *perfectly transcalescent*: able to freely transmit radiant heat (*OED, quot.* 1850).
3. *voice l'histoire*: French, meaning 'here is the story'.
4. *Phil Trans 1840*: J. Herschel, 'On the chemical action of the rays of the solar spectrum on preparations of silver and other substances, both metallic and non-metallic, and on some photographic processes', *Phil. Trans.*, 130 (1840), pp. 1–59.
5. *the late Dr Somerville*: Dr. William Somerville (1771–1860), military surgeon, (second) husband of Mary Somerville (1780–1872).

From Joseph Plateau 14 April 1862 1879

Gand, 14 Avril 1862

Mon Cher Monsieur Tyndall,

Il y a plus de quinze jours que je vous ai écrit pour vous remercier aussi que M. Frankland relativement à l'exécution de mes expériences devant la Société Royale, et pour vous demander certains renseignements; Je n'ai reçu aucune réponse, malgré votre exactitude habituelle. Ma lettre ne vous serait-elle point parvenue? Je vous priais de vouloir bien m'apprendre en premier lieu, si Mr Frankland avait énoncé, devant la Société, les lois principales qui régissent nos systèmes laminaires; en second lieu, de quel liquide il s'était servi, et quelle était la persistance des systèmes; en troisième lieu, si vous aviez reçu la caisse contenant le vase à parois planes plans en verre avec les appareils nécessaires, accessoires, aussi que la lettre dans laquelle je vous donnais des explications à cet égard et à l'égard du liquide a l'oleàte de soude; en 4ème lieu je vous exprimais quelques doutes au sujet de l'emploi de la lumière électrique pour éclairer les systèmes; enfin je vous communiquais les résultats de nos dernières études sur le liquide-glycérique au savon de Marseille, et je vous faisais connaître le procédé certain auquel je suis arrivé pour préparer un excellent liquide de cette espèce par les basses températures.

J'espère, Mon Cher Monsieur Tyndall, que vous ne me ferez pas attendre plus longtemps votre réponse, et je vous prie d'agréer l'assurance de tous mes sentiments d'affection.

J. Plateau | Professeur à l'Université | Place du Casino 22,

P.S. Mr Van Hese a reçu le montant de son compte, et il me charge de vous remercier.

Ghent, 14th April 1862

My dear Mr. Tyndall,

It has been more than two weeks since I wrote[1] to thank you as well as Mr. Frankland for the performance of my experiments in front of the Royal Society, and in order to ask you for some information. I did not receive a reply despite your usual punctuality. Did my letter not arrive to you? I ask you kindly to let me know, firstly, whether Mr. Frankland explained, in front of the Society, the main laws that govern our laminar systems. Secondly, which liquid it came from, and what was the persistence of the systems. Thirdly, whether you have received the crate containing the glass vessel with planar walls along with the additional apparatuses, as well that the letter in which I gave to you explanations in this regard and in regard to the sodium oleate liquid; Fourthly, I was telling you of some of my doubts regarding the use of electric light to illuminate the systems. And finally, I provided you with the results of our latest study on liquid glycerol in Marseille soap, and I will let you know about the particular process with which I managed to prepare an excellent liquid of this type by low temperatures.

I hope, my dear Mr. Tyndall, that you will not make me wait any longer for your answer,[2] and I ask you to please accept my warmest wishes.

J. Plateau | Professor at the University | 22 Place du Casino

P.S. Mr. Van Hese received the amount for his account,[3] and he asked me to thank you for it.

RI MS JT/1/P/94

1. *since I wrote*: see letter 1875.
2. *your answer*: letter missing or not written.
3. *the amount for his account*: see letter 1875.

To John Frederick William Herschel 15 April 1862 1880

Royal Institution of Great Britain | 15th April 1862

Dear Sir John.

Your letter[1] which I received yesterday was a source of very great gratification to me. Indeed you are one of the chosen few whose approval I think it well worth the labour of a life to obtain.

I shall prize your gift of salt exceedingly and shall try to turn it to good account.

I have directed Longman to send you a little book of mine[2] which he is just now publishing. It is totally unscientific but I do not scruple to send it to you whose mind wanders so far beyond the fringes of formal science

Believe me | Most truly yours | John Tyndall

I am on the point of running down to the coast of Devon to get up fibre for my after Easter work

RS HS 17.391

1. *Your letter*: letter 1878.
2. *a little book of mine*: *Mountaineering in 1861,* see also letter 1804, n. 9.

To George Gabriel Stokes 15 April 1862 1881

My dear Mr. Stokes

I waited till I sh[d] be able to write to you a letter fully regarding my paper.[1] The aqueous vapour question has become of such interest that I think I must ask you to publish a drawing of the apparatus[2] with which I proved its action where the rock salt plates were absent. I shall call upon my draughtsman today & if possible get the drawing from him. I think I will omit the others thro' sheer want of time to draw them. The letter press may be allowed a little accordingly.

It is not at all uncommon to use the thermo electric pile with a coating of soot exactly such as Melloni & Magnus used on the plates of rock salt. I have the very highest admiration for Melloni's ingeniousness as an experimenter & for his clearness as a reasoner; but he is not without his error & failings. He has made, as all men who work so long as he did will do—several mistakes; and this lampblack affair[3] is one of them. Of course all that I have stated is perfectly correct, but I am quite willing to publish anything you might suggest to take the edge off my remarks though I had no intention of furnishing them with an edge at all.

I should be very glad if anything struck you in my remarks regarding Magnus[4] that might be softened or improved if you would inform me. I do not wish to say a word which could unnecessarily hurt him.

Yours very since[re]ly | John Tyndall

Tomorrow evening I quit London & will be absent for 8 or 10 days.[5]

Stokes Correspondence, Cambridge University Library MS.Add.7656/ T553,1(4)–4(4)

1. *my paper*: Tyndall (1862).
2. *drawing of the apparatus*: Tyndall (1862), p. 91.
3. *lampblack affair*: Tyndall (1862), pp. 70–71.
4. *remarks regarding Magnus*: Tyndall offers extended commentary on his differences with Magnus in Tyndall (1862). For a brief account of the long running but cordial dispute with Magnus, see Introduction (this volume). For a more detailed investigation, see reference, letter 1830, n. 4.
5. *I quit London & will be absent for 8 or 10 days*: Tyndall went to Dorset and Devon. See letter 1885.

To Heinrich Debus [16 April 1862][1] 1882

Royal Institution

My dear Heins

I write in great haste. I fear your plan won't answer.[2] I have pledged myself to go down to the Isle of Portland and work my way to Torquay.[3] I wish you had come to us and then we might have made some definite arrangement.

Yours ever | John Tyndall

RI MS JT/5/11
LT Transcript Only

---

1. *[16 April 1862]*: assuming the letter was written in haste because he was about to embark on his journey, the letter can be dated as 16 April. Tyndall set off from Waterloo station at 11am to travel to the Isle of Portland. See letter 1885.
2. *I fear your plan won't answer*: no record of plan has been found.
3. *Isle of Portland and work my way to Torquay*: The Isle of Portland is a small island in the English Channel. Torquay is a seaside town in Devon, England.

From Alexander Stewart Herschel[1] 16 April 1862 1883

Royal College of Chemistry | 315 Oxford Street | April 16th 62

My dear Sir

Your welcome letter has reached me safely here this morning and it cannot fail to be the most valuable assistance that I shall receive to further my suit at the Andersonian Institution.[2] I have requested Messrs Groombridge and Sons[3] to send to you at the Museum of Practical Geology a copy of their April number of the 'Intellectual Observer'[4] in which they published a short account of observations on the heights of meteors and shooting stars made by united amateurs and professional observers last August, which I hope you will accept, with much gratitude,

from | yours very faithfully | Alexr S. Herschel.

RI MS JT/1/H/81

---

1. *Alexander Stewart Herschel*: (1836–1907). Astronomer, son of John F. W. Herschel.
2. *Andersonian Institution*: technical college based in Glasgow, Scotland founded in 1796, also known (from 1828) as Anderson's University.
3. *Messrs Groombridge and Sons*: London–based publisher.

4. *Intellectual Observer*: The full title was *Intellectual Observer: A Record and Remembrancer of Intellectual Observation in the Various Branches of Physical Science.*

From George Gabriel Stokes 16 April 1862 1884

Lensfield Cottage | Cambridge | 16 April 1862.

My dear Prof. Tyndall,

What you stated[1] about the contradictory statements on two different pages of Melloni's book[2] is of course a fact; your inference from that fact—that Melloni's mind was so preoccupied with the common notion of the absorption by lamp-black of rays of all kinds of heat that he forgot that he had found that such was not the case I think exceedingly improbable. I suggested designed suppression, but I have since thought that though the experiments of Chap. 4 could not have been made (in all probability) while the sheets were going through the press, it is not unlikely that during that interval lamp black was added to the list of substances experimented on as to their diathermancy, so that in fact when chap 2 was written the partial diathermancy of lamp-black had not yet been discovered. Both of my conjectural explanations of the discrepancy seem to me more probable than yours. To make a mistake in such a long enquiry is natural enough, but for an experimentalist to forget that he had himself discovered and corrected an important mistake, and so to fall into it again, implies an amount of carelessness or prejudice with which one should not lightly be charged.

There is nothing in what you said of Magnus as far as I recollect that is in the least discourteous, or calculated to wound his feelings.[3] If in looking over the review anything strikes me of that nature I will point it out to you.

I have no doubt the Council would authorize

<Hereafter LT Transcript Only>

the engraving of a new plate instead of the reprint of an old one. The type may be set up as the paper stands, the alteration being but trifling.

Yours very truly | G.G. Stokes

RI MS JT/1/S/231
RI MS JT/1/TYP/4/1386

---

1. *What you stated*: see letter 1881.
2. *Melloni's book*: M. Melloni, *La Thermochrose ou la Coloration Calorifique* (Naples, 1850).
3. *calculated to wound his feelings*: on Tyndall's concern, see letter 1881.

To Juliet Pollock [16 April 1862][1] 1885

Weymouth Wednesday | April 17th. 1862.

My dear Mrs Pollock.

I reached this place to-day at 4½ P.M., ordered dinner, strolled ankle deep in pebbles along the beach, returned to my whiting and chops and afterwards rambled along the seething sea edge for a couple of miles in the direction of the Portland[2] breakwater. The light gradually faded, the sky was overcast and the wind gusts which caught the wavelets by the top filled the air with a mournful music. At a little distance from me, and quite on the sea edge, with some thirty feet of cliff raising it from the water, I saw a ruined castle, gray & mournful as the evening itself: thither I wandered, and sheltered by one of its broken walls listened for a long time to the sound of the wind and of the sea. The part of human nature which comes into play under such circumstances is that which puzzles me most. That solemn unison which the soul experiences with nature, and which is a thing essentially different from the intellectual appreciation of her operations. But I will not carry you into a cloudland where I have often wandered myself without finding rest for the sole of my foot; and where probably man's spirit may wander till the end of time without becoming sensibly more instructed.

I quitted the Waterloo station to-day at 11 A.M. In the carriage with me was a maiden & a little boy—a fine, frank, fearless looking little fellow with a full brown eye, and countenance full of animation. His sister, for she must have been so, was the image of good health & good nature; and showed a grace and dignity withal that were quite charming. We were together for four hours without entering into any conversation. Still the very presence of both was pleasant to me. At Brockenhurst[3] the boy shouted 'Oh there, Pa!' A gentleman opened the door and looking at me murmured "Mr. Tyndall". It was a Mr Law[4] & the young lady was his daughter. She must be the sister of young Law whom I have often met at Barlow's[5]—Her forehead is shaped like his. Now there is gossip instead of metaphysics for you.

I shall walk over to Portland tomorrow,[6] and return here tomorrow night. On the following morn[in]g I start along the beach to Bridport,[7] and may push on as far as Lyme–Regis[8] that day before I return I hope to work my way on to Torquay[9] picking up letters at Sidmouth & Teignmouth:[10] and whenever my time is likely to be evilly occupied I will drive away Satan by taking up my pen & writing to you.

It is now bed time; Give my love to Moss, to Walt, & to Fred.[11]

goodnight | Yours always | John Tyndall

RI MS JT/1/T/1136

1. *[16 April 1862]*: 17 April 1862 was a Thursday. It is more likely that Tyndall was correct about the day—on the 15th he noted that he planned to 'quit London' on the evening of the 16th (see letter 1881)—but not the date.
2. *Portland*: Isle of Portland, Dorset, England.
3. *Brockenhurst*: Village in New Forest, Hampshire, England.
4. *a Mr Law*: not identified.
5. *Barlow's*: John Barlow (1798–1869), former Secretary of the RI.
6. *I shall walk over to Portland tomorrow*: Tyndall's account of this Thursday excursion can be found in letter 1886, which provides further evidence that he wrote this letter on Wednesday 16 April.
7. *Bridport*: Market town in Dorset.
8. *Lyme-Regis*: Coastal town in west Dorset.
9. *Torquay*: Coastal town in Devon.
10. *Sidmouth & Teignmouth*: Coastal towns in south Devon.
11. *Moss, to Walt, & to Fred*: Juliet's three sons, Maurice Emilius Pollock (1857–1932); Walter Herries Pollock (1850–1926); Frederick Pollock (1845–1937).

To Juliet Pollock [20 April 1862][1] 1886

My dear Mrs. Pollock,

From Weymouth I went on Thursday to Portland discoursing on the way with the brother of a man who had produced colour blindness by overexertion & whom I examined in London many years ago. At Portland I met my young engineer Lieutenant,[2] and rambled with him over the rocks, and over the Isle and along the breakwater. They dig splendid fossils and huge trees out of the quarries where they were buried when the world was young. By the way my friend asked me to take a glass of sherry on my arrival—He filled a glass with a most transparent amber like liquid, which when I quaffed it astonished me. Through mistake he had given me a bumper of first rate Irish whiskey;[3] but it almost cost me my breath.

From Weymouth we started along the shingly beach on Friday morning. The shingle was worse than snow & to shorten the labour we took a boat for 6 miles—we rowed past a multitude of swans; some calmly sailing on the water, and others whirring past us with a musical sound which I could not say proceeded from the bird's throat or from its wings. Onward for some hours along the beach when we turned inland to the right, and were soon comfortably housed at the Bull Hotel Bridport.

Thence we started on Saturday morning. Bridport lies splendidly: I have hardly seen anything more charming than some of the views of the town & country from points on our road—deep valleys and bold conical hills—some tree-tufted & some bare. A splendid summer scene the landscape must present. We reached the windy seacoast—up the hills, down into the gorges—The smoke of Lyme became visible & towards it we steadily walked. We reached the place at 1 P.M. had luncheon—my young friends departed for Sidmouth as they wished to spend Sunday as Christians. I stayed at Lyme, purposing to join them afterwards. During the afternoon I wandered out; three or four miles along the shore, among the remains of the dead world which is there buried in the rocks. I met some men and they warned me that the tide was coming in; but I scanned the cliffs and saw that the tide was not likely to make a prisoner of me there. So I went on beyond Pinhay Bay & turned up the Undercliff. I wandered about amid the broken knolls, unguided by a track. It was sweet; the grass was green and soft & short, and all blown over with primroses. The slow bushes were all white with blossoms, and the crags festooned with ivy. I got to the summit of the highest cliff, and as I scanned the loveliness before me I appeared to be drinking a kind of wine which made the soul slightly inebriated while it did not compromise the body's poise. It was lovely, and beyond it was the sea roaring & spreading its broken waves in seething foam upon the shore. Down again from my cliff; down among the slowbushes & primroses. I thought of you at times as I breasted the knolls & I said *[inly]*, 'This would never do for <u>him</u>, but wouldn't <u>she</u> like it.'

Yours ever | John Tyndall

RI MS JT/1/T/1228
RI MS JT/1/TYP/6/1987–88

1. *[20 April 1862]*: Juliet replies to this (see letter 1887) on Monday 21 April, so most likely Tyndall wrote it on Sunday 20 April.
2. *my young engineer Lieutenant*: Edward Wynne (see letter 1869).
3. *Through mistake he had given me a bumper of first rate Irish whiskey*: fortunately on this occasion, despite the deplorable loss of whiskey, the result of accidentally being given a mis-identified liquid was not deadly.

From Juliet Pollock — 21 April [1862][1] — 1887

59 Montagu Square W. | <u>April 21st</u>

My dear Mr. Tyndall

Your letter[2] and your message from the mountains[3] were like airs from Heaven to a cockney, whose most open skies are those of the Regents Park, and whose only attainable height is the top step of the nursery staircase which you are to climb with a sentinel, you remember; yesterday, sitting by my husband's bed, to which he was confined by a rheumatic attack; I read your Mountaineering,[4] retaining the interest I felt in it before,[5] with some additional delight: I was able more completely to enjoy the poetry of your prose and the beauty and originality of your thought without the fetter of a critic's task to bind me to that close verbal attention which sacrifices the general effect to the consideration of detail. Your miracle chapter is perfect now,[6] and is more likely to do good in its quiet, sober tone than a more impetuous protest.

We passed last Saturday at Hatton,[7] where Fred.[k] caught cold, dawdling about the garden in a cold wind, but where I enjoyed some pleasant hours walking and talking with the Chief Baron,[8] occupied with a variety of subjects equally interesting to us both: spirit rapping was one—and this led us farther on with the speculations and feelings of the highest interest.

I do not know a mind more open to truth, more candid, more perfectly liberal than his: nor do I know a heart more tender more generous in affection.

Fred.[k] is just come down and is better though not yet well: perhaps if he could join you, and do the 23 miles a day off the high road, he might be set up again. I believe on elastic turf or moss and with a variety of ups and downs, and no broad roads and no gravel I could myself accomplish 13 with advantage—you carrying the knapsack.

I have just finished the life of Pitt[9]—the life—whoever writes it—of a truly great and virtuous man, raising one's idea so often painfully degraded of humanity. Every biography tending to such a result as this of Pitt, must be esteemed as a treasure well worth the digging for.

Now—M^r^. Moss[10] summons me to a walk and he is an autocrat, and I must go. —a pavement walk. —no cuckoo which you must be hearing—no violet by a mossy stone—no primrose by a river brim—only shops and omnibuses.

But yet in its own way it may be agreeable.

It remains to me to wish you cor[dia]l inclinations as often as possible since Satan is to be combated[11] by a letter to me when he suggests any less excellent occupation.

So commending you to temptations of brandy and water without chops.

I remain most truly yours | Juliet Pollock | With love from the boys and their dad

RI MS JT/1/P/185

1. *[1862]*: reply to letter 1885 and/or 1886.
2. *your letter*: letter 1885 and/or 1886.
3. *your message from the mountains*: Juliet appears to have just received a copy of *Mountaineering in 1861,* published on 16 April and may be referring to this. It is also possible she is referring to letter 1886. Tyndall's description there of his excursion (journeying

through 'deep valleys and conical hills' etc) perhaps reimagines Dorset's hilly landscape as mountainous.

4. *your Mountaineering*: *Mountaineering in 1861,* see also letter 1804, n. 9.
5. *retaining the interest I felt in it before*: Juliet read the proofs and commented on them (especially chapter 5). See letter 1866.
6. *Your miracle chapter is perfect now*: for Tyndall's response to Juliet's comments on an earlier version of the chapter (5) discussing miracles, see letter 1866.
7. *Hatton*: Country house outside village of Hatton, Middlesex owned by Sir Jonathan Frederick Pollock, first baronet.
8. *Chief Baron*: Sir Jonathan Frederick Pollock, first baronet (1783–1870) (*ODNB*).
9. *I have just finished the life of Pitt*: Philip Henry Stanhope, *Life of the Right Honourable William Pitt* 4 vols (London: John Murray, 1861–62).
10. *Mr Moss*: Juliet's youngest son, Maurice Emilius (1857–1932).
11. *Satan is to be combated*: in letter 1885, Tyndall suggests he will 'drive away Satan by taking up my pen & writing to you'.

From Alfred Wills[1] 30 April 1862 1888

Esher, |Surrey. | 30 April 1862

Dear Tyndall

I bought a day or two ago your little Alpine book.[2] I have not had time to read it yet but in cutting open the leaves I met with something that caught my eye and read your chapter entitled 'Reflections.' It is one for w^{h}. you can hardly fail to be hauled over the coals by the religious—or so called religious world—and therefore I hope you will not take amiss a word of thanks for that chapter. It is bold, manly & suggestive & will do good—and I am glad indeed when I find a true man of science uttering straightforwardly what many think but will not say. I have thought a great deal on subjects such as you trench upon there—& have perhaps approached them from a different quarter from that from which you have come to them. The events wh. have passed over my head[3] since I saw much of you have not tended to diminish the interest of such inquiries to me nor to render less reverent the spirit in which I view them—but the more I think about them the more deeply convinced I am that pretty nearly every consideration, moral, spiritual, historical, scientific which I can understand makes against the miraculous—that the notion of it leads to an unworthy conception of the Great Supreme—and that to confound miracle with religion will become, if it is not already, one of the most fruitful sources or irreligion (w^{h}. here in a very different scene from that usually given to it in the language of polemics) and through irreligion, of immorality & misery. A great change must come—It is dawning already. The old forms of faith cannot co-exist with the ascertained results of modern investigation, whether in philosophy, science, criticism or history—and those who help to prepare the human mind for what it must have to consider & for the great change in our modes of thought & of belief which will be the consequence of such consideration do a great service to their day & to their race. Especially where their position, like yours, is such that they cannot speak without finding attentive & thoughtful hearers. Pray forgive my intruding these remarks on you, but you have touched in your book on a subject w^{h}. is to me of the deepest interest, & where you will be pretty sure to have many unfair & ingenuous attacks to meet, I felt that perhaps the sympathy of even so small a person as myself might not be unwelcome. Believe my dear Tyndall,

Yours faithfully | Alfred Wills
Prof. Tyndall F.R.S.

BL, Add MS 63092, ff. 183–84

---

1. *Alfred Wills*: Sir Alfred Wills (1828–1912), judge, mountaineer and among the founders of the Alpine Club, of which he became third president in 1863 (*ODNB*).
2. *your little Alpine book*: *Mountaineering in 1861,* see also letter 1804, n. 9.
3. *events wh. have passed over my head*: Wills's first wife died suddenly in 1860. He remarried in 1861.

To [Richard Thomas Lewis][1] [April 1862][2] 1889

Thursday night

Dear Sir

After you have looked over this proof[3] & corrected it if you find corrections necessary would you proceed & send it to the Editor of the Philosophical Magazine addressed thus

Wm Francis Esq | Red Lion Court | Fleet Street E.C. | Yours very truly | John Tyndall

RS MM/9/20

---

1. *[Richard Thomas Lewis]*: addressee extrapolated from letter 1872. Next in sequence in Royal Society archive.
2. *[April 1862]*: proof of letter 1871 (sent 25 March 1862) published in the May number of the *Phil. Mag.* This makes Thursday nights in April the most likely candidates.
3. *this proof*: published as R. Lewis, 'On the changes in the apparent size of the moon', *Phil. Mag.,* (1862), pp. 380–82. See letter 1871 and 1872.

From Hermann Helmholtz [1 May 1862][1] 1890

Heidelberg 5.1.61

Verehrter Freund

in der Vorraussetzung, dass Sie irgend ein Mal die bevorstehende große Industrie Austellung besuchen werden, möchte ich Sie bitten, dabei gelegentlich ein Exemplar des von mir construierten Myographion ansehen zu wollen, welches einer unserer geschicktesten Mechaniker, Herr Fessel aus Coeln zur Ausstellung schicken will, und nachzusehen, ob es wohl richtig zusammengesetzt und aufgestellt ist. Es ist das Instrument bestimmt Zuckungen eines Muskels auf einen rotierenden kleinen Glascylinder aufzuschreiben. Ein kurzer Bericht über die damit angestellten Versuche, die namentlich auch Messung der Fortpflanzungsgeschwingigkeit der Reizung in den Nerven bezweckten, ist einmal im Philosophical Magazine, vielleicht von Ihnen selbst gegeben worden. Herr Fessel fürchtet, daß bei der Zusammensetzung des Apparats Missverständnisse unterlaufen möchten, und hat mich deshalb gebeten einem meiner Londoner Freunde diese Bitte vorzulegen.

Mit den besten Grüßen | Ihr H. Helmholtz

Heidelberg 5.1.61

Dear Friend

Assuming that you will visit the forthcoming great Exhibition of Industry[2] at some point, I would like to ask you at some point to look at the example of the myographion[3] built by me, which one of our most skillful mechanics, Mr Fessel[4] from Cologne, will send to the Exhibition, and see if it is properly put together and exhibited. The instrument is for recording the contractions of a muscle onto a small rotating glass cylinder. A short report on the experiments conducted with it, which aimed at measuring the speed of transmission of nerve impulses, appeared in the Philosophical Magazine, perhaps arranged by yourself.[5] Mr Fessel fears that there might be misunderstandings regarding the assembly of the apparatus, and has therefore asked me to put this request to one of my London friends.

With the best greetings | Your H. Helmholtz

RI MS JT/1/H/40

---

1. *[1 May 1862]*: the letter appears to give a date of 5 January 1861, but the context of the Exhibition of Industry implies 1862. Letter 1890 appears to be the reply, making 1 May 1862 a possible date for this request. Helmholtz normally wrote the day before the month, but 1 May is more likely than 5 January for this letter.

2. *great Exhibition of Industry*: the 1862 International Exhibition was held in London from 1 May to 1 November 1862.
3. *myographion*: Helmholtz's myographion, invented in 1852, was a device that allowed the contraction of a muscle to be traced on a rotating drum.
4. *Mr Fessel*: Friedrich Fessel.
5. *appeared in the Philosophical Magazine, perhaps arranged by yourself*: Helmholtz described his experimental method in H. Helmholtz, 'On the Methods of Measuring very Small Portions of Time, and their application to Physiological Purposes', *Phil. Mag.*, 6 (1853), pp. 313–25, though he describes the instrument as made by M. E. Rekoss. Tyndall 'communicated' the paper (* on p. 313).

To Hermann von Helmholtz 5 May 1862 1891

5th. May 1862

My dear Helmholtz,

I have received your note[1] in company with a letter from Mr. Fessel[2]—When I visit the Exhibition,[3] which I hope to do in a few days, I shall make it my business to discover the instrument of Mr Fessel, and to see that it is properly arranged.

And now I have a request to prefer to you. I know you have paid a good deal of attention to the history of the Mechanical theory of heat. A short time ago I finished a course of twelve lectures on Heat,[4] and treated the subject throughout from the mechanical theory points of view. I intend to publish these lectures and should like to make myself perfectly acquainted with what Mayer has really done. Did he attempt to determine the mechanical equivalent of heat; or was Joule[5] the first? If you could give me a list of the papers of Mayer and their dates I should be extremely obliged to you.

Present my kindest regards to Professor Bunsen, and also my respects to Kirchhof.

Yours very sincerely | John Tyndall.

RI MS JT/1/T/480
LT Transcript Only

---

1. *Your note*: letter 1890.
2. *Mr. Fessel*: Friedrich Fessel.
3. *the Exhibition*: International Exhibition or Great London Exposition, South Kensington, London sponsored by the Royal Society for the Encouragement of Arts, Manufacture and Trade. The Exhibition opened on 1 May and ran until 1 November 1862. See letters 1890 and 1893 for additional references.

4. *twelve lectures on Heat*: Tyndall delivered a course of twelve lectures on heat at the Royal Institution, starting 23 January and running to 10 April 1862. They appeared in 1863 as Tyndall, *Heat.*
5. *Joule*: James Prescott Joule (1818–89), British physicist (*ODNB*).

From Rudolf Clausius 7 May 1862 1892

Zurich 7/5/62

Lieber Tyndall,

Ich habe die antwort auf Ihren freundlichen Brief bis jetzt verschoben, weil ich hoffte, Ihnen gleich die gewünschte Auskunft über Mayer's Schriften geben zu können. Da aber die Ausführung des Auftrages welchen ich einer hiesigen Buchhandlung gegeben haben, länger dauert, als ich dachte, so will ich mit meinem Brief nicht länger warten. Ich habe mir durch die Buchhandlung alle Schriften von Mayer aus Heilbronn zur Ansicht verschreiben lassen. Sobald sie kommen werde ich sie durchsehen, und diejenigen, in welchen sich etwas über die Wärme befindet, Ihnen zusenden. Sie werden aber nicht sehr viel darin finden, da Mayer über die Grundbegriffe der Mechanic nicht hinlänglich klare Vorstellungen hat. Dass Sie die Wärmelehre in Ihren Vorträgen in der Royal Institution von Standpuncte der mechanischen Wärmetheorie behandelt haben, hat mich sehr interessirt. Ich glaube dass dieses Aufgabe besonders vor einem grösseren Publicum, eine ausserördentlich schwierige ist, und sehr viel Geschicklichkeit in der Behandlung und Darstellung erfordert.

Ihre Touren in der Schweiz, besonders die Besteigung des Weisshorns, denke ich mir sehr schön. Ich habe auch von Interlaken aus einige schöne Touren gemacht, aber doch nicht so grossartig wie Ihre. Ich würde sehr gern einmal mit Ihnen zusammen eine Tour machen; Sie würden finden dass auch ich vor den Schwierigkeiten und Mühen einer solchen nicht leicht zurückschrecke, und im Bergsteigen Uebung habe. Wenn Sie einmal wieder nach der Schweiz kommen, dürfen Sie an Zürich nicht vorbeigehen, und vielleicht lässt es sich dann einrichten, dass ich Sie wenigstens auf einem Theile Ihrer Reise begleite.

Für Ihre Freundlichkeit, mit welcher Sie dafür gesorgt haben, dass meine Abhandlung übersetzt und in Ihr Phil. Mag. aufgenommen wird, danke ich Ihnen sehr.

Ichlege ein kleines photographisches Portrait von mir bei und würde mich sehr freuen eine gleiches auch von Ihnen zu erhalten.

Mit herzlichen Grüssen | Ihr | Clausius

Zurich, 7/5/62

Dear Tyndall,

I put off answering your kind letter[1] until now because I had hoped to be able to provide at the same time the information you needed about Mayer's papers. Since it is taking longer than expected for the order I made at the local book shop to be completed, I did not want to wait any longer with my letter. I have ordered all of the papers from Mayer of Heilbronn for inspection. As soon as they come, I will look through them and send you the ones that contain something about heat. However you will not find very much in them, since Mayer does not have a sufficiently clear understanding of the basic concepts of mechanics. I was very interested to hear that you covered the theory of heat from the standpoint of the mechanical theory of heat in your lectures at the Royal Institution.[2] I believe that this is an extremely difficult task, especially in front of a larger audience, and demands a great deal of skill in its treatment and presentation.

I can imagine that your tours of Switzerland, and especially climbing the Weisshorn were wonderful. I have also done a few good tours starting out from Interlaken, but not as grandiose as yours. I would like very much to tour with you sometime; you will find that I am not easily daunted by such difficulties and effort and am practised in mountain–climbing. When you come again to Switzerland, you must not pass by Zurich, and perhaps it can be arranged that I accompany you for at least a part of your journey.

I thank you very much for your friendliness in seeing to it that my paper was translated and published in your Phil[osophical] Mag[azine].[3]

I have enclosed a small photographic portrait of myself and would be very happy to receive one from you.

With fond regards, | Your | Clausius

RI MS JT/1/TYP/7/2240–41

1. *your kind letter*: letter 1877.
2. *your lectures at the Royal Institution*: Tyndall delivered a course of twelve lectures on heat at the Royal Institution, starting 23 January and running to 10 April 1862. They were published in 1863 as Tyndall, *Heat.*
3. *my paper was translated and published in your Phil[osophical] Mag[azine]*: R. Clausius, 'On the conduction of heat by gases', *Phil. Mag.*, 23 (1862), pp. 417–35.

From Hermann von Helmholtz 10 May 1862 1893

Heidelberg. 10 Mai | 1862

Verehrter Freund,

betreffs Ihrer Anfrage über J.R. Mayer's Schriften entnehme ich folgende aus Poggendorff's biographischem Handwörterbuch

1). Bemerkungen über die Kräfte der unbelebten Natur. Liebig's u. Wöhler's Annalen der Chemie Bd. XLII 1842. —Darin sind die Hauptideen erörtert, weitläuftiger

2). Die organische Bewegung in ihrem Zusammenhange mit dem Stoffwechsel. Heilbronn 1845.

3). Beiträge zur Dynamik des Himmels ; ebenda 1848 (kenne ich nicht)

4) Bemerkungen über das mechanische Äquivalent der Wärme Heilbronn. 1851. (Besitze ich selbst, enthält den wesentlichen Inhalt der früheren Schriften.) Weiter ist mir auch nicht bekannt geworden. Die Bemerkung in Poggendorff's Wörterbuch, daß Mayer im Irrenhause gestorben sei, ist unrichtig, er ist wieder geheilt, und lebt in Heilbronn, wo er sich hauptsächlich mit Weinbau beschäftigen soll, da seine ärztliche Praxis durch die Erinnerung an seine Geisteskrankheit sehr gestört ist.

Das mechanische Äquivalent der Wärme hat er berechnet, aber allerdings nicht auf einem gegen Einwürfe ganz gesicherten Wege. Er hat es nämlich berechnet aus der Erwärmung comprimirter Gase (Schallgeschwindigkeit) unter der Voraussetzung, daß alle Arbeit, die auf die Verdichtung verwendet sei, als Wärme erscheine. Diese letztere Annahme hat sich allerdings als nahehin richtig bewährt; sie schloß aber zu jener Zeit immer noch eine zweite Hypothese ein. Die Berechnung ist schon in seinem ersten Aufsatze enthalten, und giebt 365 Meter für 1 °C.

Für Ihre Absicht nach dem Myographion auf der Industrieausstellung zu sehen, sage ich Ihnen meinen Dank. Fessel's Instrumente pflegen außerordentlich sauber gearbeitet zu sein, und er ist gleichzeitig sehr intelligent, in dieser Beziehung einer der ersten Mechaniker Deutschlands, so daß ich ihm wohl einige Beachtung wünschte.

Kirchhof und Bunsen werden im Anfang August nach London kommen; ersterer im Auftrag der Badischen Regierung.

Ich hoffe, daß wenn Sie einmal wieder nach den Alpen gehen, Sie einen Halt in Heidelberg machen. Übrigens ist es von Ihnen ein sehr dankenswerthes Unternehmen, daß Sie die mechanische Wärmetheorie populär zu bearbeiten beabsichtigen. Die bisherigen Schriften über diesen Gegenstand sind alle noch zu abstract für die größere Zahl der Leute, welche Physik studiren, und Sie wissen solche Aufgaben meisterhaft zu lösen.

Mit freundlichen Grüßen | Ihr H. Helmholtz

Heidelberg. 10 May | 1862

Dear friend,

Regarding your request about J.R. Mayer's works, I take the following from Poggendorff's biographical dictionary[1]

1). Remarks on the forces of inanimate nature. Liebig's u. Wöhler's Annals of Chemistry Vol. XLII 1842.[2]—Here the main ideas are discussed more broadly

2). Organic movement in its relation to metabolism. Heilbronn 1845.[3]

3). Contributions to celestial dynamics; Ibid 1848[4] (do not know it)

4) Remarks on the mechanical equivalent of heat, Heilbronn. 1851.[5] (I have it myself, it contains the essential contents of the previous works.) I do not know anything further. The remark in Poggendorff's dictionary that Mayer died in the lunatic asylum is incorrect, he is cured and lives in Heilbronn[6] where he allegedly keeps himself busy with viniculture, as his medical practice is quite damaged by the memory of his mental illness.

He calculated the mechanical equivalent of heat but not in a way that is not quite secure against objections. As a matter of fact, he allegedly calculated it from the heating of compressed gases (speed of sound) under the assumption that all work performed during compression transforms into heat. This latter assumption proved to be approximately true; but at the time it always included still a second hypothesis. The calculation is already contained in his first article and gives 365 metres for 1 °C.

I thank you for your intention to look after the myographion at the Exposition of Industry.[7] Fessel's instruments tend to be to be extraordinarily well manufactured; at the same time, he is also very intelligent and, in this sense, one of the best mechanics in Germany, so that I would wish that he receive some notice.

Kirchhof and Bunsen are coming to London at the beginning of August; the former on behalf of the government of Baden.[8]

I hope that you will stop in Heidelberg the next time you go to the Alps. Incidentally, your intention to popularise the mechanical theory of heat is a very laudable undertaking. Previous works on this subject are all still too abstract for the greater number of people who study physics and you know how to solve such tasks in a masterly fashion.

With kind regards | your H. Helmholtz

RI MS JT/1/H/41

1. *Poggendorff's biographical dictionary*: Poggendorff, J. C. *Biographisch-literarisches Handwörterbuch zur Geschichte der exacten Wissenschaften enthaltend Nachweisungen über Lebensverhältnisse und Leistungen von Mathematikern, Astronomen, Physikern, Chemikern, Mineralogen, Geologen usw. aller Völker und Zeiten, 2 vols.* (Leipzig, Germany: J. A. Barth, 1863).

2. *1842*: J. Mayer, 'Bemerkungen über die Kräfte der unbelebten Natur', *Annalen der Chemie und Pharmacie,* 42 (1842), pp. 233–40.
3. *1845*: J. Mayer, *Die Organische Bewegung in ihrem Zusammenhange mit dem Stoffwechsel* (Heilbronn: C. Drechsler'schen Buchhandlung, 1845).
4. *1848*: J. Mayer, *Beiträge zur Dynamik des Himmels* (Heilbronn: C. Drechsler'schen Buchhandlung, 1848).
5. *1851*: J. Mayer, *Bemerkungen über das Mechanische Äquivalent der Wärme* (Heilbronn: Johann Ulrich Landherr, 1851).
6. *Heilbronn*: a city in the Kingdom of Württemberg (now incorporated in Baden-Württemberg, a state in southwest Germany).
7. *look after the myographion at the Exposition of Industry:* see letter 1892.
8. *the government of Baden*: refers to the Grand Duchy of Baden, a state that was part of the German Confederation.

To Juliet Pollock 11 May 1862 1894

11th May 1862

My dear Mrs Pollock.

An enemy goes & sows tares in his neighbour's wheat & walks off:[1] but a friend writes in the Library column of the Field.[2] I thought at first it might have been you—or Mr Pollock, but I cannot quite read either of you between the lines. Certainly this is one of the pleasantest notices of my little book[3] that has been written, and the simple reason is that the writer, whoever he is, possesses the very qualities which he ascribes to me. He is a gentleman and of polished & symmetrical intellect. Our Hall porter must have smelt the odour of the thing while cutting up the paper—and he thrust it into my hands as I passed through yesterday.

My head I am sorry to say is not quite so well as it might be at present, as I experience in a small way those symptoms which troubled me so much two years ago[4]—brought on now by one day's unwise work.

I have been long engaged to dine with Mrs Hamilton[5] on Thursday, otherwise I should attend your summons early. I do not know whether I can get away sufficiently early to join you, but if I can I will.

Yours ever truly | John Tyndall

RI MS JT/1/T/1138

---

1. *An enemy goes & sows tares in his neighbour's wheat & walks off*: biblical quotation from Matthew 13:25.
2. *Library column of The Field*: reference to *The Field, the Farm, the Garden, the Country Gentleman's Newspaper.* Review not identified.

3. *my little book*: *Mountaineering in 1861.*

4. *those symptoms which troubled me so much two years ago*: see letter 1606 for allusion to this.

5. *M*rs *Hamilton*: not identified.

To Juliet Pollock 15 May 1862 1895

15th May 1862

My dear Mrs. Pollock.

Here we have winter again in the midst of May & on this gusty morning I am off to the country. I gave Spedding[1] two days holiday yesterday and though the cause of this was the confusion of my head he rejoiced I doubt not, like schoolboys in general, over his teacher's calamity.

I do not know quite where I am going. I shall decide at the Railway Station. I sometimes think it would be jolly to have a wife to talk to when I am tired, but Oh! how very jolly it is to be thus free.

For the last week there seems to have been a bubble of air in my brain, pressing against it with elastic force and making me giddy by its rolling about. I must expel this bubble before I return:—good bye.

Your ever | John Tyndall.

I have read all through St Clements Eve:[2] The writer is a refined and able man. I read the play with intense delight. I wonder could any unpledged woman withstand the fascination of such a man as Orleans?

RI MS JT/1/T/1139

---

1. *Spedding*: unidentified.

2. *St Clements Eve*: Henry Taylor, *St Clement's Eve: A Play* (London: Chapman and Hall, 1862).

From Rudolf Clausius 17 May 1862 1896

Zürich, 17/5/62

Lieber Tyndall,

Das erste, was Mayer über die Verwandlung von Wärme in Arbeit geschrieben hat, ist eine Abhandlung in den Annalen der Chemie und Pharmacie von Wöhler und Liebig vom Jahre 1842, Band XLII Seite 233 bis 240. Diesen Aufsatz kann ich Ihnen nicht verschaffen, aber er wird Ihnen wohl andersweitig leicht zugänglich sein, und wahrscheinlich kennen Sie ihn schon.

Das zweite ist eine Brochüre von 112 Seiten vom Jahr 1845, „Die organische

Bewegung in ihrem Zusammenhange mit dem Stoffwechsel". Diese Brochüre habe ich vor einigen Tagen an Sie abgesandt und hoffentlich werden Sie dieselbe schon erhalten haben.

Die dritte ist im Brochüre von 1848 „Beiträge zur Dynamik des Himmels". Diese schicke ich gleichzeitig mit diesem Briefe an Ihre Addresse ab.

Das vierte endlich ist eine Broshüre von 56 Seiten vomjahre 1851 „Bemerkungen über das mechanische Aequivalent der Wärme". diese werde ich Ihnen in einigen Tagen auch schicken; ich möchte es nur schnell selbst noch erst lesen. Ich bitte Sie diese Broshüren zu behalten.

Nachdem ich die Broshüren von 1845 und 48 gelesen habe (welche mir früher nur dem Namen nach bekannt waren, indem ich nur die Abhandlung von 1842 wirklich gelesen hatte) muss ich das was ich in meinem vorigen Briefe geschrieben habe, dass Sie in Mayers Schriften nicht viel Bedeutendes finden würden, zurücknehmen. Ich bin erstaunt gewesen über die Menge schöner und richtiger Gedanken welche darin enthalten sind; auch ist seine Auffassung der mechanischen Principien viel klarer als ich nach der ersten Abhandlung von 1848 geglaubt hatte. Wahrscheinlich hat er sich in den Jahren von 1842 bis 1845 erst recht in die Sache hineingearbeitet. In der Schrift von 1845 finden sich eine Menge der Ideen, welche Helmholz in seinen Schriften „über die Erhaltung der Kraft" und „über die Wechselwirkung der Naturkräfte" ausgesprochen hat, und in der Schrift von 1848 findet sich mit grosser Vollständigkeit dieselbe Erklärung von der Sonnenwärme welche Thomson später gegeben hat. Ich glaube daher, dass Mayer viel mehr Anerkennung verdient, als er bis jetzt gefunden hat.

Sie werden in Poggendorfs „Biographish-literarischen Handwörterbuch" die Notiz finden: „Mayer soll vor 1858 im Irrenhause gestorben sein". Dieses ist, soviel ich weiss, unrichtig. Er ist allerdings im Irrenhause gewesen, aber ich habe gehört, dass er wieder geheilt und aus dem Irrenhause entlassen sei, und noch lebe. Ich werde versuchen einiges Nähere von seinem Leben zu erfahren da ich mich jetzt nach dem Lesen dieser Schriften für seine Person sehr interessire.

Mit herzlichen Grüssen | Ihr | Clausius

Zurich, 17/5/62

Dear Tyndall,

The first thing that Mayer wrote about the transformation of heat into work is a paper in the Annalen der Chemie und Pharmacie von Wöhler und Liebig in 1842, volume XLII page 233 to 240.[1] I cannot procure this paper for you, but you will be able to find it easily elsewhere, and you probably know it already.

The second is a pamphlet of 112 pages from 1845, 'Die organische Bewegung in ihrem Zusammenhange mit dem Stoffwechsel'.[2] I sent this pamphlet to you a few days ago and hopefully you will already have received it.

The third is a pamphlet from 1848 'Beiträge zur Dynamik des Himmels'.[3] I am sending it along with this letter to your address.

The fourth, finally, is a pamphlet of 56 pages from 1851 'Bemerkungen über das mechanische Aequivalent der Wärme'.[4] I will send it to you in the next few days; I would like to read it through quickly myself first. I ask that you keep these pamphlets.

After reading the pamphlets from 1845 and 1848 (which I had previously known only by name, but had only actually read the paper of 1842) I must retract what I wrote in my previous letter that you would find little of significance in Mayer's writings. I am amazed at the number of fine and correct ideas that they contain; his conception of mechanical principles is also much clearer than I had believed from the first paper of 1842. He probably did not immerse himself properly in the topic until the years 1842 to 1845. The paper of 1845 contains numerous ideas that Helmholtz expressed in his writings 'über die Erhaltung der Kraft'[5] and 'über die Wechselwirkung der Naturkräfte',[6] and in the essay of 1848 one finds in great detail the same explanation of the sun's heat that Thomson gave later. I believe for that reason that Mayer deserves much more recognition than he has received so far.

In Poggendorf's 'Biographish-literarischen Handwörterbuch'[7] you will find the note 'Mayer is said to have died in an insane asylum in 1858'. As far as I know, this is incorrect. He was indeed in an insane asylum, but I have heard that he was released from the asylum cured and is still alive. I will try to find out more about his life, since I have become very interested in his person after reading the writings.

With fond regards, | Your | Clausius

RI MS JT/1/TYP/7/2242–43

1. to 240: letter 1893, n. 2.
2. '... *dem Stoffwechsel*': letter 1893, n. 3.
3. '... *des Himmels*.': letter 1893, n. 4.
4. '... *der Wärme*.': letter 1893, n. 5.
5. '... *der Kraft*': Uber *die Erhaltung der Kraft. Eine physikalische Abhandlung* (Berlin: Georg Reimer, 1847).
6. *der Naturkräfte*: H. Helmholtz, Über die Wechselwirkung der Naturkräfte und die darauf bezüglichen neuesten Ermittelungen der Physik (Könisberg: Gräfe & Unzer, 1854).
7. *Biographish-literarischen Handwörterbuch*: The complete title is *Biographisch-Literarisches Handwörterbuch zur Geschichte der Exakten Wissenschaften.*

To Juliet Pollock 18 May 1862 1897

GLO'STER HOTEL. WEST COWES. |
LATE THE R.V.S CLUB HOUSE.
Sunday night | 18th May 1862.

My dear Mrs Pollock

On Thursday I started from London with Mrs. Spedding's[1] card in my pocket—I will go if my head and humour permit me. I think I should like it. Down Southwest from London, with the cold rain glancing from the windows. Low down the line the rails were dry—the rain had not stretched so far southward. I quitted the train midway between Southampton & Salisbury & then had to walk four miles into the heart of the country. I was quite alone and the road without a passenger. I enjoyed my walk extremely. I like being alone on a pleasant winding road, with hazel hedges on each side and the drooping beech branches with their tender green leaves overhead. Here and there the brightness was enhanced by the gloom of the yew trees which were scattered amid the green. At intervals the odour of the fullblown hawthorn was blown into my face, the larks were in the air and the bees humming amid the flowers—and I sauntered along free from care & filled with pleasant thoughts. or rather suffused by the warm material of thought which had not yet taken definite shape. I crossed a long avenue and though my road lay across it the extreme beauty of its woodwork tempted me down it. Pine, beech & hazel right and left, and the nightingales filling the air with music. I reached a place where the beeches expanded to greater magnitude and the floor beneath them was free of brushwood. The sun broke through the clouds and ruled straight bars of radiance along this floor, struck the boles, and turned them into pillars of a finer gilding than could be produced by artificers hand. Then into an avenue, dark as if it led to the Shades, composed entirely of yews, with their branches quite interlaced overhead, and here and there a sunflick escaping through it. Here I wander[e]d & paused & listened to the nightingales, and here the shades of evening found me, until some rain drops whisking against my cheeks warned me that it was time to seek a better shelter. That night I had a good rest & next morn[in]g my head was better.

Next day I did nothing but dawdle through the fields & read Carlyle's last volume.[2] Let the world say what it will regarding Carlyle I am not acquainted with his equal. Among my acquaintance in the literary world—which of course is but a small acquaintance he has nothing like an equal. —an intellect of the very strongest fibre horsed on an imagination of the most fiery quality—the one of them worthy of the other—Well that night I had a tolerably fair rest, but the head gave evidence of being still in a maimed condition this morn[in]g.

I started to Romsey[3]—thence walked to Southampton & took my passage to Cowes[4] with a fresh breeze in my face all the way. It did me great good, & now here I am passing the last hours of the day upon the sea fringe and listening to the faint lap of its wavelets on the beach. Tomorrow I saunter here until 4. Then to Southampton & thence to London, which I hope to reach in fair working condition.

good night | Yours always John Tyndall

RI MS JT/1/T/1137

1. *Mrs Spedding's*: probably Jane Spedding, née Headlam (c. 1809–84) sister-in-law of James Spedding (1808–81), literary editor and Francis Bacon scholar.
2. *Carlyle's last volume*: T. Carlyle, *History of Friedrich the Second, called Frederick the Great.* Vol. 3 (London: Chapman and Hall, 1862). The volume was published on 3 May.
3. *Romsey*: Market town seven miles NW of Southampton.
4. *Cowes*: Seaport town on the Isle of Wight.

To Rudolf Clausius 20 May [1862][1] 1898

20th May

My dear Clausius

Very many thanks for your last note,[2] for your portrait,[3] which now stands close to that of Faraday, and also for the pamphlet by Mayer[4] which you have been kind enough to send me.

I send you the image of my pale face in return.

I had thought of making a request of you but I am almost ashamed to do it. The request is that you would be good enough to cast your eye over the proof sheets of my lectures before they go to press. Your remarks would be of great value to me, but as I have stated, I am quite ashamed to ask you to incur so much labour. The book will not be ready until the autumn.[5] At present I am sorry to say my head is very weary and almost unfit for work.

Yours ever | John Tyndall | Kind regards to your wife. |

All your papers are sure to appear in the Phil. Mag.[6] They are most valuable to English readers

RI MS JT/1/T/234

1. *[1862]*: year added later. Year confirmed through reference to earlier correspondence with Clausius (see n. 2 and n. 3 below).
2. *your last note*: letter 1896.
3. *your portrait*: see letter 1892.

4. *pamphlet by Mayer*: see letter 1893, n. 5 for details.
5. *The book will not be ready until the autumn*: *Heat* was completed by November and was published on 4 March 1863.
6. *All your papers are sure to appear in the Phil. Mag.*: R. Clausius, 'On the application of the theorem of the equivalence of transformations to the internal work of a mass of matter', *Phil. Mag.*, 24 (1862), pp. 81–96; 201–12.

To Edward Frankland [23 May 1862][1] 1899

Royal Institution. Tuesday night

My dear Frankland

Thank you for Mr Greenwood's note.[2] The lines from Wordsworth are truly beautiful, and I believe refer to the phenomenon in question. It is curious to remark how the 'sense' of men differ as regards their appreciation of an imaginative expression. I was struck many years ago with the physical truth of the couplet from Emerson[3] which I have placed at the head of the chapter to which you refer; and until my manner of viewing the phenomena of light alters, my opinion and appreciation are likely to remain as they were.

If it were only Mrs Frankland's invitation with which I had to deal the question would be promptly solved. But I have been compelled to decline invitations from Lady Holland,[4] from Mrs Pollock, from Mrs Hamilton.[5] This evening I know I am expected near Portman Square but I cannot go. I have card from Mrs Playfair,[6] letter from Mrs Cardwell;[7] so that my only safety consists in shutting myself up; for it would entirely break me down to attempt to respond to my friends. The two soirées of the Duchess of Northumberland[8] I have allowed to pass without attending—and I shall not be able to attend our president's[9] soiree on Saturday. In fact my head requires the greatest care and Dr Bence Jones has strictly enjoined me to be perfectly quiet. Mrs Frankland will I trust see the difficulty of my position and forgive me accordingly.

Yours ever | John Tyndall

The Austrians seem disposed to allow the salt to run into water—I shall have another trial one of these days.

I have just written declining an invitation from Lady Millicent[10] for Friday.

RI MS JT/1/TYP/12/3970

1. *[23 May 1862]*: date added, probably by LT. 23 May was a Friday and the letter indicates Tuesday night. Late May, however, does make sense of reference to *Mountaineering in 1861* (published 16 April 1862). See n. 3.

2. *Greenwood's note*: probably journalist and editor Frederick Greenwood (1830–1909) (*ODNB*).
3. *couplet from Emerson*: see chapter IX, *Mountaineering in 1861.*
4. *Lady Holland*: Mary Augusta Fox, Lady Holland (1812–89), formerly wife of Henry Edward Fox (1802–59), daughter of the seventh earl of Coventry and political hostess.
5. *Mrs Hamilton*: not identified.
6. *Mrs Playfair*: Jean Ann Playfair née Millington (d. 1877), wife of Lyon Playfair.
7. *Mrs Cardwell:* Anne (Annie) Cardwell, née Parker (d. 1887), wife of Edward Cardwell (1813–86).
8. *Duchess of Northumberland*: Eleanor Percy, née Lady Eleanor Grosvenor (1820–1911).
9. *our President's*: probably referring to the President of the RS, Edward Sabine.
10. *Lady Millicent*: Lady Millicent Jones, née Acheson (d. 1887), married to Henry Bence Jones, Secretary of the RI.

From Johann Poggendorff 24 May 1862 1900

Berlin 24 Mai 62.

Mein lieber Tyndall!

Längst war es meine Absicht, an Sie zu schreiben, als ich vorgestern Ihren lieben Brief vom 19' erhielt, der mir in doppelter Hinsicht die Verpflichtung auferlegt, mit der Ausführung meines Vorsatzes nicht länger zu zögern.

Zunächst was Ihre Abhandlung betrifft, so kann ich Ihnen sagen, daß ich dieselbe in mehreren Portionen successive erhalten habe u. daß ein Theil derselben, §1 bis §7 incl., vom Dr. Becker übersetzt, bereits im Heft V gedruckt ist.

Ich würde die Abhandlung vollständig in dasselbe Heft aufgenommen haben, wenn ich sie ganz gehabt hätte. So aber schien es mir besser, eine Theilung *[an]* derselben vorzunehmen. —Die 2' Hälfte welche die Überschrift führt: On dynamic radiation and absorption würde nun in Heft VI kommen müssen, aber dem stellen sich zwei Hindernisse entgegen, welche ich Sie bitte bald möglichst aus dem Wege zu räumen.

Erstens schließt die Letzte Portion des Manuscripts nicht an die vorletzte<.>

Die vorletzte endet mit pag. 39, womit auch §9 schließt. —Die letzte Portion dagegen beginnt mit pag. 42, mitten in einem Satze mit dem Worte: accomplished *[in]* comparison with the air etc. —Es ist also klar, daß pag. 40 u. 41 fehlen!

Zweitens geht die letzte Portion bis pag. 47 (wovon der größere Theil worin auch Magnus genannt wird, wieder ausgestrichen ist)—ohne daß gesagt ist, daß damit die ganze Abhandlung geschlossen sey!—Auch darüber müsste ich schleunigst von Ihnen Nachricht erhalten, ehe ich die zweite Hälfte drucken lassen kann.

Könnten Sie die fehlenden pp. 40 u. 41 nicht ersetzen, ohne Ansicht der pp. 39 u. 42, so würde ich Ihnen diese zuschicken

Was Magnus betrifft, so können Sie ruhig seyn; er ist weit entfernt davon seinen Streit mit Ihnen persönlich zu nehmen. Er wird sich in der Sache mit der Zeit rechtfertigen, ohne Ihrer Person irgend etwas nachzutragen. Er ist übrigens noch jetzt vollkommen von der Richtigkeit seiner Resultate überzeugt, er wird Ihnen selbst noch darüber schreiben. —

Da Sie sich für meine Reise nach London so lebhaft interessiren und sogar so liebenswürdig sind, mir ein Zimmer anzubieten, so muß ich Ihnen wohl ausführlicher schreiben, wie es mit meinem Plane steht. Ich denke noch immer an diese Reise, obwohl ich manchmal as an old man etwas Furcht bekomme, ob ich noch Kräfte genug habe für das Getreibe und Getöse der großen Weltstadt,—zumal erst kürzlich Dove geschrieben hat, auch er fühle, daß er für London schon zu alt sey.

Indeß werde ich es wagen, denn wenn ich nicht in diesem Jahre nach England komme, werde ich es wohl nie wiedersehen. —Ich kann aber, wegen meiner Vorlesungen, nicht früher kommen als in den ersten Tagen des Augusts und zweitens werde ich nicht allein kommen sondern mit meiner Tochter die großen Verlangen hat, London kennen zu lernen.

Diese beiden Umstände gestatten nun leider nicht, daß ich von Ihrem gütigen Anerbieten Gebrauch machen kann; ich hoffe indeß, daß ich darum doch nicht das Vergnügen entbehren werde, Sie zu sehen,—denn vielleicht sind Sie anfangs August schon zurück gekehrt von Ihrer Reise.

Einen großen Dienst würden Sie mir erweisen, wenn Sie mir in einem passend gelegenen Boarding House zwei Zimmer verschaffen könnten; ich mache keine Ansprüche auf Eleganz, wenn sie nur so anständig sind, dass ich mit meiner Tochter darin wohnen kann,—und nicht zu übertrieben theuer. —Vielleicht ist Freund Francis bis dahin von seiner tyroler Reise zurück gekehrt; er würde gewiß (wie auch seine Schwester, die meine Töchter so gut kennt) gern behülflich seyn, meinen Wunsch erfüllen zu helfen.

Schreiben Sie mir gefälligst, was ich in dieser Beziehung zu hoffen habe, u. seyen Sie versichert, daß ich, wenn ich allein käme, Ihr gütiges Anerbieten mit dem größten Dank angenommen haben würde.

Mit schönsten Grüßen von meiner Frau

ganz | der Ihre | Poggendorff.

Berlin 24 May 62.

My dear Tyndall!

It had long been my intention to write you, when the day before yesterday I received your kind letter from the 19th,[1] which obliges me, in a double sense, to hesitate no longer with the execution of my resolution.

First, concerning your paper,[2] I can tell you that I received it successively in several portions and that a part of it, from §1 to §7 incl., has been translated by Dr. Becker[3] and printed in issue V.[4]

I would have included the complete paper in the same issue, had I received all of it. But in this way, I found it better to divide it. —The 2nd half, which carries the title On dynamic radiation and absorption would now have to go into issue VI,[5] but there are two obstacles that I ask you to clear out of the way as soon as possible.

Firstly, the last portion of the manuscript does not connect to the penultimate one.

The penultimate one ends on pag. 39, which also closes §9. —The last portion, on the contrary, starts with pag. 42, halfway into a sentence with the words: accomplished in comparison with the air etc. —It is therefore clear that pag. 40 and 41 are missing!

Secondly, the last portion goes until pag. 47 (the largest part of which, where Magnus is also mentioned, has been crossed out)—without any mention that the whole paper ends there!—About this matter also I should obtain a note from you without delay before I can send the second half to print.

In case you cannot replace the missing pp. 40 and 41 without seeing pp. 39 and 42, I would send them to you.

Concerning Magnus, you do not need to worry; he is far from taking the dispute with you personally. He will justify himself in the matter in due time, without holding a grudge against you. By the way, he is still completely convinced about the correctness of his results, he will write you about this himself. —

As you take such a lively interest in my trip to London and even were so kind as to offer me a room, I have to tell you in greater detail how my plans stand. I am still thinking about this journey, although as an old man, I sometimes wonder if I still have strength for the turmoil and the noise of the big metropolis,—especially after Dove[6] recently wrote that he also already feels too old for London.

I will risk it nonetheless, since if I do not come to England this year, I will likely never see it again. —However, due to my lectures I cannot come before the first days of August and secondly, I will not come alone but with my daughter, who has a great desire to get to know London.

Unfortunately, these two circumstances do not allow me to take advantage of your kind offer; I nevertheless hope that I will not be deprived of the pleasure of seeing you,—as you might already have returned from your journey at the beginning of August.

You would do me a great service if you could procure me two rooms in a conveniently-located boarding house; I have no demands about elegance if they are respectable enough that I can stay there with my daughter,—and not too exaggeratedly expensive. —Perhaps friend Francis[7] will be back from his trip to Tyrol; he would certainly (as much as his sister, who knows my daughters so well) gladly help to make my wish come true.

Be so kind to write to me on occasion what I can hope in this respect and rest assured that if I came alone, I would have accepted your kind offer with the greatest thanks.

With the kindest regards from my wife
entirely | Your | Poggendorff.

RI MS JT/1/P/131

1. *Your letter of the 19th*: letter missing.
2. *concerning your paper*: Tyndall (1862).
3. *Dr Becker*: the most likely contender is the chemist and photographer Dr. Ernst Becker (1826–88), Secretary and Librarian to Prince Albert from c. 1852–58.
4. *printed in issue V*: J. Tyndall, 'Ueber Strahlung und Absorption der Wärme durch gasförmige Materie', Poggen. Annal., 116 (5) (1862), pp. 1–26.
5. *into issue VI*: J. Tyndall, 'Ueber die Strahlung und Absorption der Wärme durch gasförmige Materie', *Poggen. Annal.,* 116 (6) (1862), pp. 289–307.
6. *Dove recently wrote*: Heinrich Wilhelm Dove (1803–79), German physicist.
7. *Friend Francis*: William Francis.

From Jacob Gijsbertus Samuel van Breda[1] and Wilhelmus Martinus Logeman[2] May 1862[3] 1901

XVII. *On Ampèrian Repulsion.* | *By* MM. J. G. S. van Breda *and* W. M. Logeman. | *To Dr. Tyndall, F.R.S.*

Sir,

We take advantage of the opportunity afforded by the sending of the accompanying letter, to address, through your kindness, to the readers of the Philosophical Magazine a few lines in answer to the remarks published by Mr. Croll of Glasgow,[4] in the Number of that Journal for this month.[5] pp. 365 *et seq.*, in reference to our experiments on Ampèrian repulsion.

There must be, either in the description of our experiments, or in the figure which accompanies them, some want of accuracy or of clearness; otherwise we are at a loss to understand how Mr. Croll could say that the currents in the mercury ought to repel the horizontal parts, while they attract the vertical parts of our moveable conductor. This seems to us to contradict all the received ideas of electric currents. We would willingly take trouble to make our description more clear or more accurate, but the remainder of Mr. Croll's article induces us to think that such an attempt would be quite useless.

In fact, Mr. Croll first admits that "our experiments with the reversed conductor, and some others detailed by us, certainly prove that the molecules of a conductor, while the current is passing, *do repel each other;*" and then goes on to discuss the question from a point of view which is new to us, adding that this repulsion of the molecules of a conductor proves nothing as to the repulsion of the contiguous parts of the current itself. Now we will not follow Mr. Croll into the arguments upon which he bases his opinion concerning the mutual action of the parts of the current; in the first place, because these arguments appear to us to be beyond the reach of experimental control; and in the second place, because we think it can be of but slight use to science to speculate on the properties of a current hypothetically detached from the material substratum without which we are unable really to conceive of its existence.

We have the honour to remain, &c., | J. G. S. van Breda, | W. M. Logeman.

Teylerian Laboratory, Haarlem, | May 1862.

J. G. S van Breda and W. M. Logeman, 'On ampèrian repulsion', *Phil. Mag.,* 24 (1862), pp. 126–27.

---

1. *Jacob Gijsbertus Samuel van Breda*: see letter 1873, n. 3.
2. *Wilhelmus Martinus Logeman*: see letter 1873, n. 1.
3. *[May 1862]*: published in the August number of the *Phil. Mag.*
4. *Mr. Croll of Glasgow*: James Croll (1821–90), Keeper at Andersonian Museum, Glasgow and physical geologist (*ODNB*).
5. *Number of that Journal for this month*: J. Croll, 'Remarks on Ampères experiment on the repulsion of a rectilinear electrical current on itself', *Phil. Mag.,* 23 (1862), pp. 365–67.

BIOGRAPHICAL REGISTER

This register contains the names and biographical details of people who are mentioned three or more times in the letters included in this volume. Biographical information about people mentioned only once or twice is contained in the notes appended to the appropriate letter(s).

Conventions

The sources for the biographical entries are standard biographical dictionaries, such as the *Complete Dictionary of Scientific Biography*, the *Oxford Dictionary of National Biography*, Poggendorff 's *Biographisch-literarisches Handworterbuch der exacten Wissenschaften, Historishes Lexikon der Schweiz, Neue Deutsche Biographie*, and the *Dictionary of Nineteenth-Century British Scientists.* These have not been specified in the entries unless necessary. Additionally, the editors have found material from the Tyndall correspondence and from the journals of Tyndall and T. A. Hirst very useful. Where additional sources add important material, they are given in parentheses at the end of the entry.

Agassiz, Louis (1807–73), was a Swiss naturalist. Agassiz studied medicine, botany, geology, and zoology in Germany and France before being appointed to a position at the University of Neuchâtel. In 1847 he was appointed professor of zoology and geology at Harvard in the United States and founded the Museum of Comparative Zoology in 1859. He was a key figure in developing the study of the Earth's history, proposed the theory of ice ages, and made a close study of glaciers in the Alps. In 1840 he published his important *Études sur les Glaciers* (2 vols., Neuchâtel: Gent et Gassmann), which helped establish the scientific study of glaciers. A believer in the fixity of species and a leading advocate of a polygenist form of old Earth creationism, Agassiz was a leading opponent of Darwinism in the United States.

Airy, George Biddell (1801–92), was a British astronomer. He was educated at Cambridge University and graduated senior wrangler in 1823. In 1826 he was appointed Lucasian Professor of Mathematics, and in 1828 to the Plumian Chair of astronomy and directorship of the Cambridge University observatory. In 1835 he was appointed Astronomer Royal, a position he held until 1881, and set about improving the instrumentation and research activities of the Royal Observatory, Greenwich. He vigorously pursued research on a wide range of topics, including optical diffraction, calculating the Earth's mean density, and analyzing the inequalities in the motions of the Earth and Venus. Much of his research sought to preserve the relevance of the Greenwich meridian, and its use in navigation.

Asher, Adolphus (1800–53), was a Berlin bookseller and author. Asher spent five years in England between 1820 and 1825, before he moved to Saint Petersburg, where he worked in the diamond trade before he became interested in bookselling. Due to the religious persecution of the Jewish community in Russia, Asher moved to Berlin around 1830. His shop in Linden Walk attracted many international visitors, and he also worked with the British Museum's library.

Balmat, Auguste (1808–62), was a mountaineer and guide born in Chamonix, Savoy. The grandnephew of Jacques Balmat, the first mountaineer to successfully reach the top of Mont Blanc in 1786, Auguste made his first ascent of Mont Blanc at the age of twenty-two. He met James Forbes (1809–68) in 1842 and assisted him during his studies of the movement of glaciers and survey work leading to the important map of the Mer de Glace published in Forbes's *Travels through the Alps of Savoy* (Edinburgh: A. and C. Black, 1843). In 1854 he made an early ascent of the Wetterhorn with Alfred Wills (1828–1912), who dedicated his *Wandering among the High Alps* (London: Bentley, 1856) to him. In 1858, while guiding Tyndall and Wills's ascent of Mont Blanc, he nearly lost his hands while placing thermometers at the summit. (C. D. Cunningham and W. Abney, *The Pioneers of the Alps*, 2nd ed. [London: Low, Marston, Searle and Rivington, 1888], 136–40.)

Bennen [Benet], Johann Joseph (1824–64), was a Swiss mountain guide. Born Johann Joseph Benet, he was called "Bennen" by English alpinists. He made the first ascent of the Aletschorn in 1859, and the first ascent of the Weisshorn with Tyndall in 1861. In 1862 they attempted the Lion Ridge of the Matterhorn. In the summer of 1858 he was Tyndall's guide while ascending the Finsteraarhorn. He died in an avalanche on the Haut de Cry. (C. D. Cunningham and W. Abney, *The Pioneers of the Alps*, 2nd ed. [London: Low, Marston, Searle and Rivington, 1888], 148–53.)

Brewster, David (1781–1868), was a Scottish mathematician, astronomer, editor, and educator. He was known for his work in optics and his role in founding the BAAS. Early in his career he found it difficult to cope with the stress of public speaking, and neither a career in the church nor one as a lecturer seemed viable. He became a regular contributor to numerous periodicals, and served as an editor on several, by which he mostly earned his living. He was one of three founding editors of the *Edinburgh Encyclopaedia*. He was appointed principal of the United College of Saint Salvator and Saint Leonard at the University of Saint Andrews in 1838, and principal of the University of Edinburgh in 1859.

Bunsen, Robert (1811–99), was a German chemist who received his PhD from the University of Göttingen, where he then lectured (1833–36). Bunsen subsequently held positions at the Polytechnikum at Kassel, the University of Marburg (1839–51), the University of Breslau, and, from 1852, the University of Heidelberg. He made many contributions to chemistry, including the Bunsen burner, a gas burner that could produce nonluminous flame, in 1855. With Gustav Robert Kirchhoff, he developed spectrum analysis, with which he discovered the alkali metals cesium and rubidium in 1860–61. He made extensive use of both electrolysis and spectrum analysis to analyze chemical substances. A renowned researcher, he was also an extremely popular teacher, mentoring both Tyndall and Edward Frankland at Marburg. He received the RS's Copley Medal in 1860, and its Davy Medal, with Gustav Robert Kirchhoff, in 1877.

Charpentier, Jean de (1786–1855), was a German mining engineer and glaciologist. His father was a professor at the Mining Academy of Freiberg, and Charpentier followed family tradition by studying at the Mining Academy himself. He began his career in the coal mines where his brother Toussaint worked as the director. By 1813, with the help of his friend Charles Lardy, he became director at the salt mines of Bex, Switzerland. After the 1818 Giétro Glacier disaster, where a flooding had killed dozens of people, Charpentier's attention was drawn to glaciers. He followed his friend Ignaz Venetz in considering glaciers to have been crucial in explaining the distribution of erratic blocs in Switzerland, and had Venetz's 1821 paper on the topic reprinted in 1833. His *Essai sur les glaciers* (Lausanne, 1841) was published in February 1841, months after Agassiz's *Études sur les glaciers* (Neuchâtel, 1840).

Clausius, Rudolf (1822–88), was a renowned German physicist and one of the founders of modern thermodynamics. His seminal paper on the theory of heat, "Ueber die bewegende Kraft der Warme," was published in the *Annalen der Physik* in 1850. He taught physics at the Royal Artillery and Engineering School in Berlin in 1850 before moving to a professorship in mathematical physics at the Polytechnicum in Zurich in 1855. Although he would later move on to the University of Würzburg in 1867, and the University of Bonn in 1869, his time at Zurich proved to be the most academically fertile period in his life. It was here that he developed thermodynamics further, and contributed to the kinetic theory of gases. In the 1870s his work laid more emphasis on electrodynamics.

Clausius, Adelheid (née Rimpham, d. 1875), was the wife of Rudolf Clausius, whom she married on 13 November 1859. Together they had six children. She died in childbirth in 1875.

Clowes, William, the younger (1807–83), was an English printer. The son of William Clowes, a printer, the younger William entered the trade through his father's business in 1823. At his death, the elder William left the family business to William the younger and his brothers, Winchester and George. The Royal Academy was one of the clients of the firm, as were all major book publishers. The firm experimented with different machines for mechanical composition, and became known as experts in this aspect of printing.

Cumming, William Fullarton/Fuller (1804–92), graduated in medicine from Edinburgh University in 1827 and was then employed by the Bengal Medical Establishment. His sons had been at Queenwood.

Darwin, Charles (1809–82), was an English naturalist and geologist. Having received a BA at Christ's College, Cambridge, in 1831, Darwin spent the following five years on the HMS *Beagle*, making geological observations and collecting specimens for those interested in natural history. Best known for his book *On the Origin of Species* (London: John Murray, 1859), Darwin had meant it as an abstract of a planned larger work, and *Origin* served as a starting shot for other books expanding elements of his work. From 1860 onward he started work on his *The Variation of Animals and Plants under Domestication* (1868), a massive undertaking that progressed slowly, but would expand his ideas on variation. He also intensified his study of flowers and trees, and this work culminated in his *Fertilization of Orchids* (1862), which dealt with evolutionary adaptation in the plant world.

Dawes, Richard (1793–1867), was an English educationist and dean of Hereford. Admitted to Trinity College, Cambridge, in 1813, Dawes graduated BA in 1817. From then on he served as a mathematical tutor, fellow, and bursar of the new Downing College, Cambridge, until he left Cambridge in 1837 for the rural parish of King's Somborne in Hampshire. Here he established the King's Somborne School, which educated children from different social classes, promoted the use of simple scientific experiments in a rudimentary laboratory, and used secular reading material for language development.

Dawes himself taught frequently at the school, and Edward Frankland taught agricultural chemistry as well. From 1850 onward he served as dean of Hereford, and oversaw the restoration of the cathedral.

Dawes, Mary Helen (née Gordon, 1800–92), was the wife of Richard Dawes, whom she married in 1836 (Westminster Marriages Record. The *ODNB* entry for Richard Dawes incorrectly states 1838). She was stepdaughter to the surgeon George James Guthrie.

de la Rive, Auguste-Arthur (1801–73), was a Swiss physicist and one of the founders of the electrochemical theory of batteries. He served as a professor of natural philosophy at the Academy of Geneva and worked principally on electricity, although he also investigated the specific heats of gases and calculated the temperature of the Earth's crust. His principal book was *Traité d'Électricité Théoretique et Appliquée* (3 vols., Paris: J.-B. Baillière et H. Baillière, 1854–58). C. V. Walker was translating the third volume into English when de la Rive asked Tyndall to read the proofs and correct what he called Walker's "Gallicisms." De la Rive was a member of the Society of Physics and Natural History in Geneva, of which Tyndall was elected as an honorary member in 1859.

Debus, Heinrich ("Heins") (1824–1915), was a German chemist. At the polytechnic in Kassel, he was taught chemistry by Robert Bunsen, but Bunsen left for the University of Marburg in 1839. Upon reaching adulthood, Debus joined him there, and came to work as Bunsen's personal assistant. During his time at Marburg, he finished a PhD on madder in 1848, and met Tyndall, Edward Frankland, and Thomas Archer Hirst, who were being taught by Bunsen as well. Debus moved to England in 1851, where, thanks to Frankland, he was given a post as a chemistry teacher at Queenwood College and was for a time a colleague of Tyndall's. After leaving Queenwood in 1867 he spent three years at Clifton College, Bristol, then moved to London as lecturer in chemistry at Guy's Hospital Medical School (1870–88), adding to this the position of professor of chemistry at the Royal Naval College, Greenwich (1873–88), under Hirst, who was founding director of studies at the college. Most of his research was in organic chemistry.

Drummond, Maria (nee Kinnaird, 1810–91), was a literary hostess and the adopted daughter of the wealthy MP, banker, and merchant Richard "Conversation" Sharp. She was born on the Caribbean island of Saint Vincent and orphaned by a volcanic eruption in 1812. Through Sharp's social circle she was introduced to Michael Faraday, Thomas Macaulay, Charles Babbage, and Walter Scott, among others. When Sharp died in 1835 she inherited much of his £250,000 estate and married Thomas Drummond (1797–1840), who developed the Drummond light for the Irish Survey and was undersecretary for Ireland (1835–40). Together they had three daughters, Mary, Emily, and Fanny Eleanor, before Thomas died in 1840. Maria and her daughters continued to entertain a wide circle of London's elite, including Tyndall, at their home near Hyde Park, London, and their country house at Fredley Park, in Mickleham, Surrey.

Faraday, Michael (1791–1867), was a natural philosopher, Fullerian Professor of Chemistry at the RI, and a member of the Sandemanian church. His extensive research focused on electricity and encompassed electrochemistry, electromagnetism, diamagnetism, and magneto-optics (the Faraday effect). Faraday also held the positions of director of the laboratory (1825–67) and superintendent of the house (1852–67) at the RI. He was Tyndall's patron and helped him obtain the position of professor of natural philosophy at the RI in 1853. Despite significant differences in their views on religion and on the nature of magnetism, they held the highest mutual regard for each other. Tyndall's *Faraday as a Discoverer* (London: Longmans, Green & Co., 1868) was his tribute to his mentor. By 1861 Faraday aimed to leave the RI, although his resignation was refused, and the following years he would increasingly spend more time at Hampton Court, where he died. In 1867 Tyndall succeeded Faraday as superintendent of the house and director of the laboratory at the RI

Fessel, Friedrich (1821–?), was a German optician, mechanic and instrument maker from Cologne. He was an assistant to Julius Plücker and then a teacher in the Kölner Gewerbeschule. Fessel became a full-time instrument maker from 1852 working in Cologne. (See *Zeitschrift der Deutschen Gesellschaft für Mechanik und Optik* 8 [1917]: 68). At the International Exhibition of 1862 in London, Fessel displayed his Myographion, for which he won a medal. This machine supposedly measured the speed of contraction of irritated muscles, as well as the speed by which nervous stimulation travels.

Forbes, James David (1809–68), was a Scottish physicist and glaciologist. Forbes held the chair of natural philosophy at the University of Edinburgh from 1833 until 1860.While much of his research focused on heat, he was famous for his studies of glaciers. An avid mountaineer, Forbes published *Travels through the Alps of Savoy* (Edinburgh: A. and C. Black) in 1843. His theory that a glacier is a viscous body brought him into conflict with Tyndall and Huxley in the late 1850s. Plagued by ill health, Forbes moved on to become principal of the college of Saint Salvator and Saint Leonard (the United College), at Saint Andrews University in 1860. As principal, he would occasionally lecture if able, although his main project was to draw the Scottish aristocracy back to Saint Andrews.

Foucault, (Jean Bernard) Léon (1819–68), was a French physicist. Although Foucault originally studied medicine, he switched to physics due to his fear of blood. He became the laboratory assistant of the bacteriologist Alfred François Donné, with whom he invented a photoelectric microscope in 1844. His reputation was established by the invention of an instrument used to prove the rotation of the Earth in 1850, known as the Foucault pendulum, which he had suspended from the roof of the Panthéon in Paris. For many years he also collaborated with Hippolyte Fizeau in experimental optics. While Fizeau made the first precise measurement of the speed of light, Foucault first demonstrated that light travels more quickly through air than water, which was used as verification for the wave theory of light. Tyndall translated and abridged Foucault's work for the *Phil. Mag.*

Francis, William (1817–1904), was a chemist, translator, and editor. In the mid-1830s Francis spent two years in Gera, an industrial town south of Leipzig and a center of German printing, learning the printing trade and becoming fluent in German. He was apprentice to, and natural son of, Richard Taylor, who sent him to the University of Berlin in 1839 to study chemistry. In 1841 he moved to attain his degree under Justus von Liebig and Gustav Rose in Giessen, all the while sending translations of German science back to Taylor. Francis began working for Taylor's publishing firm in 1837, became editor in 1851, and partner in 1852 when the Taylor & Francis publishing firm was established. Francis was Tyndall's most significant counselor and closest confidant as he initially maneuvered his way through the Victorian scientific scene. (William Brock and Arthur Meadows, *Lamp of Learning: Taylor & Francis and the Development of Science Publishing* [London: Taylor & Francis, 1984].)

Frankland, Edward (1825–99), was an English chemist. He began his scientific career as an apprentice apothecary. While working in Lyon Playfair's London laboratory, Frankland met Herman Kolbe, another of Playfair's assistants, who had received a PhD at Marburg under the tutelage of Robert Bunsen. It was through friendship with Frankland at Queenwood College in 1847–48 that Tyndall became interested in studying at Marburg. In 1848, Frankland again left for Marburg, where he received his PhD, and made his first observations of "organometallic chemistry," the chemistry of metals linked to carbon. In 1851, a year after taking up a position under Playfair at the

Putney College of Civil Engineering, he secured the chair of chemistry (half time) at the new Owens College, Manchester. In 1864 Frankland joined Tyndall, Thomas Archer Hirst, Joseph Dalton Hooker, Thomas Henry Huxley, and three others as members of the X Club, a monthly dining club in which they promoted science in British society.

Frankland, Sophie (née Fick, 1821–74), was the wife of Edward Frankland. During his first trip to Marburg, Edward Frankland knew little German, and was introduced by Hermann Kolbe to Sophie Fick, whose brothers Heinrich and Ludwig taught in Marburg. Her brother Adolf went on to become a professor in physiology as well, which shows the academic proclivities of the family. Their engagement was made official during Edward's second stint in Marburg in 1849, and they married in February 1851, just before they moved to Manchester. Having contracted tuberculosis, Sophie's health deteriorated in the early 1870s, and she died in Davos, Switzerland, in 1874. (Colin Russell, *Edward Frankland* [Cambridge: Cambridge University Press, 1996], 33–34, 88, 149.)

Gassiot, John Peter (1797–1877), was an electrician and wine merchant by trade. He was one of the founding members of the London Electrical Society in 1837, and served as the society's treasurer until 1841. He pursued research on electricity in the laboratory he outfitted at his home in Clapham Common, near London. During the 1860s James Clerk Maxwell worked there while establishing the unit for electrical resistance. Gassiot published papers on the functioning of the voltaic pile (battery) and showed that a spark could be produced before the circuit of a battery was closed.

Graham, Thomas (1805–69), was a Scottish chemist and physicist. Graham was only fourteen when he started at the University of Glasgow in 1819, and was inspired by the lectures of Thomas Thomson to focus on chemistry. He graduated with an MA from Glasgow in 1826, and became professor of chemistry at Anderson's College in 1930. Graham was particularly interested in gaseous diffusion, and it was at Anderson's that he first formulated what is now known as "Graham's Law" and elucidated what made the three phosphoric acids distinct from one another. After becoming the professor of chemistry at University College, London, in 1837, he published the textbook *Elements of Chemistry* (London: Hippolyte Bailliere, 1842), which was used both in England and on the European mainland. In 1854 he succeeded John Herschel as Master of the Mint, and was the last person to have this position.

Helmholtz, Hermann (1821–94), was a German physicist and physician. Educated in military medicine at the Friedrich-Wilhelms Institute in Berlin, he became the first to hold a regular post as teacher of anatomy at the Berlin Academy of Arts in 1848. His interest in physics was already developed, as in 1847, Helmholtz published his paper *Über die Erhaltung der Kraft*, which discussed the mathematical principles of the conservation of energy. This work was based on his study of muscle metabolism, and he argued that there was no such thing as a vital "force." Rather, factors such as mechanics, heat, and electricity were all manifestations of a single "force"; no energy was expended during muscle movement. He became professor of physiology at the University of Königsberg, and became professor for anatomy and physiology at the University of Bonn in 1855. It was at Bonn that he increasingly turned his attention to physics, a process that continued during his time at Heidelberg from 1858 onward. Called "an intellectual giant" by James Clerk Maxwell, Helmholtz was involved in the fields of hydrodynamics and electrodynamics.

Herschel, John Frederick William (1792–1871), was an English mathematician and astronomer. He was the son of the famous astronomer William Herschel (1738–1822) and himself became one of the most eminent scientists in Victorian Britain. He graduated from Cambridge University in 1813 as senior wrangler. He was awarded the RS's Copley Medal in 1821, and a second Copley Medal in 1847. His *Preliminary Discourse on the Study of Natural Philosophy* (London: Longmans, Rees, Orme, Brown & Green, 1830) was widely read and greatly admired. He was awarded a knighthood in 1831, followed by a baronetcy in 1838, on his return from mapping the southern skies in Cape Town, South Africa. From 1850 to 1855 he served as Master of the Mint, and on his death was buried next to Isaac Newton in Westminster Abbey.

Hese, J. van (?), was a Belgian instrument maker based in Ghent. He manufactured and supplied instruments for the Belgian physicist Joseph Plateau.

Hirst, Thomas Archer (1830–92), was an English mathematician, educator, and Tyndall's closest friend and correspondent. His friendship with Tyndall began in Halifax, Yorkshire, in 1845 when they were both working for the surveyor Richard Carter. Encouraged by Tyndall—who acted as a sort of mentor to him—Hirst studied at the universities of Marburg, Göttingen, and Berlin between 1849 and 1852. In 1853 he returned to England with a PhD and replaced Tyndall as the mathematics teacher at Queenwood College. Like Tyndall, he did not get on with the headmaster, George Edmondson. He resigned in June 1856 to look after his sick wife, Anna, who died in Paris in 1857. From June 1861 Hirst lived with Tyndall at 14 Waverly Place, a house owned by Thomas Henry Huxley (1825–95).

Hooker, Joseph Dalton (1817–1911), was a British botanist who specialized in taxonomy and the geographical distribution of plants, ultimately becoming the most famous British botanist of the Victorian era. He was the younger son of Sir William Jackson Hooker, the director of the Royal Botanic Gardens, Kew. After studying medicine at the University of Glasgow, he became a naval surgeon and spent the years 1839–43 on James Clark Ross's geomagnetic survey of the southern oceans, where he collected plants as well as serving as surgeon. Subsequently he explored and collected plants in India (1847–51), achieving considerable mountaineering success, before joining his father as deputy director of Kew Gardens in 1855. He was first introduced to Tyndall by Thomas Henry Huxley (1825–95) in November 1855, and the three became close friends.

Hopkins, William (1793–1866), was an English mathematician and geologist. He entered Cambridge University as an undergraduate at Peterhouse in 1822 after the death of his first wife. He graduated the Mathematical Tripos as seventh wrangler in 1827 and received his MA in 1830. Settling in Cambridge, Hopkins became a private mathematics tutor training undergraduates for the Tripos exams. He was one of the most successful tutors of his time and became known as the "senior wrangler maker." In his research he was one of the most significant proponents of the solidity of the Earth and came into conflict with James Forbes over their conflicting theories of glaciers.

Huxley, Thomas Henry (1825–95), was a British zoologist, educator, science popularizer, and one of Tyndall's closest friends. Following medical training at Charing Cross Hospital in London, he was a ship's surgeon aboard the HMS *Rattlesnake* expedition to Australia (1846–50) under Captain Owen Stanley. After eking a living by writing for Chapman's *Westminster Review*, in 1854 he became professor of natural history at the Royal School of Mines, London, where he remained until 1885. While in Australia he met Henrietta (Nettie) Anne Heathorn (1825–1915), whom he married in July 1855 after an eight-year engagement. He joined Tyndall at the RI as Fullerian Professor of Natural History (1855–58). The year 1860 saw the famous dispute between Huxley, who had been convinced by Darwin to support evolutionary naturalism, and Bishop Samuel Wilberforce at the Oxford meeting of the British Association for the Advancement of Science, and 1861 saw the publication of his "On the Zoological Relations of Man with the Lower Animals."

Jones, Henry Bence (1813–73), was a wealthy English physician and medical chemist. He was educated at Harrow School and Trinity College, Cambridge, before studying medicine at Saint George's Hospital, London. Attracted by the potential applications of chemistry to clinical medicine, he studied organic chemistry with Justus von Liebig at Giessen (1841–43) before assuming a successful career as a physician in private practice and at Saint George's Hospital. In May 1842 he married a cousin, Lady Millicent Acheson, daughter of the second Earl of Gosford, with whom he had seven children. In 1853 he became a manager of the RI and was elected secretary in 1860. He frequently entertained Tyndall at both his London home and at Folkestone, where he kept a summer house.

Kirchhoff, Gustav (1824–87), was a German physicist. In Königsberg, Kirchhoff studied physics under Franz Neumann and mathematics under Friedrich Julius Richelot. His formulation of the laws of electrical circuits was part of his doctoral dissertation, which he finished at Königsberg in 1847. During an associate professorship at the University of Breslau in 1850, he became acquainted with Robert Bunsen, who made sure that Kirchhoff followed his move to Heidelberg in 1854. Together with Bunsen, he worked on thermal radiation and spectrum analysis. Regarding his law of thermal radiation, he would fight a long struggle for recognition of priority with the Scottish Balfour Stewart (1828–87).

Linth, Arnold Escher von der (1807–72), was a Swiss geologist and son of Hans Conrad Escher von der Linth (1767–1823). A student of Horace-Bénédict Saussure in Geneva and Alexander von Humboldt in Berlin, he became a professor of geology at the École Polytechnique in Zurich in 1856. In 1842 he was a member of the first party to ascend the Lauteraarhorn. He produced, in collaboration with Bernhard Studer, a detailed geological map of Switzerland in 1853. His family was given the surname "von der Linth" for his father's work diverting the Linth river.

Magnus, Heinrich Gustav (1802–70), was a German chemist and physicist. A pupil of the German chemist Eilhard Mitscherlich in Berlin and of J. J. Berzelius in Stockholm, he was professor of technology and physics at the University of Berlin from 1833 until his death. In Berlin he developed a superbly equipped physics laboratory at his home to which he invited young German physicists to work, including Tyndall in 1851. In 1860 Tyndall determined that water vapor absorbed heat at a greater level compared to dry air. Magnus challenged Tyndall's findings, claiming that water vapor did not significantly absorb heat. This dispute preoccupied Tyndall in the early 1860s, as his letters during this period demonstrate.

Maskelyne, Nevil Story (1823–1911), was an English geologist and politician. Maskelyne was a professor of mineralogy at Oxford from 1856 to 1895, and Keeper of Minerals at the British Museum from 1857 to 1880. Tyndall wrote to Maskelyne from November 1861 to January 1862 requesting materials for his experiments on radiant heat.

Maxwell, James Clerk (1831–79), was a Scottish physicist and mathematician. From the late 1850s onward he developed Michael Faraday's "field theory," which had been put into mathematical expression by William Thomson some years previously. His formulation of what are now known as Maxwell's equations set the foundation for the classical theory of electromagnetism. Maxwell's work culminated in his 1873 publication, *A Treatise on Electricity and Magnetism* (Oxford: Clarendon Press).

Mayer, Julius Robert (1814–78), was a German physician and physicist. His work on the conservation of energy contributed to the foundation of thermodynamics. Previously the British physicist James Joules had been regarded as the discoverer of the mechanical equivalent of heat. Tyndall, however, in his publication *Heat as a Mode of Motion* (London: Longman, Green, Longman, Roberts, & Green, 1863) praised both men for their research, championing the significance of Mayer's work in the discovery of the conservation of energy.

Melloni, Macedonio (1798–1854), was an Italian physicist. Melloni conducted experiments in radiant heat during the 1830s and 1840s showing that heat had similar physical properties to light, as well as demonstrating the ways in which infrared radiation could pass through various materials. Tyndall, in his reevaluations of Melloni's work in the late 1850s, validated his results with the exception of Melloni's belief that infrared radiation was a different form of energy to visible rays.

Miller, William Allen (1817–70), was an English chemist, chair of chemistry at King's College London from 1845, and fellow of the Royal Society. He was the author of the popular textbook series *Elements of Chemistry, Theoretical and Practical* (London: J. W. Parker and Son, 1855, 1856, 1857), and conducted many experiments in spectrum analysis, including the effects of atmospheric water vapor on the solar spectrum.

Moore, Harriet Jane (1801–84), was a British watercolor artist who painted scenes of Faraday's apartment, study, and laboratory at the RI during the 1850s. Moore was the eldest child of surgeon James Carrick-Moore (1762–1860) and Harriet Moore (née Henderson, 1779–1866). She was elected a member of the RI in 1852. (See Frank A. J. L. James, "Harriet Jane Moore, Michael Faraday, and Moore's Mid-Nineteenth Century Watercolours of the Interior of the Royal Institution," in *Fields of Influence: Conjunctions of Artists and Scientists, 1815–1860*, ed. James Hamilton [University of Birmingham Press, 2001], 111–28.)

Moore, Julia (1804–1904), was the youngest of the three daughters of the surgeon James Carrick-Moore and Harriet Moore. The wealthy Moores had close connections with the RI and frequently entertained Tyndall at their Wimbledon home.

Mousson, Albert (1805–90), was a Swiss physicist, glaciologist, and malacologist. He studied in Bern and at the Universities of Geneva, Göttingen, and Paris. In 1832 he married Barbara Seger. After the founding of the University of Zurich in 1833 he became a lecturer, then a professor of physics there. He held the professorial chair of physics at the École Polytechnique Fédérale and the University of Zürich (1855–78). He contributed to the study of glaciers and the role of pressure on the melting of ice. In 1854 he published *Die Gletscher der Jetztzeit* (Zürich: Fr. Schulthess). He also studied meteorology and described 450 new species of molluscs.

Murchison, Roderick (1792–1871), was a geologist and geographer, who published over 180 scientific papers as well as several books. He was elected a fellow of the Geological Society in 1825, and was twice its president. In 1826 he was elected a fellow of the Royal Society. He helped found the Royal Geographical Society of London in 1830, of which he was president for sixteen years beginning in 1843. One of his important contributions to geology was his book *The Silurian System* (London: John Murray, 1839), which established the oldest fossil-bearing classification then known. In January 1859, Murchison proposed Tyndall for membership of the Athenaeum.

Murray, John (1808–92), was an English publisher and eldest child of Anne Elliot and publisher John Murray. This John Murray was the third to carry that name. He was educated at Charterhouse School (1819–26) and for a year (1827) at Edinburgh University. He began his career in the family firm at 50 Albemarle Street, London, in 1828 and became head of the firm after his father's death in 1843. Murray published the *Quarterly Review*, a popular series of *Handbooks for Travellers*, and numerous scientific publications. On 24 November 1859 Murray published both Darwin's *On the Origin of Species*, and Samuel Smiles's *Self-Help*. He became a member of the Alpine Club in 1858 and in 1860 published Tyndall's *Glaciers of the Alps*.

Parker, John William (1820–60), was a publisher and printer. He joined his father's London publishing firm in 1843 as general manager, which under his management became a leading supporter for liberal Christianity, and from 1848 for the Christian socialist movement. One of its most controversial publications was *Essays and Reviews* (1860). In July 1847 he became the editor of *Fraser's Magazine for Town and Country*, whose authors included Kingsley, Lewes, Mill, Maurice, Ruskin, Tennyson, and Tyndall.

Plateau, Joseph (1801–83), was a Belgian physicist. Born in Brussels, Plateau studied at the University of Liège, where he graduated in 1829 with a doctorate in mathematics and physics. In 1832 Plateau invented a device that could give the illusion of a moving image, which he called the phenakistoscope. In 1835 he was appointed professor of experimental physics at the University of Ghent. Plateau is best known for his experimental work in which he investigated what is now known as the Plateau problem. From 1840 he conducted experiments to demonstrate that a minimal surface could be formed from a soap film, made of a solution of soapy water and glycerine, which was stretched on a wire frame.

Playfair, Lyon (1818–98), was a Scottish chemist who worked in Justus von Liebig's laboratory in Giessen (1839–41). In 1842 he became professor of chemistry at the Royal Manchester Institution. In 1845 he became chemist to the Geological Survey of Great Britain and professor of chemistry at the School of Mines, London, and in 1848 was elected a fellow of the Royal Society. In 1853 Playfair was appointed secretary for science in the Department of Science and Art. In this position, he attempted to establish a national system of science education. Playfair maintained a close relationship with the royal family and especially Albert, the Prince Consort, and for a time educated the Queen's sons Albert and Alfred. In 1858 he became professor of chemistry at Edinburgh University.

Playfair, Jean (née Millington, d. 1877), was a daughter of Crawley Millington of Crawley House. She married Lyon Playfair on 17 December 1857, after the death of his first wife in August 1855. They had one daughter, Ethel Mary Lyon, born 6 January 1862.

Poggendorff, Johann Christian (1796–1877), was a German physicist at the University of Berlin. His major contribution to science was as editor for fifty-two years (1824–76) of *Annalen der Physik und Chemie*, which became known as *Poggendorff's Annalen* during his editorship. He used his broad knowledge of science and scientific men in producing a

Biographisch-literarisches Handworterbuch zur Geschicte der exacten Wissenschaften (Leipzig: Johann Ambrosius Barth), the first two volumes of which were published in 1863.

Pollock, Juliet (née Creed, 1819–99), was a daughter of Catherine Herries and the Reverend Henry Creed, vicar of Corse, Gloucestershire. In 1844 she married William Frederick Pollock, with whom she had three sons. She and her husband were loyal supporters of the RI, attended Tyndall's lectures, frequently entertained Tyndall to dinner, and were among Tyndall's closest friends during the period covered by this volume. Juliet was the author of several books, including *Julian and his Playfellows* (London: Grant and Griffith, 1852), *New Friends* (London: John W. Parker and Son, 1858), and *Macready As I Knew Him* (London: Remington and Co., 1885).

Pollock, Maurice Emilius (1857–1932), was the third son of Juliet and William Frederick Pollock. He became a sculptor and married Lydia Helen Roberts (1853–84) in 1880, and Mabel Mary McPherson (1860–1937) in 1889.

Pollock, Walter Herries (1850–1926), was the second son of Juliet and William Frederick Pollock. He was educated at Eton College (1862–65) and was admitted to Trinity College, Cambridge, where he obtained a BA (1871) and MA (1875). He was admitted to the Inner Temple in 1870 and called to the bar in 1873. Pollock became a writer, poet, and journalist, and was the editor of the *Saturday Review* (1884–94). Letters in this volume show a close affinity between Walter and Tyndall. Juliet referred to Tyndall as Walter's "especial friend" and related how he took a photograph of Tyndall with him when leaving for school in Brighton at the age of eight.

Pollock, William Frederick (1815–88), second baronet, was a lawyer and author. He was the eldest son of Jonathan Frederick Pollock, first baronet (1783–1870), and Frances Rivers (d. 1827). He married Juliet Creed in 1844. He was admitted to Trinity College, Cambridge, in 1832, obtaining his BA (1836) and MA (1840). While there, he was admitted to the Inner Temple in 1833 and called to the bar in 1838. He became a master of the Court of Exchequer in 1846 and was appointed Queen's Remembrancer in 1874. Pollock's *Personal Remembrances* (2 vols., London: Macmillan and Co., 1887), illustrate how the family circulated among London's elite and entertained at their Montagu Square home. Both William and Juliet were loyal supporters of the RI, and Tyndall was frequently among their guests.

Rendu, Louis (1789–1859), was a French glaciologist and Roman Catholic bishop of Annecy. In 1840 he published *Théorie des glaciers de la Savoie* (Chambéry: Chez Puthod), in which he explained glacier motion in terms of flowing ice. The Scottish glaciologist and physicist James David Forbes (1809–68) was later accused of borrowing work from Rendu in his suggestion that glaciers behaved like viscous bodies. Tyndall was one of Forbes's most outspoken opponents, arguing that ice, being a hard, crystal-like solid, could not have viscous properties.

Ruhmkorff, Heinrich Daniel (1803–77), was a German engineer and leading scientific instrument maker based in Paris. He held a reputation for high-quality electrical apparatus, and for his commercialization of the induction coil, which was often known as the Ruhmkorff coil.

Sabine, Edward (1788–1883), was an Anglo-Irish astronomer, geophysicist, and military officer who led the British effort to establish magnetic observatories all over the globe. A graduate of the Royal Military Academy at Greenwich, he reached the rank of general in the 1850s. He accompanied J. C. Ross's Arctic expedition (1818) and W. E. Parry's expedition (1819). He received the Copley Medal from the RS in 1821 for his magnetic observations in the Arctic. He was closely connected with the administration of the RS, and served as its treasurer (1850–61) and president (1861–71). He was an outstanding lobbyist for science, an effective administrator, and a valuable patron to Tyndall from 1851.

Saussure, Horace-Bénédict de (1740–99), was a Swiss geologist, botanist, and mountaineer, educated at the University of Geneva. He held the chair of philosophy at the Academy of Geneva in 1762 and married Albertine Amélie Boissier in 1765. During the 1770s he began the research that was published in his four-volume *Voyages dans les Alpes* (Neuchâtel: Samuel Fauche, 1779–96). Saussure made the third ascent of Mont Blanc in 1787, a year after Jacques Balmat and Michel Paccard. Spending four and a half hours on the summit, he conducted a number of scientific experiments. His expedition caught the popular imagination and earned him international fame.

Simpson, James (1811–70), was a Scottish physician and obstetrician. Simpson held the chair of midwifery at Edinburgh University, and was elected president of the Royal College of Physicians in Edinburgh in 1850. Simpson pioneered the use of chloroform for anaesthetic purposes.

Stokes, George Gabriel (1819–1903), was an English mathematician and physicist. He graduated from Cambridge University as senior wrangler and Smith's Prizeman in 1841 and served as Lucasian Professor of Mathematics at Cambridge University (1849–1903). He augmented his income as a lecturer at the Royal School of Mines (1849–56). As secretary of the RS (1854–85) he was responsible for refereeing papers for the *Phil. Trans.*, a task for which he frequently sought Tyndall's assistance.

Thomson, William (1824–1907), was a distinguished Irish-Scottish mathematician, physicist, and inventor. He was educated at the University of Glasgow and Peterhouse College, Cambridge University, where he was second wrangler and Smith's Prizeman in 1841. He was professor of natural philosophy in Glasgow from 1846, paying frequent visits to London. He first met Tyndall at the Edinburgh meeting of the BAAS in 1850 and from the start the two physicists had a strained relationship. For much of his career Tyndall regarded Thomson as a scientific rival.

Tyndale, Hector (1821–80), was an American Union Army general during the American Civil War, and cousin of John Tyndall. In 1865 Tyndale was brevetted major general. During Tyndall's successful lecture tour of the United States in 1872–73 he appointed Tyndale a trustee of the tour proceeds, which were devoted to the establishment of a fund to promote scientific education in physics.

Tyndale, William (*c.* 1494–1536), was a religious reformer during the Protestant reformation and a translator of the Bible. A Protestant martyr, he was burned at the stake. It was a family tale that Tyndall was descended from "The Martyr," and he wrote in his journal that he "was struck by the remarkable reproduction of his temper amongst us;—give us the same circumstances and I am sure either Hector or myself would act as he did tomorrow" (Journal, RI MS JT/2/13c/1187). In 1869 Tyndall visited the supposed birthplace of William Tyndale, and the monument to him erected at Nibley Knoll in Gloucestershire.

Whately, Arthur Pepys (1829–1910), was a practicing barrister and member of the Alpine Club from 1858. He was placed on the club's committee the same year, and appointed honorary secretary in 1861–62. Tyndall was a member of the club from November 1858, and elected vice-president in 1861 but resigned the same year. He was made an honorary member in 1887.

White, Walter (1811–93), was a librarian and writer. He was appointed assistant secretary of the RS in 1861, where he spent much time taking minutes for its various committees until his resignation in 1884. It was in this role that he was introduced to many of the men who attended the lectures, and who lectured at the society, including Tyndall. From 1850 he wrote extensive travel narratives on his expeditions to various places, including Mont Blanc and the Tyrol.

Wilson, George (1818–59), was a Scottish chemist and museum director. In 1840 he received a license from the Royal College of Surgeons of Edinburgh to teach chemistry in its degree program. He was appointed lecturer in 1843 at the Edinburgh Veterinary College, the School of Arts, and the Scottish Institution, and was elected fellow of the Royal Society of Edinburgh in 1845. The following year he was appointed director of the Industrial Museum of Scotland, and held the position of Regius Chair of Engineering at the University of Edinburgh. His most popular works were a chemistry textbook published in the Chambers Education Course series, and his account of the senses in the *Five Gateways of Knowledge* (London: Macmillan & Co., 1856), which went through at least seven editions.

With, George Henry (1827–1904), was an English chemist and schoolmaster trained at Saint John's College in Battersea. He taught at Alverstoke National School in Gosport, where he was recognized for his teaching abilities by Richard Dawes, the rector of King's Somborne, who encouraged his interest in science. From 1851 to 1876 he was master of Blue Coat boys' school in Hereford, and later taught science at Hereford Ladies' College, Hereford Cathedral School, Barr's Court School, and Ludlow Grammar School. As a chemist he specialized in soil science, and worked with the Industrious Aid Society. Interested in astronomy, With produced around two hundred mirrors, which he sold as parts to be used in the construction of telescopes.

INDEX

The index refers to the letters using their serial numbers. The letter 'n' denotes a footnote reference and appears if an item is not otherwise mentioned in the letter to which the footnote pertains. A number in **bold** signifies a letter to or from a correspondent.